The Cambridge Handbook of Environmental Sociology Volume 2

The Cambridge Handbook of Environmental Sociology is a go-to resource for cutting-edge research in the field. This two-volume work covers the rich theoretic foundations of the sub-discipline, as well as novel approaches and emerging areas of research that add vitality and momentum to the discipline. Over the course of sixty chapters, the authors featured in this work reach new levels of theoretical depth, incorporating a global breadth and diversity of cases. This book explores the broad scope of crucial disciplinary ideas and areas of research, extending its investigation to the trajectories of thought that led to their unfolding. This unique work serves as an invaluable tool for all those working in the nexus of environment and society.

KATHARINE LEGUN is Assistant Professor in Communication, Philosophy and Technology at Wageningen University in the Netherlands. Her work considers how plants, measurement systems, and new artificial intelligence technology shapes ecological and economic agency in agri-food systems and the political dynamics of farming.

JULIE C. KELLER is Assistant Professor in the Department of Sociology and Anthropology at the University of Rhode Island. Her research and teaching focus on rural inequality, agricultural labor, farmers, and immigration. She is the author of *Milking in the Shadows: Migrants and Mobility in America's Dairyland* (Rutgers University Press, 2019).

MICHAEL CAROLAN is Professor of Sociology and Associate Dean for Research and Graduate Affairs at Colorado State University. His research focuses on issues related to food, agriculture, sustainability, and justice. Some of his recent books include *Society and the Environment: Pragmatic Solutions to Ecological Issues*, third edition (Routledge, 2020) and *The Food Sharing Revolution: How Start-Ups, Pop-Ups, and Co-Ops are Changing the Way We Eat* (Island Press, 2018).

MICHAEL M. BELL is Chair and Vilas Distinguished Achievement Professor of Community and Environmental Sociology at the University of Wisconsin–Madison. He is the author of *City of the Good: Nature, Religion, and the Ancient Search for What Is Right* (Princeton, 2018) and the widely used environmental sociology textbook, *An Invitation to Environmental Sociology*, now in its sixth edition.

The Cambridge Handbook of Environmental Sociology

Volume 2

Edited by

Katharine Legun
Wageningen University

Julie C. Keller
University of Rhode Island

Michael Carolan
Colorado State University

Michael M. Bell
University of Wisconsin–Madison

CAMBRIDGE
UNIVERSITY PRESS

University Printing House, Cambridge CB2 8BS, United Kingdom

One Liberty Plaza, 20th Floor, New York, NY 10006, USA

477 Williamstown Road, Port Melbourne, VIC 3207, Australia

314–321, 3rd Floor, Plot 3, Splendor Forum, Jasola District Centre,
New Delhi – 110025, India

79 Anson Road, #06–04/06, Singapore 079906

Cambridge University Press is part of the University of Cambridge.

It furthers the University's mission by disseminating knowledge in the pursuit of
education, learning, and research at the highest international levels of excellence.

www.cambridge.org
Information on this title: www.cambridge.org/9781108429337
DOI: 10.1017/9781108554558

© Cambridge University Press 2020

This publication is in copyright. Subject to statutory exception
and to the provisions of relevant collective licensing agreements,
no reproduction of any part may take place without the written
permission of Cambridge University Press.

First published 2020

A catalogue record for this publication is available from the British Library.

ISBN 978-1-108-42933-7 Hardback

Cambridge University Press has no responsibility for the persistence or accuracy of
URLs for external or third-party internet websites referred to in this publication
and does not guarantee that any content on such websites is, or will remain,
accurate or appropriate.

Contents

List of Figures	*page* viii
List of Tables	ix
List of Contributors	x
Foreword RILEY E. DUNLAP	xix
Introduction KATHARINE LEGUN, JULIE C. KELLER, MICHAEL CAROLAN, AND MICHAEL M. BELL	1

Part I Methods

1	Re-compos(t)ing the Ghosts of Sociologies Past: Towards More Cosmoecological Sociologies MARTHA MCMAHON AND CHELSEA POWER	11
2	On Discourse-Intensive Approaches to Environmental Decision-Making: Applying Social Theory to Practice STEVEN E. DANIELS AND GREGG B. WALKER	29
3	Community-Based Research RANDY STOECKER	47
4	Using Geographic Data in Environmental Sociology RACHEL A. ROSENFELD AND KATHERINE J. CURTIS	61

Part II Embodied Environmental Sociology

5	Strangers on the Land? Rural LGBTQs and Queer Sustainabilities JULIE C. KELLER	87
6	Masculinity and Environment KATHRYN GREGORY ANDERSON	103
7	Toxicity, Health, and Environment JENNIFER S. CARRERA AND PHIL BROWN	117

8	The Environment's Absence in Medicine: Mainstream Medical Coverage of Leukemia MANUEL VALLÉE	137

Part III Beyond the Human

9	Interventions Offered by Actor-Network Theory, Assemblage Theory, and New Materialisms for Environmental Sociology KATHARINE LEGUN AND ABBI VIRENS	161
10	Plants *and* Philosophy, Plants *or* Philosophy MICHAEL MARDER	176
11	Animals and Society: An Island in Japan MARGO DEMELLO	188

Part IV Sustainability and Climate Change

12	Possibilities and Politics in Imagining Degrowth VALÉRIE FOURNIER	205
13	Sustainable Consumption EMILY HUDDART KENNEDY	221
14	Sustainability Cultures: Exploring the Relationships Between Cultural Attributes and Sustainability Outcomes JANET STEPHENSON	236
15	Socio-Ecological Sustainability and New Forms of Governance: Community Forestry and Citizen Involvement with Trees, Woods, and Forests BIANCA AMBROSE-OJI	249
16	Carbon Markets and International Environmental Governance JOHN CHUNG-EN LIU AND MARK H. COOPER	267
17	The Multi-Level Governance Challenge of Climate Change in Brazil LEILA DA COSTA FERREIRA	285

Part V Resources

18	Enclosing Water: Privatization, Commodification, and Access DANIEL JAFFEE	303
19	Speech Is Silver, Silence Is Gold in the Fracking Zone DEBRA J. DAVIDSON	324

20	Environmental Sociology and the Genomic Revolution VALERIE BERSETH AND RALPH MATTHEWS	342
21	The Future Is Co-Managed: Promises and Problems of Collaborative Governance of Natural Resources NATHAN YOUNG	360

Part VI Food and Agriculture

22	Future and Food: New Technologies, Old Political Debates MICHAEL CAROLAN	377
23	Eating Our Way to a Sustainable Future? JOSÉE JOHNSTON AND ANELYSE M. WEILER	390
24	Neoliberal Globalization and Beyond: Food, Farming, and the Environment GEOFFREY LAWRENCE AND KIAH SMITH	411
25	The Sociology of Environmental Morality: Examples from Agri-Food PAUL V. STOCK	429

Part VII Social Movements

26	Alternative Technologies and Emancipatory Environmental Practice CHELSEA SCHELLY	447
27	The Global Fair Trade Movement: For Whom, By Whom, How, and What Next ELIZABETH A. BENNETT	459
28	Possibilities for Degrowth: A Radical Alternative to the Neoliberal Restructuring of Growth-Societies BARBARA MURACA	478
29	Achieving Environmental Justice: Lessons from the Global South PEARLY WONG	497
30	Conclusion: Envisioning Futures with Environmental Sociology JULIE C. KELLER, MICHAEL M. BELL, MICHAEL CAROLAN, AND KATHARINE LEGUN	515
	Index	523

Figures

2.1	Collaborative learning foundations	*page* 34
14.1	Sustainability cultures framework	239
17.1	Brazilian GHG emissions, by activity sector (1990–2014)	288
17.2	Brazilian states with approved climate change policies	293
19.1	Alberta oil production by type over time	330
19.2	Alberta natural gas production by reserve over time	331
19.3	Fracking fluid composition as depicted by *FracFocus*	333
22.1	"Good citizen" according to big data industry	382
22.2	"Good citizen" according to Farm Hack activists	383

Tables

2.1	Discourse-intensive approaches and four theorists	*page* 41
8.1	Toxicants associated with adult-onset leukemias	142
8.2	WebMD articles on leukemia	145
15.1	Innovative forest governance in Britain: community and citizen involvement in different kinds of forest practice with different facilitating/leading actors	253

Contributors

KATHRYN GREGORY ANDERSON is a researcher and instructor of Sociology and Environmental Studies at UC-Berkeley and UW-Madison. She's consulted on climate, biodiversity, and land use for the United Nations Development Program and the Brazilian Government, and she currently researches U.S. environmental politics in the farm sector. Focusing on how race, gender, nation, and other social hierarchies create barriers for collective action to promote economic democracy and environmental justice, her research identifies leverage points for increasing diversity and inclusion and she develops evidence-based organizational interventions. Professor Kate is an award-winning college teacher with experience in online teaching and the psycho-cognitive aspects of teaching and learning.

BIANCA AMBROSE-OJI is an environmental sociologist and currently holds the position of Senior Social Scientist with Forest Research in the UK, and is a Visiting Professor with the Forest and Nature Conservation Policy group at Wageningen University in the Netherlands. She has twenty-five years' experience researching the socio-ecological dimensions of forestry in the UK, Europe, Africa, and Asia with a particular focus on community and public engagement in forest governance, stakeholder representation in policy making processes, and socio-economic evaluation of programs and projects. Prior to working with Forest Research she was based at the Centre for Arid Zone Studies at Bangor University, where as well researching, she also directed and taught an MSc in International Natural Resources and Development.

MICHAEL M. BELL is Chair and Vilas Distinguished Achievement Professor of Community and Environmental Sociology at the University of Wisconsin-Madison, where he is also a faculty associate in Environmental Studies, Religious Studies, and Agroecology. He has published more than ten books, including *City of the Good: Nature, Religion, and the Ancient Search for What Is Right* and the widely used environmental sociology textbook, *An Invitation to Environmental Sociology*, now in its sixth edition. Mike is as well a prolific composer and performer of grassroots and classical music. For more on his work and passions, see his website: www.michael-bell.net.

ELIZABETH A. BENNETT is Associate Professor of International Affairs and Director of Political Economy at Lewis & Clark College in Portland, Oregon (USA). She holds a PhD in Political Science from Brown University and a MALD (Master of

Arts in Law and Diplomacy) in political economy and international development from The Fletcher School at Tufts University. Her research focuses on fair trade, living wages, textile supply chains, sustainability, and the cannabis sector. Her work is published in the *American Journal of Sociology, World Development, Sustainable Development, Agriculture and Human Values*, and *Globalizations*, among other journals, and she is the co-editor of *The Handbook of Research on Fair Trade* (with Laura Raynolds) and co-author of *The Civic Imagination: Making a Difference in American Political Life*.

VALERIE BERSETH is a PhD Candidate in Sociology at the University of British Columbia. Her research examines legal, scientific, and political conflicts in natural resource governance. Her current work is focused on climate change adaptation and genetic re-wilding strategies in Pacific Salmon management.

PHIL BROWN is University Distinguished Professor of Sociology and Health Science at Northeastern University, where he directs the Social Science Environmental Health Research Institute, and the NIEHS T-32 training program, "Transdisciplinary Training at the Intersection of Environmental Health and Social Science." He is the author of *No Safe Place: Toxic Waste, Leukemia, and Community Action; Toxic Exposures: Contested Illnesses and the Environmental Health Movement*; and co-editor of *Social Movements in Health* and *Contested Illnesses: Citizens, Science and Health Social Movements*. He studies biomonitoring and household exposure and reporting back data to participants, chemical policy and health effects of flame retardants and per- and polyfluoroalkyl substances (PFAS) (https://pfasproject.com/), and health social movements.

MICHAEL CAROLAN is Professor of Sociology and Associate Dean for Research and Graduate Affairs at Colorado State University. His research focuses on issues related to food, agriculture, sustainability, and justice. Some of his recent books include Society and the Environment: Pragmatic Solutions to Ecological Issues, 3rd edition (Routledge, 2020) and The Food Sharing Revolution: How Start-Ups, Pop-Ups, and Co-Ops are Changing the Way We Eat (Island Press, 2018).

JENNIFER S. CARRERA is an assistant professor of Sociology and Environmental Science and Policy at Michigan State University. She has a PhD in Sociology from the University of Illinois at Urbana-Champaign, an MS in Environmental Engineering from the University of Illinois at Urbana-Champaign, and an MS in Biostatistics from Emory University. Her research focuses on water access and quality, with particular attention to the politics of knowledge in environmental justice struggles. In 2018 she received an NIH K01 career award from the National Institute of Environmental Health Sciences to study community-driven low-cost technology development for improving knowledge production and sharing around water quality and public health in Flint, Michigan.

JOHN CHUNG-EN LIU is an associate professor of sociology at National Taiwan University. His scholarship draws from economic and environmental sociology to study climate change governance, and his work was featured in media outlets such as *Foreign Policy*, *Guardian*, and Public Radio International. Previously, he was an assistant professor at Occidental College and a Postdoctoral Fellow at the Harvard Kennedy School. Liu received his PhD from the University of Wisconsin-Madison, master's degrees from Yale University, and a BS from National Taiwan University.

MARK H. COOPER is an assistant professor in the Department of Human Ecology and the Department of Animal Science at the University of California, Davis. His research includes work in political ecology, economic geography, science and technology studies, and environmental sociology. His current research examines the governance of greenhouse gas emissions from agriculture, the politics and policy of climate change in California, and the role of measurement and metrics in regulation. He was previously a Mellon Postdoctoral Fellow at the College of William & Mary and a Postdoctoral Researcher at Lund University.

KATHERINE J. CURTIS is Professor of Community and Environmental Sociology and Director of the Applied Population Laboratory at the University of Wisconsin–Madison. Curtis's research is centered in spatial demography with a strong program in population–environment interactions, most centrally the relationship between migration and environmental change.

STEVEN E. DANIELS is on the faculty of Utah State University, where he serves as a Professor in the Sociology, Social Work, and Anthropology Department and as a Community Development Extension Specialist. His career focuses on the nexus of communities, land management agencies, and decision processes. While widely published (over sixty peer-reviewed articles), he also has extensive experience as a trainer/process coach and facilitator. His interest is in developing and employing innovative discourse systems that enable participants to improve situations that they regard as intractable. Most of his projects are fairly brief, but he has served for nine years as the facilitator of a landscape-level multi-party collaborative process in central Utah that has received a number of state and national accolades and accomplished far more than anyone might have predicted.

DEBRA J. DAVIDSON is Professor of Environmental Sociology in the Department of Resource Economics and Environmental Sociology at the University of Alberta, having received her PhD from the University of Wisconsin (1998). Her key areas of teaching and research include impacts and adaptation to climate change, and crises and transitions in food and energy systems. Dr. Davidson is President of the Research Committee on Environment and Society in the International Sociological Association, and she was a Lead Author in Working Group II of the Intergovernmental Panel for Climate Change's 5thAssessment Report. Her work is featured in several journals, including *Science, Nature, Global Environmental Change, British Journal of Sociology,* and *International Sociology,* among others. She is co-editor of *The Oxford Handbook of Energy and Society* (2018) and *Environment and Society: Concepts and Challenges* (Palgrave, 2018), and co-author of *Challenging Legitimacy at the Precipice of Energy Calamity* (Springer, 2011).

MARGO DEMELLO holds a PhD in Cultural Anthropology from UC Davis. She is an adjunct professor at Canisius College in the Anthrozoology Masters Program, and serves as the Program Director for Human–Animal Studies at the Animals and Society Institute. She has published over a dozen books, most within the field of human–animal studies, and dozens of articles and book chapters.

VALÉRIE FOURNIER is Senior Lecturer in Organization Studies at Leicester University. Her research interests range from Critical Management Studies to alternative organizations and economies. Her recent work has explored ideas of utopia, degrowth and commoning, and she has co-authored *The Dictionary of Alternatives: Utopianism and Organization* as well as the *Routledge Companion to Alternative Organization*. Her current research focuses on the ways in which the idea of the commons can reshape our understanding of, and relationships to, the environment, and in particular to biodiversity.

LEILA DA COSTA FERREIRA is a Full Professor in Environmental Sociology at Campinas University, Center for Environmental Studies and Research and Sociology Department. She is a member of the International Sociological Association (ISA) and the Earth System Governance Lead Project (Future Earth). She has been a Visiting Professor at Jiao Tong University in Shanghai, China (2009) and a Visiting Professor at University of Texas at Austin, USA (1998). She is the author of ten books and about sixty papers in environmental sociology. Her expertise includes environmental sociology, interdisciplinary social theory, and climate change.

JOSÉE JOHNSTON is Professor of Sociology at the University of Toronto. She is the co-author of *Foodies* with Shyon Baumann, as well as *Food and Femininity* with Kate Cairns. She has published recent articles in venues such as *Social Forum, Journal of Consumer Culture, Theory and Society*, and *British Journal of Sociology*. Her major substantive interest is the sociological study of food, which is a lens for investigating questions relating to consumer culture, gender, sustainability and inequality.

EMILY HUDDART KENNEDY is Assistant Professor in the Department of Sociology at the University of British Columbia. Her research is focused on sustainable consumption, including motivations and barriers to adopting sustainable practices, and an examination of how these practices intersect with gender and class inequality. In addition to co-authoring the edited volume, *Putting Sustainability into Practice: Applications and Advances in Research on Sustainable Consumption* with Maurie Cohen and Naomi Krogman (Edward Elgar Publishing), her research is published in venues such as the British Journal of Sociology, Environmental Sociology, and Social Forces.

DANIEL JAFFEE is an Associate Professor of Sociology at Portland State University. His research examines social contestation over the commodification of water, food and water justice movements in the global South and North, and the politics of fair trade and agri-food certification. He is the author of *Brewing Justice: Fair Trade Coffee, Sustainability, and Survival* (University of California Press), which received the C. Wright Mills Book Award. He is also a recipient of the Fred Buttel Scholarly Achievement Award from the Rural Sociological Society.

JULIE C. KELLER is Assistant Professor in the Department of Sociology and Anthropology at the University of Rhode Island. Her research and teaching focus on rural inequality, agricultural labor, farmers, and immigration. She is the author of *Milking in the Shadows: Migrants and Mobility in America's Dairyland* (Rutgers University Press, 2019).

GEOFFREY LAWRENCE is Emeritus Professor of Sociology at The University of Queensland, Australia. His work spans the areas of agri-food restructuring, neoliberal globalization, financialization, and food security. During his academic career Geoff has raised some $15 million in research grants and has published 25 books and over 500 journal articles and book chapters. Recent books include *THINK Sociology* and *The Financialization of Food Systems: Contested Transformations*. He was President of the International Rural Sociology Association (2012–2016). He is a Life Member of the Australian Sociological Association and an elected Fellow of the Academy of Social Sciences in Australia.

KATHARINE LEGUN is Assistant Professor in Communication, Philosophy and Technology at the University of Wageningen. Her work considers more-than-human approaches to understanding economic and environmental practice, focusing particularly on the role of plants and technology in shaping human agency within agriculture. Her work has looked at aesthetics and politics in the apple industry, hop geopolitics and craft beer, digital sustainability programs in wine and dairying, and the social implications of automation in horticulture. She has published in *Society and Natural Resources*, *Economy and Society*, *Geoforum*, *The Journal of Rural Studies*, *Agriculture and Human Values*, and *Environment and Planning A*, and has co-edited a special issue of *the Journal or Rural Studies* on the post-human turn in agri-food research.

MICHAEL LÖWY was born in Brazil in 1938 and has lived in Paris since 1969. He is presently Emeritus Research Director at the CNRS (National Center for Scientific Research). His books and articles have been translated into 29 languages. He is co-author with Joel Kovel of the *International Ecosocialist Manifesto* (2001). Among his main publications: *Romanticism against the Current of Modernity* (with Robert Sayre), Duke University Press, 2001; *Fire Alarm: Reading Walter Benjamin's "On the Concept of History,"* Verso, 2005; *On Changing the World; Essays in Political Philosophy from Karl Marx to Walter Benjamin*, Haymarket, 2013; and *Ecosocialism; A Radical Alternative to the Capitalist Ecological Catastrophe*, Haymarket, 2015.

MICHAEL MARDER is IKERBASQUE Research Professor in the Department of Philosophy at the University of the Basque Country (UPV-EHU), Vitoria-Gasteiz. He holds a PhD in Philosophy from The New School for Social Research in New York. Prior to coming to the Basque Country, he taught at universities in the United States (Georgetown, George Washington, and Duquesne Universities) and in

Canada (University of Toronto & University of Saskatchewan). In addition, Marder has been a visiting professor at the University of Bristol (UK), University of Lisbon and University of Coimbra (Portugal), University of Sichuan (China), Diego Portales University (Chile), and Forum on Contemporary Theory (India). His writings span the fields of phenomenology, political thought, and environmental philosophy. He is the author of numerous scientific articles and thirteen monographs, including *Plant-Thinking: A Philosophy of Vegetal Life* (2013); *Phenomena–Critique–Logos: The Project of Critical Phenomenology* (2014); *The Philosopher's Plant: An Intellectual Herbarium* (2014); *Pyropolitics: When the World Is Ablaze* (2015); *Dust* (2016); *Energy Dreams* (2017); *Heidegger: Phenomenology, Ecology, Politics* (2018); and *Political Categories* (2019) among others.

MARTHA MCMAHON is an associate professor in the department of Sociology at the University of Victoria, in British Columbia, Canada. She teaches and does research in the areas of environmental sociology, ecological feminism, feminist theorizing, local food and farming, agri-food governance. She is currently the director of the human dimensions of climate change undergraduate interdisciplinary minor degree and certificate programs. She is also a part-time farmer.

RALPH MATTHEWS is Professor of Sociology at the University of British Columbia in Vancouver, and Professor Emeritus of Sociology at McMaster University in Hamilton, Ontario. The author of five books and over 100 papers and reports, he has written extensively on issues of community sustainability, viability, and resilience. His recent work explores these issues in the context of climate change and resource management, particularly with regard to fisheries and aquaculture. He is also involved in research on the environmental implications of genomics and social aspects of neuroscience.

BARBARA MURACA is Assistant Professor of Environmental Philosophy at the Department of Philosophy and Environmental Studies of University of Oregon and co-director of the International Association of Environmental Philosophy (IAEP). Her research interests encompass social philosophy, sustainability and degrowth research, environmental ethics, process philosophy, feminist philosophy, and ecological economics. She has published several articles on degrowth, both in English and in German, and the book *Gut Leben: Eine Gesellschaft jenseits des Wachstums* (Wagenbach, 2014) where she explores the importance of concrete utopias and social experiment for a social-ecological transformation.

CHELSEA POWER is an MA student in Sociology at the University of Victoria in Victoria, British Columbia, where she also attained her BA in Sociology. She has a strong interest in ecofeminist theory, human and nonhuman–nature relationships, and sustainable energy futures. Her thesis research is in ecofeminist theory, environmental sociology, and current ecological crises. She is also an experienced cave explorer, rock climber, and outdoor enthusiast on Vancouver Island.

RACHEL A. ROSENFELD is a doctoral candidate in sociology and community and environmental sociology at the University of Wisconsin–Madison. Her research examines the connections between environmental changes and demographic processes. Rosenfeld received a BA in international studies from Middlebury College and a Master of Public Policy from Georgetown University. She previously worked as a health, nutrition, and population consultant at the World Bank.

KIAH SMITH is a Research Fellow at the University of Queensland in the sociology of environment and development, with a focus on critical agrifood studies. Her research agenda contributes new understandings of the social dimensions of food system transformation, through examining contestation within the corporate food regime and the discourses and practices of food justice in Australia and globally. She has researched and written on fair and ethical trade, resilience, land speculation and financing, governance, social solidarity economy, gender empowerment, livelihoods, and the green economy. Smith is also a Future Earth Fellow, co-convenor of the Brisbane Fair Food Alliance, and has recently been awarded a three-year Australian Research Council award for her study on "Civil Society, Fair Food Futures and the SDGs."

JANET STEPHENSON is Associate Professor and Director of the Centre for Sustainability, an interdisciplinary research center at the University of Otago, New Zealand. A social scientist, her research interests include Indigenous resource management, the interconnections between people and their local environments, and the role of individuals and organizations in the transition to a sustainable future. Much of her work is undertaken in interdisciplinary teams working on energy and climate change challenges.

RANDY STOECKER is a professor in the Department of Community and Environmental Sociology at the University of Wisconsin with an affiliate appointment in the University of Wisconsin-Extension Center for Community and Economic Development. He practices, publishes, conducts trainings, and speaks frequently on community organizing and development, community-based participatory research/evaluation, higher education community engagement strategies, and community information technology. His complete vita may be found at http://comm-org.wisc.edu/stoeckerfolio/stoeckvita.htm.

PAUL V. STOCK is an Associate Professor of Sociology and the Environmental Studies Program at the University of Kansas. As a rural and environmental sociologist, Paul is concerned with the practices farmers and their families employ to sustain their families, identity, livelihoods, and ecosystems in the context of changing political, economic, and social pressures. Paul has written extensively on the global food system and how it impacts people's everyday life including the books *New Farmers* (2019), *Food Utopias* (2015), and *Food Systems Failure* (2011).

MANUEL VALLÉE is a Senior Lecturer in Sociology at the University of Auckland, New Zealand. His research focuses on the relationship between environmental pollution, disease, and ideology. One project examines how governments contribute to the production of disease by producing knowledge that normalizes the use of pesticides and other toxicants. Another project examines the way governments and other hegemonic institutions obscure the relationship between pollution and disease, through the production of disease information that systematically obscures the relationship between pollution and disease. He is also the co-editor of the edited volume *Resilience, Environmental Justice, and the City*.

ABBI VIRENS is a PhD candidate at the University of Otago, New Zealand. Her current research explores the more-than-human dynamics within urban foraging spaces and how these affective interactions can produce post-capitalist possibilities. She is also interested in the intersections of post-humanism and Indigenous ontologies and how productive collaborations in these disciplines could participate in decolonization. Abbi holds a BA in anthropology from Laurentian University in Canada.

GREGG B. WALKER is a professor of communication and an adjunct professor in the environmental sciences, forestry, geosciences, and public policy programs at Oregon State University where he teaches courses in conflict and dispute resolution processes, international negotiation, environmental conflict, and science communication. Gregg conducts training programs, designs and facilitates collaborative processes about natural resource and environmental policy issues, and researches community-level collaboration efforts. He consults with the National Collaboration Cadre of the USDA-Forest Service and is on the roster of the US Institute for Environmental Conflict Resolution. Gregg also leads teams from two NGOs at the UN climate change negotiations.

ANELYSE M. WEILER is a PhD candidate in the Department of Sociology at the University of Toronto, a 2015 Pierre Elliott Trudeau Foundation Scholar, and a Culinaria Research Centre Fellow. Her teaching and research explore the convergence of social inequalities and environmental crises across the food chain, with a particular focus on migrant farm workers. She actively contributes to several organizations advocating for food security, migrant justice, and decent work. You can read her academic writing and public scholarship at anelyseweiler.com

PEARLY WONG is a Fulbright doctoral student pursuing a dual degree in Cultural Anthropology and Environment and Resources at the University of Wisconsin-Madison. Prior to coming to Madison, she had worked for more than three years in community development and sustainable natural resource management with UNESCO and United Nations University in Nepal, Cameroon, Sri Lanka, and Indonesia. Her research interests include critical development studies, environmental justice, and decolonizing knowledge in Asia. Her dissertation examines changing and conflicting understandings of development and well-being by community actors, in the context of rapid transformation in Nepal.

NATHAN YOUNG is Professor of Sociology at the University of Ottawa, Canada. His recent research includes studies of public discourses about climate change, conflicts over fisheries and aquaculture, the role of different types of knowledge in environmental management and decision-making, and community resilience in the face of environmental changes. He regularly collaborates with natural scientists on research into the human dimensions of environmental issues. His latest book is entitled *Environmental Sociology for the Twenty-First Century* (Oxford).

JULIA N. ALBRECHT is a senior lecturer at the University of Otago (New Zealand). She is an editor of *Pacific Geographies* and has recently joined the Editorial Board of the *Journal of Hospitality & Tourism Research*. Much of her recent research concerns visitor management, nature-based attractions and tourism in winegrowing regions.

CHELSEA SCHELLY received her PhD from the Department of Sociology at the University of Wisconsin–Madison and is an associate professor of Sociology in the Department of Social Sciences at Michigan Technological University. Her work explores how the sociotechnological systems used to support residential dwelling shape the organization of social life and conceptions of human–nature relationships. She researches and writes about a wide array of alternative sociotechnological systems.

Foreword
The Evolving Field of Environmental Sociology
Riley E. Dunlap

Although it had predecessors in sociological work on natural resources, primarily by rural sociologists, and on housing and the built environment, primarily by urban sociologists, environmental sociology as a distinct field emerged in the 1970s in the United States. It arose in the midst of high societal interest in environmental matters prompted by mobilization for the first Earth Day (April 22, 1970), which put environmental quality firmly on the nation's agenda. By the mid-1970s sociologists were engaged in a wide variety of work on environmental issues, from studying public opinion and environmentalism to exploring the social impacts of energy shortages and the possibility of "limits to growth," while also engaged in a range of applied work such as conducting social impact assessments. These diverse interests coalesced with the formation of the Section on Environmental Sociology within the American Sociological Association, established in 1976 and often viewed as the "official" birth of the field.

While it has had some ups and downs, environmental sociology has clearly grown into an established and reputable sociological specialty, with undergraduate courses commonly taught across the nation, graduate training programs established in several universities, and scholarship appearing in leading disciplinary journals. Importantly, it has also spread well beyond its early North American confines, now reaching large parts of the world and institutionalized within the International Sociological Association with the Research Committee on Environment and Society. Along the way, it has become far more diverse, as the range of topics investigated and theoretical and methodological perspectives employed have expanded rapidly and continue to do so.

Early on William Catton and I defined environmental sociology narrowly as the "study of societal–environmental interactions," distinguishing it from a "sociology of environmental issues" that studied phenomena such as environmentalism and environmental attitudes employing traditional sociological perspectives. We did so purposefully to create intellectual space and justification for studying such interactions (e.g., the impacts of resource scarcities on society), because doing so was rather problematic given conditions in the larger discipline. In the 1970s mainstream sociology tended to view the biophysical environment as little more than a (nonsignificant) context for the study of modern human societies and had an understandable hesitancy to invoke environmental conditions given past uses of environmental, geographical, and biological factors to justify racist, sexist, and nationalist ideologies. Avoiding these prior excesses, however, led sociology to endorse a socio-cultural determinism which looked askance at consideration of environmental conditions.

So strong was the "taboo" against considering biophysical conditions, that as late as 1980 sociologists of agriculture were arguing that *sociological* analyses of energy intensity of farms and farmers' adoption of innovations ought not take into account aridity and soil type when criticized for ignoring such biophysical conditions. Echoing a Durkheimian emphasis on explaining social facts with other social facts, they suggested that broadening their focus beyond social factors such as farm structure (size and ownership) and farmer characteristics by including biophysical conditions as explanatory variables was "non-sociological."

Disciplinary neglect of the biophysical environment began to change rapidly in the 1980s when environmental hazards attracted sociological attention, including their impacts on community dynamics, on residents' health, and especially their unequal distribution across racial and socioeconomic sectors. These negative impacts of environmental pollution and contamination were observable (albeit sometimes only with the help of scientific measurements), and typically inequitably distributed across social strata, making them ideal topics for a discipline long concerned with inequality. The emergence of global environmental problems in the 1990s, and the growing availability of national-level data on deforestation, air and water pollution, energy consumption, pesticide use, greenhouse gas emissions, and ecological footprints sparked cross-national analyses of the societal factors (or "driving forces") giving rise to these environmental conditions. These studies, often highlighting disparities between rich and poor nations, have rapidly proliferated, and combined with continued work on environmental justice at the community level have helped make the sociological investigation of environmental conditions commonplace. Indeed, both qualitative and quantitative studies of the social processes that generate environmental problems and the societal impacts of these problems now abound – and researchers need not fear having such work labeled as non-sociological.

Of course, there was always far more going on within environmental sociology than empirical studies of societal–environmental interactions, as research on topics such as environmental attitudes, environmental movements, environmental politics and governance, sustainability, and resistance to environmental reforms – among others – continued and evolved. In particular, such work has often broadened from a local or national focus to the cross-national and international levels, as with multinational comparisons of environmental attitudes and investigations of Northern versus Southern hemisphere and even global environmental movements.

Furthermore, new conceptual and theoretical perspectives have arisen, often induced by intellectual trends (in sociology and beyond), perhaps most notably the "cultural turn" stimulated by postmodernism. This led some environmental sociologists to problematize environmental conditions, highlighting the societal processes that led to some being labeled as problems while others were neglected, and others to interrogate the scientific practices and knowledge giving rise to "environmental problems." Even more fundamentally, the concepts of "environment" and "nature" attracted scrutiny, with analysts emphasizing their inherent ambiguities, cultural contingencies, gendered natures, and the like. Work along the above lines provided vital insights, such as the importance of examining the discourse, narratives, and frames employed in environmental controversies. Yet, such work sometimes attracted criticism from those who felt that "relativizing" scientific knowledge of environmental problems played into the hands of powerful interests opposed to ameliorating them, and that erasing (or at least blurring) the human–environment distinction made empirical analyses of the causes of such problems impossible. While the vigorous "constructivist-realist" debates of the 1990s have largely subsided, schisms persist – including over the role of scientific knowledge in our field (a topic of increasing complexity given the growing dismissal of such knowledge in our "post-truth" era).

In short, the evolution of environmental sociology over the past four-plus decades has been influenced by changes in real-world conditions – from the continual emergence of new environmental problems often, as in the case of climate change, at larger geographical scales, to changing societal conditions such as globalization, the rise of social media and the spread of neoliberal hegemony – and by intellectual and disciplinary trends. The latter include new theoretical perspectives (postmodernism, feminist theory, intersectionality, etc.), methodological tools (Discourse Analysis, Geographic Information Systems, Multi-Level Modeling, Computational Topic Modeling, etc.) and ever-growing data sets on environmental conditions. It is little wonder then that the environmental sociology of today differs substantially from that of the early days, and is constantly changing.

The diverse and evolving nature of environmental sociology is captured by this large, two-volume handbook. Indeed, the editors have taken an exceptionally expansive view of environmental sociology, referring to it as a "careful, committed conversation" about matters concerning the intersection of society and the environment – rather than a field with readily discernible characteristics that can be clearly distinguished from other fields. They have ensured the implementation of their expansive perspective in several ways: (1) by soliciting contributions from scholars in a range of environmental fields beyond sociology per se, capturing the inherent, interdisciplinary nature of environmental sociology; (2) by inviting contributions from scholars at various career stages to capture intergenerational variation in perspectives and ensure new voices are being heard; (3) by soliciting contributions that focus on the Southern hemisphere to ensure geographical variation (although Asian environmental sociology, due to linguistic barriers, is under-represented); and (4) by covering a vast range of topics, from well-established concerns in the field (e.g., economy and environment, environmental politics, social justice, and sustainability) to newer foci (e.g., sexuality, bodies, animals, plants, and even outer space).

The result departs from the traditional notion of handbooks as summaries of existing knowledge about a set of key topics deemed central to an area of inquiry, including a prior handbook focused on U.S. environmental sociology (R. E. Dunlap and W. Michelson, eds., *Handbook of Environmental Sociology*, Greenwood Press, 2002) and two editions of another with a more international focus (M. R. Redclift and G. Woodgate, eds., *The International Handbook of Environmental Sociology*,

Edward Elgar, 1997, 2010). I believe many scholars will welcome this new, more expansive perspective on environmental sociology, as it certainly captures the rapid diversification and permeability of the field. Some of the newer, as yet less central, concerns will no doubt blossom into productive lines of inquiry – just as rather recent work on food systems and food justice has done and gender and the environment appears to be doing. Some others may prove to have been rather short-lived intellectual explorations that fail to spark sustained attention and research. While the academic marketplace of ideas is far from perfect, with many built-in inequities, it does over time nonetheless tend to sort out those that prove especially insightful. Only time will tell which of the many newer topics and perspectives offered in these two volumes will "catch on," and the editors are to be congratulated for giving them a chance to do so by inclusion in this significant platform.

As far back at the 1980s, during the Ronald Reagan era, I was fond of saying "this is a bad time for the environment, but a good time for environmental sociology" (the latter in the sense that our work was highly relevant). I could never have imagined then that we would be looking back at the anti-environmental actions of the Reagan Administration as moderate compared to those of the George W. Bush Administration, nor as meager relative to the current Trump Administration's all-out assault on environmental science and regulations. Nor could I have imagined that the successful diffusion of neoliberal ideology and more recent rise of right-wing populism (itself spawned in large part by the success of neoliberalism in undermining economic progress for the working class) would spread anti-environmentalism to many parts of the world, including Northern Europe where a decade ago some sociologists were proclaiming the success of ecological modernization in creating more sustainable nations.

Clearly the need is now greater than ever for environmental sociology to both shed light on the nature of our environmental predicaments and offer insights for trying to move beyond them into a more just and sustainable future. Our field has a long history of engaging with society at large, nowadays labeled "public sociology," most obvious in the work of environmental justice scholars but increasingly apparent in other emphases from environmental health to climate change. Coupled with an ever-growing knowledge base, and continually expanding theoretical and methodological tools, environmental sociology is poised to help overcome current threats and barriers to sustainability while also pointing to alternative, more ecologically sustainable, futures. Many will find the wide range of insights offered in this handbook of considerable help in these endeavors.

Introduction

Katharine Legun, Julie C. Keller, Michael Carolan, and Michael M. Bell

As with any edited collection, when planning *the Cambridge Handbook of Environmental Sociology* we imagined who would be using the text, and when they would be using it. We considered a researcher, starting on a new project and looking for approaches to better understand a complex environmental problem; a student, having been exposed to environmental sociology, excited by some ideas and looking to become better oriented with the field; a teacher, looking for readings to assign to students in the upcoming semester; or a practitioner, whose interest lies somewhere in that liminal space straddling town and gown. We thought of the purpose of handbooks, in a world where a quick search on the Internet can generate an article to answer any question, and sometimes an article to seemingly support just about any belief. In this context, a handbook can act as a reliable reference point that includes a broad, but not boundless, survey of ideas and a quick, but not superficial, snapshot of some of the empirical work that supports and elaborates those ideas. A handbook should provide some grounding and some coherence in what can feel like a world of shaky information. It should support the kind of depth and rigor that can come from physical, textual interactions, and contribute to a sense of intellectual continuity and community as part of the enduring work of universities.

Environmental sociologists can play a strong role in contemporary public life, and we often find ourselves feeling obliged to do so. The worst effects of climate change loom heavily on the horizon, while deforestation and energy use seem to only accelerate, and the worst effects compile most strongly on the least advantaged. At the same time, political debates and growing populism internationally seem to often carry a distinct environmental flavor. Debates over the use of public funds for environmental protection and energy transitions, or over reduced access to resources for conservation purposes, have reinvigorated old divisions between economic and ecological wellbeing. These are not new: they align with some of the earliest work in environmental sociology that discussed how our economic paradigms became incompatible with, and even hostile to, more environmentally focused paradigms (see Catton and Dunlap 1978). Political economy approaches, like the "treadmill of production" (Schnaiberg 1980; Gould, Pellow, and Schnaiberg 2015), further illustrate how our economic system is dependent on the increased exploitation the environment. The inherent, capitalist pressures to expand production and consumption and to reduce costs are linked to environmental degradation through the search for cheaper, more accessible resources and easier ways to dispose of increasing waste, with little regard to issues of inequality. This exploitative economic model being enacted is not the only approach our society can take, and many environmental sociologists have focused their careers trying to dislodge it and shift political discourses that are grounded in its assumed inevitability. Clearly the economy, in a holistic sense, suffers with ecological collapse and rampant environmental injustice.

Politicians advocating for the pursuit of economic growth through the relaxing of environmental and labor regulations, and thus reinforcing these assumed dichotomies, is not novel, and yet the deployment of this logic in new contexts contains new dynamics and new implications. Better understanding historical patterns and their distinct operation in our current setting is an important task for us. For example, social media increasingly plays a forceful role in shaping the circulation of ideas and information, and can be seen to spur populism by spreading dubious information and supporting it in an echo chamber (Groshek and Koc-Michalska 2017). Social media can also be a powerful force in social movements, bringing masses of diverse people together around environmental issues (see Steinhardt and Wu 2016). The growth of social media as a dominant form of communication can be seen broadly to influence inter-generational social and political dynamics while also connecting people quickly over significant geographical spaces. Social media is just one of the many factors that has changed the world in which policy debates about the environment occur.

We embrace the continuity we see in the subjects we pursue, and we also recognize the disjuncture and the need for work firmly situated within those new contexts, able to consider their unique contemporary characters. After all, discontinuity and disjuncture are, in many ways, characteristics of contemporary social and environmental life: where the rate of environmental change and extent of environmental injustice challenges our ability to adapt (see Barnett et al. 2015), we are facing an increasing rate of species extinction (Ceballos et al. 2015), and we see high rates of mass global migration as a result of conflict, but also increasingly as a result of environmental change (Mortreaux and Barnett 2009). In their article in *Nature*, Black et al. (2011) suggest that migration from climate change can provide opportunities, and that reluctance to move can generate the greatest suffering in relation to climate change. We hear similar comments in response to hurricanes: those that suffered the most were those that stayed rather than leaving their homes. Change – the embrace of change, and the normalization of significant continuous change – is a defining feature of our era. Yet, the meaningful change we want to see toward environmental reform and justice, and away from unfettered economic growth, can sometimes seem beyond our capacity, and this is certainly true if we're looking at an individual level.

It is perhaps this feeling of discontinuity and disjuncture that also compels us to try to build broad and diverse coalitions across our field, and outside our field. We do this particularly in the context of an academic space that is also shifting. The neoliberal influences on universities have led many institutions to be run like businesses, with academic work becoming more precarious. These changes pose a number of challenges for our work: funding for research is increasingly dependent on investment from private businesses, exerting pressure on us to make our work palatable to the private sector's aims and logics; we must produce work that is measurable according to the metrics of our institutions, so that we are encouraged to prioritize rapid academic publications over other types of slow and engaged forms of scholarship (Berg et al. 2016; Ball 2012); and, the changes in university staffing and university priorities can make it challenging to maintain enduring, meaningful connections with communities outside the university (Brackmann 2015). These connections are particularly important given the work we do in environmental sociology: we learn about society, but we also participate in it and hope that our work can lead to a better future. Developing and maintaining strong relationships with people outside of the university is necessary if we are to be relevant and meaningful.

A handbook is also a device for disciplines, which are changing. Universities seem to be constantly shifting to more businesslike forms of organization, where degrees are being designed to cater to the labor market, rather than a coherent disciplinary foundation, and departments are being managed in ways that better align with these degrees. In this context, handbooks are playing a role in reproducing the idea of a *discipline*. This can sometimes be a way to draw boundaries around a field and thus further valorize a limited set of founding figures. Alternatively, disciplines can also be a powerful way to create a sense of continuity and support through an intellectual community; a way to bring together without disciplining out heterogeneity and the willingness to experiment.

We use the term "discipline" to recognize a community of environmental sociologists, drawn together by interests in phenomena occurring at the intersection of society and the environment. We also share some general conceptual foundations, and tend to have an underlying commitment to questions around social and ecological justice. And yet, while we may share some foundations, and may steer our ship toward some similar kinds of beacons, it would be a disservice to many of our colleagues to draw thick disciplinary boundaries, or ignore the vibrant evolution of ideas that has proliferated over the past fifty years. Indeed, there may be more diversity of ships and beacons within those who identify as environmental sociologists then overall differences with allied environmental humanities and social sciences, such as political ecology, environmental anthropology, science and technology studies, environmental history, ecological economics, environmental philosophy, and food studies. We do not use the word discipline to reproduce a strong artificial boundary around what may be considered environmental sociology, or to insist on rigorous adherence to a foundational canon. To do so would implicitly exclude a lot of brilliant work being done by those we consider part of our community.

Rather, we envision the notion of a discipline more in the sense of careful, committed conversation about common concerns, than in the sense of boundaries around participation in that conversation. Instead, the idea of a discipline can be a way of imagining a community at a collective level, and which welcomes those "environmental borderlands" (Zimmerer 2007) that often defy conventional disciplinary classifications. Conceptualizing the "discipline" as careful, committed conversation allows us to be reflexive about the work we are doing, our role in our communities, and the hopes we have about our work in the future. Along these lines, developing this Handbook was an exercise in both trying to recognize some important currents in environmental sociology broadly, but also thinking about how to participate in the discipline through a handbook. We wanted to have chapters that reflected major traditions in environmental sociology and a strong and clear sociological grounding, but we also wanted chapters that illustrated a creative breadth of work. We also wanted to include authors who were writing from a range of different places about those places, who were at different career levels, and those who had some ways of approaching environmental sociology from different angles.

Textbook editors have a particular snapshot of a topic or area, and this influences the book they put together. We had four editors, but we are all from a predominantly North American academic tradition and English speaking. We knew this would limit what we know of environmental sociology. In preparing the text proposal, we explored some areas of research that we had not encountered before and became familiar with new bodies of work. We were also fortunate in that many of the contributing authors, and many of our colleagues, gave us suggestions about people to include. We tried to balance what we knew with an exploration of the field, which, while not exhaustive, led us to some exciting corners and edges.

In keeping with our view of the environmental sociological project as careful, committed conversation, a lot of the authors included in the Handbook are not, in the bounded sense of a discipline, environmental sociologists. And yet, they are part of a shared community as their foundations, approaches, and methods so strongly align with what it means to *do* environmental sociology that we would be remiss to exclude their work. In other cases, we can see how other disciplines may pointedly speak to particular problematics in environmental sociology, and felt it valuable to bring that work explicitly into this conversation. Environmental concerns are necessarily interdisciplinary, but that doesn't mean being anti-disciplines. To be interdisciplinary, and even transdisciplinary, there have to be disciplines to integrate. What we are against is the bounded sense of discipline that prevents connecting our conversations. The challenges we face as humans on a struggling planet require a collective effort. Building a strong community within our discipline in a shifting academic landscape, and building coalitions outside our discipline with those who share our concern for the future and our commitment to a better one, is not just enjoyable as it has been in developing this Handbook, but it is absolutely necessary.

Conceptualizing and Practicing Environmental Sociology

The Handbook is divided into two volumes: one focuses on conceptualizing environmental sociology, and the other on practicing environmental sociology. Of course, drawing some kind of boundary between more conceptual work and more practical work is fraught, and many of the chapters included in the conceptual volume are quite practical, and vice versa. Regardless, we chose to make this rough distinction for organizational purposes, and include chapters that were tackling conceptual issues more directly in the first volume, covering the broad areas of theory, economy, culture, politics, and justice. In the second volume, we included chapters that focused on a particular substantive area of environmental sociology, a particular subfield, or a particular kind of environmental problem. This volume includes sections on methods, embodiment, research beyond-the-human, sustainability, natural resources, food and agriculture, and social movements.

The first volume begins by tackling foundational concerns of the field, tracing and challenging environmental sociology's emergence in relation to classical social theory (Holleman, Chapter 1) and centering globalization and the need for a globalized approach (Lidskog and Lockie, Chapter 2). New theoretical directions are found in the reformulation of nature/"nature" in connection with labor and sex-gender (Salleh, Chapter 3), and in relation to the divine and the human and the material and ideological conflict of bourgeois and pagan (Bell, Chapter 4). Another promising theoretical avenue is the burgeoning scholarship on environmental microsociology (Brewster and Puddephatt, Chapter 5).

Part II brings together theories of the economy and environmental sociology, emphasizing again the importance of critical materialist frameworks (Longo and York, Chapter 6) and ecosocialism specifically (Löwy, Chapter 9), as well as a call to "emplace" sustainability (Barron, Chapter 11). The substantive issues addressed in this section include analyses of "green economies" (Li and Green, Chapter 7; Bresnihan, Chapter 8), community economies (Barron, Chapter 11), as well as the commons and "communing" practices (García-López, Chapter 10) as possibilities for contesting and transforming the capitalist economy.

In Part III, the theme of culture unites topics as disparate as media and environmental activism (Hannigan, Chapter 12), colonialism and parks (Ramutsindela, Chapter 13), nature-based tourism and environmental philosophy (Zhang, Higham, & Albrecht, Chapter 16), outer space as environment (Ormrod, Chapter 15), as well as "futuring" and transition pathways (White and Roberts, Chapter 14).

In Part IV, authors focus squarely on politics, power, and the state. Whether through development, enclosure, or restoration, the heavy hand of the state in land management is made clear in John Zinda's chapter, as he makes a case for the transfer of knowledge between political ecologists and environmental sociologists (Chapter 19). Cock identifies contested "environmental imaginaries" in the fight over land rights in South Africa, arguing for indigenous frameworks that challenge the state's view of nature as commodity (Chapter 17). Campbell's chapter examines the emergence of the audit culture, the mechanisms responsible for the eco-labeling of goods, identifying a new form of governance in this recent phenomenon (Chapter 18). Michael A. Long, Michael J. Lynch, and Paul B. Stretesky cover the area of green criminology, arguing for uniting the treadmill of production approach with ecological Marxism to further understand environmental problems (Chapter 20). Philip Macnaghten (Chapter 21) discusses the role of governance in shaping the practices and outcomes of science, outlining four different governance models including a co-production model, and suggesting that this may be applied through a responsible research and innovation framework. Noah Feinstein (Chapter 22) delves into the relationship between public knowledge and democracy, and proposes some solutions to the enduring challenges around the development and exercise of public knowledge for decision making. Lastly, Hale and Carolan consider how equity and diversity can contribute to an understanding of relational resilience, and explore relational resilience through an analysis of two food cooperatives (Chapter 23).

In Part V, the authors bring social justice to the forefront. David N. Pellow (Chapter 24) expands traditional understandings of environmental justice (EJ) through examining four pillars of "Critical EJ Studies." Henri Acselrad (Chapter 25) discusses the struggle for environmental justice in Brazil, while John C. Canfield, Karl Galloway, and Loka Ashwood (Chapter 26) consider challenges to environmental justice in rural communities in the USA. Leslie King (Chapter 27) traces the relationship between capitalism and socioecological harm, highlighting corporate power and signaling opportunities for disrupting this control to work toward environmental justice. Phillip Warsaw (Chapter 28), looks at environmental inequality from the perspective of environmental economics, through a critical approach to neoclassical economics.

In the second volume, focusing on practicing environmental sociology, we start with some new epistemological and methodological insights particular to the field of environmental sociology. Martha McMahon and Chelsea Power (Chapter 1) explore the tensions and potential intersections between more-than-human approaches and strong sociological traditions attending to power structures. Steven E. Daniels and Gregg B. Walker outline the epistemological foundations of discourse-intensive approaches (Chapter 2), while Randy Stoecker describes the history, aims, and challenges associated with community-based research (Chapter 3), and Katherine J. Curtis and Rachel A. Rosenfeld elaborate on the strategies and benefits of drawing from spatial data in conducting social research (Chapter 4).

In the second section of the second volume, we focus on embodied work in environmental sociology, focusing on sexuality and gender, as well as health and wellbeing. These chapters are connected in the ways that the environment shapes social wellbeing through its relationship with our bodies. Julie C. Keller (Chapter 5) discusses the discursive exclusion of queer people from an imagination of rurality, along with the rich history of queer rural social movements and spaces. Kathryn Gregory Anderson considers how masculinities are constructed with and through the environment, and how particular kinds of masculinities can be hostile to the environment (Chapter 6). Jennifer S. Carrera and Phil Brown (Chapter 7) discuss the role of sociology in bringing a structural analysis to work on public health, while Manuel Vallée (Chapter 8) describes how the environment gets excluded from popular accounts of health issues.

In Part III, we look at the role of non-humans in environmental sociology, focusing particularly on plants and animals. Katharine Legun and Abbi Virens (Chapter 9) describe the interventions offered by more-than-human approaches, and discuss what it can offer environmental sociology. Michael Marder (Chapter 10) considers our limited philosophical understanding of plants, and describes how a better philosophy of plants might open up radical new ways of thinking. In Margo DeMello's chapter focusing on human–animal studies, she describes the field's growth as well as the methodological value of multispecies ethnography (Chapter 11).

We focus on sustainability and climate change in Part IV, beginning with a look at the burgeoning area of degrowth through the work of Valérie Fournier (Chapter 12).

Emily Huddart Kennedy takes us through inequalities and sustainable consumption (Chapter 13), and Janet Stephenson discusses the relationship between culture and sustainable outcomes (Chapter 14). The remaining three chapters deal with questions of governance and sustainability or climate change. John Chung-En Liu and Mark Cooper (Chapter 16) consider the challenges and limitations to carbon markets as a method of international environmental governance. Bianca Ambrose-Oji considers the practice of community governance of foresty in the UK (Chapter 15), while Leila da Costa Ferreira tackles multi-level governance in Brazil, considering how action around climate change scales up from the city, to state, to federal level (Chapter 17).

The section on resources, Part V of volume two, covers water, petroleum, as well as the role of genomics and adaptive co-management in conservation and the management of resources. Daniel Jaffee discusses the privatization of water resources (Chapter 18), while Debra J. Davidson discusses the politics around hydraulic fracking for petroleum and the silencing of dissent (Chapter 19). Valerie Berseth and Ralph Matthews discuss how the application of genomics is changing the way we understand the natural world (Chapter 20). Nathan Young explores the challenges and opportunities of adaptive co-management structures which aim to involve communities in the management of natural resources (Chapter 21).

The penultimate section of the handbook, Part IV, covers food and agriculture. Michael Carolan discusses the development and adoption of precision tools and big data in agriculture (Chapter 22). Josée Johnston and Anelyse Weiler consider whether the food system can be changed through changing consumption patterns (Chapter 23). Geoffrey Lawrence and Kiah Smith look at the effects of neoliberal globalization on the food system (Chapter 24), and Paul V. Stock considers environmental morality from an agri-food approach (Chapter 25).

The last section of the two volumes, Part VII, is meant to offer some critical but insightful, and potentially even inspiring, thoughts on how to achieve a more just and beneficial approach to the environment. The section on social movements includes a chapter by Chelsea Schelly who describes different kinds of technologies and their social implications, and how alternative technologies can lead to more emancipatory practices (Chapter 26). Elizabeth A. Bennett outlines recent developments in the fair trade movement, and considers how the movement can best develop in the future (Chapter 27). Barbara Muraca (Chapter 28) extends a discussion of degrowth, elaborating on the history of the movement, its contemporary relevance, and its possibilities for radical transformation. Lastly, Pearly Wong outlines what lessons can be learned from environmental justice movements of the global South (Chapter 29).

Over the course of the text, we see some similar themes popping up, and some notable absences, which we outline in our conclusion (Chapter 30). It was challenging to decide where chapters should fit, and what would make a coherent, or at least sensible, section. We spent considerable time shifting chapters around and renaming sections, and mulling over whether, for example, it made sense to have a more-than-human section given the extent to which so many chapters touch on more-than-human approaches. We ultimately decided to break up a section on science and technology studies (STS) for this very reason – it was simply too difficult to see what chapters to include and which belonged elsewhere, and we decided that STS would easily be incorporated into other sections due to the prevalence of the approach in all areas of environmental sociology. In short, we made some choices about how to organize the two volumes, but as most scholars will know, these categories are porous and fluid in reality, even if they are inscribed on paper. We hope that readers will see connections and continuities outside of those we have chosen to develop organizationally, and may find some personal inspiration and creative insight through those unwritten threads – inspiration and insight they ultimately contribute back to the careful, committed conversation of environmental sociology.

References

Ball, S. J. (2012). Performativity, commodification and commitment: An I-spy guide to the neoliberal university. *British Journal of Educational Studies*, 60(1), 17–28.

Barnett, J., Evans, L., Gross, C., et al. (2015). From barriers to limits to climate change adaptation: Path dependency and the speed of change. *Ecology and Society*, 20(3), 5.

Berg, L. D., Huijbens, E. H., & Larsen, H. G. (2016). Producing anxiety in the neoliberal university. *The Canadian Geographer/le géographe canadien*, 60(2), 168–180.

Black, R., Bennett, S. R., Thomas, S. M., & Beddington, J. R. (2011). Climate change: Migration as adaptation. *Nature*, 478(7370), 447.

Brackmann, S. M. (2015). Community engagement in a neoliberal paradigm. *Journal of Higher Education Outreach and Engagement (TEST)*, 19(4), 115–146.

Catton, W., & Dunlap, R. (1978). Environmental sociology: A new paradigm. *The American Sociologist*, 13(1), 41–49.

Ceballos, G., Ehrlich, P. R., Barnosky, A. D., et al. (2015). Accelerated modern human-induced species losses: Entering the sixth mass extinction. *Science Advances*, 1(5), e1400253.

Gould, K. A., Pellow, D. N., & Schnaiberg, A. (2008). *Treadmill of production: Injustice and unsustainability in the global economy*. New York: Routledge.

Groshek, J., & Koc-Michalska, K. (2017). Helping populism win? Social media use, filter bubbles, and support for populist presidential candidates in the 2016 US election campaign. *Information, Communication & Society*, 20(9), 1389–1407.

Lave, R. (2012). Neoliberalism and the production of environmental knowledge. *Environment and Society*, 3(1), 19–38.

Mortreux, C., & Barnett, J. (2009). Climate change, migration and adaptation in Funafuti, Tuvalu. *Global Environmental Change*, 19(1), 105–112.

Schnaiberg, A. (1980). *The environment: From surplus to scarcity*. New York: Oxford University Press.

Steinhardt, H. C., & Wu, F. (2016). In the name of the public: Environmental protest and the changing landscape of popular contention in China. *The China Journal*, 75(1), 61–82.

Zimmerer, K. S., (2007). Cultural ecology (and political ecology) in the "environmental borderlands": Exploring the expanded connectivities within geography. *Progress in Human Geography*, 31(2), 227–244.

PART I

Methods

1 Re-compos(t)ing the Ghosts of Sociologies Past: Towards More Cosmoecological Sociologies

Martha McMahon and Chelsea Power

Given current environmental challenges, is it perverse of feminist environmental sociologists to seem to abandon the sociological ancestors and appeal instead to mushrooms, to call for making kin with other critters, to turn to telling different stories or to compos(t)ing ourselves differently as humans (Haraway, 2016; Tsing, 2015)?

We begin by asking what is sociology and who is it for? One answer is that sociology is the study of people 'doing things together' (Becker, 1986). Feminist sociology would say that it is both about and also 'for people' (Smith, 2005). We hold it must also be for and about 'more-than-people'.[1] Environmental sociology's future lies in attending to new ways of being and of doing things together in an enlarged understanding of being together. It includes studying how people can 'make kin' with our multi-speciesed others and the arts, not just the sciences, of how to live well in the ruins of capitalism, as some ecologically minded feminists describe how we must learn to live (Alaimo, 2016; Haraway, 2016; Tsing, 2015, 2017; Shotwell, 2016). That requires learning with and from, and not just about, what many post-humanists call 'the more-than-human' and some Indigenous peoples call Mother-Earth (Mendez, 2018; McGregor, 2016) and where knowledge is better understood as a verb (activities of knowing) and is embodied in transformative relationships. We must become different in order not just to live well with others; but to stay alive at all. Living respectfully with the more-than-human offers new (and old) possibilities to ways of being human, or as sociologists would say, being social. New words like cosmoecology (Despret & Mueret, 2016) or alter-politics (Hage, 2015) come to mind. Haraway's book *Staying with the Trouble: Making Kin in the Chthulucene,* is a lot about the connections, concepts and the stories with which we think and know in these urgent times alternatively named the Anthropocene, Capitalocene and Chthulucene. Knowing, being and doing are inseparable. Can environmental sociologists study people and the 'more-than-people' doing things together without reproducing prophesies of anthropogenic environmental or climate apocalypse? This

---

1 The term in the literature is more usually 'the more-than-human' which presents a problem for this chapter as the concept of humanity is challenged. However, because it is so widespread in the feminist literature, we will sometimes use the term more-than-human even though the term 'more-than-people' might be more sociological. Similarly we use the term 'nature' at times in ways inconsistent with our conceptual disruption of the term but we have no other short-hand referent.

means understanding how people, their institutions, their stuff, their comings and goings are implicated in ecologically and socially destructive outcomes without falling into the kinds of mystifications about humanity and society that abound in the popular and policy talk about climate change.

Sociology teaches us that we are produced and sustained in our ongoing social relationships. Most of these are invisible to us as individuals and even as collectivities. Environmental sociology stretches the concept of significant relationships within which to understand 'people doing things together' and in which we become individuals and collectivities to include the more-than-people.

The ontological turn towards post-humanism in the humanities and social sciences disrupts the older binaries of nature/culture and subject/object central to sociology's own history. This ontological turn raises new challenges as it works to overcome old problems. It is critiqued as being inadequate to engage the realities of globalized structures and power-relations. By contrast environmental sociology's strengths lie precisely in attending to power, structure and inequality, in exposing harms, debunking the politics of false stories, and offering analytical critique that show how inequalities and multiple forms of harm are done (Dunlap and Brulle, 2015). This is what sociologists are really good at doing.

So one can envision environmental sociology as offering hybrid political-intellectual spaces between post-humanism and critical sociological traditions. To shed the US/Eurocentricism and anthropocentricism of modernist binaries, as post-humanism proposes, does not mean we must shed politics or stop thinking sociologically. The ontological contributions of post-humanism can be knitted into environmental sociology's political critique and vice versa.

If, as many now warn, climate change imaginaries operate with an untheorized concept of humanity and are symptomatic of a 'post-political condition', clearly more sociological work is urgently needed. Our sociological politics must include the more-than-human, but not as objects. They must be recognized as central and as co-participants in the stories we tell and the worlds about which they are told. Without that, we as sociologists are not up to the challenges of our times. We fail our discipline and our times.

Environmental sociology makes clear that neither environmental problems generally nor climate change in particular can be understood without attending to structure, power and culture (Dunlap & Brulle, 2015). It sees injustice and inequality as being as central to climate change as are green house gas (GHG) emissions. Thus the popular and common natural science referent of humanity as an historical actor becomes conceptually and empirically untenable (Malm, 2016) and ideologically dangerous. It cannot be justified even in the service of the 'good cause' of trying to explain climate change to wider populations. Neither can one continue to justify sociological conceptualizations of power that are anthropocentric (or racialized, or androcentric). Time to recompose if not re-compost our diversely embodied selves and our stories differently.

We believe environmental sociology needs to move to the centre of sociology. Sociology needs the kinds of conceptual tools to allow engagement with environmental crises and climate change in ways that are deeply respectful of the diversity of people(s) and of the more-than-human. The sociology 'ancestors' may not have bequeathed the tools to accomplish this, but we need these tools.[2] Even the word environment creates the illusion that the environment is something other to the knower, separate from 'us' (Lykke, 2009). Of course there is no 'us' in the sense implied, and many people(s) do not experience themselves as separate. The term environment itself contains a problematic political epistemic stance. Without conceptual 're-tooling', particularly in the context of a-political climate change public discourses (Swyngedouw, 2013), there is a real danger that environmental sociology will 'morph' into a discipline that enables and intensifies the control of people and nature. Weber's iron cage writ large. We argue that this is a real danger as long as environmental sociology remains too imprisoned in and by sociology's origin stories[3] to be able to engage the kinds of gendered geo-politics and ontological politics that are buried in the discipline's foundations and from which it must escape.

This might help us understand why, as Latour (2004) pointed out, some well-established sociological conceptual tools such as critical construction are not apt for navigating the new waters of fake news and climate change debates. They seem, he believes, to have 'run out of [critical] steam'.

Too often the role assigned to the social sciences (and humanities) with respect to climate change is of figuring out and encouraging the general population to 'get with the program'. There is no 'program' in the conventional sense, just a mix-and-match grab bag of discourses circulating around valuable yet limited knowledge traditions and assumptions about forms of climate governance that focus mostly on GHG emission reduction, too often limited to commodifying carbon. In these discourses carbon emissions are fetishized and much of the humanities and social sciences are rendered marginal. A reduction in GHG emissions is urgent but must not be conflated with the deeper challenges with which emissions are inseparably entangled. Sociological analysis can change climate change discourses as climate change forces sociology to change.

A call for foundational disciplinary change isn't new. Not very long ago, feminism pushed sociology to engage with the bio-politics of sexed and gendered lives. Feminist socio-ecological analysis now calls for even more deeply transformative conceptual change in sociology. These are more than good ideas. By building on environmental sociology's unique appreciation of the role of social organization and institutionalization, new kinds of sociological hybrids can help bring into being and institutionalize far deeper respect for people and the more-than-people. This won't come easily to a discipline and university departments that can bureaucratically incorporate calls for change with business as usual and still treat feminist analysis as primarily for feminist scholars and environmental sociology as being for the environmentalists in the department, or see post-colonial thought as a non-core-program course option, or Indigenous studies as a make-up for colonial pasts, and understand nature as an object of theoretical and discourse analysis, not as referencing agency or different forms of subject-ness. Only the privileged can believe that life and

---

2 Some argue Marx and Weber did bequeath such tools but they were not recognized.
3 We are responding to the kinds of oversimplified origin stories popular in sociology. The reality of sociology's origins is more complex, diverse and messy.

scholarly business will go on as usual. It won't. Climate change is indifferent to sociology and its projects. It won't wait, nor will the misery of so many others. Sea levels will rise, the gulf-stream will shift course, people and other creatures will die (are dying, have gone extinct) and new ones emerge, all sorts of 'stuff' will and is happen(ing). It is time to revisit foundational stories. It seems to us that the discipline of sociology is founded on multiple repressions. These are intergenerational repressions reproduced by the origin stories of sociology and from which environmental sociology needs to be liberated if it is to contribute to ecologically liveable and socially just worlds. That said, we believe sociology is an indispensable resource for liveable futures in which people, the more-than people, and justice can flourish.

Don't Simply Abandon Sociological Founding Fathers, but Wriggle Out of Their Shrouds

Origin stories (including sociology's one) tell us about how people understand their place in the universe and their relationship to other living things. They enable the development of collective identities grounded in time and place.

Reading origin stories differently can be freeing. One does not have to abandon the sociological ancestors but it is time to revisit origin stories and wriggle out of their shrouds lest their shrouds become ours.

One origin story is that sociology was preoccupied with tensions between a focus on social order versus emancipation. Substantive sub-specializations often sidestepped that tension but it remains both pressing and instructive. Often read in terms of political tension between Left and Liberal, such foundational tensions in our discipline can now be re-read in terms of ontological politics of appreciating the inseparable co-productions and entanglements of nature and society. This enables us to escape epistemologies of objectification and control and the limitations within Enlightenment thinking about nature, freedom and justice

Not a conceptually easy task. The political tensions in sociology have always been entangled with the politics of nature although this was not recognized. Even feminist sociology with its promise to offer conceptually and politically enabling tools for social change found itself constrained (albeit only partially so) by Enlightenment tradition of nature/society binaries and distributive notions of justice. The innovative concept of gender soon proved inadequate for engaging embodied differences and multiple sexualities. Similarly the traditional sociological focus on distributive justice proved inadequate for the politics of recognition (Fraser, 1995). It also doesn't work well for understanding the lives of those who live more directly and culturally entangled with locally varied relationships with the more-than-human (de Sousa Santos, 2015; Tsing, 2015). Sociology developed a great conceptual took kit, but it is no longer enough.

We draw on feminist analysis of planetary troubles, work on re-peasantisation, and post-colonial and post-humanist insights to argue that environmental sociology as taught in what might be called the English-speaking Global North would do well to abandon any universalizing aspirations, re-visit sociology's origin stories and confront tensions and erased histories in the discipline in order to be able to usefully engage the catastrophes of environmental degradation and climate change. In visiting the ancestors' graves we can discover new relations and new self-understandings.

The Metamorphosis of the World: Climate Change as Transformative

It was tempting to start this chapter with the title of Ulrich Beck's (2016) posthumously published book: *The metamorphosis of the world: how climate change is transforming our concept of the world*. Can environmental sociology herald such a metamorphosis? Sociology's general engagement with 'nature' has been enigmatic. It has been unreflective about the degradations of the more-than-human in the name of progress which it used to call modernization, but more often now calls 'the economy' (Mitchell, 2008). Sociology used to advise the powerful on how to 'modernize' people and places. Some conclude that sociology actively colluded with the destruction of the more-than-human. At the same time sociology drew neat conceptual boundaries in which sociology's concern is with the social. Like a lot of political boundary drawing, borders are ways of keeping potential threats outside. Troubling information was externalized. Its flagship journals still tend to relegate interest in the lives of peoples who lives seem culturally embedded in 'nature' to anthropology or geography or minority sociological fields of inquiry. Mainstream sociology has treated peasants as historical left-overs: nostalgia-inspiring remnants of vanishing worlds (Handy, 2009). The body and reproduction have only relatively recently entered sociology text books and journals. Many sociology programs no longer offer courses in 'rural sociology' and attend almost exclusively to the experience of urbanized and virtual lives.

The title of Beck's book on a metamorphosing concept of the world seems to imply an unwarranted universalism. There are many worlds. Metamorphoses of worlds are not new. Sociology has witnessed violent metamorphoses of social and natural worlds and their catastrophic ends. Worlds have been disappeared. Their loss sanitized and rendered intelligible by stories of progress, economic growth or modernization (Scott, 1998). Sociology's roots are often invisibly in bloody cataclysms (Connell, 2014). We believe that there are far too many contained and troubling shadows in the history of sociology's sense-making frameworks to allow confidence in moving along to engage climate change.

Critically engaging some of the shadow stories of sociology's origins is one way to help ensure that masculinist, US/Euro or anthropocentric stories and the associated colonization of natural and social worlds are not endlessly reproduced. Too often politically inscribed epistemologies and ontologies worked to repress respectful relationships, reduce others to objects, to natural resources, to de-subjectified foci of analysis and kinds of data to be known, disciplined and managed. Knowledge is political. Politics is not a dirty word and we cannot pretend that science somehow allows us to escape hard political questions (Latour, 2004). Much depends on the politics. Feminist scholarship proposes very different political ontologies. Can knowledge facilitate reverse-colonization (Hage, 2016) and what politics of knowledge would be emancipatory for social and more-than-human worlds? If one can think of knowing the more-than-human as an embodied dialogue (Bell, 2011) perhaps a respectful eye will replace an imperial eye/I ?

Outside Epistemic Hegemony

Some from the Global South doubt sociology's capacity for self-transformation. They see sociology's core concepts, categories and methodologies as having been created for a particular time and place. These were then transposed to help render legible, reconfigure and manage other realities 'in their own image' – kind of like a God, we are told, who made the world in His (sic) own image. Haraway termed this way of knowing 'the God trick'. Although Marx and Weber may have engaged the tensions of nature/society, social control/emancipation (Dunlap & Brulle, 2015), twentieth-century interpretations kept those sociological traditions embedded in the 'society' side of that false nature/society conceptual divide. Both assumed that the important social and political action lay with social control/emancipation tensions and not nature/society tensions. Thus neither traditions adequately appreciated colonial, peasant or feminist struggles which now animate so much political and intellectual changes. Consequently, whether Weberian or Marxian, the stories of modernity or capitalism respectively still centre accounts in which it is the logic of changes emanating in origin from European experience that is considered analytically significant. For scholars like Bhambra (2011), although the social relations of colonialism, imperialism and slavery are coextensive with capitalism, each sociological foundational tradition renders them peripheral to the development of capitalist modernity. Even today, high-status Western knowledge institutions continue to act as gate-keepers of the discipline (Connell, 2014), deciding what sociology is and what is significant in the discipline. Mignolo (2014) explains that scholars from outside the centre know the other stories of sociology past and present only too well. The kind of sociology that was born in Europe, he explains, was developed to either understand (or allow elites to manage) its transitions to urbanized industrial social (some would say capitalist) formations. These stories about change (more accurately about cataclysmic upheaval) being embedded in imperial/colonial imaginaries imposed themselves on much of the rest of the world as the only ways in which the *Other* could be rendered intelligible and more importantly legible.[4] Feminist scholars would add that even such post-colonial counter-readings of sociology's origin stories do not always fully appreciate the gendered natures of those imperial imaginaries and ontologies.

---

4 This theme can be found in work that seeks to relativize and challenge traditional paradigms associated with a 'Western' or 'modern' ontology, whether the natural science, Eurocentric social sciences or andro- and anthropocentrism humanism. At core of much of this work is the understanding that the Enlightenment view of nature is inextricably tied to colonial European ambitions to dominate the world. We add that much of life in what is called Europe was itself characterized by internal colonial-like relations and ask if that matters.

Some scholars suggest that there is little use now for the social sciences, and sociology in particular, outside the Global North: its day is done (de Sousa Santos, 2015; Go, 2017). Others continue to find themselves constrained by it as the frame of reference in terms of which marginal scholars must explain themselves. Feminist and Indigenous scholars are constantly called on to translate themselves and their work into such terms of conventional intelligibility. Some others see sociology's intellectual future to lie in border zones and connected histories. Evocative of feminist work (Lugones, 2010) a border thinker is someone who doesn't reject Western epistemology but who also doesn't bend to it. Border thinkers have epistemic potential that territorial/imperial thinkers do not (Mignolo, 2014). Remedies such as de-Westernization and decolonization are recommended to and for sociological work focused within the Global North. In practice almost everywhere right now a range of different knowledge formations are emerging from the ontologies, epistemologies, experiences and traditions of marginalized and non-US/European centric worlds and their long and often unhappy engagements with epistemic hegemonies (de Sousa Santos, 2015; Mignolo, 2014).

We are arguing that ontological politics, and not only the geo-politics of sociological origin stories, matter. Sociological knowing was grounded in the massive transformations associated with capitalism, industrialization, the development of the nation state, bureaucracies and professions and processes unhelpfully explained by the term "modernization." The entanglements of the changes in the Centre with those of colonialism and slavery are seldom made visible. However significant erasures were (are) also internal to Europe as well as external. It was the rapidly growing urban centres, industrial factories and workers in nineteenth-century England or Germany and new biopolitics of managing populations in the nineteenth century that came to occupy sociology's centre stage. It was not the bogs of Ireland, the out-ports of Newfoundland, the dairy barns or crop fields of the European countryside, just as it was not grain fields of the settler-colonies, the diversities of Indigenous people's lives, the dailiness of most women's lives or the realities of the more-than-human which were also entangled with the massive transformations that came to shape sociology's founding stories. These appeared as the backdrop to the making of history, passive, exploitable enablers. Those worlds became Anthropology's office. The misery of the industrial working classes was theorized, although their lived-experience was often objectified even when it evoked sympathy and political solidarity. The lived experiences of those living outside the centre of action of industrializing capitalism and nation/imperial state formation and whose identities or lives seemed entangled with nature were erased or left to literature and the arts. These were worlds and peoples who were 'lost', as though simply misplaced by history. Industrialization and bureaucratization seemed to be where the significant sociological action, if not the march of history, was at.

Many sociology students now speak with disturbing ease of the West as though the West were all of a piece. They don't realize that it is the sociological and other kinds of story-telling that makes things 'neat'. There are few sociologist story-tellers who were peasants, mineworkers, laundry workers, homeless women, woodworkers or gravediggers. Or rather there probably were but their work did not enter the canon.

The significance of the diverse peasantries, gypsies, women land-workers and other social strata of regions now called Europe seemed to have been collapsed into a story of loss vis-a-vis the growth of industrialization, modernity and capitalism. As though all of a sudden there is now an explosion of research on re-peasantization and the growth of social movement networks around agriculture and food and exploitative trade regimes as well as Indigenous resurgence. These political and intellectual eruptions are exposing false 'end of history' stories about people(s) whose lives have always been centred by the more-than-human. If sociology's origin stories had been grounded in peasant lives the stories would have been framed to express the realization that human lives are utterly interdependent with the rich flourishing of the more-than-human and not with nature understood as an object or resource. Haraway (2016) characterized such different lives as being organized around non-innocent bonds of respect. Non-innocent lives of respect are not always neat, sweet or easy ways of living and dying. There are no places of purity (Shotwell, 2016) or non-complicity. If such experiences of life do not become central to the new stories of the world it will really be the end of history for many.

How different would new sociological stories be if different actors and different agents, human and the more-than-human, moved to centre stage? Feminist and environmental sociological unsettlings of sociology must now centre the more-than-people in ways that also continue to centre new understandings of social justice.

What if Things Were Understood Differently?

Post-colonial critiques of sociology might be different if it were recognized that many in what is now called Europe were not really European in the geo-political sense or epistemic sense. Many had their lives re-organized or disappeared by the hubris of modernist ontologies or fantasies of progress. Put another way, the lives, ontologies and experiences of many were very different to sociological stories. Feminism explains that including the excluded isn't a matter of 'adding and stirring' but changes everything. We think that the critiques of sociology should be read not so much about the geographical location from where sociology emerged and more to do with for whom and about whose lives sociology was developed. It is no accident that feminist sociology's early challenge to mainstream sociology explored the idea of a sociology 'for women'. Although later this epistemic strategy was abandoned and the concept of woman was troubled, the idea of knowledge as being a relationship and as being situated remains central to feminism. Just as Indigenous voices and others are unsettling mainstream sociology, it is helpful also to recuperate the untold stories of those at the geo-political centre who were and are marginal to the main sociological story-lines, including the more-than-human. Mignolo (2014) argues that the roots of much of colonial harms were ontological and epistemic and that it was because of this that they could become political and economic. This situates ontology at the centre of questions about power in environmental sociology.

Re-interpreting sociology's origin stories with a focus on the relationships between power and ontologies is more faithful to the dead and more adequate for the living in these crises of our times. It is more open to productive kinds of border-work around climate change than most of what we read about in the media, climate change policy statements or mainstream sociology texts. We agree with the feminist scholars who talk of making kin in multi-speciesed worlds because in re-compos(t)ing ourselves via our entanglements with the more-than-human we have opportunities to re-compose ourselves diversely as people(s).

Our argument is that stories told of sociology's origins are partial truths that help keep environmental sociology marginal in the discipline and perpetuates sociology's inadequacy for engaging the multiple kinds of crises in which climate change is embedded.

Good Humans, Dangerous Nature: Bad Humans, Mortally Wounded Nature

One reads almost everywhere that humanity is causing epoch-level climate change. The identification of humanity as the agent or actor rather than natural cycles helps explains the appeal of the term 'Anthropocene' to counter climate change denial. In contrast, environmental sociology doesn't really work with a concept of humanity. Sometimes it starts by using the word 'humanity' but quickly moves to examine the institutions that organize the actions, events, biophysical, atmospheric and other changes and events under inquiry. It explores the ways in which fossil fuel is enmeshed in contemporary social formations and cultural arrangements or the relations between the fossil fuel corporate sector and States and so on. It analyzes connections between post-World War 2 growths in consumption of energy and natural resources (the great acceleration), new basis of political stability and issues of identity in late modernity. These process and happenings cannot be understood as expressions of some phenomenon labeled humanity. Because sociology conceives of the person as an assemblage of social relations, sociological focus typically is on the complexities of socio-economic, cultural arrangements, and inequality in explanations of environmental destruction, whether deforestation, pollution, species extinction or loss of topsoil and climate change. What sociology has been less good at doing has been theoretically and empirically engaging with what it calls nature and what we here call the more-than-human(/people).

In many ways the concept of humanity is nonsense. Humanity is not an actor or agent in any sociologically meaningful sense of the word. Thus, the concept of humanity cannot be invoked in sociology in the same way it is in popular or natural science discourses of climate change. In contrast, sociology offers analyses of institutional arrangements and of how things (including the production of knowledge of climate change) are socially organized. This works as analytical antidotes to essentialist or universalizing notions such as humanity.

There is a great deal of pressure to create a false political consensus around the concept of the Anthropocene and mainstream climate change discourses because the issues are so urgent. If environmental sociology succumbs to this pressure it is at risk of becoming non-sociological and therefore not much use in really engaging with climate change issues. Even the concern with inequality in mainstream climate change policy-related discourse can easily give a false sense of being on the same page. The policy implications of much of climate change talk is that protecting us all, and the poor in particular, requires a shift from democratic to technocratic decision making, a price highly unlikely to deliver protection to the poor and very likely to deliver greater concentration of power and inequality. The underlying connections between the sources of inequality, injustice and climate change are too often ignored in Anthropocene-talk. Similarly, sympathy and concern for women's greater vulnerability may be displayed. But the masculinist nature of the institutional arrangements associated with climate change emissions is left unaddressed. So although the extent of environmental degradation, the miseries of fellow creatures, the sickening excesses of consumption and the toxicities of plastic worlds might make one want to 'throw in one's lot' with a those expressing profound disgust with 'humanity', such a-political readings of climate change have deadly consequences (Hornborg, 2015; Stehr, 2013).

Another Short Trip to the Graveyard

Emerging governance of climate change should encourage us to revisit the nineteenth-century political intellectual tensions between British neo-classical economics with its focus on the self-interested individual as the basic unit of a well ordered society and the more social continental European sociological traditions. It will help us in staying alive.

Contemporary neoliberalism represents the resurrection and triumph of this 'British'[5] concept of economic man (McMahon, 1997) over the more social being of sociological theories. You can't keep a bad man down, it seems. Economic man as individual rational actor is a key social policy analytical concept behind much climate change policy and is woven into a great deal of policy and techno-managerialism (Shove, 2010). Impediments to economic man's economic self-interested rationality, whether faulty market signals or other distortions, are to be removed, according to much climate change policy. The efficiency of a perfectly working competitive market is understood to allow the expression of the aggregated rationality of individuals. Carbon pricing offers to make the market work more efficiently to regulate society according to the rational decisions of individuals by internalizing the environmental and climate costs of fossil fuels, costs that had hitherto been externalized and/or deferred. Such policies, it is claimed, remove barriers to the future efficient working of the market, conceived of the best institutional arrangement to allow economic man to act rationally and the allocations of resources to be done with maximum efficiency. Technologies such as proxy pricing and so on can make the unworkable workable. In this market-myth the old belief in the universe-ordering divinity of God seems replaced by the rationality of economic man. One can see why some people would be attracted to post-humanism over this kind of story of 'man'. Recognizing the real costs of fossil fuel or other stuff that gets

---

5 We are breaking our own guidelines about not confounding geography and power.

used is usually a good idea and so using the market to address environmental problems can make sense. For many political liberals and liberal lefts, the market can work if it is subordinate to broader social, ethical and ecological relationships (Polanyi, 1957).[6] Ecological modernization theory with its faith in ecological rationality co-joining economic rationality appeals to many sociologists, while for others it suggests that sociology has lost out to economics in the wider competition of environmental policy ideas.

In hegemonic climate change discourses humanity, like economic-man, is typically conceptualized as the aggregate of human individuals, historically and currently. Once identified as anthropogenic in origin, climate change is treated as a social problem as well as a geo/biophysical one. Unsurprisingly it calls forth definitions of social problems and solutions framed within the usual kinds of neoliberalizing biopolitics of dominant governance. This concept of humanity and the idea of humanity as having the kind of agency that natural scientists, economists, politicians and media ascribe to it is an oxymoron for a discipline that attends to the complexities and diversities of social organization to explain social realities. The bounded individualism embedded in hegemonic thinking is incompatible with the relational, 'becoming-with-others' understanding of the person in sociological and feminist thought. It is incompatible with a liveable planet (Haraway, 2016).

Sociologists typically problematized conventional framings of social problems. Sociological thinking questions the use of universalizing account of humanity as causal agency because it attends to the diversity of social organization. Sociology is empirical as well as analytical. It recognizes that that neither those who lived in the past nor those who are alive today were or are 'we' in any easy sociological sense. In the context of massive inequality and diversity what exactly is accomplished by the use of the term 'we' in so much climate-change talk? De-politicization, mystification rather than some hoped-for new cosmopolitics. The production of GHG emissions is profoundly entangled with the production and reproduction of inequality. Few of us, as everyday kinds of individuals, have the power to effect significant change in the political and economic systems that organize their lives. Few make choices in the ways assumed in economic and behavioural psychology models (Shove, 2010). Lives and actions are institutionally organized, their meanings are culturally shaped and, in the context of complex globalized socio-economic arrangements with transnational supply chains, aggregate consequences of individual action are rarely visible to actors either in time or space. To theorize climate change through a model of aggregate individual choice is non-sense.

Although politically dominant and science-led policy analysis of climate change may appear to be taking the social into greater account and acknowledge that the poor, women, marginalized communities and some regions of the world suffer disproportionately, the real differences between environmental sociological analysis and other accounts of climate change are conceptual, political and profound.

---

6 However, in the not too distant future there will be overt political conflict from both the Left and from more authoritarian political interests that will both argue for very different reasons that the liberal faith in the market is not necessarily the only or the best way to accomplish change. Old tensions between authoritarian and emancipatory images of the future will once again become the subject matter of sociological inquiry and of everyday life.

For many sociologists, environmental destruction is not some unintended consequences of modernity or by-product of industrialization. It is not the unanticipated cost of progress. Nor is it an expression of untamed human greed nor an outcome of market failure. Most environmental sociologists recognize the central roles of the logics and history of colonialism and capitalism and many also see racialization, culture and patriarchy as analytical and historical keys to understanding the political, cultural and socio-economic arrangements that are so entwined with ecological destruction. Capitalism is not used as a single causal variable. Most recognize the environmental destruction in the former USSR. Scott identifies the modern State, centralized planning and the rise of resource managing professions and associated bureaucracies as implicated in many socio-ecological disasters. In terms of climate change, theorists like Parenti (2015) help identify the central role of the State in making nature available for capital accumulation.

These kind of sociological analyses leave one with little confidence in the capacity of the State or international networks of States to address climate change. Equally discouraging of hope are insightful theories such as Schnaiberg and Gould's treadmill of production which hold out little prospect of change initiated by the corporate sector in capitalism (Schnaiberg & Gould, 2000). Although sociological ecological modernization theory argue that ecological rationality can emerge to counter the dominance of the economic rationality of capitalism and the narrow bureaucratic rationality of the State there is widespread fear among many social theorists that authoritarian futures legitimated by appeals to climate change are real threats while offering little hope in preventing ecological disaster (Stehr, 2018).

Not surprisingly some see support for ecological modernization type policies as environmental sociology's most pragmatic option. It seems like a hybrid of the European sociological traditions and neoclassical economics. Because sociology seems kind of stuck in political and conceptual space we conclude that burying the foundational conflicts of sociology's origins won't turn out well. In the abstract economic rationality may be conjoined with ecological and bureaucratic rationality but the deeper more imprisoning gendered and nature/society framing, the de-subjectification of the more-than-human and the ontologies and epistemologies of domination remain intact. This is not how to make kin, nor a path to social justice and multi-species flourishing

The Politics of Nature: The Issue Is Not Bringing Nature Back into Politics, or Politics into Nature. They Have Always Been Entwined

In Soci 100 most of us learned that sociology grew as a distinctive discipline by carving out the social as its intellectual space for analysis and expertise. Nature belonged to the natural sciences. We may also have learned that because sociology emerged in a world characterized by racism, social Darwinism and eugenics, and from which its more liberal and progressive practitioners sought to distance themselves, the discipline itself came to eschew sociological explanations that invoked nature to account for social phenomena, especially those around differences and inequalities. From early on, therefore, sociologists who positioned themselves on the side of justice often (but not always) tended to de-politicize rather than politicize nature.[7] This political-epistemic move was useful in early feminism and much anti-racist work. Feminist sociology took up de Beauvoir's insight that women *are made, not born*. "Biology is not destiny," the slogan read. The sociological concept of gender allowed one to see that the gender inequalities that were attributed to nature or to body difference were in fact socially constructed differences that justified oppression and injustice. Similarly, the concept of race was re-conceptualized as racialization in resistance to the naturalization of attributed body differences invoked in the service of inequality. Among sociologists, for much of the later part of the twentieth century it seemed that de-politicizing talk of nature by exposing its subtexts or distancing oneself from invoking nature was universally on the side of justice. That is to say, sociologists challenged the ways in which nature was used to justify vested interests or inequality and became generally uncomfortable with talk of nature in sociological explanation. The binary separation of nature and society that is the hallmark of modernist thinking thus seemed employed by sociologists in the service not just of progress but of justice, whether of gender or racialization, but also of class justice.

Things change. Climate change and the concept of the Anthropocene disturbed the neat binaries entangled with the progressive promises of modernity even among people who had never felt a need to engage with post-humanism, post-colonialism or feminism. The emerging subtext of the Anthropocene seems to carry a new political subtext about nature and society. That subtext is that the modern age is over and *now* nature in the form of climate change is in control. With that newly declared abandonment and inversion of modernity's foundational promise of control over nature will shortly go (and are already going) the other modernist promises of emancipation, equality and justice.

However, one should not believe all that one reads or is told. The Anthropocene story does not really abandon its modernist binaries of society versus nature. Rather it appears that it will use nature for political ends in new ways (Stehr, 2013). It may well, however, abandon its ideological historical travelling companions of progress, liberation and justice. Grove and Chandler (2017) argue that the Anthropocene imaginary destabilizes what is called the modern subject by bringing nature back into politics. It is in this sense that, according to Grove and Chandler (2017), the Anthropocene imaginary signals the end of both nature and humanism. The issue, however, is not that nature is brought back into politics and vice versa or that the appeal to science can no longer be a way of evading or silencing political debate and avoiding value differences (Latour, 2004). It is about *how* nature and politics are and will be connected and how that will be understood.

One may agree that the Enlightenment separation of nature and society has been entangled with European colonial ambitions of control. It does not follow however, that all efforts to disrupt that binary are emancipatory or positive. Rather than being subversive of hegemonic world views and power relations, such ontological moves

---

[7] Sociological de-bunking the taken-for-granted sexist and racist political uses of nature, of course, could be seen as a different kind of politicization of nature.

can have the opposite outcomes (Hornborg, 2017). Grove & Chandler (2017) caution that discourses of the Anthropocene now carry new forms of governance that deepen and extend the neoliberal reach on human and more-than-human worlds. This extended neoliberal reach is accomplished by constructing nature no longer as our home nor as something that can be rendered hospitable to humans or managed by modernist knowledge practices,[8] 'nature' is increasingly represented as the volatile, unpredictable enemy from which the population must be protected (Grove and Chandler, 2017; Stehr, 2013). If modernity naively promised to secure the individual from the harms of nature and if modernist social and political institutions claimed to subordinate nature to the power of science and society, now Anthropocene-talk represents nature as being in charge. One can see power as being re-naturalized. Or de-politicized in a profoundly political way. The technologies of remediation, adaption, and technocratic led social policies to build resilience or to de-carbonization the economy typically still represent nature as a technical challenges in which the certainties of science are not subject to democratic debate. In the Anthropocene imaginary a new epoch of a man-made pathological earth (the Anthropocene) legitimates new regimes of security, protection and power. For many feminists this sounds more like making war than making kin.

Don't Phone 911 or Call in the Army

In times of crisis people in uniform (more often than not men) are called on to come to the rescue: firefighters, police, and other emergency personnel. Hegemonic climate change imaginaries promise a future of perpetual emergency: like an endless War-Measures Act. These Acts come with the suspension of the normal operation of civil society, justified by 'the emergency'. In the discursive context of climate change emergency help may include people in white lab coats, business attire of experts, academics or maybe even carbon traders. Evocative of Agamben's (in Hage, 2015) use of the concept *État de siège*, which connotes the sense of a permanent state of exception or siege or a kind of perpetual war, the appeal to emergency or crises functions as a mode of governmentality that constantly hovers between legitimacy and coercion. This permanent sense of siege will likely be the new normal (Hage, 2015) in the Anthropocene.

How perverse, therefore, of feminist-inspired analyses of climate change to appeal instead to mushrooms, to call for making kin with other critters, to turn to telling different stories or to compos(t)ing ourselves differently as humans. They call for new kinds of ontological politics (onto-politics) not new wars of survival or old class wars, or endless states of emergency.

Some on the left reject such feminist onto-politics as inadequately political (Hornborg, 2017a, 2017b; Grove and Chandler, 2017). These critics may be conflating diverse post-humanist moves to 'de-centre man' with a slippery slope of relativism.

---

8 The contrast between this new scientific ownership of the meanings of nature and place and space and those of Indigenous peoples and peasants is profound. The new stories now impose a new meaning on the place-based meanings of being alive with others.

Revisiting tensions in sociology's origin-stories helps us recognize these concerns as being like the unnecessary realism versus social constructionism debates. The case for new kinds of onto-politics is not a feminist or other argument for relativism or for biased or unscholarly work. It is based on an awareness that all knowledge is relational, always political, partial and situated. Historically sociology eschewed the naturalization of social and political arrangements because of many sociologists' political commitments to justice and equality. Now sociology finds itself on the side of politicizing accounts of nature in order to avoid the naturalization of social control and inequality that is coming most particularly with climate change stories. Sociology can caution us about new kinds of politics of nature and the growing policy invitations to techno-managerialism and repressive biopolitics, done in the name of nature. The apparent destabilization of the human subject in Anthropocene imaginaries is not post-modernism as many assume it is. It can be modernism repackaged. Repackaged now without its historic commitments to equality, justice and freedom in which the objectification of the more-than-human and most of humanity still remains intact. To escape this trap feminist environmental sociology advocates for new conceptual tools that embody and enable new ways of being people(s) in more-than-people worlds.

Ecological feminism goes further than simply politicizing or reading the politics embedded in accounts of nature. It proposes new ontological politics of nature to replace the dominance of modernist ones of capitalism. An offer, as the saying goes, one can't refuse. This is a key distinction for future governance in times of climate change and it should calm the fears of those on the left who fear feminist ontological politics and post-humanism mean the end of the old emancipatory and social justice commitments. One must remember that what it means to govern is co-constituted with the 'nature' of the life to be governed. Ontologies shape both epistemologies and methodologies. They are co-constituted. Ontologies carry politics. Despret and Meuret (2016) add a kind of twist to the meaning of ontology by suggesting that the ontology question is not about the status or nature of this or that being. Rather, they argue, it lies in asking how does this or that being hold onto the task of its existence and what does this achievement and maintenance of being require? They almost counter-intuitively suggest that it is less the amount of natural resources extracted from the earth that will make lives unliveable than it is the disruption of the arts and ways creatures (including people) figure out how to stay alive: to continue *to be*. This disruption of the creative art of life sadly is what the 'security' ethos in so many policies being developed to manage climate change is accomplishing. No wonder many feminists are coming to think of staying alive and flourishing as art.[9] New ways of living and flourishing as part of the more-than-human call for new ontological relationships that open the possibilities of different ways of *being* and doing together. Isn't the study of being and doing things together what sociology is about?

---

9 Yet where we work there are efforts to replace the arts and humanities with the sciences in the name of climate change.

Doing Together Differently: Towards a Sociological Cosmopolitics of Respect, Obligation and Flourishing

Feminist thinkers such as Haraway, 2016; Méndez, 2018; Tsing, 2015 and others invite us not simply to know differently and think differently but more importantly, to *be* differently. To diversely re-embody ourselves as part of the more-than-human in ways that allow ourselves and others to flourish. For sociology, this is not a call to post-humanism but a call to be differently human; different kinds of people(s). (A welcome release from being neoliberal subjects.) Sociologist would add that this kind of change requires different kinds of institutions, whether to do with governance, food, health, transport, knowledge, etc.: different technologies of being and doing. Unlike some post-humanist work that simply dissolves boundaries between subject and object, nature and culture and put everything on the same ontological plane, feminists who work in this area recognize differences within and among others (peoples and more-than-people) and often attend in particular to animals and other living organisms for sources of inspiration. For Anna Tsing learning the 'arts of living on a damaged planet' means new kinds of entangling of humans and more-than-human animals that are contributing to a new cosmo-ecology, creating new connections. For Haraway both the Anthropocene and Capitalocene are *dead-end* stories. While Capitalocene rather than Anthropocene might better name the villains of the 'double-death-loving epoch' (Haraway, 2016, 47), both tell stories or actually story unlivable worlds. We need, according to Haraway, "a new story to carry lives of multispecies flourishing entangling myriad temporalities and spatialities of intra-active entities-in-assemblage – including the more-than-human, other-than-human, inhuman, and human-as-humus'. Sociologists have long understood that relationships, interactions and meanings produce identities. Expanding worlds of respectful interaction produces different identities, differently embodied beings. Living differently with the more-than-humans co-produces different kind of people differently. For Haraway this means recognizing that people and other animals are tied together in non-innocent bonds of respect. The term respect here means to hold in regard, to pay attention in a context where each bears the consequences of the other's ways of living and dying.

A true sociological cosmopolitics does more than simply take the other into account. Rather it embodies the scope of how (and where) other beings obligate us: people are multiply obligated. Like symbolic interactionism's concept of meaning, others and situations have the power to obligate and *call out* (Blumer, 1969) actions and responses in us and shape our identities. The more-than-human become among our significant others. Perhaps our most significant other if we understood ourselves differently. A feminist sociological cosmopolitics would also do politics differently. Like Hage's (2015) notion of alter-politics this would-be politics grows not from opposition to, or critique of, our current systems. It is a politics grown from attention to another way of being, one that involves and appreciates the centrality of other kinds of living beings and social justice. Environmental sociology has established that social justice is central to engaging environmental challenges.

What environmental sociology can contribute analytically to some of this innovative feminist work is the indispensable recognition that how we do things together with others, however understood, however embodied and however related in time and space, however storied, is socially organized – albeit now we recognize this as organized with more-than-human others. The importance of sociologically understanding institutions and organization remains key to learning the arts of living well on a damaged planet.

References

Alaimo, S. (2016). *Exposed: Environmental politics and pleasures in posthuman times*. Minneapolis, MN: University of Minnesota Press.

Beck, U. (2016). *The metamorphosis of the world: how climate change is transforming our concept of the world*. Hoboken, NJ: John Wiley & Sons.

Becker, H. S. (1986). *Doing things together: selected papers*. Evanston, IL: Northwestern University Press.

Bell, M. M. (2011). *An invitation to environmental sociology*. Thousand Oaks, CA: Sage Publications.

Bhambra, G. K. (2011). Talking among themselves? Weberian and Marxist historical sociologies as dialogues without 'others'. *Millennium*, *39*(3), 667–81.

Blumer, H. (1969). *Symbolic interactionism: perspective and method*. Upper Saddle River, NJ: Prentice-Hall.

Connell, R. (2017). In praise of sociology. *Canadian Review of Sociology/Revue canadienne de sociologie*, *54*(3), 280–96.

Connell, R. (2014). Using southern theory: Decolonizing social thought in theory, research and application. *Planning Theory*, *13*(2), 210–23.

de Sousa Santos, B. (2015). *Epistemologies of the South: justice against epistemicide*. Abingdon, UK: Routledge.

Despret, V., & Meuret, M. (2016). Cosmoecological sheep and the arts of living on a damaged planet. *Environmental Humanities*, *8*(1), 24–36.

Dunlap, R. E., & Brulle, R. J. (eds). (2015). *Climate change and society: sociological perspectives*. Oxford: Oxford University Press.

Fraser, N. (1995). From redistribution to recognition? Dilemmas of justice in a 'post-socialist' age. *New Left Review*, *212*, 68.

Go, J. (2017). Decolonizing Sociology: epistemic inequality and sociological thought. *Social Problems*, *64*(2), 194–99.

Grove, K., & Chandler, D. (2017). Introduction: resilience and the Anthropocene: the stakes of 'renaturalising' politics. *Resilience*, *5*(2), 79–91.

Hage, G. (2016). État de siège: A dying domesticating colonialism? *American Ethnologist*, *43* (1), 38–49.

Hage, G. (2015). *Alter-politics: critical anthropology and the radical imagination*. Melbourne: Melbourne University Publishing.

Handy, J. (2009). 'Almost idiotic wretchedness': a long history of blaming peasants. *The Journal of Peasant Studies*, *36*(2), 325–44.

Haraway, D. J. (2016). *Staying with the trouble: making kin in the Chthulucene*. Durham, NC: Duke University Press.

Hornborg, A. (2017a). Artifacts have consequences, not agency: toward a critical theory of global environmental history. *European Journal of Social Theory, 20*(1), 95–110.

Hornborg, A. (2016). Dithering while the planet burns: anthropologists' approaches to the Anthropocene. *Reviews in Anthropology, 46*(2–3), 61–77.

Latour, B. (2004). Why has critique run out of steam? From matters of fact to matters of concern. *Critical Inquiry, 30*(2), 225–48.

Lugones, M. (2010). Toward a decolonial feminism. *Hypatia, 25*(4), 742–59.

Lykke, N. (2009). Non-innocent intersections of feminism and environmentalism. *Kvinder, Køn & Forskning, 3–4,* 36–44.

Malm, A. (2016). *Fossil capital: the rise of steam power and the roots of global warming.* New York, NY: Verso Books.

McGregor, D. (2016). Living well with the Earth: Indigenous rights and the environment. In C. Lennox & D. Short (eds.), *Handbook of Indigenous Peoples' Rights* (pp. 167–80). New York, NY: Routledge.

McMahon, M. (1997). From the ground up: ecofeminism and ecological economics. *Ecological Economics, 20*(2), 163–73.

Méndez, M. J. (2018). 'The River Told Me': rethinking intersectionality from the world of Berta Cáceres. *Capitalism Nature Socialism, 29*(1), 7–25.

Mignolo, W. D. (2014). Spirit out of bounds returns to the East: the closing of the social sciences and the opening of independent thoughts. *Current Sociology, 62*(4), 584–602.

Mitchell, T. (2008). Rethinking economy. *Geoforum, 39*(3), 1116–21.

Parenti, C. (2015). The 2013 ANTIPODE AAG lecture: The environment making state: territory, nature, and value. *Antipode, 47*(4), 829–48.

Polanyi, K. (1957). *The great transformation: the political and economic origin of our time.* Boston, MA: Beacon Press.

Schnaiberg, A., & Gould, K. A. (2000). *Environment and society: the enduring conflict.* Caldwell, NJ: Blackburn Press, 2000.

Scott, J. C. (1998). *Seeing like a state: how certain schemes to improve the human condition have failed.* New Haven, CT: Yale University Press.

Shove, E. (2010). Beyond the ABC: climate change policy and theories of social change. *Environment and Planning A: Economy and Space, 42*(6), 1273–85.

Shotwell, A. (2016). *Against purity: living ethically in compromised times.* Minneapolis, MN: University of Minnesota Press.

Smith, D. E. (2005). *Institutional ethnography: a sociology for people.* Toronto: AltaMira Press.

Stehr, N. (2013). An inconvenient democracy: knowledge and climate change. *Society, 50*(1) 55–60.

Stehr, N. (2018). Climate change: what role for Sociology? In M. T. Adolf (ed.), *Nico Stehr: Pioneer in the Theory of Society and Knowledge* (pp. 343–54). Berlin: Springer.

Swyngedouw, E. (2013). The non-political politics of climate change. *ACME: An International Journal for Critical Geographies, 12*(1), 1–8.

Tsing, A. L. (2015). *The mushroom at the end of the world.* Princeton, NJ: Princeton University Press.

Tsing, A. L., Bubandt, N., Gan, E., & Swanson, H. A. (eds.). (2017). *Arts of living on a damaged planet: ghosts and monsters of the Anthropocene.* Minneapolis, MN: University of Minnesota Press.

2 On Discourse-Intensive Approaches to Environmental Decision-Making: Applying Social Theory to Practice

Steven E. Daniels and Gregg B. Walker

1 Introduction

Writing in *The Good Society* almost three decades ago, the late Robert Bellah and colleagues addressed "the patterned ways Americans have developed for living together, what sociologists call institutions." (1991, p. 4). They sought to understand "how much of our lives is lived and though institutions, and how better institutions are essential if we are to live better lives (p. 5). The chapter considers a particular set of institutional practices that emerged in the early 1990s in response to environmental and natural resource management controversies; practices typically associated with environmental governance and collaborative, community-based organizations. Neither then nor now are these organizations tightly defined or standardized; rather they emerged and endure in varied forms across the globe. Regardless of their forms and functions, collaborative organizations and practices designed to foster environmental innovation reflect institutional reform – a commitment to inclusivity and diversity.

This chapter examines the recent rise of more inclusive approaches to environmental decision-making from a sociological perspective. In diverse locales there are coalitions of citizens, non-governmental organizations, and the traditional resource management agencies forging agreements in the face of deeply seated value differences and vexing technical complexity. These innovative efforts are making progress when conventional processes seem paralyzed, and this increased democratization of natural resource/environmental decision making has arguably been perhaps the biggest shift in the field in the past twenty years. Often arising spontaneously outside of long-standing administrative structures, they have increasingly become respected and institutionalized, even as they operate in the face of competitive political incentives.

One of initial challenges this chapter faces is the lack of a standard terminology. A huge number of terms are used to describe these processes: advisory groups, advocacy coalitions, appreciative inquiry, charettes, citizen involvement, citizen juries, civic science, consensus conferences, collaborative learning, co-management, community-based collaboration, collaborative public management, deliberative decision-making, facilitated dialogue, grassroots environmental management, negotiated rule making, participatory governance, partnerships, platforms, pluralism, principled negotiation, public issues education, public participation, search conferences, social learning, strong democracy, watershed councils, and working groups – to mention a selection of commonly used terms. This panoply of terms has arisen because the literature on this paradigm shift in environmental management has been studied through many different disciplinary lenses – each with their favorite terminology – and in many different geographic regions – each with their preferred labels.

At the risk of adding yet another term to the already excessive inventory, this chapter refers to all of these approaches collectively as discourse-intensive (DI) approaches. A subsequent section will explain DI approaches more fully, but the first-level explanation is that they are approaches that devote considerable effort to designing and enacting a process that is transparent, inclusive, and thorough in order to arrive at decisions that are technically sound as well as socially legitimate.

The range of contexts within which these innovative discourse-intensive approaches have appeared is immense. By way of illustration (and by no means comprehensive) it ranges from land use planning in the Northwest Territories in Canada (Caine, 2013), biodiversity conservation in Australia (Hill et al., 2015), Chilean fisheries management (Ramirez and Hernandez, 2005), and wetlands protection in Italy (Celino and Concillio, 2011). The book length treatments of this phenomenon are bounded by the now-ancient Bingham (1986) and Amy (1987) and the more contemporary Clarke and Peterson (2015).

This chapter follows a straightforward path. The key features of DI approaches are briefly laid out. Four perspectives from sociology are then applied to the emergence of DI approaches: Weber's timeless writing on bureaucracies, Bourdieu's work on the interaction of people and their context through the notions of habitus and field, Luhmann's focus on communicative systems, and Sherif's classic work on intergroup competition. A particular DI approach – Collaborative Learning – that is broadly representative of the range of techniques that are employed around the globe – is then presented to provide a more applied understanding of how to promote discourse around complex natural resource/environmental decisions. Finally, the chapter concludes by juxtaposing the theoretical perspectives of the four theorists with the common attributes of discourse-intensive approaches.

2 Understanding Discourse-Intensive Approaches in General

It is necessary at the outset to acknowledge that this chapter uses discourse in a somewhat looser way than communication theorists (e.g., Gee, 1990, and Trappes-Lomax, 2004) might prefer. To them, it is a very specific and nuanced term, but we shall use it as broad descriptive category. Discourse can be both a verb (people engage in discourse) and a noun (the text is a discourse). In either case, it is a narrative construction that defines terms, expresses values, analyzes (or ignores) choices, and argues for outcomes. It is a rhetorical construction of reality that frames how the situation ought to be examined, what issues ought to matter, and who ought to be considered relevant. It may be fairly technical intellectual argument or it may be much more an appeal to emotion: either way, it functions as persuasive argumentation through the way it represents reality. It is richly communicative, although it does not always require face-to-face interaction (e.g., a mass media public health campaign). Discourses can be undertaken so as to promote understanding and broad participation, but can also exert influence to control public opinion or policy outcomes.

Discourse-intensive approaches to natural resource/environmental management can therefore be thought of in several different ways (this list draws upon Clarke and Peterson (2015):

1. The designers/conveners of DI approaches are intentionally trying to create an alternative discourse that is typically more inclusive than a traditional agency-centric discourse. Rather than a policy discourse that an agency develops and imposes (often with the intention to control and dominate), DI approaches are participatory, with stakeholders sharing discourses (some of which may be competing) in an effort to generate a new discourse with common ground.
2. This new discourse is inclusive of the various discourses that the participants bring into the process; it does not attempt to define whose discourse is "right" but strives to construct a shared discourse that incorporates/honors the individual discourses.
3. The approach is implemented with considerable attentiveness to how seemingly minor details might privilege various groups or discourses.
4. The use of language, technical jargon, mental models, etc. will be scrutinized and designed in order to promote a broad and inclusive space in which the full range of participants can variously discover, articulate, and pursue their interests.
5. The process will develop norms of information sharing rather than withholding information to be used competitively. This may include joint fact-finding/monitoring activities when the science on an issue is contested.
6. The process discourse is not conducive to coalitions of stakeholders forcing the adoption of their discourse through the domination or exclusion of others' interests.
7. The process of constructing a collective discourse will be fair; various participants will all have an adequate opportunity to embed their values and interests in it. It should resonate with norms about what constitutes a legitimate process, particularly when participants from indigenous or high context cultures are involved.
8. DI approaches go beyond single issues; participants view the situation holistically and think critically in terms of a system.

So to the extent that conventional natural resource decision-making processes can accurately be seen as highly technical, agency-centric, and prone to political escalation, discourse-intensive approaches strive to be much the opposite. They are more humane, inquisitive, respectful interactions that attempt to acknowledge the legitimacy of competing points of view and craft decisions that are innovative and respectful of the full range of participants and their values. They are much less "me versus you" and more "us versus this complex situation." They tend not to be quick, are rarely easy, and demand much of the participants. But the collective experience from around the world is that they are able to yield real progress in situations that seem to confound more conventional management models.

3 Examining a Particular Discourse-Intensive Approach: Collaborative Learning

To illustrate how a discourse-intensive approach can be designed and applied in order to be as responsive as possible to the features of a problem domain, this chapter now focuses on Collaborative Learning. As an approach to facilitating dialogue and deliberation around natural resource management issues, Collaborative Learning is a relatively mature technique, having been formulated and disseminated in the late 1990s (Daniels and Walker, 1996; Daniels and Walker, 2001). It began with a desire to develop a methodology that could deal successfully with decision situations that simultaneously exhibit these three dimensions: complexity, controversy, and uncertainty.

3.1 Common Features of Environmental Decisions

3.1.1 Complexity

There may be no conflict setting with more inherent complexity than natural resource and environmental policy situations. Faure and Rubin (1993) highlight a number of "distinctive attributes" of environmental policy negotiations: multiple parties and multiple roles, multiple issues, meaningless boundaries, scientific and technical uncertainty, power asymmetry, joint interest, negative perceptions of immediate outcomes, history, long time frame, changing actors, public opinion, institutionalization of solutions, and new regimes and rules (pp. 20–26). Moreover, beyond these lists of attributes, the trend in recent decades is for government agencies to employ natural resource management schemes with increasingly complex formulations: managing for more natural functions, social values, and economic benefits across more acres and across longer time horizons. Simplistic or solely technical solutions are no longer acceptable, if they ever were.

3.1.2 Controversy

In addition to their complexity, natural resource and environmental conflict and decision situations are typically controversial. The controversial nature of these situations emerges from a variety of factors. First, many different viewpoints exist concerning the array of issues in the situation. Second, tension or incompatibility in the situation may relate to one or more of the following: facts, culture, values, jurisdiction, history, personality, relationship, procedure (Wehr, 1979; Daniels and Walker, 2001). Third, given the number of parties, varied sources of tension, and deeply held values, consensus may be very hard to achieve. Fourth, parties may hold strong emotional ties to the issues and the landscape, including strong attachments to place (Cheng et al., 2003; Cantrill and Senecah, 2001). Fifth, parties may display cognitive biases such as overconfidence, fixed-pie, and attribution that contribute to competitive frames (Bazerman, 2002; Daniels and Walker, 2001).

This complexity is compounded because each stakeholder group tends to ask for their issue, on all acres, for all time without any consideration for either the constraints of the natural system or the legitimate interests of other stakeholders. This puts public lands managers into the "Solomon Trap" that Carpenter and Kennedy referred to in their classic book (1988) on managing public disputes.

3.1.3 Uncertainty

Natural resource management and environmental policy situations draw on knowledge from varied sources and in many forms. Such knowledge, including that from scientific inquiry, is inherently uncertain. "Scientists treat uncertainty as a given," Costanza and Cornwall (1992, 14), assert: "a characteristic of all information that must be acknowledged and communicated." They add that "uncertainty should be accepted as a basic component of environmental decision making at all levels and be better communicated" (Costanza and Cornwall 1992, 15). The presence of uncertainty in complex and controversial situations draws attention to the importance of learning and methods for adaptation, such as adaptive management (Ascher, 2004; Lee, 1993). Consequently, "a new role for scientists will involve the management of the crucial uncertainties: therein lies the task of assuring the quality of the scientific information provided for policy decisions" Funtowicz and Ravetz (2001, 178).

3.2 The Emergence of Collaborative Learning

As noted above, the "Collaborative Learning" methodology was intentionally designed to respond to complexity, controversy, and uncertainty. Theories, concepts, and techniques from diverse fields and disciplines not typically integrated in the natural resource and environmental policy literature were integrated in a manner that had previously not been attempted.

Work in systems – systems theory, systems thinking, systems design – gave Collaborative Learning the capacity to cope with complexity. Soft systems theory (e.g., Checkland and Scholes, 1990; Wilson and Morren, 1990; Flood and Jackson, 1991) and systems thinking (e.g., Senge, 1990; Senge et al., 1994; Meadows, 2008) were fundamental to designing ways stakeholders to more fully grasp the inherent complexity of the situation at hand, and develop innovative ideas that emerged from that that richer understanding.

The conflict and dispute resolution literature – the extensive research and practice in negotiation, mediation, facilitation, and conflict management – provided the foundation for addressing controversy. Fisher, Ury, and Patton's (1991) ideas regarding interest-based negotiation, for example, inspired Collaborative Learning's emphasis on concerns and interests as a means for discovering common ground.

Scholars' emphasis on conflict management rather than resolution (e.g., Deutsch, 1973; Folger et al., 2008; Wilmot and Hocker, 2018) related to the Collaborative Learning concept of "improving the situation" and the correspondent "situation improvement" technique.

Scholarship in experiential and adult learning theories offered guidance on how to address uncertainty. Kolb's (1984) "learning cycle," featured in Wilson and Morren's work on soft systems, provided groundwork for the Collaborative Learning phases circle (Daniels and Walker, 2001). Senge's (1990) commentary on "the learning organization" and "learning teams" linked active learning and systems thinking.

Lastly, the integration of systems, conflict management, and learning ideas relied on participatory communication. Participatory communication covers "a wide range of communication functions that cross the development spectrum," Ramirez and Quarry point out, such as "information, public relations, social marketing, community voice and so on" (2004, p. 13). Participatory communication is "a dynamic, interactional, and transformative process of dialogue between people, groups, and institutions that enables people, both individually and collectively, to realize their full potential and be engaged in their own welfare" (Singhal and Devi, 2003, p. 2, citing Singhal, 2001). Regardless of the context, channel, or function, the core of communication activities are the meanings parties create and co-create (Walker, 2007). Figure 2.1 features the integrated dimensions of the Collaborative Learning approach.

Over the past twenty-five years Collaborative Learning has been applied to a range of projects in a variety of ways. Examples include recreation area planning, fire recovery efforts, forest management plan revision, and river sediment management (see Daniels & Walker, 2001; Walker, Senecah, & Daniels, 2006). The CL method has evolved, with new tools to foster discourse-intensive activities and effective and enduring environmental policy decisions. In recent years the CL method has become central to the work of the National Collaboration Cadre of the United States Forest Service agency (www.fs.fed.us/emc/nfma/collaborative_processes/default.htm).

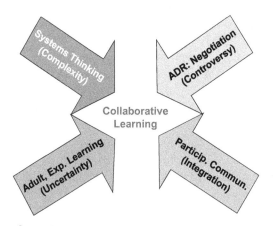

Figure 2.1 *Collaborative learning foundations*

The fundamental contention of this chapter is that discourse-intensive approaches to natural resource management are social institutions that are peoples' best efforts to make progress on complex and contentious issues. As grand as they may sound, more often than not they begin as *ad hoc* "pick-up teams" of people who are sufficiently frustrated by their lack of success in other venues that they are willing to try something new.

4 Four Sociological Perspectives on Discourse-Intensive Approaches

The various discourse-intensive approaches to natural resource decision making that are the focus of this chapter have arisen largely outside of traditional agency processes. Agencies have embraced them incrementally and it is important to reflect on the forces that continue to shape the rate and character of this adoption. They are also hugely social processes, and as such social theorists offer us many ways to understand both their potential and pitfalls. With that in mind this chapter now turns to applying the theoretical insights of four influential thinkers to the emergence of discourse-intensive approaches.

4.1 A Weberian Understanding the Nature of the Agency/Citizenry Relationship

One factor that must be kept in mind when reviewing natural resource management agencies' comfort with DI methods is that those organizations are typically Weberian bureaucracies and, as such, will not come naturally to collaboration with external parties. Max Weber was a German sociologist active in the late nineteenth and early twentieth centuries.[1] Some of his influential writing was on the nature of bureaucracies, which were emerging simultaneously with – and not entirely independent from – the decline in the power of European monarchies and their patronage-based approach to governance (e.g., the king's mistress's brother is appointed judge despite a lack of training or disposition for the job). Several features described the ideal Weberian bureaucracy, and they continue to be visible in contemporary organizations: clearly defined roles and hierarchical structures in which subordinate people are expected to comply with the authority of their superiors, people holding positions due to professional qualifications and certification rather than favoritism, a measure of isolation from political pressure (sometimes through lifetime tenure), decisions are reached by objectively comparing alternatives to optimize social benefit rather than personal gain, a stronger sense of loyalty to the bureaucracy than to external parties, emblems/uniforms that clearly display one's membership and status within the organization, etc. Weber viewed bureaucracy as a modernizing force that allowed government processes to mirror the increasingly complex and competent industrial-era

---

1 This passage relies heavily on the Gerth and Mills' (1946) edited anthology of Weber's essays, notably pp. 196–240.

private sector organizations of the day, while at the same time recognizing that the rigid structure and role specialization created an "Iron Cage" aspect for employees. And even as bureaucracies operate on behalf of the citizenry, there is a sense of their superiority over the everyday citizen who does not have the same extensive training and understanding of the issues as the professional bureaucrat (an elitism that existed in the ideal type that Weber was describing rather than advocating/promoting). Just as there is an assumption that bureaucrats will defer to their superiors, the citizenry is similarly expected to defer to the bureaucracy.

Many natural resource management agencies are archetypical Weberian bureaucracies. Indeed, the federal Forest Service in the United States was created in the early twentieth century by foresters who had studied in Germany and were greatly influenced by the Germanic attitudes of the day (Dana and Fairfax, 1980). As such, the agency epitomized Weber's model: military-appearing uniforms, credential/examination-based hiring and promotion, clearly- defined ranks and career ladders, frequent transfers between locations that prevent employees from becoming embedded in local communities (thereby ensuring their primary loyalty was to the agency), etc.

Simply stated, operating collaboratively with outside parties is not in the DNA of Weberian bureaucracies. It flies in the face of the sources of power and legitimacy upon which their authority has historically been derived. A Weberian bureaucrat would find it very difficult to be open to the ideas and critique from outside parties, which creates the following dynamic:

1. Agency people tend to approach their jobs through a technical frame; they rely on their scientific understanding of the resources being managed to guide their decisions.
2. Citizens may assign a large number of additional values to landscapes and resources that may have some combination of personal, historical, cultural, or aesthetic roots.
3. Agency planning documents tend to be quite long, technical, and more-or-less impenetrable to people without applicable technical training.
4. In many locations there is a significant lack of trust in agency processes, although that distrust can emerge from very different sources in different contexts.
5. Agency planning processes tend to be formal and rigid, and participants with high human capital are better able to protect their interests and achieve their goals.

4.2 Bourdieu and the Design of Discourse

It is instructive to apply Bourdieu's concept of social practice to the design of DI approaches. His conceptualization of social practice locates an individual's actions and interactions with others within a context of social conditions and constraints (Bourdieu [1980](1990)). The notion of habitus is integral to understanding how individuals navigate this social space; it is the internalized and embodied sensibility that goes beyond an actor's mental faculties. It is attitudes, values, perceptions, tastes and dispositions that are anchored in one's experience in social institutions and a particular cultural milieu. A discourse-intensive approach to crafting decisions about natural resource management would constitute what Bourdieu would refer to as a field, in that it is a social space within which social practice occurs. A field is constituted by actors and their interests, the power relationships among them, and the institutional structures that define the process. Bourdieu would conclude that people whose habitus is closely aligned with the characteristics of a particular field would be both at ease in that field, and also able to exhibit a high degree of agency in terms of being able to pursue and protect their interests Bourdieu (1991). Stated another way, a meeting in which a New York lawyer would be effective would be alien to an indigenous person from the Amazon, while a community ceremony in the Amazon would confound the lawyer.

The fundamental nugget that emerges from this brief Bourdieusian analysis is that a discourse-intensive approach can be understood as an intentional effort to achieve alignment between the field and the habitus of the participants (Bourdieu and Wacquant, 1992). To a sociologist engaged in passive observation of social processes, the field is the result of historical, cultural, political, and social influences. Someone trying to design a DI process would see the situation substantially differently. While the field is undeniably shaped by macro-level cultural and social forces, it is also the result of intentional design choices enacted specifically to increase the participants' efficacy. Suppose a federal agency is conducting a planning process in Puerto Rico – should it be conducted in English or Spanish? Choosing the former will merely reinforce existing power relationships and perpetuate the social distance between agencies and the residents. Choosing the latter will allow the residents to be far more effective in conveying nuanced preferences and they will feel far more legitimized by the choice.

4.3 Luhmann and Communicative Systems

Sociologist Niklas Luhmann's contributions to social theory are also relevant to DI methods and techniques. In a number of works, Luhmann has emphasized the importance of communication in social systems (e.g., 1979, 1989, 1992, 2000). Peterson and colleagues observe that Luhmann "defined society as a self-organizing subsystem that is distinguished from its environment by communication" (2004, p. 17). An environment can be described objectively (e.g., in terms of its natural attributes, species communities, ecological conditions), while society as a system is constructed dynamically through communication behavior and interaction.

Hall and colleagues (2017) explain that "Luhmann's macro-level social theory departs from his predecessors' social stratification theories (of Marx, Weber, Mills, etc.)." It features functional differentiation; "how society organizes itself by its actions and how these become self-reinforcing" (p. 2, citing Luhmann, 2000). Hall and colleagues posit that Luhmann's theory "offers language well-suited for integrating social and ecological system functioning" for problem solving and decision making, and is "useful for explaining how societies become aware of ecological problems and ... how social systems can react by aligning multiple sectors of society into sustainable solutions" (p. 2). Hall and colleagues note further, that according to Luhmann's theory, "social systems only become aware of how they are coupled to environmental systems at points, where the environment communicates with – disrupts – the value logics that organize social functioning" (p. 4).

Luhmann's commentaries on communication can be challenging to read and comprehend, but his writing reveals ideas germane to environmental conflict resolution and participatory discourses that foster collaborative, mutually beneficial decisions. To illustrate the relevance of Luhmann's work, imagine a municipality dealing with the issue of how to manage its solid waste with specific concerns about maintaining a landfill and providing a recycling center.

First, Luhmann recognizes the importance of paradox. He asserts that "when the world is included in communication, it appears as the paradox of the unity of difference, a paradox that requires a solution [Aufldsung] if things are to continue at all" (1994, p. 27). When siting a landfill, if a single (e.g., municipality-centric) discourse dominates the decision situation, citizen views are paradoxically excluded, minimized, or dismissed. In contrast, DI approaches provide opportunities for parties to work through any paradox constructively, by providing and evaluating, for example, a range of alternatives.

Second, Luhmann notes that an individual's observations are limited; each person has a "blind spot." Consequently, the observer, as she or he is "ploughing through the world . . . is exposed to the observation of observations. There is no privileged point of view, and the critic of ideology is no better off than the ideologue" (1994, p. 28). When considering landfill development, stakeholders' ideas about what is desirable and feasible should generate a more robust decision and a sense of shared ownership – both of the process and outcome.

Third, as a societal order evolves, its history "can be reconstructed if one considers that all communication depends on the cooperation of conscious systems, and that must therefore assume a perceptible form. Communication must transform complexity; "it must transform infinite informational loads into finite ones" (1994, p. 29). When a municipality's stakeholders contribute ideas from varied perspectives and develop a shared discourse about waste disposal, they develop a story of achievement and success through collaborative governance. The story can provide guidance for future decisions.

Fourth, Luhmann's theoretical approach highlights the importance of meaning; for Luhmann a social system is to a significant extent a communication system that creates and transmits meaning (Luhmann, 1989). According to Leydesdorf (2000), Luhmann "considered the core concept of symbolic interactionism, that is, the interactive construction of social meaning, as the unit of operation of social systems" (p. 274). What stakeholders (both government and non-government) consider to be a sustainable waste management policy, both through landfill maintenance and recycling facilities, will vary. Through an innovative DI process, parties will share their meanings and interpretations of a viable waste management system. Decision authorities – such as municipal governments – gain support by understanding the importance of DI approaches that feature participatory and dynamic communication systems.

An essential question, then, in environmental conflict resolution and decision making is – how does society structure its capacity for processing environmental and ecological information, particularly in light of citizens' blind spots? (Luhmann, 1989, 1994; Hall et al., 2017). While Luhmann did not contemplate public participation in environmental policy decision making, his ideas featured here relate directly to discourse-intensive approaches. DI efforts address paradox; particularly the paradox between technical expertise and citizen engagement (Daniels & Walker, 2001). DI work fosters shared learning; overcoming individual blind spots, managing information, and attending to the meanings that people construct. Some DI methods, such as the Collaborative Learning approach outlined above, draw on systems thinking to address complexity.

4.4 Sherif and In-Group/Out-Group Competition

The notion that people readily and often unconsciously assign other people to being members of either an in-group (i.e., a group in which you feel you belong) or out-group (i.e., a group in which you do not feel you belong) is hugely influential in social psychology, and it is so widely applied that it is difficult to assign its origins to any one researcher. At the risk of being arbitrary, this chapter sees its origins in experiments of Muzafer Sherif and colleagues in the late 1940s and early 1950s, in which boys were assigned membership of one of two different groups that were then placed in competition with one another (the well-known Robbers Cave experiment (1954/1961)). The extent to which the groups coalesced into unified bodies that then acted very negatively toward members of the other group was striking in this research as well as in countless variations on it that followed. In-group/out-group cognitive frames have been closely aligned with social identity and linked to acts of discrimination and prejudicial attitudes (Tajfel, 1959; Tajfel and Turner, 1969).

It is easy to see how in-group/out-group dynamics might be triggered and reinforced by macro variables such as race, religion, or nationality. The research on these processes indicate that they can also be activated by far more minor distinctions (there was a famous informal experiment conducted by a school teacher in the 1960s who divided her class by eye color and thereby generated strongly discriminatory attitudes and interpersonal conflict among children who had previously been friends (janeelliot.com; www.youtube.com/watch?v=KHxFuO2Nk-0). In-group/out-group assignments can be based on characteristics like peoples' hobbies, place of birth/residence, or professions. Once in-group/out-group assignments are established, a suite of attitudinal biases are activated. We tend to regard in-group members as more intelligent, trustworthy, and ethical than others. If an in-group member fails to meet a deadline, we are more likely to attribute it to external factors beyond her control, but the identical behavior from an out-group member might be met with a "see, you can't trust any of them" classist attribution. People are far more likely to behave cooperatively (often generously) toward in-group members and competitively (often viciously) toward out-group members (see Fiske and Taylor (2017) for a more complete explanation of these processes that this chapter can provide).

The divisive power of this "us versus them" social psychological framing provides a core motivation for the use of discourse-intensive approaches in natural resource management. Traditional agency-centric processes position the agency as the arbiter of competing claims between interest groups, and arguably can exacerbate intergroup competition by prompting rigid positional rhetoric (Daniels and Walker, 2001; Daniels, 2009). To the extent that DI approaches can challenge and soften those negative attributions and reduce the incentives to for the various parties to compete in order to achieve their goals. Returning to the original Sherif et al. experiments mentioned above, they found that giving the teams superordinate goals shifted the orientation from "us v. them" to "all of us v. the problem" (Sherif, 1958). The designers of DI approaches have the opportunity to pursue just such a reframing, thereby reducing social polarization.

5 Connecting Theory to Practice

Even as discourse-intensive approaches become an important component of environmental policy decision making, they remain social institutions appropriately subject to critical scrutiny. Staunch empiricists might argue that the most reliable way to identify those DI practices that promote innovative environmental decision making would be to observe these processes in action, and correlate outcome with process features. Indeed there is considerable research that attempts to do just that (e.g., Imperial, 2005; Pagdee, Kim, and Daugherty, 2006; Brown, Buck, and Lassoie, 2008; Ducrot, 2009). An alternative approach would be to rely upon and extend the implications that emerge from theory.

To that end, Table 2.1 illustrates connections between theory and practice; specifically between key ideas of the theorists featured in this chapter and the eight observations about discourse-intensive approaches presented earlier in Section 2. The table highlights the theorists' contributions that relate to the development of discourse-intensive approaches to public participation and stakeholder engagement processes like Collaborative Learning.

6 Conclusion

Decisions about how to manage natural resources and address environmental problems are inherently social because they are institutionalized patterns of behavior that are deeply embedded in socially constructed notions of legitimacy and competence. Who has a voice in the process, who has rights that must be protected, whose data matter, and how private benefits and public costs will be balanced are all socially constructed choices. Not only are they socially derived, their outcomes subsequently shape both physical and social reality. As such, there are many significant thinkers whose work could contribute to this domain: Plato, Aristotle, Bentham, James, Argyris, Foucault, Giddens, Ostrom, etc. With that in

Table 2.1 Discourse-intensive approaches and four theorists

DI Propositions	Weber	Bourdieu	Luhmann	Sherif
The designers/conveners of DI approaches intentionally try to create an alternative discourse that is typically more inclusive than a traditional agency-centric discourse.	Agency processes are often rigid	Habitus may reinforce conventions but can also change		
This new discourse includes the various discourses that the participants bring into the process; it does not attempt to define whose discourse is "right" but strives to construct a shared discourse that incorporates/honors the individual discourses.			Communication fosters cooperation and overcoming blind spots	Trying to impose the "right" discourse will prompt competition
The approach is implemented with considerable attentiveness to how seemingly minor details might privilege various groups or discourses.		Habitus may encourage privilege		The process ought not create a competitive advantage
The use of language, technical jargon, mental models, etc. will be scrutinized and designed in order to promote a broad and inclusive space in which the full range of participants can variously discover, articulate, and pursue their interests.		Habitus is subjective; as fields change so to can individual agency	Ecological communication provides a language for integrating social and ecological systems	
A DI approach will develop norms of information sharing rather than withholding information to be used competitively. This may include joint fact-finding/monitoring activities when the science on an issue is contested.	Agencies should encourage information sharing			Groups will tend trust their own data over others'

Table 2.1 (cont.)

DI Propositions	Weber	Bourdieu	Luhmann	Sherif
A DI approach is not conducive to coalitions of stakeholders forcing the adoption of their discourse through the domination or exclusion of others' interests.	Traditional approaches encourage groups to lobby the agency			Competitive negotiation involves proving the superiority of your position
The process of constructing a collective discourse will be fair; various participants will all have an adequate opportunity to embed their values and interests in it. It should resonate with norms about what constitutes a legitimate process, particularly when participants from indigenous or high context cultures are involved.		A field is both objective and subjective; guiding social relations and action	Communication confronts paradox and respects different constructions of meaning	Notions of fairness need to go beyond "fair is what allows my side to prevail"
DI approaches go beyond single issues; participants view the situation holistically and think critically in terms of a system.			Communication integrates social and ecological systems	Learning that people legitimately care about different issues prompts innovation

mind, the four whose relevance was specifically examined in this chapter should been seen as illustrative rather than exhaustive.

The notion that "a problem cannot be solved with the same level of thinking that gave rise to it" is generally attributed to Albert Einstein in reference to the existential risk that nuclear weapons pose to humanity.[2] If we apply this maxim to natural resource management, we can paraphrase it into "you cannot solve a problem in the institutions that have given rise to it." Discourse-intensive approaches represent an institutional innovation; grassroots efforts to solve problems that have confounded the conventional intelligence of the organizations assigned to deal with them. Arguably the trickiest problems that any generation faces are those that the previous generation could not conquer *in the context of the institutions at their disposal.* The rise of discourse-intensive approaches offers the very real prospect of meaningful progress on previously vexing problems. But that does not mean that the previous institutions nor the actors within them were somehow wrong or ill-intended. Increasing the discourse intensity of environmental management approaches supplements our existing set of institutional frames, and need not been seen as replacing or rejecting them. By the same token, it would be arrogant to conclude that it represents the last generation of institutional innovations that will be necessary.

As an example of a discourse-intensive approach, Collaborative Learning has been designed, applied, adapted, and refined over twenty-five years. Like any DI approach, its resilience and effectiveness rely on its evolution, remaining organic (e.g., being responsive to the uniqueness of each situation) rather than formulaic (e.g., becoming part of an operations manual). Consequently, Collaborative Learning and other DI innovations (e.g., appreciative inquiry, citizen juries, consensus conferences, facilitated dialogue, mediated modeling, scenario mapping, world cafes) face the challenge of avoiding entrenched and static institutionalization, even as they become embedded in bureaucratic processes themselves.

And to close with a caveat: when reflecting on the ideas of the theorists featured in this chapter and the twenty-five years of applying Collaborative Learning strategies and tools, the 1990s movie "Groundhog Day," comes to mind. For a television reporter, the same day occurred over and over again. Advocates of DI approaches may at times wonder, "haven't we tried this – haven't we done this before?" Given the complexity, controversy, and uncertainty in natural resource and environmental management decision situations, well-applied DI methods should generate progress, and while at times seemingly slow and incremental, make important progress nonetheless.

---

2 Sourced from the following: "Our world faces a crisis as yet unperceived by those possessing power to make great decisions for good or evil. The unleashed power of the atom has changed everything save our modes of thinking and we thus drift toward unparalleled catastrophe. We scientists who released this immense power have an overwhelming responsibility in this world life-and-death struggle to harness the atom for the benefit of mankind and not for humanity's destruction. We need two hundred thousand dollars at once for a nation-wide campaign to let people know that **a new type of thinking is essential if mankind is to survive and move toward higher levels**. This appeal is sent to you only after long consideration of the immense crisis we face. . . . We ask your help at this fateful moment as a sign that we scientists do not stand alone." (Source: *New York Times* – May 25, 1946, p. 13 – "Atomic Education Urged by Einstein." Emphases in original) (http://icarus-falling.blogspot.com/2009/06/einstein-enigma.html

References

Amy, D. J. (1987). *The Politics of Environmental Mediation*. New York: Columbia University Press.

Ascher, W. (2004). Scientific information and uncertainty: challenges for the use of science in policymaking. *Science and Engineering Ethics*, *10*: 437–455.

Bazerman, M. H. (2002). *Judgment in Managerial Decision Making*, 5th ed. New York: John Wiley.

Bellah, R. N., Madsen, R., Sullivan, W.M., Swindler, A., & Tipton, S.M. (1991). *The Good Society*. New York: Alfred A. Knopf.

Bingham, G. (1986). *Resolving Environmental Disputes: A Decade of Experience*. Washington, DC: The Conservation Foundation.

Bourdieu, P. [1980] (1990). *The Logic of Practice*. Cambridge, UK: Polity Press.

Bourdieu, P. (1991). On symbolic power. In J.B. Thompson, ed., *Language and Symbolic Power*, Cambridge, MA: Harvard University Press, pp. 163–170.

Bourdieu, P., & Wacquant, L. (1992). *An Invitation to Reflexive Sociology*, Chicago, IL: University of Chicago Press.

Brown, H. C., Buck, L., & Lassoie, J. (2008) Governance and social learning in the management of non-wood forest products in community forests in Cameroon. *International Journal of Agricultural Resource, Governance and Ecology*, *7*(3): 256–275

Caine, K. J. (2013). Logic of land and power: The social transformation of northern natural resource management. In J.R. Parkins and M.G. Reed, eds., *Social Transformation in Rural Canada: Community, Cultures, and Collective Action*. Vancouver: UBC Press, pp. 169–188.

Cantrill, J. G., & Senecah, S. L. (2001). Using the "sense of self in place" construct in the context of environmental policy making and landscape planning. *Environmental Science and Policy*, *4*: 185–204.

Carpenter, S., & Kennedy, W. J. D. (1988). *Managing Public Disputes: A Practical Guide to Handling Conflict and Reaching Agreements*. San Francisco, CA: Jossey-Bass.

Celino, A., & Concillio, G. (2011). Explorative nature of negotiation in participatory decision making for sustainability. *Group Decision and Negotiation*, *20*: 255–270.

Checkland, P., & Scholes, J. (1990). *Soft Systems Methodology in Action*, New York: John Wiley & Sons.

Cheng, A. S., Kruger, L. E., & Daniels S. E. (2003). "Place" as an integrating concept in natural resource politics: Propositions for a social science research agenda. *Society and Natural Resources*, *16*(1): 87–104.

Clarke, T. L., & Peterson, T. R. (2015). *Environmental Conflict Management*, Los Angeles, CA: Sage Publications.

Costanza, R., & Cornwall, L. (1992). The 4P approach to dealing with scientific uncertainty. *Environment*, *34*(9): 12–22.

Dana, S. T., & Fairfax, S.K. (1980). Forest and Range Policy, 2nd ed. New York: McGraw-Hill.

Daniels, S. E. (2009). Exploring the feasibility of mediated final offer arbitration as a technique for managing "gridlocked" environmental conflict. *Society and Natural Resource*, *22*(2): 261–277.

Daniels, S. E., & Walker, G. B. (1996). Collaborative learning: improving public deliberation in ecosystem management. *Environmental Impact Assessment Review*, *16*: 71–102.

Daniels, S. E., & Walker, G. B. (2001). *Working Through Environmental Conflict: The Collaborative Learning Approach*. Westport, CT: Praeger.

Deutsch, M. (1973). *The Resolution of Conflict*. New Haven, CT: Yale University Press.

Ducrot, R. (2009). Gaming across scale in peri-urban water management: Contributions from two experiences in Bolivia and Brazil. *International Journal of Sustainable Development and World Ecology*, *16*(3): 240–252.

Faure, G.-O., & Rubin, J. (1993). Organizing concepts and questions. In G. Sjostedt, ed., *International Environmental Negotiation*, Newbury Park, CA: Sage Publications. pp. 17–26.

Fisher, R., Ury, W., & Patton, B. (1991). *Getting to Yes: Negotiating Agreement without Giving In*, 2nd ed. New York: Basic Books.

Fiske, S. T., & Taylor, S. E. (2017). *Social Cognition: From Brains to Culture*, 3rd ed. Newbury Park, CA: Sage Publications.

Folger, J. P., Poole, M. S. & Stutman, R. K. (2008). *Working Through Conflict: Strategies for Relationships, Groups, and Organizations*, 6th ed. Boston, MA: Allyn and Bacon.

Flood, R. L., & Jackson, M. C. (1991). *Creative Problem Solving: Total Systems Intervention*. Chichester, UK: John Wiley.

Funtowicz, S., (2001). Global risk, uncertainty, and ignorance. In J. Kasperson and R. Kasperson, eds., *Global Environmental Risk*, London: Earthscan, pp. 173–194.

Gee, J. P. (1990). *Social linguistics and Literacies: Ideology in Discourses, Critical Perspectives on Literacy and Education*. London: Routledge.

Gerth, H. H., & Mills, C. W. (1946). *From Max Weber: Essays in Sociology*. New York: Oxford University Press.

Hall, D. M., Feldspausch-Parker, A., Peterson, T. R., Stephens, J. C., & Wilson, E. J. (2017, April). Social-ecological system resonance: A theoretical framework for brokering sustainable solutions. *Sustainability Science*. Published online: https://link.springer.com/journal/11625

Hill, R., Davies, J., Bohnet, I. C. et al. (2015). Collaboration mobilises institutions with scale-dependent comparative advantage in landscape-dependent biodiversity conservation. *Environmental Science and Policy*, *51*: 267–277.

Hocker, J. L., & Wilmot, W. (2018). *Interpersonal Conflict*, 10th ed. New York: McGraw-Hill.

Imperial, M. T. (2005). Using collaboration as a governance strategy: Lessons from six watershed management programs. *Administration and Society*, *37*(3): 281–320.

Kolb, D. A. (1984). *Experiential Learning: Experience as the Source of Learning and Development*. Englewood Cliffs NJ: Prentice-Hall.

Lee K. (1993). *Compass and Gyroscope: Integrating Science and Politics for the Environment*. Washington, DC: Island Press.

Leydesdorff, L. (2000). Luhmann, Habermas, and the theory of communication, *Systems Research and Behavioral Science*, *17*: 273–288.

Luhmann, N. (1979). *Trust and Power*. H. Davis, J. Raffan, & K. Rooney, trans. Chichester, UK: Wiley.

Luhmann, N. (1989). *Ecological Communication*. J. Bednarz, trans. Chicago, IL: University of Chicago Press.

Luhmann, N. (1992). What is communication? *Communication Theory*, *2*: 251–259.

Luhmann, L. (1994). Speaking and silence. *New German Critique*, *61*: 25–37.

Luhmann, N. (1995). *Social Systems*. J. Bednarz & D. Baecker, trans. Palo Alto, CA: Stanford University Press.

Meadows, D. H. (2008). In D. Wright, ed., *Thinking in Systems: A Primer*. Hartford, VT: Chelsea Green Publishing.

Pagdee, A., Kim, Y.-S., and Daugherty, P. J. (2006). What makes community-based forestry successful: A meta-study from community-based forests around the world. *Society and Natural Resources*, *19*(1): 33–52.

Peterson, T. R., Peterson, M. J., & Grant, W. E. (2004). Social practice and biophysical process. In S. L. Senecah, ed., *The Environmental Communication Yearbook*, Vol. 1. Mahwah, NJ, USA: Lawrence Erlbaum Associates, pp. 15–32.

Ramirez, R., & Fernandez, M. (2005). Facilitation of collaborative management: Reflections from practice. *Systemic Practice and Action Research*, *18*(1): 5–20.

Ramirez, R., & Quarry, W. (2004). *Communication for Development: A Medium for Innovation in Natural Resources Management*. Ottawa and Rome: International Development Research Centre and the Food and Agriculture Organization of the United Nations.

Senge, P. M. (1990). *The Fifth Discipline: The Art and Practice of the Learning Organization*. New York: Doubleday/Currency.

Senge, P. M., Kleiner, A., Roberts, C., Ross, R., & Smith B. (1994). *The Fifth Discipline Fieldbook: Strategies and Tools for Building a Learning Organization*. New York: Currency/Doubleday.

Sherif, M. (1958). Superordinate goals in the reduction of intergroup conflict. *American Journal of Sociology*, *63*: 349–356.

Sherif, M., Harvey, O. J., White, B. J., Hood, W. R., & Sherif, C. W. (1954/1961). *Intergroup Conflict and Cooperation: The Robbers Cave Experiment* Norman, OK: University Book Exchange.

Singhal, A. (2001). *Facilitating Community Participation through Communication*. New York: UNICEF.

Singhal, A. & Devi, K. (2003). Visual voices in participatory communication. *Communicator*, *38*(2): 1–15.

Tajfel, H. (1959). Quantitative judgment in social perception. *British Journal of Psychology*, *50*: 16–29.

Tajfel, H. & Turner, J. C. (1979). An integrative theory of intergroup conflict. In W. G. Austin & S. Worchel (eds.), *The Social Psychology of Intergroup Relations*. Monterey, CA: Brooks-Cole, pp. 47–61.

Trappes-Lomax, H. (2004). Discourse analysis. In A. Davies and C. Elder, eds., *The Handbook of Applied Linguistics*. Malden, MA: Blackwell Publishing, pp. 133–164.

Walker, G. B. (2007). Public participation as participatory communication in environmental policy decision-making: From concepts to structured conversations. *Environmental Communication: A Journal of Nature and Culture*, *1*: 64–72.

Walker, G. B., Senecah, S. L., & Daniels, S. E. (2006). From the forest to the river: Citizen views of stakeholder engagement. *Human Ecology Review*, *13*: 193–202.

Wehr, P. (1979). *Conflict Regulation*. Boulder, CO: Westview Press.

Wilson, K. & Morren, G. E. B. (1990). *Systems Approaches to Improvements in Agriculture and Resource Management*. New York: Macmillan.

3 Community-Based Research

Randy Stoecker

What Is Community-Based Research?

Community-based research, or CBR, has a history that is distinct from environmental sociology. Environmental sociology developed in part from the growing environmental activism of the 1970s, but it became a mostly academic pursuit of attempting to explain either the influence of the environment on society, the influence of society on the environment, or the dialectical interaction of the two (Dunlap and Catton, 1979; Catton, 1980; Schnaiberg, 1980). Especially in the early days of the subfield, the connection between research and action was not central.

CBR has a longer history and multiple origins in the global post-World War II period. In North America the practice was first known as action research, and is attributed to Kurt Lewin (1948). In the Global South the practice was first known as participatory research, mostly attributed to Rajesh Tandon (Brown and Tandon, 1983), with related practices such as Paulo Freire's (2000) popular education (Myles Horton (1997) also practiced popular education in North America) and Augusto Boal's Theater of the Oppressed (1985). There were, in those early days, some theoretical and political distinctions between the two practices, with action research a more top-down academic-controlled process and participatory research and its related practices focused on anti-oppression organizing (Brown and Tandon, 1983; Stoecker, 2003).

Today, the definitional milieu within which the term CBR is embedded is a mess. Chandler and Torbert (2003) counted at least twenty-seven terms referring to CBR, and there are probably more now. Some of the most commonly used, besides the original terms, are community-based research, community-based participatory research, and participatory action research. The terms are often associated with disciplines. Action research appears more in the education disciplines, community-based participatory research in public health, community-based research in the liberal arts, and participatory research and participatory action research tend to cross disciplines. It is tempting to try and divide the labels up into camps, but the wide variations of practice within each label today are greater than any differences in practice between the labels. Minkler and Wallerstein's (2008) thick volume on community-based participatory research shows a range of practices varying from what appear to be old-school action research to activist forms of participatory research. Whyte's (1990) book *Participatory Action Research* reads very much like Lewin's action research approach, and Stringer's (2013) book *Action Research* reads much more like a participatory research approach. Many of the authors cited in this review will use the terms participatory action research, community-based participatory research, and others. I will refer to them all as CBR, simply for convenience.

Regardless of what term we use, we none the less need to define it. I still prefer the definition that comes from Strand et al (2003a: 6, also see 2003b). "1. CBR is a collaborative enterprise between academic researchers (professors and students) and community members. 2. CBR seeks to democratize knowledge by validating multiple sources of knowledge and promoting the use of multiple methods of discovery and dissemination. 3. CBR has as its goal social action for the purpose of achieving social change and social justice."

Let's parse out this definition a bit. When we talk about CBR as a collaborative enterprise we mean that neither the research question nor the research method is determined solely by academics. In fact, one of the most important principles for Strand et al. is that "the community" (variously defined as informal groups or formal organizations and sometimes stretched to include various institutional bodies) chooses the research problem. In my interpretation of this principle, a professional researcher should always be a part of that collaboration but student involvement is not a requirement. By democratizing knowledge we mean that our collaboration extends the forms and processes of knowledge that get counted as legitimate, including oral traditions (Bruce, 2003; Hoare et al., 1993) and experiential knowledge. And when we say that the ultimate goal of CBR is social action for social change, we mean that the research process and research results need to both support and inform social action. CBR can support social action by designing the research process itself to organize people. It can inform social action by providing knowledge groups can use to inform their strategies or policy proposals.

The definition proposed here is a higher standard than most, particularly in its emphasis on action for social change and social justice. Much of the literature still limits itself to a focus on the relationship between, usually, some kind of service provider and an academic to do a research project without strategically directing it toward social change (Stoecker, 2016). In other cases, people count themselves as doing CBR when they only have token community involvement, or when their research has no direct connection to action (Stoecker, 2009). One of the reasons that all of the labels have become so meaningless is that people have subverted them to include work that should be categorized elsewhere. This review will hold tight to all three parts of the strict CBR definition in discussing CBR in environmental sociology.

Challenges Facing CBR

Aside from the inability of many CBR researchers to meet the standards of the strict definition, CBR also faces challenges from the broader academic and scientific cultures in which its practitioners often operate. Many scholars remain concerned about whether CBR can produce valid, reliable knowledge. To a large extent, however, this issue has been put to rest. CBR articles and books are published across the disciplines. And there are strong arguments that CBR is not just a legitimate form of research, but may even produce better knowledge than traditional scientific research methods (Balazs and Morello-Frosch, 2013). The main problem with much traditional science is that people don't trust the scientists. This distrust has been most pronounced among indigenous peoples, who have been subjected to science that has informed genocide at worst, and has misrepresented indigenous cultures at best. Smith (2012) may be the most visible, but certainly not the only, scholar to show science's complicity in colonization of the mind and body. The consequent distrust is not limited to indigenous peoples, however. Ross and Stoecker (2016) have shown that the distrust of researchers felt by residents of the lower 9th Ward of New Orleans extended even to CBR researchers and remained years after Katrina. As we will also see, however, as researchers work to rebuild trust, they are producing examples of CBR that are both valid and impactful.

In this discussion of the validity of CBR we must also attend to a common association of CBR with a qualitative research approach. But CBR is actually non-aligned when it comes to research methods. We can think of CBR as occupying the liminal space where epistemology and methodology touch each other. Epistemologically, CBR values the synergy created when abstract expert knowledge combines with specific concrete community knowledge (Nyden and Wiewal, 1992). From this standpoint we can already see why it is logical for CBR to value diversity of knowledge forms, inter- or trans-disciplinarity, and a mosaic of ways of knowing. Methodologically, then, CBR will be inclusive of whatever approaches best address a given research question, and happily includes even quantitative methods when the research question calls for such tools. It may be that CBR gets its association with qualitative methodology because it operates with a qualitative logic. That is, CBR values the production of knowledge through relationships rather than through the social distance created by positivist objectivity. So even when CBR researchers deploy quantitative research methods, they do so through a strong relationship with a community group that is involved in every step of the research process (Stoecker and Brydon-Miller, 2013).

Another concern about CBR is whether the products of CBR are given the same emphasis as the process. There is a large literature on the importance of building relationships in CBR that is heavily influenced by the larger literature on service learning and community-higher education partnerships (Stoecker, Reece, and Konkle, 2018). That literature focuses on the relationship itself, strategizing ways to create and maintain it, rather than on developing a strategy that connects research and action. While we will see that CBR does in fact produce outcomes on environmental issues, even those examples do not often focus on building models of how to produce outcomes.

Stoecker's (2013) project-based approach to CBR is one attempt to build a model linking research and action. The model begins with research diagnosing some social condition – understanding its dimensions and causes. The next step involves research focused on designing a prescription for changing that condition – producing a policy proposal or an action strategy of some kind. The next steps are to implement the prescription and engage in research to evaluate its effects and thus maximize its outcomes. Hidayat et al. (2013) and Calderón and Cadena (2007) have shown the applicability of such a model in their CBR projects.

A final concern is that CBR is as subject to neoliberal co-optation as any other form of higher education community engagement. This concern shows that CBR, by itself, is not automatically insulated from colonizing tendencies, but must be explicitly designed to resist such tendencies by engaging with organized groups that are explicitly anti-colonial and anti-capitalist (Jordan and Kapoor, 2016). In addition, CBR must be designed more in accord with the Strand et al. (2003) definition offered above and, in an elaboration of the principles, to put social change first rather than last in the priority of higher education community engagement (Stoecker, 2016)

CBR in Sociology

CBR has not occupied a prominent place in sociology in general. Instead, the discipline has gravitated toward an approach called public sociology. In public sociology the emphasis is overall on communicating traditionally created and constructed sociological knowledge to various "publics" (Burawoy, 2004), and not so much about co-constructing knowledge in collaboration with those publics. This may be because, in contrast to a discipline like public health, which has a strong practice paradigm as well as being an academic discipline, sociology has not as readily embraced a practice that would allow it to focus attention on the empowerment of grassroots people, especially those who are oppressed, exploited, or excluded.

The closest sociology has come, in any institutionalized form, to engaged scholarship like CBR, has been through what is usually called applied sociology. The works on applied sociology, such as Steele and Price (2007) and Zevallos (2009), however, read more as a form of expert-directed Lewinian action research that is easily exploitable by governments and corporations, rather than a collaborative form of CBR that can be put to use by grassroots people and their organizations. So the theory and practice of CBR has lagged in sociology.

Additionally, environmental sociology is a very young subfield, only about fifty years old. And like many subfields in sociology, it is quite interdisciplinary and, unusually, invokes not just other social sciences but also natural sciences (Lutzenhiser, 1994; Heinrichs and Grosse, 2010). The subfield itself also covers wide-ranging topics that themselves require diverse forms of interdisciplinary expertise such as limnology, medicine, physics, geology, and many others. Thus, this review will include work that is sociological in its approach even when the authors of the work may not identify as sociologists.

CBR in Environmental Sociology

Because environmental sociology is so wide ranging in its foci, CBR applications will also be substantively diverse. They can to some extent be grouped, so here we will cover a few of the major topics where CBR has been widely used: climate change, food systems, sustainable behavior change, and environmental justice.

Climate Change

CBR focused on climate change in environmental sociology has lagged in contrast to disciplines like public health. Health Canada, for example, funded thirty-six CBR projects over a three-year period to focus on the health effects of climate change and strategies for managing those effects, particularly in Inuit communities (Peace and Myers, 2012). Canada also has an advantage in conducting such research because of their more developed ethics guidelines for CBR, set out in Chapter 9 of the Tri-Council Policy Statement (Government of Canada, 2018) that focuses on research with indigenous people.

In some ways it also "natural" that CBR around climate change would appear prominently in Canada as the effects and their consequences are being felt so intensely in the Arctic region. CBR (usually called community-based participatory research around climate and health issues in Canada) projects have also informed a model called community-based adaptation or CBA that integrates research and planning ways to respond locally to climate change. Such projects are not free of critiques that they are still distorted by the legacy of colonialism and its colonizing knowledge processes (Ford et al., 2016). Some of these projects focus on transportation and mobility issues arising on land and water that is rapidly thawing (Tremblay et al., 2008). Similar projects have also occurred in the United States, such as the comprehensive study of Oakland on the effects of climate change and planning a response (Garzón et al., 2012). Interestingly, in this study, and showing the chaos of definitions, the authors use the label "community engaged" rather than "participatory," arguing that the old participatory label was about getting grassroots people to collect data while their "engaged" label was about getting grassroots people to have power in decision making about the research and planning processes.

Some of these projects are also about knowledge exchange – bringing together scientists' knowledge with community-based knowledge (Galvin et al., 2015).

Food Systems

The issue of food equity has exploded onto the scene in the past decade. Many of those working on the issue of food have also employed a CBR approach. In a Masters-level UK course, for example, students engage in a variety of interdisciplinary projects that range from more natural science projects on land use to marketing studies, to financial studies (Moragues-Faus, Omar, and Wang, 2015). In Guelph, Canada, the local university supported a local food system planning process engaging grassroots people using a CBR approach (Hayhurst et al., 2013)

Food systems may be the area where the link between research and action is strongest. Guzmán et al. (2013). describe a set of case studies in Spain following a five-step agroecology-informed model of preliminary exploration, diagnostic design, research, action, and evaluation and adjustment. So synergistic are the issue of food systems and the practice of CBR that their combination has catalyzed new organizational forms such as Activating Change Together for Community Food Security (2015) or Food First (n.d.). And agroecology seems to be an especially good fit for CBR, with explicit support for combining experiential and credentialed knowledge (Méndez et al., 2017).

An increasing focus of environmental sociology-related CBR links two areas of concern in environmental sociology – climate change and food. Loo (2014) advocated for a CBR approach to improve our understanding of how climate change affects food. Loo argued that CBR, because of its ability to access local knowledge, could better inform interventions. A CBR project in Malawi, with farmers leading the process, focused on strategies to achieve food security in the context of climate change (Soils, Food and Healthy Communities, 2018). Similar projects in Ghana and Zimbabwe showed how a form of CBR that incorporates principles of popular education in planning-action-reflection cycles, improved farmers' responses to climate change (Mapfumo et al., 2013).

Natural Resource Management

Connected to food systems, but defined more broadly, CBR also plays an increasingly powerful role in natural resource management, particularly in the developing world. CBR, in such contexts, becomes attractive through the lens of community-based natural resource management (CBNRM), where the focus is on understanding the needs of farming communities and then attempting to meet those needs while also protecting the environment (Gruber, 2010; Yamanoshita, 2013). Since CBNRM starts from the interests of professional natural resource managers, the focus of CBR in such contexts is mostly on shifting the perspectives and practices of scientists rather than building the capacity of communities. Programs such as that of the Center for People and Forests, for example, pose CBR specifically in contrast to top-down natural resource management, and train professionals in working collaboratively with communities (RECOFTC, 2011). But doing so is easier said than done. Scientists using a Lewinian action research version of CBR which, as we discussed above, is the most researcher-driven model, were never able to fully build a partnership with farmer's organizations whose social objectives did not easily integrate with the scientists' environmental objectives (Castellanet and Jordan, 2002).

The focus of CBR in natural resource management is not solely on managing natural resources. It is also on the development of communities in poverty (Tyler, 2006). Consequently it also confronts existing government policies and practices (John and Phalla, 2006; Van Tuyen et al., 2006), and promotes new forms of cooperative and collaborative local community relationships (Nong and Marschke, 2006; Ykhanbai and Bulgan, 2006).

Sustainable Behavior Change

One of the problems that US researchers have concerned themselves with has been individual behavior and attitude change. Perhaps it is because of the massively individualist culture in the United States, along with the rising power of neoliberal hegemony but, in any event, US researchers seem most prominent in the environmental sociology literature when it comes to promoting individual behavior change. But in some ways the challenge seems intractable, and researchers bemoan the lack of a solid successful strategy (Eilam and Trop, 2012; Kinzig et al., 2013).

Intriguingly, however, CBR may show a way forward. Bywater (2014) used PAR as a tool to support college student behavior change. Cachelin, Rose, and Lee (2016) showed CBR as a way to get people to think beyond individual behavior change to think about how their sustainability behaviors affect their place. Ballard and Belsky (2010) found that CBR enhanced the environmental learning of undocumented Latinx salal harvest workers. These findings fit with the popular education form of CBR, where people develop new behaviors and attitudes because they did their own study of the problems they were facing, and came up with their own solutions. CBR may be able to help sociology get past its own vanguardist tendencies.

Environmental Justice

Environmental justice has developed as an alternative to the mainstream environmental movement, focusing on the intersection of environmental pollution and class and especially race inequalities. Environmental justice movements tend to manifest themselves at the local level, led by grassroots people suffering the ill effects of local toxic air, water, or soil pollution (Skelton and Miller, 2016; Perez et al., 2015). An important task of these groups is documenting the physical effects of the toxins being foisted upon their communities, and this is where CBR comes in. In addition, in contrast to the big environmental lobbying organizations, because environmental justice organizing emphasizes the importance of building the leadership of grassroots people, it has a natural affinity with the best practices of CBR (Bacon et al., 2013).

A great deal of that CBR work is epidemiological, but it has sociological characteristics because the research is done with an awareness of class and race exploitation and oppression, rather than simply gathering up health statistics. Thus, environmental justice is another example of the advanced status of public health in relation to CBR (Shepard et al., 2002; McOliver et al., 2015).

Some of the most highly respected work, which has also been the most attacked by state and capitalist interests, has been that of the late Steve Wing. His work with rural

African American communities suffering the ill effects of concentrated animal feeding operations – CAFOs – has led to court cases and legislative battles (Wing, Cole, and Grant, 2000; Wing et al., 2008). Some of the controversies spurred on by Wing and colleagues' work involves the challenges of measuring exposure. Harrison's (2011) work on air pollution caused by pesticide drift directly confronts the challenges involved in doing CBR with relatively untrained grassroots people when sample collection and measurement are fraught with technical difficulties. And Petersen and colleagues (2007) describe a CBR project focused on assessing lead levels in children linked to pollution caused by mining operations in rural Oklahoma.

Urban areas are equally beset by environmental justice issues, including lead exposure. Garcia et al. (2013) describe a CBR project supporting organizing efforts to reduce air pollution around major industrial transportation hubs. And some of the measurement issues inherent in documenting the health effects of pollution may actually be addressed by high quality CBR work. Jordan and colleagues (2000) show how CBR leads to better data in a CBR project measuring the effects of lead exposure in the Phillips neighborhood of Minneapolis. Because residents could feel some ownership and influence over the research methods in ways that seemed more sensitive to their children's needs, they were more likely to have their children participate in the research.

Conclusion and Future Directions

CBR has achieved a prominent role in environmental scholarship and movements, including the broadly interdisciplinary scholarship that is included in environmental sociology. There is not a specifically environmental sociology CBR. The interdisciplinary nature of environmental issues, which often invoke natural science-based knowledge at the input end and health knowledge at the output end, make an environmental sociological CBR too narrow for practical benefit. And the approach of too many scholars, who privilege academic forms of knowledge over community forms of knowledge, will further limit the potential of environmental sociological CBR.

If environmental sociological CBR is to have a future in helping to create a better world for all life from microbes to sequoias (which, whether we admit it or not, are all inextricably bound together) it must shift its emphasis from privileging academic knowledge to the principles of CBR posed at the beginning of the chapter: collaboration, democratization, and promotion of justice. Those three principles, in fact, question the path that environmental sociological CBR is on. Oppressed, exploited, and excluded people rarely start with environmental issues. They start with economic, housing, food, safety, health and other immediate life issues. Climate change, deforestation, pollution and other environmental issues are well down the list. Most immediate issues will eventually point to environmental issues as well, but it's not the starting point.

So collaboration in environmental sociological CBR means starting where people are, and that usually means not starting with the environment. The closest model we have to such an approach is with environmental justice. But the collaboration needed even in environmental justice requires strong community organizing. So CBR is also not the starting point. Issue-identification is the starting point. We need to build strong community-controlled organizations that can develop their own social justice campaigns and engage academics in collaborating with them on the research they need to win those campaigns. When and if those campaigns point to environmental issues, academics should stand prepared to contribute.

Democratization then means that decision-making does not privilege elites – whether they be political elites, economic elites, or educational elites. Thus, and again invoking community organizing, we need to practice two forms of social change with people – acting as research allies to support community groups working on immediate issues, and doing so in such a way that it builds the people's capacities to find, create, use, and amplify knowledge (Stoecker, 2012).

But perhaps the hardest part for academics is committing ourselves to social justice as both professionals and persons. With academic freedom being crushed by neoliberalism and an expanding fascism focused on producing a master class of rich white men who can dismantle democracy unchecked, we in the academy seem to cling ever more fervently to a morally adrift empiricism – the very thing which alienates much of the population from the very idea of research. In such an atmosphere, academics making explicit commitments to social justice not only feels risky but is in fact risky. But doing so is not without precedent, as the Community Learning Partnership (n.d.) – a consortium of mostly community colleges and social justice groups in numerous locations across the United States – is showing us.

The greatest innovation that environmentally sociological CBR can engage in is to integrate community organizing into its practice. As academics, we already understand the research methods and processes. What we by and large do not understand is how our research can be used in community organizing campaigns. The need for community organizing is illustrated by the absence of truly impactful environmentally sociological CBR. We do the research, but because community groups are not organized to guide our research and use it strategically, it becomes shelf research as much as the typical journal article.

For those academics uncomfortable with the thought of making an explicit commitment to justice, one role they can play is to research the research. There are many questions about how to make our research matter. The focus in most of the published scholarship reviewed here, and in general, has indeed been on the research methods and processes or the relational processes. When outcomes have been discussed, they have not been theorized. The field remains sorely lacking in carefully theorized discussions of the relationship between research, action, and outcomes. Why and how does research support, or not support, action? What kinds of research support what kinds of action? How important are what kinds of research, under what kinds of circumstances, in informing action, or in producing outcomes? These, for the moment, will remain questions for further study and further advancement of CBR practice in environmental sociology, environmental studies in general and, indeed, environmental movements. These are not just, or even primarily, academic questions. As the Pacific Institute (2018) notes "Communities facing persistent environmental health and justice issues have the right to excellent research." Knowledge justice is part of the path to environmental justice.

References

Activating Change Together for Community Food Security. (2015). Making Food Matter: Strategies for activating change together. A participatory research report on community food security in Nova Scotia. https://foodarc.ca/wp-content/uploads/2014/11/Making-Food-Matter-Report_March2015rev.pdf

Bacon, C., deVuono-Powell, S., Frampton, M. L., LoPresti, T., & Pannu, C. (2013). Introduction to empowered partnerships: Community-based participatory action research for environmental justice. *Environmental Justice* 6(1), 1–8. https://doi.org/10.1089/env.2012.0019

Balazs, C. L., & Morello-Frosch, R. (2013). The Three Rs: How community-based participatory research strengthens the rigor, relevance, and reach of science. *Environmental Justice* 6 (1), 9–16.

Ballard, H. L., & Belsky, J. M. (2010). Participatory action research and environmental learning: implications for resilient forests and communities. *Environmental Education Research* 16(5–6), 611–627.

Boal, A. 1985. *Theatre of the Oppressed*. London: Pluto.

Brown, L. D., & Tandon, R. (1983). Ideology and political economy in inquiry: Action research and participatory research. *Journal of Applied Behavioral Science* 19, 277–294.

Bruce, H. E. (2003). Hoop dancing: Literature circles and Native American storytelling. *The English Journal* 93(1), 54–59.

Burawoy, M. (2004). Public sociology: Contradictions, dilemmas and possibilities. *Social Forces* 82(4), 1603–1618.

Bywater, K. (2014) Investigating the benefits of participatory action research for environmental education. *Policy Futures in Education* 12(7). http://dx.doi.org/10.2304/pfie.2014.12.7.920

Cachelin, A., Rose, J., & Rumore, D. L. (2016). Leveraging place for critical sustainability education: The promise of participatory action research. *Journal of Sustainability Education* 11. www.jsedimensions.org/wordpress/wp-content/uploads/2016/03/Cachelin-Rose-Rumore-JSE-February-2016-Place-Issue-PDF-Ready.pdf

Calderón, J. Z., & Cadena, G. R. (2007). Linking critical democratic pedagogy, multiculturalism, and service learning to a project-based approach. In J. Z. Calderón (ed.), Race, poverty, and social justice: Multidisciplinary perspectives through service learning. Sterling, VA: Stylus, pp. 63–80.

Castellanet, C., & Jordan, C. F. (2002). *Participatory Action Research in Natural Resource Management: A Critique of the Method Based on Five Years' Experience in the Transamozonica Region of Brazil*. New York: Taylor and Francis.

Catton, W. R. (1980). *Overshoot: The Ecological Basis of Revolutionary Change*. Urbana, IL: University of Illinois Press.

Chandler, D, & Torbert, B. (2003). Transforming inquiry and action: interweaving 27 flavors of action research. *Action Research* 1, 33–52.

Community Learning Partnership. (n.d.). http://communitylearningpartnership.org/

Dunlap, R. E., & Catton Jr.,W. R. (1979). Environmental Sociology. *Annual Review of Sociology* 5, 243–273.

Eilam, E., & Trop, T. (2012). Environmental attitudes and environmental behavior – which is the horse and which is the cart? *Sustainability* 4, 2210–2246.

Food First. (n.d.) Institute for Food & Development Policy. https://foodfirst.org/about-us/our-work/.

Ford, J. D., Stephenson, E., Cunsolo Willox, A., Edge, V., Farahbakhsh, K., et al. (2016), Community-based adaptation research in the Canadian Arctic. *WIREs Climate Change* 7, 175–191.

Freire, P. (2000). *Pedagogy of the Oppressed*. 30th anniversary ed. Translated by M. B. Ramos. New York: Continuum.

Galvin M., Wilson J., Stuart-Hill S., Pereira T., Warburton M. et al. (2015). Planning for adaptation: Applying scientific climate change projections to local social realities. South African Water Research Commission. WRC Report No. 2152/1/15. http://wrc.org.za/Knowledge%20Hub%20Documents/Research%20Reports/2152-1-15.pdf

Garcia, A. P., Wallerstein, N., Hricko, A., Marquez, J. N., Logan, et al. (2013). THE (Trade, Health, Environment) Impact Project: A community-based participatory research environmental justice case study. *Environmental Justice* 6(1), 17–26.

Garzón, C., Cooley, H., Heberger, M., Moore, E., Allen, L. [...] and The Oakland Climate Action Coalition. (Pacific Institute). (2012). Community based climate adaptation planning: case study of Oakland, California. California Energy Commission. Publication number: CEC-500–2012-038. www.energy.ca.gov/2012publications/CEC-500–2012-038/CEC-500–2012-038.pdf.

Guzmán, G. I., López, D. Román, L., & Alonso, A. M. (2013). Participatory action research in agroecology: Building local organic food networks in Spain. *Agroecology and Sustainable Food Systems* 37(1), 127–146.

Government of Canada. (2018). *Tri-Council Policy Statement: Ethical Conduct for Research Involving Humans*. https://ethics.gc.ca/eng/policy-politique_tcps2-eptc2_2018.html

Gruber, J. S. (2010). Key principles of community-based natural resource management: A synthesis and interpretation of identified effective approaches for managing the commons. *Environmental Management* 45(1), 52–66.

Harrison, J. L. (2011). Parsing "participation" in action research: Navigating the challenges of lay involvement in technically complex participatory science projects. *Society & Natural Resources* 24(7), 702–716,

Hayhurst, R. D., Dietrich-O'Connor, F., Hazen, S., & Landman, K. (2013). Community-based research for food system policy development in the City of Guelph, Ontario. *Local Environment: The International Journal of Justice and Sustainability* 18(5), 606–619.

Heinrichs, H., & Gross, M. (2010). Introduction: New trends and interdisciplinary challenges in environmental sociology. In M Gross and H. Heinrichs, eds., *Environmental Sociology: European Perspectives and Interdisciplinary Challenges*. New York: Springer, pp. 1–16.

Hidayat, D., Stoecker, R., & Gates, H. (2013). Promoting community environmental sustainability using a project-based approach. In K. O. Korgen and J. White, eds., *Sociologists in Action: Sociology, Social Change and Social Justice*. Thousand Oaks, CA: Pine Forge Press, pp. 263–268.

Hoare, T., Levy, C., & Robinson, M. P. (1993). Participatory action research in Native communities: Cultural opportunities and legal implications. *The Canadian Journal of Native Studies* 13(1), 43–68.

Horton, M. (1997). *The Long Haul: An Autobiography*. New York: Teachers College Press.

Jordan, C., Lee, P, & Shapiro, E. (2000). Measuring developmental outcomes of lead exposure in an urban neighborhood: The challenges of community-based research. *Journal of Exposure Analysis and Environmental Epidemiology*, 10(6 Pt 2), 732–742.

Jordan, S., & Kapoor, D. (2016). Re-politicizing participatory action research: Unmasking neoliberalism and the illusions of participation. *Educational Action Research*, 24 (1), 134–149. https://doi.org/10.1080/09650792.2015.1105145

John, A. J. I., & Phalla, C. (2006). Community-based natural resource management and decentralized governance in Ratanakiri, Cambodia. In S. Tyler, ed., *Communities, Livelihoods, and Natural Resources Action Research and Policy Change in Asia*. Warwickshire, UK: Intermediate Technology Publications; and Ottawa, CA: International Development Research Centre, pp. 33–56.

Kinzig, A. P., Ehrlich, P. R., Alston, L. J., Arrow, K., Barrett, S., et al. (2013). Social norms and global environmental challenges: The complex interaction of behaviors, values, and policy. *Bioscience* 63(3), 164–175.

Lewin, K. (1948). Resolving Social Conflicts: Selected Papers on Group Dynamics. G. W. Lewin, ed. New York: Harper & Row.

Loo, C. (2014). The role of community participation in climate change assessment and research. *Journal of Agricultural and Environmental Ethics* 27(1), 65–85.

Lutzenhiser, L. (1994). Sociology, energy and interdisciplinary environmental science. *The American Sociologist* 25(1), 58–79.

Mapfumo, P., Adjei-Nsiah, S., Tambanengwe, F. M., Chikowo, R., & Gillerd, K. E. (2013). Participatory action research (PAR) as an entry point for supporting climate change adaptation by smallholder farmers in Africa. *Environmental Development* 5, 6–22. https://doi.org/10.1016/j.envdev.2012.11.001

Minkler, M., and Wallerstein, N., eds. (2008). *Community-Based Participatory Research for Health: From Process to Outcomes*, 2nd Edition. San Francisco: Jossey-Bass.

McOliver, C. A., Camper, A. K., Doyle, J. T., Eggers, M. J., Ford, T. E., et al. (2015). Community-based research as a mechanism to reduce environmental health disparities in American Indian and Alaska Native communities. *International Journal of Environmental Research and Public Health* 12(4), 4076–4100.

Méndez, V.E., Caswell, M., Gliessman, S. R., & Cohen, R. (2017). Integrating agroecology and participatory action research (par): lessons from Central America. *Sustainability* 9(5), 705. www.mdpi.com/2071–1050/9/5/705

Moragues-Faus, A., Omar, A., & Wang, J. (2015), Participatory Action Research with local communities: Transforming our food system. November 18. Food Research Collaboration Policy Brief. DOI:10.13140/RG.2.1.4304.7766.

Nong, K., & Marschke, M. (2006). Building networks of support for community-based coastal resource management in Cambodia. In S. Tyler, ed., *Communities, Livelihoods, and Natural Resources Action Research and Policy Change in Asia*. Warwickshire, UK: Intermediate Technology Publications; and Ottawa, CA: International Development Research Centre, pp. 151–168.

Nyden, P., & Wiewal, W. (1992). Collaborative research: Harnessing the tensions between researcher and practitioner. *American Sociologist* 23(4), 43–55.

Pacific Institute. 2018. Participatory Action Research. https://pacinst.org/issues/empowering-people-and-communities/participatory-action-research/

Peace, D. M., & Myers, E. (2012). Community-based participatory process – climate change and health adaptation program for Northern First Nations and Inuit in Canada. *International Journal of Circumpolar Health*, 71. www.ncbi.nlm.nih.gov/pmc/articles/PMC3417663/

Perez, A. C., Grafton, B., Mohai, P. et al. (2015). Evolution of the environmental justice movement: Activism, formalization and differentiation. *Environmental Research*

Letters, 10(10). http://iopscience.iop.org/article/10.1088/1748–9326/10/10/105002.

Petersen, D. M., Minkler, M., Vásquez, V. B. et al. (2007). Using community-based participatory research to shape policy and prevent lead exposure among Native American children. *Progress in Community Health Partnerships* 1(3), 249–256.

RECOFTC. 2011. Participatory Action Research for Community-Based Natural Resource Management. www.recoftc.org/basic-page/participatory-action-research-community-based-natural-resource-management.

Ross, J. A., & Stoecker, R. (2016). The emotional context of Higher Education Community Engagement. *Journal of Community Engagement and Scholarship.* 9, 7–18.

Shepard, P. M., Northridge, M. E., Prakash, S., & Stover, G. (2002) Preface: Advancing environmental justice through community-based participatory research. *Environmental Health Perspectives*, 110(supplement 2), 139–140. www.ncbi.nlm.nih.gov/pmc/articles/PMC1241155/pdf/ehp110s-000139.pdf.

Skelton, R., & Miller, V. (2016). The Environmental Justice Movement. National Resources Defense Council. www.nrdc.org/stories/environmental-justice-movement.

Schnaiberg, A. (1980). *The Environment: From Surplus to Scarcity.* New York: Oxford University Press.

Smith, L. T. (2012). *Decolonizing Methodologies: Research and Indigenous Peoples.* London: Zed Books.

Soils, Food and Healthy Communities. (2018). Participatory Climate Change Adaptation Research. Farmer-led Research to Improve Food Security and Nutrition in Malawi. http://soilandfood.org/research-results/participatory-climate-change-adaptation-research/.

Steele, S. F., & Price, J. (2007). *Applied Sociology: Terms, Topics, Tools and Tasks*, 2nd ed. Belmont, CA: Thomson Higher Education.

Stoecker, R. (2016). *Liberating Service Learning and the Rest of Higher Education Civic Engagement.* Philadelphia, PA: Temple University Press.

Stoecker, R. (2013). *Research Methods for Community Change: A Project Based Approach.* Thousand Oaks, CA: Sage.

Stoecker, R. (2003). Community-based research: from theory to practice and back again. *Michigan Journal of Community Service Learning* 9, 35–46.

Stoecker, R. (2009). Are we talking the walk of community-based research? *Action Research* 7, 385–404.

Stoecker, R. (2012). CBR and the two forms of social change. *Journal of Rural Social Sciences* 27, 83–98.

Stoecker, R., & Brydon-Miller, M. (2013). Action research. In A. A. Trainor and E. Graue, eds., *Reviewing Qualitative Research in the Social Sciences*, New York: Routledge, pp. 21–37.

Stoecker, R., Reece, K., & Konkle, T. R. (2018). From relationships to impact in community-university partnerships. In M. Seal, ed., *Participatory Pedagogic Impact Research: Co-Production With Community Partners in Action.* London: Routledge, pp. 214–237.

Strand, K., Marullo, S., Cutforth, N., Stoecker, R., & Donohue, P. (2003a). Principles of best practice for community-based research. *Michigan Journal of Community Service Learning* 9, 5–15.

Strand, K., Marullo, S., Cutforth, N., Stoecker, R., & Donohue, P. (2003b). *Community-Based Research and Higher Education: Principles and Practices.* San Francisco, CA: Jossey-Bass.

Stringer, E. T. (2013). *Action Research*. Thousand Oaks, CA: Sage.
Tremblay, M., Furgal, C., Larrivée, C., Annanack, T., Tookalook, P., et al. (2008). Climate Change in Northern Quebec: adaptation strategies from community-based research. *Arctic*, 61(supplement 1), 27–34.
Tuyen, T. V., Chat, T. T., Hanh, C. T. T., Tinh, D. V., Thanh, N. T., et al. (2006). Participatory local planning for resource governance in the Tam Giang lagoon, Vietnam. In S. Tyler, ed., *Communities, Livelihoods, and Natural Resources Action Research and Policy Change in Asia*. Warwickshire, UK: Intermediate Technology Publications; and Ottawa, CA: International Development Research Centre, pp. 57–84.
Tyler, S. (2006). Community-based natural resource management: a research approach to rural poverty and environmental degradation. In S. Tyler, ed., *Communities, Livelihoods, and Natural Resources Action Research and Policy Change in Asia*. Warwickshire, UK: Intermediate Technology Publications; and Ottawa, CA: International Development Research Centre, pp. 13–30.
Whyte, W. F. (1990). *Participatory Action Research*. Thousand Oaks, CA: Sage.
Wing S, Cole D, & Grant, G. 2000. Environmental injustice in North Carolina's hog industry. *Environmental Health Perspectives* 108(3), 225–231.
Wing, S., Horton, R. A., Muhammad, N. et al. (2008). Integrating epidemiology, education, and organizing for environmental justice: community health effects of industrial hog operations. *American Journal of Public Health* 98(8), 1390–1397.
Yamanoshita, M. (2013). Participatory Action Research for Community Based Natural Resource Management Workshop in Vietnam Forestry University: Workshop Report. Institute for Global Environmental Strategies. https://pub.iges.or.jp/pub/participatory-action-research-community-based
Ykhanbai, H., & Bulgan, E. (2006). Co-management of Pastureland in Mongolia. In S. Tyler, ed., *Communities, Livelihoods, and Natural Resources Action Research and Policy Change in Asia*. Warwickshire, UK: Intermediate Technology Publications; and Ottawa, CA: International Development Research Centre, pp. 107–128.
Zevallos, Z. (2009). What Is Applied Sociology? A brief introduction on applied sociology. Sociology at Work. www.sociologyatwork.org/about/what-is-applied-sociology/.

4 Using Geographic Data in Environmental Sociology

Rachel A. Rosenfeld and Katherine J. Curtis

In an era of increasingly accessible spatially- referenced environmental data, environmental sociologists have the opportunity to use spatial data and spatial analytical techniques to unpack the complex relationships between humans and the environment. The goal of this chapter is to provide an overview of how spatial data and analyses can help researchers and policymakers advance two defining features of environmental sociology: (1) understand the characteristics of the interconnectedness of human and environment patterns and processes; and (2) focus on the role of social inequality and power in molding human–environment interactions (Pellow and Nyseth Brehm 2013). We suggest that environmental sociology benefits from explicitly integrating space and spatial relationships in theoretical and methodological approaches because the environment is fundamentally spatial.

Effective use of spatial data allows social scientists to incorporate the characteristics of place and the relational nature between places while maintaining analytical focus on the role of individual decision-making and behavior. Focusing on the individual or household level alone often ignores the influence of circumstances and surroundings in which individuals exist and, thus, neglects a potentially important dimension of human–environment relationships. Additionally, exclusive attention toward aggregated data ignores that individuals, not groups or aggregates in space, are influential actors (Voss 2007). In this chapter, we begin with the premise that individuals are embedded within specific environmental contexts, and spatial data and analyses are mechanisms that help identify key dimensions through which humans organize their societies. These dimensions, in turn, generate outcomes at the finer individual scale up to the broader societal level.

We begin this chapter with an overview of general definitions and foundational theoretical and methodological concepts in Sections 1 and 2. In Section 3, we highlight compelling spatial work already occurring in the environmental sociology literature, and we conclude with future directions for the incorporation of spatial data to enhance the theory on human–environment interactions in Section 4. Many of the examples we cite in this chapter reference quantitative spatial data and methods, yet our overarching aim is to present a broader argument for the potential advantages of spatially-informed theories and analyses. Ultimately, we aim to elucidate the relational nature between locations, the environment, and the human–environment processes occurring in these spaces.

1 What Are Spatial Data and Spatial Analyses?

Defining "Spatial" Data and Spatial Thinking

Spatial data are widely used in geography, economics, sociology, regional science, demography, and environmental studies. In one of the earliest examples of spatial analysis, during the first half of the nineteenth century, Briton Dr. John Snow generated a map linking the location of communal water pumps and cholera deaths in a London neighborhood. Careful analysis of the maps helped Snow conclude that one water pump was the source of cholera deaths (de Castro 2007). While Snow's spatial analysis was more rudimentary than many of the tools and techniques used today, Snow demonstrated the importance of understanding spatial relationships to highlight the origins of a disease and the environmental conditions leading to the cholera outbreak.

The relational quality of spatial phenomena – the explicit consideration of one location relative to another or others – is a key, distinguishing feature of "spatial" studies. In theorizing space, contemporary scholar Doreen Massey (1994) conceptualizes space and place in terms of social relations, namely through class and gender. Space is "constructed out of the multiplicity of social relations across spatial scales" from the global to national to local levels, in which we can observe the dynamic, shifting "geometry of social/power relations" (Massey 1994: 4). Massey contrasts space with place by emphasizing that the notion of place is reduced to a nationalist, regionalist, localist, or enclosed identity, whereas space transcends this enclosed, limited, and static definition. Similar to the manner in which social classes do not exist as discrete entities, spatial entities exist because they are in relation to each other at a given point in time and across time (i.e., longitudinally).

A considerable share of contemporary spatial social science research engages in spatial awareness but relatively little spatial analysis. The latter is mostly considered quantitative, whereas the former refers to a spatially-integrated way of thinking rather than formal data analysis. For instance, spatially-informed research takes into account that social and demographic patterns may vary at different rates in different spatial contexts (Weeks 2004). Social science researchers embrace a spatial orientation in a variety of different methodologies, including those following more qualitative traditions. For example, in her work on sidewalk use by street vendors in Ho Chi Minh City, Vietnam, Annette Kim uses spatial ethnography to describe her "iterative process" of using GIS to identify the location of sidewalk vendors, and then interviewing vendors and participating on the sidewalk as customers and loiterers (Kim 2015: 218). In another example, Akkan and colleagues (2018) adopt a spatial orientation as a methodological tool to understand children's well-being as part of their physical environment (including the relationships between spaces they navigate) and their social environment. By integrating basic mapping with in-depth interviews and focus groups, the study is an example of how an underlying spatial awareness more fully illuminates the relationships within and between environments (Akkan et al. 2018).

Adopting a more historical view, until the mid-twentieth century many social science disciplines were essentially spatial in terms of both conceptual awareness and analytical techniques, most notably demography (Weeks 2004). Shortly after the mid-twentieth century, much of demographic research shifted focus away from spatial units of analysis and onto the individual (Voss 2007). This shift emerged because of three factors: (1) development and widespread accessibility of microdata at the individual and household levels; (2) a desire by social scientists to avoid aggregation bias or what sociologists refer to as the ecological fallacy – inaccurately using aggregate data to draw inferences about individuals; and (3) an attempt to avoid statistical and computational challenges that are common at the aggregate level. Voss asserts there has been a reemergence of spatial thinking in population science in part owed to computational advances and data availability, and he discusses the ways in which spatial analyses elucidate the patterns and behavior among spatial units at various aggregate levels rather than the patterns and behaviors of individual actors.

Not all fields within the social and environmental sciences, and not all demographers, shifted focus away from aggregate social data. Rural sociologist Charles Galpin and The Chicago School of urban and community development of the 1920s and 1930s incorporated many themes of biology and ecology to the study of humans and, in these instances, ecological processes emphasized the study of individuals in the context of the structure, change, and behavior of aggregate groups. For example, in the mid-twentieth century, studies in both rural and urban sociology paid attention to social area analysis and the ways that social heterogeneity mapped itself onto ecological patterns and processes (Porter and Howell 2012; Voss 2007). Galpin introduced the isotropic map as a tool for social research in 1915 and, furthermore, the role of space along with the ecological environment flourished throughout the sociological literature during mid-century motivated by Amos Hawley's *Human Ecology: A Theory of Community Structure* (1950) (Porter and Howell 2012). Porter and Howell discuss geographical sociology (geo-sociology) as a "synergy between the ecologically centered macro-theory and the application of spatially-centered research methods" to sociological questions (Ibid.: 1). They discuss the ways in which scholars have increased their attention to identifying relationships between geographic contexts and ecological settings in which social processes occur, and have developed better tools to operationalize and test the core spatial concepts of containment, proximity, and adjacency. Nonetheless, despite the Chicago School's emphasis on human–environment interactions, its analysis was largely limited to high-income regions and was rarely embraced for lower-income settings (Balk and Montgomery 2015).

One of the ways in which spatial analysis has played an increasingly theoretical role in the demographic literature is through the environmental context or human ecological perspective (Weeks 2004; See also Hunter et al. 2015; Montgomery and Balk 2011; Redman and Foster 2008; de Sherbinin et al. 2007; Entwisle 2007; Rindfuss et al. 2004). Taken as a whole, and due to factors Voss discusses, demographers have historically spent more time focusing on the interactions between demographic characteristics and social organizations, and less time investigating the natural and built environments and their interactions with demographic processes. This focus has changed in more recent years, giving rise to a new literature studying the population–environment nexus and, in turn, expanding the analysis of spatial data. Weeks highlights the importance of thinking about space in the local context or environment to conceptualize the complex social relations and activities that occur in a built environment situated within a specific natural environment.

This type of thinking, and the analytical approach that it fosters, is what makes a spatial analysis. Cressie defines a spatial analysis "as a quantitative data analysis in which the focus is on the role of space and which relies on explicitly spatial variables in the explanation or prediction of the phenomenon under investigation" (Cressie 1993 as cited in Weeks 2004: 383). As with Massey's theorizing of space, Cressie's definition is centered on devoting analytical attention to the relational aspect of spatial data. In classic statistical analysis, locational attributes are positioned as statistical "nuisances" that must be corrected rather than as useful data points that inform the conceptual framework guiding the study (Weeks 2004: 383). In spatial analysis, the researcher emphasizes the location in the spatial arrangement as a driver of observed trends. In spatial data, everything is relational, and nearer places tend to have more similarities than more distant places (i.e., [positive] spatial autocorrelation) (Tobler 2002, 1970). If spatial autocorrelation is present in the data, the analyst might discover a spatial process at play (e.g., spatial dependence, spatial heterogeneity), one of conceptual relevance and not merely a statistical artifact of the data (Curtis et al. 2012; LeSage and Pace 2009).

Indeed, Weeks reminds us that place-based characteristics and information on a physical location do not automatically create a spatial analysis, and highlights that spatial analysis is typically divided into two categories: (1) analyses that focus on local or neighborhood attributes; and (2) analyses that examine the connections between locations or spaces (Weeks 2004). Weeks also offers useful warnings for researchers using spatial data including: the tendency to neglect the ecology of a phenomenon and, instead, overemphasize individual level analyses that do not incorporate neighborhood or regional influences; and the risk of masking local patterns by analyzing counties or provinces or combinations of these larger spatial units. Weeks also identifies exemplars of spatial analyses that have focused on the spatial patterns of demographic outcomes. For example, using data from the 1911 census for England and Wales, Reid demonstrates that infant and child mortality were associated with a father's education and also where the child lived (Reid 1997 as cited in Weeks 2004).

Types and Examples of Spatial Data

Spatial data, also known as geospatial data and geographic information, refer to measurements or observations on specific characteristics at specified locations. Within the social sciences, spatial data tend to correspond with areal aggregates – of or pertaining to social science attributes in the context of an area or space – and include researching economic, demographic, social, and environmental attributes aggregated to some level within a geographic hierarchy (Voss 2007). Generally, spatial data represent the location, size, or shape of a place or feature on earth. Referencing Massey's and Weeks' positions, these locational data are relational data since spatial data reference relative positions of locations by identifying boundaries, either natural or constructed, or exact locations (i.e., geographic coordinates) (Haining 1990). Indeed, spatial data do not necessarily correspond with place-based data, the latter of which in fact can be aspatial (e.g., average educational attainment). At a basic level, spatial data are "spatial" because they refer to the *relationships* that they have to other points or features on a map. In other words, data are spatial when the analyst is interested in the spatial arrangement or relational position of the data on a map. In contrast, place-based data refer to the characteristics of a place, neighborhood, or region, and they are aspatial because the location on the map and the relationship to other places on the map is not of primary analytical concern.

The origins of most spatial data reflect the perspectives and goals of the agencies creating them. For example, the US Census Bureau focuses on data collection and compilation in a format best suited to understand human geographies, whereas the US Geological Survey gathers and synthesizes data in a manner intended to best capture natural environments. Spatial data come in two general forms: geometric and geographic. Geographic spatial data are the form that is most commonly used by environmental sociologists, and these data typically are represented by latitudinal and longitudinal coordinates and boundary lines. Spatial data can be discrete or continuous (de Castro 2007; Haining 1990), and examples of discrete data are point data (e.g., geographic coordinates) and lattice data (e.g., data aggregated at an administrative level such as a county, district, or province), whereas continuous data do not have well-defined breaks between the values or features (e.g., temperatures, precipitation, or topographical features). Of course, as with data in general, continuous spatial data can be transformed into discrete (or categorical) spatial data as the analyst sees fit and, and yet while possible, converting discrete spatial data into a continuous form poses a larger analytical challenge.

The availability and quality of spatial data are continually changing, both as a result of technological advances that enable the capture, processing, and analysis of new data, and of shifting priorities among agencies that fund data collection, compilation, and dissemination. Traditional sources of environmental spatial data include archival sources, such as maps, census materials, aerial photographs, field observation from survey data and from direct observation, and simulation work in a laboratory environment. For example, Tian et al. (2014) use historical archives to follow the history of land use in India from 1880 to 2010. The archives include information on changes in croplands, forests, grasslands, and the constructed environment over the study period. More modern sources incorporate satellite imagery and newer government and commercial spatially-referenced databases (Haining 1990). As early as the 1970s, satellite photos of land cover in West Africa showed significant deforestation (Vittek et al. 2014), and satellite imagery highlighted overgrazing in the Western Sahel countries as well (Schmidt 2001). Another modern and widely used data source is the Normalized Difference Vegetation Index (NDVI), which is used to estimate biophysical variables of temperate vegetation through remote sensing measurements (Hunter et al. 2014; Foody et al. 2001). For rainfall data, the Climate Hazards Group InfraRed Precipitation with Station dataset (CHIRPS) is a 30-plus-year global precipitation dataset that incorporates 0.05 degree resolution satellite imagery with in-situ station data to generate gridded rainfall time series to monitor seasonal droughts (Funk et al. 2015). When thinking about rural development, nighttime light images may be a proxy for levels of urbanization or development in less developed country settings (Dorélien et al. 2013; Weeks 2004). Nighttime light can also be used to highlight urbanization in developed country contexts (Dorélien et al. 2013). At the global scale, IPUMS Terra marks a recent scientific effort to integrate population and environment data including population censuses and surveys; land use and land cover from satellite imagery, censuses, and surveys; and temperature, precipitation, and climate data from weather stations (Minnesota Population Center 2020). Given its spatial and temporal expanse, this data enterprise has the potential to dramatically lower the barriers to spatial analysis among environmental sociologists.

2 Spatial Analytical Approaches

Identifying Spatial Patterns

A spatial concentration describes an observed pattern of a phenomenon or behavior, and developing a sense of when and how to look for spatial data patterns helps analysts select spatial models that effectively illustrate such patterns (Haining 1990). For most social science research, space often is not defined as a point but instead as an area or region (e.g., lattice data). Awareness about spatial differentiation across a study site is essential, whether it be differences in an outcome or differences in a relationship between key variables among the spatial regions that make up the study site (e.g., county, province). For example, space has long been factored into rural sociological analyses focusing on differentiation across the rural-urban continuum (recent examples include Lee and Sharp 2017; Ward and Shackleton 2016; Winkler et al. 2016). Increasingly, researchers in other fields within sociology are thinking more about spatial continuity and heterogeneity, including environmental sociologists (we discuss some ways in which spatial environmental and social data are used in environmental sociological research in Section 3).

The concept of a region in spatial analysis is common, yet it is important to recognize that regional variation (or between-unit variation) is only one possible type of spatial differentiation. When studying a spatial region, two major problems can arise. First, an attribute (e.g., precipitation or soil quality) in a distinct spatial region is usually subject to continuous change rather than discrete change; rarely is something internally uniform in a designated region whether it be a county, province, or country. Similarly, sharp regional boundary delimitations are generally features of political boundaries. However, environmental considerations do not neatly conform to the boundaries of most political and social spheres (e.g., wildfire smoke or other air pollutants). Moreover, spatial structures do not emerge independently, but spatial processes generate spatial structures. In the classic text, *Spatial Data Analysis in the Social and Environmental Sciences* (1990), geographer Robert Haining identifies four spatial processes that generate spatial structures: diffusion; exchange and transfer; interaction; and dispersal. This chapter does not aim to specify formal links between these processes and their mathematical properties, but we emphasize the importance of understanding that there are relationships between space, populations, soil, economies, etcetera that give rise to the spatial patterns observable in spatial data (for a detailed discussion of these spatial processes, see Haining 1990: 24–26).

Inherently, most spatial data have a certain amount of spatial autocorrelation. Tobler's First Law (Tobler 1970) establishes that neighboring values are more similar to one another than are values among those farther apart. While most spatial patterns abide by this law of positive spatial autocorrelation, there are instances of negative spatial autocorrelation – neighboring values are systematically different from one another. The key is that when there is a nonrandom spatial distribution of values across the study site, there is a spatial pattern in the data. In testing for autocorrelation, researchers test a hypothesis that the data are autocorrelated in space relative to the null hypothesis of randomly distributed values between spatial regions (Ward and Gleditsch 2008; Haining 1990). If spatial autocorrelation is identified, researchers can use it to inform the types of analyses pursued in the investigation of a given phenomenon or behavior.

Spatial Analysis

Simply using spatial data does not ensure a spatial analysis, as discussed in the early section of this chapter. One of the key questions in spatial analysis concerns how methods of exploratory data analysis can best detect spatial structures in datasets (e.g., local spatial outliers – isolated values or clusters that diverge with the rest of the values or clusters; uneven spatial coverage; and spatial dependence or heterogeneity processes) (Ward and Gleditsch 2008; Haining 1990).

Beginning in the 1980s as computing power increased, researchers embraced more complex statistical analyses using multilevel modeling for conducting spatially-informed analyses. Multilevel modeling (e.g., hierarchical linear models, nested models, and random effects models) is one class of statistical modeling of parameters that vary at several levels. For example, a model of migration intentions might contain measures on individual households as well as measures on communities within which the households are grouped. Of course, multilevel modeling strategies introduce a new set of distinct methodological issues that require extensive computing power and statistical packages that offer tools to specify and model hierarchical data. A defining feature of these statistical packages is that they avoid focusing too heavily on the individual or the aggregated data by modeling social processes at both the individual and aggregate levels; the associations between variables are estimated at one level and are then interacted with the second level, and thus can incorporate both individual or household data with spatial data (Ward and Gleditsch 2008; Voss 2007).

In reflecting on our earlier discussions of spatial and aspatial, multilevel models are not inherently spatial and, in fact, are aspatial unless they explicitly incorporate the relational quality of spatial data. Incorporating place-based characteristics at the aggregate level does not make a spatial analysis, whereas incorporating the spatial reference of locations between one another at the aggregate level does. Today, a variety of advanced spatial analytical models are freely available as a result of continuous development and open-source sharing of code. The R Project is one, but not the only, outlet, and developments like GitHub have helped put myriad spatial analytical tools in the hands of analysts (GitHub 2018; R Project 2018).

The utility of advanced spatial analysis techniques hinges on the structure of the data and the quality of the underlying exploratory analysis. Exploratory spatial data analysis (ESDA) can inform more advanced statistical analyses or, in some instances, make up the full extent of an analysis. ESDA also may influence the direction for more qualitative approaches by helping the researcher identify a location to study or where differences emerge that are in need of further exploration. Indeed, rigorous exploratory analysis is essential for the identification of outliers, interpretation of the distribution (e.g., location, skew, spread, and tail properties), identification of spatial properties (e.g., spatial outliers and spatial relationships), and identification of spatial relationships between observational units (Wooldridge 2013; Haining 1990).

Conceptually, analysts are best served by remembering that spatial data are rarely the outcome of a well-defined experiment; rather, they are observational data. Consequently, data of this nature violate one of the classical notions of statistical inference: random sampling. Therefore, analysts are compelled to ask: under what conditions can one assume that the data are representative? A wide variety of test statistics exist, each calibrated for a particular type of data (for more information, see Haining 1990: 228). The central point is that researchers are best served if cognizant of the various pathways through which empirical errors might occur since empirical errors often lead to incorrect theoretical conclusions. The analysis of spatial data does not negate these general empirical errors but builds onto them errors specific to spatial data.

Key Challenges and Limitations

All data analyses confront certain empirical problems, and all of these apply to spatial datasets in addition to problems specific to spatial data. In broad strokes, these include, but are not limited to, problems that emerge when statistical and modeling assumptions are not met and those that arise from a particular dataset. In terms of problems with the data themselves, at the most basic level, researchers may find both measurement error and missing data. When data describe a region, scholars often encounter what is known by quantitative geographers as the modifiable areal unit problem (also known as MAUP). Over time, the data may answer the same questions but the boundaries of a particular area might change as a result of zoning changes or because different agencies collect similar data (de Castro 2007; Haining 1990). This modifiable areal unit problem also is a concern in terms of the boundaries that are drawn at a given point in time. Furthermore, aggregation up to a certain, cruder level might mask relationships occurring at a more local, finer level.

In addition to changing boundaries, a major challenge in using spatial data (and, more generally, aggregate data) is that researchers discover different relationships between variables when looking at the same variables at different scales or levels of aggregation. For instance, analysts tend to uncover different findings when focusing on attributes at the US census tract level rather than at the county level (typically, the former biases the individual rather than the spatial qualities). Even when physical boundaries are not observed, spatial boundaries often point to social and political boundaries that limit interactions and social relationships (Irwin 2007). Traditional statistical analyses at the individual level tend to use dichotomous classifications of places (i.e., metropolitan or nonmetropolitan) as contextual attributes presumed to illustrate place attributes. Yet these groupings only work when assuming that they match the spatial cohesion of the social relationships. Dorélien et al. (2013) propose that a dichotomous urban definition fails to adequately characterize the full range of human settlement patterns across the globe (Dorélien et al. 2013; See also Lichter and Ziliak 2017; Balk and Montgomery 2015). Furthermore, they argue that there is a lack of a common framework on the global level for understanding the spatial boundaries of urban areas, and this lack of a common conceptual framework makes cross-country comparisons and aggregations challenging.

In the ecological context, alternative spatial scales (including those not considered administrative boundaries, e.g., watershed) can reveal more about a place's characteristics than scales that reflect administrative or governance boundaries (e.g., municipality) (Irwin 2007). This challenge is difficult to handle when researchers draw conclusions based on different levels of aggregation, which are often arbitrarily determined boundaries for an administrative unit because the spatial scale analyzed often depends on the availability of data produced for institutional purposes (e.g., US Census Bureau) (Fussell et al. 2014; Voss 2007). Researchers often use US Census data when thinking about spatial and sociological concepts, but we often observe a mismatch between the different types of census units and the sociological concepts we aim to explore. Census data generally are constructed to facilitate administration and data gathering. For example, zip codes can be used as a level of analysis, but these codes were originally developed to aid in postal service delivery. There also are many different types of census units from small areas (e.g., census block) to relatively large areas (e.g., state), and most groupings tend to bias urban areas as opposed to more sparsely populated, geographically isolated areas (Irwin 2007).

Generally, confidentiality is a concern when using data that includes humans, and this issue extends to spatial data. For example, a dataset might reveal a location for an individual or household, in which case the confidentiality of that individual or household might be violated (de Castro 2007; VanWey et al. 2005). Many funding and data sources, such as the National Institutes of Health, National Science Foundation, the Internal Revenue Service, the Social Security Administration, to name a few, publish spatial data and follow rigorous protocols to ensure that these data are released in a way that guarantees the confidentiality of the respondents (Ibid.). Depending on the specific method, preserving confidentiality can make it difficult to link social data to fine-scale environmental data; the fine-scale might be lost by the need to aggregate up to the cruder scale of the social data; or the "swapping" of respondents between spatial units to preserve confidentiality might create erroneous associations between social and environmental attributes (Shlomo et al. 2010).

Newer spatially-explicit data sources also bring with them challenges to preserving confidentiality. Volunteered geographic information (VGI) is an example of participatory approaches to geographic information systems (GIS), in which large websites provide a general base map and allow users to generate their own content by marking new attributes on the maps that are not included in the base map (e.g., OpenStreetMap). Due to the open-source nature of VGI, confidentiality is a hurdle data scientists have yet to solve (Goodchild 2007). Geographic methods are often used to certify confidentiality (VanWey et al. 2005), but there is no simple answer or method to the challenging issue of confidentiality. As data collection and analytical approaches continue to increase in complexity and accuracy, researchers will need to develop new techniques to secure confidentiality. Moreover, the increased access to restricted federal data through the Federal Statistical Research Data Centers has opened new possibilities to generate spatially-referenced data and conduct spatial analyses, but not without some modification in order to adhere to federal privacy and confidentiality guidelines.

3 Spatial Data and Analyses in Environmental Sociology

This section highlights the ways in which environmental sociologists currently utilize spatial data and analyses. Geographers have long discussed the concepts of space and place, but only more recently have sociologists and environmental sociologists explicitly explored space in their research (Lobao 2007; Massey 1994; Haining 1990). For the past forty years, a core tenet of environmental sociology is that individuals – their outcomes, behaviors, and processes – are not isolated from their community, environment, and place broadly defined. Sociologists Pellow and Nyseth Brehm (2013) suggest two defining features of the field of environmental sociology: (1) the interconnectedness of human–environment patterns and processes (what Pellow and Nyseth Brehm call human and nonhuman); and (2) the focus on the role of social inequality and power in molding human–environment interactions. We organize this section accordingly.

Scholars, public policy practitioners, and community activists within environmental sociology advocate for a paradigm that focuses more heavily on contextual, environmental, and social structural factors, as well as economic and technological influence. In its early years, environmental sociology occupied a space on the fringe of traditional sociology (Pellow and Nyseth Brehm 2013; Dunlap and Michelson 2002). Therefore, early environmental sociologists often collaborated with scholars in a wide range of disciplines (e.g., economists, climate scientists, geographers, limnologists, political scientists, urban planners, rural demographers, anthropologists, and biologists). This emphasis on interdisciplinarity has enabled environmental sociologists to grapple with the nuances of the interactions between humans and their built and non-built environments, and to generate a "robust and defensible account of sociological reality" (Pellow and Nyseth Brehm 2013: 231). While this cooperation across disciplines has been a rich part of environmental sociology, sociology as the home discipline ensures that inequality and inequity remain in the foreground. Inequality and power more broadly are key to sociological inquiry, and environmental sociology can provide a unique and innovative lens to understand how social and environmental systems benefit certain groups and disadvantage others. Accordingly, environmental sociology encourages scholars and activists to broaden the sociological view of human inequality by incorporating the interactions between humans, nonhuman species, and ecosystems into their scholarship (Pellow and Nyseth Brehm 2013; Bell 2012; Costanza 1989).

Research has established that rapid and slower onset environmental disasters are forcing people to change their household migration and fertility decisions in both more developed and less developed countries (DeWaard et al. 2016; Curtis et al. 2015; Hunter et al. 2015; Nobles et al. 2015; Fussell et al. 2014). The average warmer weather throughout the world increases malaria, dengue, diarrhea, Lyme disease, tick-borne encephalitis, and food-borne pathogens to name a few (Bell 2012). Additionally, increased pollutants in the atmosphere and the ozone hole result in higher incidences of respiratory illness, certain cancers, and premature death (Malley et al. 2017; Zhang et al. 2010; Hudson et al. 2006). Sub-Saharan Africa is one of the world's regions most greatly impacted by droughts, and this change in precipitation has huge implications for global social inequality (FEWS 2016). Incorporating spatial data into environmental sociological research can provide a unique link between the individual and community levels for a number of these pressing concerns. Indeed, spatial data in environmental sociology can advance monitoring, evaluating, and implementing development, environmental, and population policies (de Castro 2007; Wachter 2005).

When environmental sociology emerged in the 1970s and 1980s, there was growing awareness about spatial data analyses in geography, regional science, economics, and applied and rural demography. As computationally demanding modeling tools for spatial analysis were catching up to spatial theory on the interactions between individuals and their environment or the space in which they live, social science embraced or re-embraced spatial analysis to varying degrees. Urban, rural, and applied demographers, spatial econometricians, and geographers were essential in closing the gap between macro- and micro-level associations and used spatial data and techniques to do so (Voss 2007). Understanding the spatial context and relationships between places has been instrumental as part of data analyses for environmental sociology, yet more opportunities exist for environmental sociologists to incorporate spatial data and, subsequently, advance theories on human–environment interactions.

Interconnectedness of Human–Environment Patterns and Processes

Both domestic and international scholars think about the relationships between places and changing environmental conditions, and migration is a large part of this story. Demographer Kenneth Wachter writes that "'migration' is the usual label for the spatial subfield of demography, and movement is its preoccupation" (Wachter 2005: 15,299). Furthermore, sociologist and demographer Paul Voss suggests that "migration is an area of research often located within a broader field of spatial data analysis dealing specifically with 'spatial interaction data'" (Voss 2007: 458). Fussell et al. (2014) offer examples of incorporating spatial awareness through the environmental dimensions of human migration. Fussell and colleagues discuss how drought and rainfall cause variability in local natural resources, such as wood for fuel, building materials, and medicinal plants, and this variability affects the likelihood of out-migration for resource-dependent households. Natural resource variability can decrease a family's agricultural productivity or ability to produce and sell goods created from natural resources (e.g., wicker baskets). Thus, families may be forced to send a family member to migrate to diversify their household livelihoods, causing the family to rely more heavily on remittances (i.e., the new economics of labor migration [Massey et al. 1998]) (Fussell et al. 2014). Another complicating factor for the family may be that the natural resources used in their origin community may drastically differ from their new destination.

Hunter et al. (2014) examine changes in spatial and temporal associations between temporary out-migration and natural resource availability in Agincourt, South Africa (see also Leyk et al. 2012). Hunter and colleagues examine migration and environment associations using demographic survey data along with satellite imagery capturing natural resource availability (i.e., NDVI which identifies vegetation). By employing this combination of data, they are able to analyze associations between out-migration (typically for income-generating activities) and environmental variables that vary across the study space. Both Hunter et al. and Leyk et al. use NDVI to quantify spatial differences in resource availability for twenty-one villages in Agincourt (for more information on NDVI, see Foody et al. 2001). Specifically, they are able to focus on differences in associations between variables when using the whole population, a village, versus a sub-village, and they find that socioeconomic status and NDVI produce high degrees of spatial variation across the study site. While in many cases natural resource availability and changes in environmental conditions more broadly induce out-migration, particularly disadvantaged households are often restricted in their migration options because of the upfront costs migration entails (Fussell et al. 2014; Gray and Mueller 2012).

In Mexico, Hunter et al. (2013) use the Mexican Migration Project to link environmental factors to commonly understood migration predictors, such as employment opportunities, political instability, and family ties. This work highlights temporal and social dimensions that give rise to spatial differentiation in human–environment interactions by showing that rural Mexican households experiencing drought at least two years prior and with relatively deeper migration networks and histories are more likely to send a family member to the United States compared to households experiencing recent droughts and with weaker migration networks (Hunter et al. 2013).

Of course, while well documented in the literature, environmentally induced migration is not the only pathway through which environmental sociologists consider space and spatial relationships. Mena and colleagues (2011) use agent-based modeling to understand the impact of land use change on agriculture in the Ecuadorian Amazon (see also Hunter et al. 2015). While agent-based modeling is used to simulate decision-making processes at the household level, the authors combine this with GIS (mapping) and Landsat Thematic Mapper to determine spatial relationships between farms and their environment. From an applied geography perspective, this work is key to understanding environment–population interactions for recognizing not only individual farm decision-making processes, but also the spatial connections between farms, such as labor sharing among adjacent farms, and how spatial decision-making can play a role within land use decisions (Mena et al. 2011).

Few studies have linked rainfall and vegetation data with location-specific health and nutrition data. However, in a recent study, López-Carr et al. (2016) examine climate-related child undernutrition in the Lake Victoria Basin (including regions within Kenya, Tanzania, and Uganda). The study features a unique combination of datasets including NDVI and multidecadal remotely sensed precipitation data which López-Carr and colleagues integrate with cluster level data from the Demographic and Health Surveys (DHS). Their analysis is motivated by the limited amount of literature linking population dynamics with the environment and climate change outside the associations between migration and the environment (López-Carr et al. 2016). In their work, they create a framework that maps climate-related chronic childhood undernutrition (i.e., as measured by child stunting) by using weighted and standardized climate exposure variables. The authors find that child stunting vulnerability is higher in areas experiencing a negative change in NDVI along with an increase in precipitation, and the rainfall change has a greater association with child undernutrition. This study offers a unique perspective on climate change interactions with the local environment and how these interactions affect health and nutrition.

Social Inequality in Molding Human–Environment Interactions

In this section, we focus on how social inequality manifests at the intersection of human and environment processes. The work of environmental sociologists is intimately linked to the environmental justice movement, which elevated the issue of environmental inequality to the US national policy stage in the late 1980s and early 1990s (Downey 2003). Environmental justice is defined by Presidential Executive Order 12898 as "the fair treatment and meaningful involvement" of all people regardless of race, color, sex, national origin, or income with respect to the development, implementation, and enforcement of environmental laws, regulations, and policies (Schlosberg 2009). Environmental inequality most certainly still exists, and this section highlights the work of some environmental sociologists whose effective use of spatial data has shown that populations of color and lower income populations are affected disproportionately by environmental degradation and climate unpredictability. Environmental sociologists are positioned to draw connections between social and economic inequality and environmental hazards, which in turn informs how or whether society moves closer toward protection from environmental degradation, prevention of adverse health impacts from worsening environmental situations before harm happens, and restoration of the environment in collaboration with human rights principles for all people (Cutter 1995).

Environmental justice principles are part of global development initiatives, and a large part of economic growth and sustainable development centers on supporting sustainable jobs and education systems that promote equality (United Nations 2015). Sustainable economic development incorporates protecting the environment from degradation to support the needs of future generations and, accordingly, reducing poverty, expanding access to quality education, increasing climate action, and responsible consumption and production are several key targets of the United Nations Sustainable Development Goals (SDGs). An important task for the United Nations and other, similarly directed agencies is to design policies that help shield vulnerable populations against the negative economic and health impacts of climate change.

Recent environmental sociological research has demonstrated the interlinkages between poverty and climate action and, in a way, supports the SDG initiatives. Conceptually, poverty is not just the lack of income or resources that guarantee a sustainable livelihood, but includes limited access to food, water, nutrients, basic health, and education services. By linking systematic inequality in educational attainment to rainfall data and spatial variation in climate shocks, Randell and Gray (2016) show that the rural poor in sub-Saharan Africa are especially vulnerable to barriers to early childhood development and to school participation since the region faces more frequent droughts and heat waves in the summer months. Furthermore, in Hunter and colleagues' (2014) study on migration and spatially-varying natural resource use in South Africa, they illustrate that natural resource usage changes based on educational attainment, and relatively higher levels of human capital achieved through education may contribute to the advancement and intensification of agricultural production and the extraction of natural resources (Hunter et al. 2014; See also Gray 2009).

More domestically, a sizeable and growing literature focuses on the spatial dimensions of environmental inequalities within the United States. In a call to action, Barbara Entwisle's Presidential Address to the Population Association of America (2007) challenges population scientists to focus attention on the intersection of human agency, neighborhoods, and health to advance scholarship on inequality. At the core of her argument, Entwisle emphasizes that individuals operate within specific social, spatial, and biophysical environments, and at the same time individuals are constrained by these environments. Yet research does not reflect nor address the influence of spatial and environmental contexts on human health. Too often the neighborhood effects literature is too narrowly focused on poverty, "a category that groups together measures of structural disadvantage, inequality, and concentrated affluence" (Entwisle 2007: 692). Out of a survey of 503 published studies on local contexts and health, Entwisle reports that less than 25 percent of the articles focus on health services, the built environment, and toxins – the domains of space and the environment – whereas more than 70 percent of published articles focus on poverty. Highlighting another spatial dimension ripe for theoretical and analytical advancement, Entwisle points out that neighborhood change over time is rarely incorporated into these studies; a mere 2.4 percent of the 503 studies highlight neighborhood change, whereas the vast majority focus on individual change (Entwisle 2007: 693).

As an early exemplar of Entwisle's vision for future research, Downey (2003) advances the study of environmental and racial inequality in the United States by pursuing an explicit spatial analysis of manufacturing plants (known to emit environmental toxins) and racial segregation between 1970 and 1990 in Detroit, Michigan. Leveraging US Census data, Michigan manufacturer directories, US Environmental Protection Agency (EPA) Toxic Release Inventory data, and GIS (mapping) techniques, Downey derives measures of the number of manufacturing facilities within a quarter-kilometer radius of the center of GIS-generated grid cells across the study site, plus the distance in meters from the center of each grid cell. Through this approach, Downey avoids the assumption that facilities are evenly distributed within census tracts (i.e., the typical measure of neighborhood). Downey's work infuses an environmental inequality dimension into the classic work of Massey and Denton (1993) on residential segregation by leveraging spatial data and analytical tools, and shows that black and white households with similar incomes do not share the same spatial exposure to environmental hazards.

Partly addressing Entwisle's call, Sharp et al. (2015) investigate how the structure of health inequalities is shaped by residential neighborhoods and by additional activity spaces (i.e., the areas within which people move or travel or work during the course of their daily activities). By expanding their spatial lens, Sharp and colleagues find evidence of spatial effects on health outcomes characteristic of a relative deprivation process. The authors show that spatial effects extend beyond the traditionally measured local neighborhood and include other spaces with(in) which people interact, and that the nature of the extra-local spatial effect is conditional on how the spaces rank relative to one another. Chief among their findings is that adults living in the most disadvantaged neighborhoods are more likely to report worse health when they spend more time working or commuting in more advantaged neighborhoods, and adults who live in more advantaged neighborhoods report worse health when they spend more time in disadvantaged neighborhoods (Sharp et al. 2015). This study and others exploring the measurement of the spatial context (e.g., Jones and Pebley 2014) highlight the potential of probing human activities within multiple spatial units, perhaps with shared and overlapping boundaries, and by identifying which populations systematically follow which activity space pathways (i.e., treating the activity space system as the object of inquiry).

4 Future Directions

While environmental sociologists have embraced spatial awareness and the use of spatial data, there remains much room for development in the application of spatial thinking, spatial data, and spatial analyses. In this section, we highlight some of the opportunities for future directions within environmental sociology. First, more theorizing and identifying different mechanisms through which spatial data can be used for understanding the rapidly changing global climate would benefit scholarship. Archival sources, mapping, and satellite data can be incorporated for understanding both rapid onset and slower onset environmental changes, yet slower onset events and climate change are less well understood and are conceptually more complex since they emerge without an exogenous shock (Neumann and Hilderink 2015). In terms of data collection, embracing citizen science – participation in scientific research by the public (Riesch 2015) – and increased VGI usage could be a part of this advancement, particularly in the case of rapid onset environmental shocks and in more remote areas of the world. Conditions on the ground may prevent the capture of satellite digital imagery or systematic precipitation monitoring because of power outages, a lack of Internet connection, or no centralized computer hardware and software to track patterns. Yet, the local population may be able to report changes in the environment through mobile phones via text, voice, or photos (Goodchild 2007). Capturing widespread data from immediate reports has the potential to give researchers more tools and better measures to generate research that could increase the capacity for international and national organizations to implement improved disaster risk management and climate adaptation and mitigation strategies. Nevertheless, even in a world of increased VGI usage and engagement with citizen science, slower onset environmental changes will be more difficult to measure and study. Incorporating VGI and traditionally gathered spatial data in tandem might yield more realistic scenarios (Goodchild 2010).

Second, researchers can invest in explicitly and systematically linking environmental justice to spatial inequality, and the US EPA offers data tools of potential use to researchers and policymakers. For example, in 2015, the EPA released the initial version of the Environmental Justice Screening and Mapping Tool (EJSCREEN) to better combine environmental and demographic indicators to identify areas and populations facing environmental risks (Hansman 2015). Little scholarly work has been published using this new tool, yet it could provide new opportunities for environmental sociologists to theorize and measure inequality at the intersection of social and environmental disadvantage. Much of the research using EPA spatial data to date has focused on the spatial distribution of air toxins (Ard 2015; Gilbert and Chakraborty 2011). The EJSCREEN geocoded datasets aim to help researchers better link the spatial distribution of air toxins in addition to water toxins, hazardous waste, and other environmental indicators to various demographic indicators (e.g., percent low-income, percent minority, educational attainment, linguistic isolation, age) (EPA 2014). This spatial database creates a promising opportunity for environmental sociologists interested in incorporating spatial data in their scholarship on environmental justice.

Third, advancements in spatial concepts and measures could enable researchers and policymakers to reconsider international definitions of aggregation that would reduce analytical barriers to cross-country comparisons and subsequently may contribute to global and regional policy recommendations (Dorélien et al. 2013). There are a number of data compilation sources that are used by both domestic and international scholars to think geographically about settlement patterns, population, and the environment. For example, IPUMS Terra synthesizes microdata (data at the individual level) with area-level data describing geographic units along with other spatial information such as climate and land use data (Minnesota Population Center 2020), and the Global Rural-Urban Mapping Project (GRUMP) offers researchers geo-referenced data that can contribute to their understanding of the full range of human settlement patterns from densely populated to sparsely populated and isolated areas through a variety of measurements including population counts, population density, geographic unit area grids, among other data (SEDAC 2018). IPUMS Terra and GRUMP are potentially useful sources for research within a country and in cross-country comparisons, since these data projects incorporate types of standardized, aggregate measures for international geographic comparisons, including spatially-harmonized measures (i.e., spatially-consistent measures over time despite administrative boundary changes) and more thorough classifications along the rural-urban continuum. However, neither IPUMS Terra nor GRUMP offers a complete solution to the quest to discover a consistent set of international definitions of aggregation, and scholarship would be best served by incorporating improved ways of classifying at the aggregate level whether through population density, primary economic activity, types of government, or along other dimensions (Dorélien et al. 2013).

Fourth, although research on migration and the environment is among the most spatially-developed scholarship, more research could focus on the role of education in identifying environmental adaptive capacity. Such focus is consistent with the SDGs emphasis on ensuring inclusive and equitable quality education. On the international scale, rural populations, especially in sub-Saharan Africa, still are the fastest growing, poorest populations, and future studies could explore mechanisms that incorporate education as a tool through which to enhance resilience for these communities. From a livelihoods perspective, the human capital acquired through educational attainment provides an additional resource from which households can draw when faced with hardships or uncertainty. Education is strongly related to contraceptive knowledge and usage (Ainsworth et al. 1996), meaning that educated households that wish to limit fertility in response to environmental uncertainty may be more successful in achieving those goals as one form of adaptation to fewer natural resources. Migrants are more likely to be more educated than those who do not migrate, especially when migrating across country borders (Feliciano 2005), and future research could continue to consider the relationships between migration as adaptive capacity and the spatial differences in climatic variability and migration (Hunter et al. 2013). Spatial data have the potential to highlight areas of highest risk for environmental degradation using precipitation, land cover, sea level rise data, etcetera, and these can be linked to data about human settlement patterns. Using these resources, environmental sociologists can consider how education has the potential to inform adaptation to various agricultural strategies because of changing climatic conditions and considerations of livelihood diversification strategies more broadly for both those who do and do not migrate (Verner 2010).

Finally, our aim is to encourage environmental sociologists to continue to integrate spatial data and analyses where appropriate, and to engage a new way of thinking about space and spatial relationships across a variety of research methodologies. Environmental sociologists already successfully embrace spatially-informed thinking and spatial data to answer pressing questions. In the spirit of making further strides, collaborations across disciplines in which the use of spatial data is more widespread would allow environmental sociologists to incorporate novel expertise, experience, and analytical methods of researchers from other fields (e.g., geography, ecology, climate science, and demography).

What is most striking about the promise of spatial data is the theoretical and analytical potential to incorporate spatial analyses in environmental sociology. While this chapter has focused on quantitative spatial data and analyses, relationships within and between locations have a robust qualitative nature. Thus, adopting a variety of research methodologies that incorporate spatial thinking and data promises to advance scholarship in innovative and meaningful ways. There are myriad possibilities for integrating spatial data in environmental sociology, and future research depends on nurturing a spatial way of thinking to effectively incorporate spatial data and analyses into the inquires and pursuits at the center of the field.

Acknowledgments

This chapter was supported by center Grant #R24 HD047873 and training Grant #T32 HD07014 awarded to the Center for Demography and Ecology at the University of Wisconsin-Madison, and by funds to Curtis from the Wisconsin Agricultural Experimental Station and the Wisconsin Alumni Research Foundation.

References

Ainsworth, Martha, Kathleen Beegle, and Andrew Nyamete. 1996. "The Impact of Women's Schooling on Fertility and Contraceptive Use: A Study of Fourteen Sub-Saharan African Countries." *The World Bank Economic Review* 10(1): 85–122.

Akkan, Başak, Serra Müderrisoglu, Pınar Uyan-Semerci, and Emre Erdogan. 2018. "How Do Children Contextualize Their Well-Being? Methodological Insights from a Neighborhood Based Qualitative Study in Istanbul." *Child Indicators Research* 12: 443–460.

Ard, Kerry. 2015. "Trends in Exposure to Industrial Air Toxins for Different Racial and Socioeconomic Groups: A Spatial and Temporal Examination of Environmental Inequality in the U.S. from 1995 to 2004." *Social Science Research* 53(September): 375–390.

Balk, Deborah L., and Mark R. Montgomery. 2015. "Guest Editorial: 'Spatializing Demography for the Urban Future.'" *Spatial Demography* 3(2): 59–62.

Bell, Michael M. 2012. *An Invitation to Environmental Sociology.* Fourth edition. Thousand Oaks, CA: Pine Forge Press.

Costanza, Robert. 1989. "What Is Ecological Economics?" *Ecological Economics* 1(1): 1–7.

Cressie, Noel A.C. 1993. *Statistics for Spatial Data: Revised Edition.* New York: John Wiley & Sons.

Curtis, Katherine J., and Annemarie Schneider. 2011. "Understanding the Demographic Implications of Climate Change: Estimates of Localized Population Predictions under Future Scenarios of Sea-Level Rise." *Population and Environment* 33(1): 28–54.

Curtis, Katherine J., Elizabeth Fussell, and Jack DeWaard. 2015. "Recovery Migration After Hurricanes Katrina and Rita: Spatial Concentration and Intensification in the Migration System." *Demography* 52(4): 1269–1293.

Curtis, Katherine J., Paul R. Voss, and David D. Long. 2012. "Spatial Variation in Poverty-Generating Processes: Child Poverty in the United States." *Social Science Research* 41(1): 146–159.

Cutter, Susan L. 1995. "Race, Class and Environmental Justice." *Progress in Human Geography* 19(1): 111–122.

de Castro, Marcia Caldas. 2007. "Spatial Demography: An Opportunity to Improve Policy Making at Diverse Decision Levels." *Population Research and Policy Review* 26(5/6): 477–509.

de Sherbinin, Alex, David Carr, Susan Cassels, and Leiwen Jiang. 2007. "Population and Environment." *Annual Review of Environment and Resources* 32: 345–373.

DeWaard, Jack, Katherine J. Curtis, and Elizabeth Fussell. 2016. "Population Recovery in New Orleans after Hurricane Katrina: Exploring the Potential Role of Stage Migration in Migration Systems." *Population and Environment* 37(4): 449–463.

Dorélien, Audrey, Deborah Balk, and Megan Todd. 2013. "What Is Urban? Comparing a Satellite View with the Demographic and Health Surveys." *Population and Development Review* 39(3): 413–439.

Downey, Liam. 2003. "Spatial Measurement, Geography, and Urban Racial Inequality." *Social Forces* 81(3): 937–952.

Dunlap, Riley E. and William Michelson, eds. 2002. *Handbook of Environmental Sociology*. Westport, CT: Greenwood Press.

Entwisle, Barbara. 2007. "Putting People into Place." *Demography* 44(4): 687–703.

Famine Early Warning Systems Network (FEWS). 2016. "Famine Early Warning Systems Network." Accessed February 25, 2018. www.fews.net.

Feliciano, Cynthia. 2005. "Educational Selectivity in U.S. Immigration: How Do Immigrants Compare to Those Left Behind?" *Demography* 42(1): 131–152.

Foody, Giles M., Mark E. Cutler, Julia McMorrow et al. 2001. "Mapping the Biomass of Bornean Tropical Rain Forest from Remotely Sensed Data." *Global Ecology and Biogeography* 10(4): 379–387.

Funk, Chris, Pete Peterson, Martin Landsfeld et al. 2015. "The Climate Hazards Infrared Precipitation with Stations – a New Environmental Record for Monitoring Extremes." *Scientific Data* 2(December): 1–21.

Fussell, Elizabeth, Katherine J. Curtis, and Jack DeWaard. 2014. "Recovery Migration to the City of New Orleans after Hurricane Katrina: A Migration Systems Approach." *Population and Environment* 35(3): 305–322.

Fussell, Elizabeth, Lori M. Hunter, and Clark L. Gray. 2014. "Measuring the Environmental Dimensions of Human Migration: The Demographer's Toolkit." *Global Environmental Change* 28(September): 182–191.

Gilbert, Angela, and Jayajit Chakraborty. 2011. "Using Geographically Weighted Regression for Environmental Justice Analysis: Cumulative Cancer Risks from Air Toxics in Florida." *Social Science Research* 40(1): 273–286.

GitHub. 2018. "GitHub." Accessed February 27, 2018. https://github.com.

Goodchild, Michael F. 2010. "The Role of Volunteered Geographic Information in a Postmodern GIS World." *ArcUser.* 20–21. Accessed November 14, 2018. www.esri.com/news/arcuser/0410/vgi.html.

Goodchild, Michael F. 2007. "Citizens as Sensors: The World of Volunteered Geography." Santa Barbara, CA: National Center for Geographic Information and Analysis, UC Santa Barbara.

Gray, Clark L. 2009. "Environment, Land, and Rural Out-Migration in the Southern Ecuadorian Andes." *World Development* 37(2): 457–468.

Gray, Clark L., and Valerie Mueller. 2012. "Natural Disasters and Population Mobility in Bangladesh." *Proceedings of the National Academy of Sciences* 109(16): 6000–6005.

Haining, Robert. 1990. *Spatial Data Analysis in the Social and Environmental Sciences.* New York: Cambridge University Press.

Hansman, Heather. 2015. "The EPA Has a New Tool for Mapping Where Pollution and Poverty Intersect." *Smithsonian.* Accessed February 27, 2018. www.smithsonianmag.com.

Hawley, Amos H. 1950. *Human Ecology: A Theory of Community Structure.* New York: The Ronald Press.

Hudson, Robyn, Aline Arriola, Margarita Martínez-Gómez, and Hans Distel. 2006. "Effect of Air Pollution on Olfactory Function in Residents of Mexico City." *Chemical Senses* 31(1): 79–85.

Hunter, Lori M., Jessie K. Luna, and Rachel M. Norton. 2015. "Environmental Dimensions of Migration." *Annual Review of Sociology* 41(1): 377–397.

Hunter, Lori M., Raphael Nawrotzki, Stefan Leyk et al. 2014. "Rural Outmigration, Natural Capital, and Livelihoods in South Africa." *Population, Space and Place* 20(5): 402–420.

Hunter, Lori M., Sheena Murray, and Fernando Riosmena. 2013. "Rainfall Patterns and U.S. Migration from Rural Mexico." *International Migration Review* 47(4): 874–909.

Irwin, Michael D. 2007. "Territories of Inequality." In *The Sociology of Spatial Inequality*, edited by Linda M. Lobao, Gregory Hooks, and Ann R. Tickamyer, 85–109. Albany, NY: State University of New York Press.

Jones, Malia, and Anne R. Pebley. 2014. "Redefining Neighborhoods Using Common Destinations: Social Characteristics of Activity Spaces and Home Census Tracts Compared." *Demography* 51(3): 727–752.

Kim, Annette M. 2015. "Critical Cartography 2.0: From 'Participatory Mapping' to Authored Visualizations of Power and People." *Landscape and Urban Planning* 142: 215–225.

Lee, Barrett A., and Gregory Sharp. 2017. "Ethnoracial Diversity across the Rural-Urban Continuum." *The ANNALS of the American Academy of Political and Social Science* 672(1): 26–45.

LeSage, James P. and Robert Kelley Pace, 2009. *Introduction to Spatial Econometrics.* Boca Raton, FL: CRC Press.

Leyk, Stefan, Galen J. Maclaurin, Lori M. Hunter et al. 2012. "Spatially and Temporally Varying Associations between Temporary Outmigration and Natural Resource Availability in Resource-Dependent Rural Communities in South Africa: A Modeling Framework." *Applied Geography* 34(May): 559–568.

Lichter, Daniel T., and James P. Ziliak. 2017. "The Rural-Urban Interface: New Patterns of Spatial Interdependence and Inequality in America." *The ANNALS of the American Academy of Political and Social Science* 672(1): 6–25.

Lobao, Linda M., Gregory Hooks, and Ann R. Tickamyer, eds. 2007. *The Sociology of Spatial Inequality*. Albany, NY: State University of New York Press.

López-Carr, David, Kevin M. Mwenda, Narcisa G. Pricope et al. 2016. "Climate-Related Child Undernutrition in the Lake Victoria Basin: An Integrated Spatial Analysis of Health Surveys, NDVI, and Precipitation Data." *IEEE Journal of Selected Topics in Applied Earth Observations and Remote Sensing* 9(6): 2830–2835.

Malley, Christopher S., Daven K. Henze, Johan C.I. Kuylenstierna et al. 2017. "Environmental Health Perspectives – Updated Global Estimates of Respiratory Mortality in Adults ≥30 Years of Age Attributable to Long-Term Ozone Exposure." *Environmental Health Perspectives* 125(8): 1–9.

Massey, Doreen. 1994. *Space, Place, and Gender*. Minneapolis, MN: University of Minnesota Press.

Massey, Douglas S. 1993. *American Apartheid: Segregation and the Making of the Underclass*. Cambridge, MA: Harvard University Press.

Massey, Douglas S., Joaquin Arango, Graeme Hugo et al. 1998. *Worlds in Motion: Understanding International Migration at the End of the Millennium*. New York: Oxford University Press.

Mena, Carlos F., Stephen J. Walsh, Brian G. Frizzelle, Yao Xiaozheng, and George P. Malanson. 2011. "Land Use Change on Household Farms in the Ecuadorian Amazon: Design and Implementation of an Agent-Based Model." *Applied Geography* 31(1): 210–222.

Minnesota Population Center. 2020. "IPUMS Terra: Integrated Population and Environmental Data." Accessed March 5, 2020. https://terra.ipums.org/.

Montgomery, Mark R., and Deborah Balk. 2011. "The Urban Transition in Developing Countries: Demography Meets Geography." In *Global Urbanization*, edited by Eugenie L. Birch and Susan M. Wachter, 89–106. Philadelphia, PA: University of Pennsylvania Press.

Neumann, Kathleen, and Henk Hilderink. 2015. "Opportunities and Challenges for Investigating the Environment-Migration Nexus." *Human Ecology* 43(2): 309–322.

Nobles, Jenna, Elizabeth Frankenberg, and Duncan Thomas. 2015. "The Effects of Mortality on Fertility: Population Dynamics After a Natural Disaster." *Demography* 52(1): 15–38.

Pellow, David N., and Hollie Nyseth Brehm. 2013. "An Environmental Sociology for the Twenty-First Century." *Annual Review of Sociology* 39(1): 229–250.

Porter, Jeremy R., and Frank M. Howell. 2012. *Geographical Sociology: Theoretical Foundations and Methodological Applications in the Sociology of Location*. Dordrecht, Netherlands: Springer.

R Project. 2018. "R: The R Project for Statistical Computing." Accessed February 27, 2018. www.r-project.org.

Randell, Heather and Clark Gray. 2016. "Climate Variability and Educational Attainment: Evidence from Rural Ethiopia." *Global Environmental Change* 41(November): 111–123.

Redman, Charles L. and David R. Foster, eds. 2008. *Agrarian Landscapes in Transition: Comparisons of Long-Term Ecological and Cultural Change*. New York: Oxford University Press.

Reid, Alice. 1997. "Locality or Class? Spatial and Social Differences in Infant and Child Mortality in England and Wales, 1895–1911." In *The Decline of Infant and Child*

Mortality. The European Experience: 1750–1990, edited by C.A. Corsini and P. P. Viazzo, 129–154. Dordrecht, Netherlands: Martinus Nijhoff.

Riesch, Hauke. 2015. "Citizen Science." In *International Encyclopedia of the Social & Behavioral Sciences (Second Edition)*, edited by James D. Wright, 631–636. Oxford: Elsevier.

Rindfuss, Ronald R., Stephen J. Walsh, B. L. Turner, Jefferson Fox, and Vinod Mishra. 2004. "Developing a Science of Land Change: Challenges and Methodological Issues." *Proceedings of the National Academy of Sciences of the United States of America* 101(39): 13,976–13,981.

Schlosberg, David. 2009. *Defining Environmental Justice: Theories, Movements and Nature.* Oxford: Oxford University Press.

Schmidt, Laurie J. 2001. "From the Dust Bowl to the Sahel: Feature Articles." National Aeronautics and Space Administration (NASA). Accessed February 26, 2018. https://earthobservatory.nasa.gov/Features/DustBowl.

Sharp, Gregory, Justin T. Denney, and Rachel T. Kimbro. 2015. "Multiple Contexts of Exposure: Activity Spaces, Residential Neighborhoods, and Self-Rated Health." *Social Science & Medicine* 146(December): 204–213.

Shlomo, Natalie, Caroline Tudor, and Paul Groom. 2010. "Data Swapping for Protecting Census Tables." In *Privacy in Statistical Databases*, edited by Josep Domingo-Ferrer and Emmanouil Magkos, 41–51. Berlin: Springer.

Socioeconomic Data and Applications Center (SEDAC). 2018. "Global Rural-Urban Mapping Project (GRUMP), v1 | SEDAC." Accessed January 17, 2018. http://sedac.ciesin.columbia.edu/data/collection/grump-v1.

Tian, Hanqin, Kamaljit Banger, Tao Bo, and Vinay K. Dadhwal. 2014. "History of Land Use in India during 1880–2010: Large-Scale Land Transformations Reconstructed from Satellite Data and Historical Archives." *Global and Planetary Change* 121 (October): 78–88.

Tobler, Waldo R. 2002. "Global Spatial Analysis." *Computers, Environment and Urban Systems* 26(6): 493–500.

Tobler, Waldo R. 1970. "A Computer Movie Simulating Urban Growth in the Detroit Region." *Economic Geography* 26: 234–240.

United States Environmental Protection Agency (EPA). 2014. "Understanding EJSCREEN Results – Data and Tools." Accessed March 4, 2018. www.epa.gov/ejscreen/understanding-ejscreen-results.

United Nations. "Sustainable Development Goals – United Nations." 2015. *United Nations Sustainable Development* (blog). Accessed February 25, 2018. www.un.org/sustainabledevelopment/sustainable-development-goals/.

VanWey, Leah K., Ronald R. Rindfuss, Myron P. Gutmann, Barbara Entwisle, and Deborah L. Balk. 2005. "Confidentiality and Spatially Explicit Data: Concerns and Challenges." *Proceedings of the National Academy of Sciences* 102(43): 15,337–15,342.

Verner, Dorte. 2010. *Reducing Poverty, Protecting Livelihoods, and Building Assets in a Changing Climate: Social Implications of Climate Change in Latin America and the Caribbean.* Herndon, VA: World Bank Publications.

Vittek, Marian, Andreas Brink, Francois Donnay, Dario Simonetti, and Baudouin Desclée. 2014. "Land Cover Change Monitoring Using Landsat MSS/TM Satellite Image Data over West Africa between 1975 and 1990." *Remote Sensing* 6(1): 658–676.

Voss, Paul R. 2007. "Demography as a Spatial Social Science." *Population Research and Policy Review* 26 (5/6): 457–476.

Wachter, Kenneth W. 2005. "Spatial Demography." *Proceedings of the National Academy of Sciences* 102(43): 15,299–15,300.
Ward, Catherine D., and Charlie M. Shackleton. 2016. "Natural Resource Use, Incomes, and Poverty Along the Rural–Urban Continuum of Two Medium-Sized, South African Towns." *World Development* 78(February): 80–93.
Ward, Michael Don, and Kristian Skrede Gleditsch. 2008. *Spatial Regression Models*. Thousand Oaks, CA: Sage Publications.
Weeks, John R. 2004. "Role of Spatial Analysis in Demographic Research." In *Spatially Integrated Social Science*, edited by Michael F. Goodchild and Donald G. Janelle, 381–399. New York: Oxford University Press.
Winkler, Richelle L., and Kenneth M. Johnson. 2016. "Moving Toward Integration? Effects of Migration on Ethnoracial Segregation Across the Rural-Urban Continuum." *Demography* 53(4): 1027–1049.
Wooldridge, Jeffrey M. 2013. *Introductory Econometrics: A Modern Approach*. Fifth edition. Mason, OH: South-Western Cengage Learning.
Zhang, Junfeng, Denise L. Mauzerall, Tong Zhu et al. 2010. "Environmental Health in China: Progress towards Clean Air and Safe Water." *The Lancet* 375(9720): 1110–1119.

PART II

Embodied Environmental Sociology

5 Strangers on the Land? Rural LGBTQs and Queer Sustainabilities

Julie C. Keller

Introduction

In an episode aired on August 16, 2016 on the Premiere Radio Network, host Rush Limbaugh denounced the US Department of Agriculture's Rural Pride campaign, an initiative to spread awareness of LGBT-identified rural people and offer programs to support marginalized farmers. "It's not about lesbian farmers," he told his listeners. "What they're trying to do is convince lesbians to become farmers ... They are trying to bust up one of the last geographically conservative regions in the country; that's rural America." The always-colorful New York Daily summed up the host's concerns with the provocative headline, "Rush Limbaugh Warns of Alleged Lesbian Farmer Invasion." Social media was momentarily abuzz with critical and comical responses to Limbaugh's dramatic warning. Facebook groups popped up proclaiming their solidarity with queer farmers, as well as T-shirts emblazoned with the (slightly tongue-in-cheek) urgent message, "America Needs Lesbian Farmers." The Huffington Post and NBC's LGBTQ arm, NBC Out, took the opportunity to inform readers of the diversity of rural places and people, pointing out that LGBTQs are not new to rural places. Indeed, they've always been there (Erbentraut, 2016; Moreau, 2016).

Though focusing too much on Limbaugh may create a space for his homophobic rant to settle deeply into the valleys of our consciousness, I argue that there is value in examining these remarks in a discussion of how sexuality and environmental sociology come together. I see the threat of the lesbian farmer – and her backlash – as an opportunity to explore several key ways in which sexuality and the environment inform one another. First is Limbaugh's characterization of the countryside as the last bastion of conservative values. Second, as the articles above highlighted, Limbaugh assumes that LGBTQs are not already rural residents. The third facet in the intersection of sexuality and environment is what caused Limbaugh's ire in the first place: the US Department of Agriculture's (USDA) new campaign to target rural LGBTQs in their outreach efforts. Finally, the T-shirt that was born in response to Limbaugh's homophobic rant is symbolic of resistance efforts and offers an opportunity to highlight social movements that push back against various forms of oppression.

What assumptions about the environment, rural spaces, and sexuality are embedded in these assertions? Why is the countryside viewed as conservative territory? Why are gender and sexual minorities framed as foreign invaders to the countryside? What are the complex realities of rural LGBTQs, particularly those who are farmers? I explore these questions and more to highlight avenues for theoretical and empirical work on the intersections of sexuality and the environment within the field of environmental sociology.

Though not central to most analyses of environmental issues, sexuality has nonetheless surfaced in quite a few theoretical approaches. From deep ecology's unsettling focus on fertility rates to the enduring tenets of ecofeminism to the promising new framework of total liberation, sexuality has certainly played an important role in environmental studies, and more narrowly, within environmental sociology. This chapter is not a comprehensive review of scholarship on sexuality in environmental sociology. Rather, I use the case of queer rural folks – and queer farmers more specifically – to sketch out opportunities for strengthening the analytical lens used to examine the sexuality-environment connection. I highlight several areas that show particular promise for investigating sexuality and the environment, and conclude by arguing that perhaps lesbian farmers are not only what America needs, but also what environmental sociology needs.

Conservative Countryside?

In his rant on lesbian farmers, Limbaugh unwittingly touched on several theoretical and empirical concerns that have preoccupied scholars who study rural places and people. Characterizing rural America as "one of the last geographically conservative regions," he invoked an old dualism linking place with politics. The political left is also guilty of painting rural politics with broad brushstrokes, evident, for instance, fifteen years ago on the front page of Seattle's *The Stranger*, after the re-election of George W. Bush: "Do not despair ... Don't think of yourself as a citizen of the United States. You are a citizen of the urban archipelago. The United Cities of America." What these narratives confirm is the worn-out urban/rural binary that pits the menacing city against the treasured countryside, or intellectuals against rubes.

Media scholar Mary L. Gray, together with historian Colin R. Johnson and anthropologist Brian J. Gilley, illustrates in *Queering the Countryside: New Frontiers in Rural Queer Studies* (2016) that the political underpinnings of rural and urban areas are in reality far more complicated. As Gilley pointed out in a recent interview, "The notion that the right owns rurality – guns, animal husbandry, country music. All of that stuff is a new development" (Moreau, 2016). In terms of political identification, the percentage of voters in rural counties who identify as Republican is greater than that of urban counties, and has increased in recent years (Pew, 2018). Yet, as Gilley and Johnston point out, the characterization of rural places as a last bastion for conservative values actually misrepresents the history of agrarian politics in the United States and beyond. For instance, social movements such as the Grange and the Farmers' Alliance worked toward establishing agricultural cooperatives to challenge the growing power of corporations (e.g., Loose, 2014). Globally, rural people have started countless progressive movements, including the Landless Rural

Workers Movement of Brazil, and France's *Confederation Paysanne* (e.g., Woods, 2008). As Bell and colleagues (2010) argue, rural places are not stable: they are dynamic, active, and mobile. Evident in studying social movements, migration, economies, or demographics, rural places in reality cannot possibly live up to a stagnant version of a rural idyll that many long for.

The notion of the countryside as "one of the last" remaining conservative strongholds brings to mind the state slogan of Montana: "The Last Best Place." The message draws upon national pride, conjuring up what legal scholar Lisa R. Pruitt describes as "popular perceptions of the rural that persist in our national consciousness" (2006, p. 159). As Schmalzbauer (2014) skillfully makes clear in her ethnography of Mexican immigration in Big Sky Country, as a state with such a high percentage of whites, Montana's "The Last Best Place" slogan is rife with racist undertones. Unpacking rural places as "one of the last geographically conservative regions," reveals not only an inaccurate understanding of rural political history, but in a discussion of queer migration, it also signals an assumption of widespread homophobia in rural places. In reality, however, the politics of the countryside are not necessarily defined by discriminatory attitudes toward LGBTQs.

As Gray, Johnson, and Gilley (2016b) say, consider that Vermont was the first state legislature to approve same-sex marriage. And, in the same year the Iowa Supreme Court unanimously upheld a district court's ruling that recognized same-sex marriage (Richburg, 2009). More recently, we've seen states embroiled in battles over "bathroom bill" legislation, which would prevent transgender children and adults from using the bathroom that aligns with their gender identity. Some of the most rural states in the United States – Montana, Kentucky, and South Dakota – have failed to pass or sign into law bills such as these (Lutey, 2018; NCSL, 2017). And, in 2018, Vermont Democrats nominated the country's first transgender gubernatorial candidate of a major party (Bidgood, 2018). On the other hand, in the 2016 presidential election, rural voters were far more likely than their urban counterparts to vote for Donald Trump, who pledged loyalty to LGBTQ voters during the campaign season but chose Mike Pence as his running mate, the Indiana governor who signed into law a religious liberty bill that permitted discrimination against LGBTQs (Drabold, 2016). Rural politics are indeed complex. What is clear is that we cannot expect rural places to fit easily into a binary understanding of politics and geography.

The Queer Threat to the Rural Idyll and Queers in Rural Places

The simplistic sketch of the countryside as untouched by sexual minorities again relies on an age-old construction of rural places as pure and unadulterated, as places chock-full of wholesomeness and so-called "family values." Urban places, by contrast, are seen as dirty and crime-filled places to satisfy any number of vices. These binary understandings of the rural and the urban have long been a focus for scholars studying culture and place (e.g., Bell, 2006; Bell, 2007; Williams, 1973). As some have argued (Bell, 2000; Keller & Bell, 2014), embedded in this dualistic breakdown of places are assumptions about sexuality, in which the wholesome procreative sexuality of rural places is the foil to the pariah sexuality of urban locales. The threat that LGBTQs pose to rural areas, then, is not just about the supposed political conspiracy of turning red districts blue, but it is also about what the queer body is supposed to represent to rural America – a foreign invasion of "natural" (read: heterosexual) spaces, and a threat to a rural way of life. Furthermore, the framing of this threat also ignores any diversity in the political identification of LGBTQs. Although small in number (Pew, 2016) we cannot ignore the existence of conservative LGBTQs, such as Log Cabin Republicans.

Framing rural places as territories to be protected against the encroachment of gender and sexual minorities not only promotes a fear that the sacred institution of heterosexuality will be trampled, but ignores rural realities. Queers are part of rural places, and not due to a leftist conspiracy of relocation. As Wypler (2018) pointed out, responding to Limbaugh's comments, there is actually a rich history of queers in the American countryside, exemplified by the lesbian land (or landdyke) movement that emerged in the 1970s. These queer women created various communities in rural spaces where they could live off the land in a way that challenged patriarchy and heterosexism, while also protecting the environment (Unger, 2010).

Accusing the USDA of recruiting sexual minorities to live and work in rural areas reveals a belief that queers are not already part of rural communities. It is this inability to *see* rural LGBTQs that Philo (1992) referred to as the marginalization of "neglected rural others." In circulating this invasion rhetoric, Limbaugh was promoting a fear-based reaction to this supposed imminent threat. But he was also challenging the government's attempts to make this marginalized group *legible* – to be seen as both rural residents (perhaps as farmers, too) and as LGBTQs. Limbaugh did this not by an outright denial of their existence, but through his inability to render rural queers legible. By his logic, the only way LGBTQs could be rural residents was through counter-urbanization.

Although population estimates of LGBTQs in rural places are hard to pin down, researchers have done well in documenting the experiences of this group. Since the mid-1990s or so, an impressive body of literature on rural LGBTQs has greatly contributed to our understanding of sexuality in rural spaces and small towns. Ushered in by D'Augelli and Hart's "Gay Women, Men and Families in Rural Settings" (1987), Bell and Valentine's "Queer Country" (1995), and Kath Weston's "Get Thee to a Big City" (1995), a steady stream of work on rural sexual minorities continues to flow, coming from scholars in geography, anthropology, sociology, psychology, history, media studies, and other disciplines. As Sachs (2014) points out, the empirical work in this area is rich and complex, revealing the suffering of rural LGBTQs on the one hand, but also the ways in which urban LGBTQs misunderstand and misrepresent these struggles. Sexual minorities may face heteronormativity and homophobia in rural places, yet they also suffer from the harms of "metronormativity" within LGBTQ social movements (Gray, 2009; Sachs, 2014).

Historical approaches have highlighted the lives of rural queer men. E. Patrick Johnson's (2011) *Sweet Tea: Black Gay Men of the South* presents rich oral histories of queer black men, many from rural origins in the American South. Similarly, John Howard's (1999) *Men Like That* includes oral histories from 100 queer men in

Mississippi that reach as far back as the 1940s up to the 1980s. Contrary to metro-centric research and common belief, Howard argues that in Mississippi queer sex was not unheard of, and that with the improvement of roads and transportation, queers could travel to other towns to find companions, but also found each other close to home in their rural communities.

In terms of queer migration, counter-urbanization trends do include queers coming to the countryside to live for the first time, but they also include queers returning to their rural roots. Studies focusing on back-and-forth rural–urban migration among LGBTQs have looked at the implications for finding community and for identity development, both in terms of sexuality and a sense of self grounded in rural culture (e.g., Annes & Redlin, 2012; Preston & D'Augelli, 2013). Although making up just a slice of the scholarship on rural queers, this research offers critical insights into the complex meanings involved for LGBTQs in leaving the countryside for the city and then returning to live in a rural community.

The wide range of experiences that LGBTQs have in rural places challenges the assumptions of urban sexual minorities that rural life must be quite depressing and isolating for folks who are perceived as "stuck" in their countryside environs (Halberstam, 2005, p. 36). There are actually numerous examples of rural spaces that host thriving queer cultures. There is the so-called "lesbian capital" of Hebden Bridge, a small town in West Yorkshire in the UK (Robehmed, 2012; Smith and Holt, 2005), the Connecticut River Valley in Massachusetts, U.S. that encompasses the college towns of Amherst and Northampton (Kirkey & Forsyth, 2001), not to mention the multiple options for gay-friendly getaways in non-metropolitan areas. These include Kings Beach in New South Wales, Australia, Guerneville in the Russian River Valley in California, Provincetown in Massachusetts, Ogunquit in Maine, Key West in Florida, and Fire Island in New York, to name a handful. These are places with decades-long histories of queer vacationers as well as year-round residents. As early as the 1930s, for instance, Cherry Grove on Fire Island offered a respite from the bustling city life for lesbians and gay men, as well as an "escape from straight domination" (Newton, 2014).

LGBTQ-affirming vacation destinations in non-metropolitan areas offer the opportunity to theorize the appeal of these places, bringing to light the concept of the *queer rural idyll*. This version of the rural idyll is an imagined view of natural spaces and rural places as potential utopias for queers (Keller & Bell, 2014). It's about locating the rural in the sexual – discovering how the rural is invoked in the queer sexual imaginary. The queer rural idyll operates on the symbolic level; it is made up of images and idyllic associations that rural spaces may conjure up for LGBTQs, whether it means exploring nature with fellow queers or experiencing sexual liberation in a space unfettered by the constraints of urban life. To acknowledge the material realities of queers living in rural spaces as well as the imagined queer rural idyll is what Michael M. Bell and I refer to as a *plural rural sexual* approach (Keller and Bell, 2014).

Others have similarly highlighted this complexity in the relationship between material and ideal understandings of rurality and sexuality. In a study of the Michigan Womyn's Music Festival, for example, a yearly celebration that was held on rural land in the upper Midwest, Browne (2011, p. 21) found that lesbian utopias challenge "hegemonic heterosexual ruralities." These utopias included outdoor nudity, rural sex spaces, and a spirituality connected to the land (Browne, 2011). The concept of a queer rural idyll can also be seen in the formation of year-round queer rural spaces, such as the lesbian separatist communities of Oregon. In these rural communities, Sandilands (2002) found that women constructed queer safe spaces, carved out rural lesbian identities, and experienced natural spaces as erotic sites. That said, Sandilands emphasized that, "After 27 years of Oregon women's lands, not a single lesbian I spoke to in the course of my research subscribed to a view of the women's lands as a utopia on earth" (p. 140). Tensions unfolded in these intentional communities due to various issues, from racism and classism to difficulties farming the land and financial pressures (Sandilands, 2002). The queer rural idyll is, after all, an ideal rural; the reality of queer rural spaces does not always live up to the imagined promise (Keller & Bell, 2014).

Taken together, the construction of rural places as exclusively heterosexual, and the framing of LGBTQs as foreign invaders to a rural way of life are not only harmful, but render invisible the experiences of rural LGBTQs. Rural queers exist. The challenges this group faces are complex, and we must pay attention to the various ways that LGBTQs create meaningful spaces in rural areas. Though not often the focus of environmental sociologists, the lives of rural queers — their experiences with nature, connections to the land and agriculture, as well as rural ways of being — offer much promise for investigating connections between sexuality and the environment.

Queering the Rural State?

A critical piece in understanding the source of Limbaugh's discontent is the fact that the USDA, during the Obama administration, had officially recognized LGBTQs as part of rural communities by offering directed programing to this marginalized group. The assumption operating on the political right is that the USDA had a partisan agenda that depended on encouraging the urban-rural migration of LGBTQs. Never mind that queers are already living and working in rural America, as discussed above. Besides tapping into conspiracy theories beloved by talk show hosts looking to engage and enrage listeners, this claim introduces the notion of special privileges granted to certain groups and not others.

The Rural Pride campaign was born in 2014, the result of a collaboration between the USDA and the National Center for Lesbian Rights. From 2014 to 2016, the campaign launched a series of summits in partnership with local rural-based organizations across the country. On the front page of their 2016 Outreach Report, the Natural Resources Conservation Service (NRCS) of California featured photos of events geared toward veterans, Latinos, and LGBTs. The report notes that, "As part of USDA's civil rights initiative, field-based agencies such as NRCS were asked to support a series of Lesbian, Gay, Bisexual [and] Transgender (LGBT) events to engage rural families and inform them of USDA programs and services. NRCS took the lead in California, hosting a full-day Rural Pride Summit in Visalia on Thursday, July 21st." The article goes on to say that the NRCS-California event was the best attended, with roughly 250 people, among the twelve of such events held across the nation. Although it appears that multiple USDA offices were asked to participate in the initiative, it is notable that the Natural Resources Conservation Service made such a prominent effort to include rural LGBTs.

This campaign, as part of the LGBT Emphasis Program at the NRCS, appeared to signal that the USDA viewed LGBTQ identity as a minority status, perhaps even worthy of including under the umbrella of "minority and women farmers and ranchers" (USDA, n.d.). For those who have been conducting research on queer farmers, and on rural queers generally, it is astounding that the rural government giant had recognized the gender and sexual diversity of its employees and of the countryside. This initiative seemed to signal that the USDA valued the voices and concerns of LGBTQs in rural spaces. Including LGBTs in USDA programming is a move toward legibility for sexual and gender minorities. The institution communicated the message that they *see* this marginalized rural group as clients that deserve to be served, and are in need of special assistance.

President Trump's appointment of Sonny Perdue as Secretary of Agriculture, however, reversed the gains made. In 2018, Perdue's chief of staff asked the National Institute of Food and Agriculture, the department that oversees 4-H (a network of organizations offering life skills to young people in agriculture and beyond) to rescind a policy protecting LGBT youth that some local groups had adopted. This resulted in the forced resignation of a 4-H director in Iowa who defended the policy supporting LGBT youth (Crowder & Clayworth, 2018). And, the Rural Pride campaign seems to have been dismantled. There is no mention of LGBTQs on the main page of the NRCS national website, yet the LGBT Special Emphasis Program is still listed under the Contact Us tab, alongside programs for other minorities, including Hispanic, Native American, Black, people with disabilities, veterans, women, and Asian Americans.

The effort to include rural LGBTQs by a governmental institution should be applauded. Yet Limbaugh spun this attempt at inclusion into a zero-sum game, warning that sexual minorities were receiving special privileges from the government, and that it was happening right in his viewers' backyards. Invoking the rhetoric of "special rights" is of course a well-worn strategy for stoking the insecurities of the working-class and increasingly precarious middle-class. In Arlene Stein's (2001) study of a small logging town in Oregon, *The Stranger Next Door*, she found that opposition to "special rights" for lesbians and gays was a central theme for residents supporting antigay ballot initiatives. Yet it wasn't a hatred of gays that fueled their politics, according to Stein. Rather, Stein argues, it was the closing of mills and growing economic insecurity that drove residents to vote against civil rights for LGBTs. Residents were also infuriated with environmentalists who were against logging, perceiving this group as a threat to the community's economic livelihood.

In the era of Trump, this is a familiar story. The fear of minority groups attaining "special rights" is especially resonant. As sociologist Arlie Hochschild (2016) found in *Strangers in Their Own Land*, working-class whites who perceived themselves as last in line after "Affirmative Action blacks, immigrants, [and] refugees" found their concerns not only validated, but fueled, by Trump the candidate, now Trump the president. When it comes to the concerns of rural LGBTQs, queering the rural state may be awhile down the road. But that doesn't mean we should lose focus on the potential for the government to recognize and reach out to rural LGBTQs.

Queer Sustainabilities and Eco-Queer Social Movements

It may be a stretch to suggest that lesbian *sustainable* farmers were the target of Limbaugh's remarks. However, it's not so farfetched to suggest that implicit in the cries of a lesbian farmer conspiracy is not only fear of the invasion of contaminated sexual others into rural places, but also fear of changing the land itself. This offers the chance to explore connections between sexuality and sustainable agriculture.

Since the 1990s, scholars have pointed to the intersections of gender and sustainable agriculture. Those working in this area have found that women's sustainable agriculture organizations provide members with opportunities to exchange farming knowledge, a women-only space where personal agency could be realized, and the chance to challenge "conventional notions of femininity" (Hassanein, 1999; Trauger, 2004). Considering that farming in the United States has historically been male-dominated, and that the public face of farming is often male, these spaces are critical for women to assert themselves as "farmers" (see Keller, 2014; Sachs, Barbercheck, Braiser, Kiernan, & Terman, 2016). Sachs and colleagues put forward what they call feminist agrifood systems theory, which explains why women face obstacles in farming, how they access land, labor, and capital through innovative means, and how women "shape new food and farming systems by integrating economic, environmental, and social values" (p. 2).

Turning from gender to sexuality, scholars have begun asking LGBTQs who farm sustainably similar questions about their pathways into agriculture. In a recent study of thirty sustainable farmers in New England, Isaac Leslie (2017) found that LGBTQ farmers entered sustainable agriculture for a variety of reasons, some related to gender and sexuality. Queer women and trans farmers described the appeal of gender-neutral farm clothing, as well as anticapitalist values drawing them to sustainable agriculture in particular. For some queer farmers interviewed, heterosexism was a barrier that affected access to land and training opportunities. Queer farmers in Leslie's study described the importance of relationships in sustainable agriculture, whether depending on others in the community for sustainable fertilizer or for marketing purposes, and some of these farmers experienced heterosexist microaggressions from those in the community. Notably, the majority of queer farmers Leslie interviewed did not experience overt instances of heterosexism, but even the expectation that they might experience heterosexism in a rural space was enough to act as a deterrent to farming.

Leslie argues convincingly that engaging with queer approaches offers the sustainability movement an inclusive strategy for recruiting more farmers into its ranks.

But the movement has to be willing to challenge its heterosexism. Leslie observes that there are plenty of anticapitalist queers who do not see themselves represented in the mainstream LGBTQ movement, and the sustainability movement will have to reconfigure itself to create space for this new group of potential and beginning farmers. Building off of feminist and queer theory, Leslie (2017, p. 274) proposes "queering sustainable agriculture" and the "sustaining of rural queer people." This framework parallels Sachs et al.'s (2016) feminist agrifood systems theory by emphasizing how heteronormativity structures rural life and sustainable agriculture, and demonstrating the various ways farmers are creating queer agricultural spaces.

Stepping toward a broader theoretical framework for understanding sexuality and environment, Catriona Sandilands and Bruce Erickson's (Mortimer-Sandilands & Erickson, 2010; Sandilands, 2002) concept of queer ecologies offers much to consider, beyond the research and practice of queer farmers:

> Specifically, the task of a queer ecology is to probe the intersections of sex and nature with an eye to developing a sexual politics that more clearly includes considerations of the natural world and its biosocial constitution, and an environmental politics that demonstrates an understanding of the ways in which sexual relations organize and influence both the material world of nature and our perceptions, experiences, and constitutions of that world. (Mortimer-Sandilands & Erickson, 2010, p. 5)

A framework such as this allows for a multilevel analysis of how sexuality and environment inform one another. Akin to environmental justice's focus on race and ecofeminism's focus on gender, queer ecology pushes us to interrogate how heterosexism organizes natural spaces, and understand, for instance, forms of resistance to this domination found among lesbian separatist communities in rural spaces (Sandilands, 2002). By placing politics at the center, Sandilands invites us to imagine the power of uniting sexual minority interests with environmental concerns, encouraging us to envision the possibility of political collaborations.

Building off of this framework is Joshua Sbicca's (2012) conceptualization of what he terms the "eco-queer movement." Emerging social movements grounded in LGBTQ rights and environmentalism can be spotted in both rural and urban spaces.

For these LGBTQ activists and their allies, it isn't difficult to see the connections between sexuality, gender, and the environment, and they see the creation of spaces and movements as central to forming communities around the shared principles of sustainability and anti-oppression. In urban spaces, there are the numerous examples of food-related queer projects listed by Sbicca (2012), including Queer Food for Love and Rainbow Chard Alliance in San Francisco, California, and even more have emerged recently. The queer-operated Side Yard Farm & Kitchen in Portland, Oregon is an urban farm specializing in organic herbs and vegetables, as well as a catering business that follows a "seed-to-plate" food philosophy. Nearby is Sweet Delilah Farm, an organic flower and herb farm that began when the owner was looking for a calming space for LGBTQ youth transitioning out of corrections facilities. Both farms offer inclusive, community spaces for gathering and healing (Feldman, 2018).

Turning back to rural spaces, it may be surprising to some that so many examples of queer land projects and communities in the countryside exist. Similar to the landdyke movement that began in the 1970s, queer people today have found solace, freedom, and creative possibility in countryside intentional communities that emphasize sustainable practices. One of these is the Idyll Dandy Arts (IDA) community, located in rural Tennessee in the United States. As a "rural community land project," IDA provides an opportunity for those who are queer, trans, and gender nonconforming to lead a communal life by working and living together, while also offering an educational space to learn farming skills and a creative space for artists (IDA, 2019). Another example is Fancyland in Northern California, a "queer land project" that acts as a "small-scale rural resource for learning and sharing useful rural living skills such as alternative building, appropriate technology, gardening, and land stewardship" (Fancyland, n.d.).

In Millerton, New York, Wildseed Community Farm and Healing Village is led by a Black and Brown "queer-loving" rural community on nearly 200 acres, with an organic farm and healing sanctuary. Not only does Wildseed provide an empowering space for people of color and queers, but they offer healing spaces for "those impacted by the criminal (in)justice system, and other communities on the frontlines of ecological disruption" (Wildseed, n.d.). Operating on Wildseed land is Linke Fligl, a queer and Jewish-led chicken farm and land-based cultural project. Nearby, Rock Steady Farm is women and queer-owned, operating twelve acres of farmland and supplying sustainably grown vegetables and flowers to over 200 CSA members and restaurants. Rock Steady Farm is a worker-owned cooperative that is guided by social justice and ecological principles.

Other farms that aren't queer-founded emphasize the inclusion of queer and trans people, bringing various marginalized groups together under anti-oppression principles. Soul Fire Farm in upstate New York, for instance, co-founded by Leah Penniman, lists in its mission statement to "bring diverse communities together on this healing land to share skills on sustainable agriculture, natural building, spiritual activism, health, and environmental justice" (Penniman, 2018).

Alliances that bring queer farmers together, whether in urban or rural locations, emphasize the need for even more spaces to connect, share skills, and build community. In 2018, Rock Steady Farm, Wildseed, and Linke Fligl co-hosted the first Northeast Regional Queer Farmers Alliance gathering. These groups invited LGBTQ farmers from both urban and rural locales to attend a meeting and social gathering to discuss myriad issues, such as protections for trans and gender nonconforming people, developing a giving circle, and sharing skills, to name just a few. The invitation explicitly welcomes indigenous people and people of color, and "all farmers within the spectrum of LGBTQ." In forging a sense of community across divisions, the emerging Northeast Regional Queer Farmers Alliance seeks to bring both rural and urban queers together around shared principles of social justice and environmental sustainability.

The emerging spaces and networks described above offer queer farmers and their allies an alternative way of participating in agriculture, engaging with food, and living in rural spaces. The mission statements and organizing principles of many of these groups are practicing what Leslie, Wypler, and Bell (2019) call *relational agriculture*. As they argue, heteropatriarchy is a major organizing principle in our food system. Agricultural institutions that we take for granted, such as the "family farm," carry with it certain assumptions about gender and sexuality (Keller, 2015; Leslie, 2017). Entangled in these gendered and sexualized meanings are messages about what is traditional, wholesome, and fundamentally good. A relational agriculture lens, as Leslie and colleagues explain (2019), is an approach that challenges and reframes heteropatriarchal relationships in food and agriculture, and it has implications for sustainability. As the examples above make clear, queer farmers are carving out spaces to enact different kinds of relationships in agriculture, and forging new possibilities for social justice and sustainable practices. These queer sustainabilities and eco-queer movements have much to offer to environmental sociology, as they point toward new and creative configurations in the nexus between environment and sexuality.

A Queer Environmental Justice Future?

An analysis of the 2016 anti-lesbian-farmer moment, as troubling as that moment was, offers multiple paths for more deeply integrating sexuality into environmental sociology. Through a focus on rural LGBTQs, queer sustainabilities, and eco-queer movements, this chapter makes a case for deeply considering how gender and sexual minorities figure into discussions of countryside spaces, nature, agriculture, and sustainability.

Analyzing how binary framings of rural places and sexual minorities operate, and what those who deploy them are trying to achieve, is necessary work if we want to understand the political dimensions of rural queer issues, and queer sustainabilities in particular. Simplistic at best, and hostile at worst, binary framings of rural versus urban spaces, and the sexual identities and practices found therein, such as those found in Limbaugh's commentary, are particularly harmful to rural queers. These have implications for sustainable agriculture when, as Leslie (2017) suggested, perceptions of homophobia or transphobia act as a deterrent for some LGBTQs considering living in rural places and pursuing a career in agriculture. Through using a *relational agriculture* lens (Leslie et al., 2019), we can better name and unpack the traditional gender and sexual relationships in food production that prevent the participation of all members of society, and constrict the possibilities of sustainable practices.

There are numerous examples of queer intentional communities, Do It Yourself (DIY) spaces, artistic projects, and alternative food cultures in both urban and rural areas that are carving out possibilities for community and sustainable living. As I've argued elsewhere (Keller, 2014), the growing number of queer spaces dedicated to food production and food issues – many of which involve explicit principles of sustainability – require a re-framing of the urban/rural assumptions about sexuality to see the increasingly complex ways in which place and sexuality come together. But these spaces are more than just about the intersection of sexuality and place, or sexuality and environment. As their mission statements make clear, many of these communities endeavor to move beyond a single-axis focus on sexuality in order to tackle multiple forms of injustice.

Related to these efforts, an additional avenue for deepening the connection between sexuality and the environment is through critically integrating sexuality as an important dimension of environmental justice (EJ). As Leslie King writes in this collection, environmental justice offered new ways of understanding the meaning of the environment and environmentalism. Far beyond protecting lands and oceans from pollution, and saving wildlife from extinction, EJ puts social inequality at the center of the conversation, focusing on the uneven distribution of environmental privilege and harm. EJ approaches have tended to focus on inequality as a result of race and class. However, some scholars have expanded the EJ paradigm to include sexuality as a dimension of ecological inequality. Sandilands (2004, p. 123), for instance, argued that the separatist communities she studied were working toward a "queer environmental justice" by creating explicitly lesbian spaces on rural land and thereby challenging traditional sexual relations, as well as through introducing ideas about nature into lesbian communities. More recently, Pellow (2018) offered a Critical Environmental Justice framework, which includes a focus on additional axes of difference, such as gender, sexuality, ability, and species. This expansion of categories of difference and an intensified commitment to intersectionality when it comes to environmental injustice deepens the EJ paradigm, offering further interdisciplinary connections.

Not only does a focus within EJ on sexuality open the door to further research on gender and sexual minorities and environmental harms (Collins, Grineski, & Morales, 2017), but from a social movements perspective, a queer re-framing of EJ offers the chance for new organizations to critique and reject exclusionary social movements centered on the environment and/or sexuality. A queer EJ perspective demands a conversation about the meaning of justice for multiple marginalized communities, including people of color, LGBTQs, and low-income people. Many of the land-based projects discussed above are built around the understanding that social movements centered on sustainable agriculture, environmentalism, and gay rights have often erected exclusionary barriers to keep those with fewer resources from participating. As just one example, farmer Leah Penniman (2018) and sociologist Monica White (2018) point out that the historical contributions of people of color who developed groundbreaking farming techniques and alternative practices have been largely erased from the contemporary sustainable agriculture movement. African American soil scientist George Washington Carver was a proponent of cover cropping (Penniman, 2018) and African American farmer Fannie Lou Hamer developed strategies for community-based food systems (White, 2018). Many of the queer-founded or queer-inclusive organizations and spaces I highlighted in the section above are committed to confronting erasures such as these and combatting various oppressions using an intersectional approach to food and environmental injustices.

While EJ scholars do not typically place countryside issues at the center of their research, Pellow (2016) has traced a number of opportunities for strengthening the connections. One of these is Food Justice Studies, which brings urban and rural people together to tackle issues of food access and food sovereignty. As I've made clear here, examples abound of queer farmers and activists in both urban and rural locales working toward food justice for their communities. Both environmental sociologists and rural sociologists have common ground here in understanding how these growing food justice movements frame their objectives, and the extent to which intersectionality is an organizing feature of their work. On a practical level, these queer farmers and their allies are key agents in challenging our unequal food system, and both environmental sociology and rural sociology would benefit through further analysis of their efforts.

In sum, there is plenty of territory to explore as environmental sociologists consider how sexuality could be more deeply integrated into the field. This chapter has shed light on a few particular junctures – the connections between rural spaces, LGBTQs, queer sustainabilities, and eco-queer movements. The work of practitioners on the ground, combined with recent theoretical and empirical contributions of scholars centering the connections between sexuality and the environment, offer much hope for an even more rich and complex area of study ahead.

References

Annes, A., & Redlin, M. (2012). Coming out and coming back: Rural gay migration and the city. *Journal of Rural Studies*, *28*(1), 56–68.

Bell, D. (2000). Farm boys and wild men: Rurality, masculinity, and homosexuality. *Rural Sociology*, *65*(4), 547–561.

Bell, D. (2006). Variations on the Rural Idyll. In P. Cloke, T. Marsden, & P. H. Mooney (eds.), *Handbook of Rural Studies* (pp. 149–160). London: SAGE.

Bell, D., & Valentine, G. (1995). Queer country: Rural lesbian and gay lives. *Journal of Rural Studies*, *11*(2), 113–122.

Bell, M. M. (2007). The two-ness of rural life and the ends of rural scholarship. *Journal of Rural Studies*, *23*(4), 402–415.

Bell, M. M., Lloyd, S. E., & Vatovec, C. (2010). Activating the countryside: Rural power, the power of the rural and the making of rural politics. *Sociologia Ruralis*, *50*(3), 205–224.

Bidgood, J. (2018, August 14). Christine Hallquist, a transgender woman, wins Vermont Governor's Primary. *The New York Times*. Retrieved from www.nytimes.com/2018/08/14/us/politics/christine-hallquist-vermont.html

Browne, K. (2011). Beyond rural idylls: Imperfect lesbian utopias at Michigan Womyn's music festival. *Journal of Rural Studies*, *27*(1), 13–23.

Collins, T. W., Grineski, S. E., & Morales, D. X. (2017). Sexual orientation, gender, and environmental injustice: Unequal carcinogenic air pollution risks in greater Houston. *Annals of the American Association of Geographers*, *107*(1), 72–92.

Crowder, C., & Clayworth, J. (2018, November 18). 4-H: Trump agency push to dump LGBT policy led to Iowa leader's firing. *Des Moines Register*. Retrieved from www.desmoinesregister.com/story/news/investigations/2018/11/18/4-h-transgender-lgbt-iowa-john-paul-chaisson-cardenas-iowa-state-university-civil-rights/1572199002/

D'Augelli, A. R., & Hart, M. M. (1987). Gay women, men, and families in rural settings: Toward the development of helping communities. *American Journal of Community Psychology*, *15*(1), 79–93.

Drabold, W. (2016). Mike Pence: What He's Said on LGBT Issues Over the Years. *Time Magazine*, July 16. http://time.com/4406337/mike-pence-gay-rights-lgbt-religious-freedom/

Erbentraut, J. (2016). These lesbian farmers aren't here to take over America. They want to grow it. *Huffington Post*, September 4. www.huffpost.com/entry/lesbian-farmers-rush-limbaugh_n_57c879d6e4b0e60d31ddf5c0

Fancyland. (n.d.). About. https://fancylandy.wordpress.com/

Feldman, L. (2018). 'We created this village': Breaking bread with the women farmers of Portland. *Time Magazine*, July 27. https://time.com/longform/women-urban-farmers/

Gray, M. L. (2009). *Out in the Country: Youth, Media, and Queer Visibility in Rural America*. New York City: New York University Press.

Gray, M. L., Johnson, C. R., & Gilley, B. J. (2016a). *Queering the Countryside: New Frontiers in Rural Queer Studies*. New York City: New York University Press.

Gray, M. L., Johnson, C. R., Gilley, B. J. (2016b). Introduction. In *Queering the Countryside: New Frontiers in Rural Queer Studies* (pp. 1–21). New York City: New York University Press.

Halberstam, J. J. (2005). *In a Queer Time and Place: Transgender Bodies, Subcultural Lives*. New York City: New York University Press .

Hochschild, A. R. (2016). I Spent 5 Years with Some of Trump's Biggest Fans. Here's What They Won't Tell You. *Mother Jones*. Retrieved from www.motherjones.com/politics/2016/08/trump-white-blue-collar-supporters/

Howard, J. (2001). *Men Like That: A Southern Queer History*. Chicago, IL: University of Chicago Press.

IDA. (2019). About IDA. Idyll Dandy Arts. https://idylldandyarts.tumblr.com/about

Johnson, E. P. (2011). *Sweet Tea: Black Gay Men of the South*. Chapel Hill, MD: University of North Carolina Press.

Keller, J. C. (2014). "I Wanna Have My Own Damn Dairy Farm!": Women farmers, legibility, and femininities in rural Wisconsin, U.S. *Journal of Rural Social Sciences*, *29*(1), 75–102.

Keller, J. C. (2015). Rural Queer Theory. In B. Pini, B. Brandth, & J. Little (eds.), *Feminisms and Ruralities* (pp. 155–166). Lanham, MD: Lexington Books.

Keller, J. C., & Bell, M. M. (2014). Rolling in the Hay: The Rural as Sexual Space. In C. Bailey, L. Jensen, & E. Ransom (eds.), *Rural America in a Globalizing World* (pp. 506–522). Morgantown, WV: West Virginia University Press.

Kirkey, K., & Forsyth, A. (2001). Men in the valley: Gay male life on the suburban–rural fringe. *Journal of Rural Studies*, *17*(4), 421–441.

Leslie, I. S. (2017). Queer Farmers: Sexuality and the transition to sustainable agriculture. *Rural Sociology*, *82*(4), 747–771.

Leslie, I. S., Wypler, J., & Bell, M. M. (2019). Relational agriculture: Gender, sexuality, and sustainability in U.S. farming. *Society & Natural Resources*, *32*(8), 853–874.

Loose, S. K. (2014). History from below: Connecting rural Oregon to its social movement history. *Oregon Historical Quarterly*, *115*(2), 244–251.

Lutey, T. (2018, June 29). Transgender bathroom initiative about 15 K signatures short of qualifying for Montana ballot. *Billings Gazette*. Retrieved from https://billingsgazette.com/news/state-and-regional/govt-and-politics/transgender-bathroom-initia tive-about-k-signatures-short-of-qualifying-for/article_c66255f2-1240-505b-b5b8-756da0987eb9.html

Moreau, J. (2016, August 27). Why is Rush Limbaugh so afraid of lesbian farmers? *NBC News*. Retrieved from www.nbcnews.com/feature/nbc-out/why-rush-limbaugh-so-afraid-lesbian-farmers-n638736

Mortimer-Sandilands, C. (2010). *Queer Ecologies: Sex, Nature, Politics, Desire*. Bloomington, IN: Indiana University Press.

NCSL. (2017). Bathroom Bill Legislative Tracking. Washington, DC Retrieved from www.ncsl.org/research/education/-bathroom-bill-legislative-tracking635951130.aspx

Newton, E. (2014). *Cherry Grove, Fire Island: Sixty Years in America's First Gay and Lesbian Town* (2nd ed.). Durham, NC: Duke University Press.

Pellow, D. N. (2016). Environmental justice and rural studies: A critical conversation and invitation to collaboration. *Journal of Rural Studies*, 47, 381–386.

Pellow, D. N. (2018). *What Is Critical Environmental Justice?* Cambridge: Polity Press.

Penniman, L. (2018). *Farming while Black: Soul Fire Farm's Practical Guide to Liberation on The Land*. White River Junction, VT: Chelsea Green Publishing.

Pew. (2016). *Lesbian, Gay, Bisexual Voters Remain Solidly Democratic*. Washington, DC: Pew Research Center. Retrieved from www.pewresearch.org/fact-tank/2016/10/25/lesbian-gay-and-bisexual-voters-remain-a-solidly-democratic-bloc/

Pew. (2018). *What Unites and Divides Urban, Suburban and Rural Communities*. Washington, DC: Pew Research Center. Retrieved from www.pewsocialtrends.org/2018/05/22/what-unites-and-divides-urban-suburban-and-rural-communities/

Philo, C. (1992). Neglected rural geographies: A review. *Journal of Rural Studies*, *8*(2), 193–207.

Preston, D. B., & D'Augelli, A. R. (2013). *The Challenges of Being a Rural Gay Man*. New York: Routledge.

Pruitt, L. R. (2006). Rural rhetoric. *Connecticut Law Review*, *39*(1), 159–240.

Richburg, K. B. (2009, April 4). Iowa Legalizes Same-Sex Marriage. *Washington Post*. Retrieved from www.washingtonpost.com/wp-dyn/content/article/2009/04/03/AR2009040300376.html

Robehmed, S. (2012, February 9). Why is Hebden Bridge the lesbian capital? BBC News Magazine. www.bbc.com/news/magazine-16962898

Sachs, C. E. (2014). Gender, Race, Ethnicity, Class, and Sexuality in Rural America. In C. Bailey, L. Jensen, & E. Ransom (eds.), *Rural America in a Globalizing World: Problems and Prospects for the 2010s* (pp. 421–434). Morgantown, WV: West Virginia University Press.

Sachs, C. E., Barbercheck, M., Braiser, K., Kiernan, N. E., & Terman, A. R. (2016). *The Rise of Women Farmers and Sustainable Agriculture*. Iowa City: University of Iowa Press.

Sandilands, C. (2002). Lesbian separatist communities and the experience of nature: Toward a queer ecology. *Organization & Environment*, *15*(2), 131–163.

Sandilands, C. (2004). Sexual Politics and Environmental Justice. In R. Stein, M. Knopf Newman, A. Lucas, W. LaDuke, B. Berila, & G. Di Chiro (eds.), *New Perspectives on Environmental Justice: Gender, Sexuality and Activism* (pp. 109–126). New Brunswick: Rutgers University Press.

Sbicca, J. (2012). Eco-queer movement(s): Challenging heteronormative space through (re)imagining nature and food. *European Journal of Ecopsychology*, *3*, 33–52.

Schmalzbauer, L. (2014). *The Last Best Place? Gender, Family, and Migration in the New West*. Stanford, CA: Stanford University Press.

Smith, D. P., & Holt, L. (2005). 'Lesbian migrants in the gentrified valley' and 'other' geographies of rural gentrification. *Journal of Rural Studies*, *21*(3), 313–322.

Stein, A. (2001). *The Stranger Next Door: The Story of a Small Community's Battle Over Sex, Faith, and Civil Rights*. Boston, MA: Beacon Press.

Trauger, A. (2004). 'Because they can do the work': women farmers in sustainable agriculture in Pennsylvania, USA. *Gender, Place & Culture*, *11*(2), 289–307.

Unger, N. C. (2010). From Jook Joints to Sisterspace: The role of nature in lesbian alternative environments in the United States. In C. Mortimer-Sandilands & B. Erikson (eds.) *Queer Ecologies: Sex, Nature, Politics, Desire* (pp. 173–198). Indiana University Press,

USDA. (n.d.). Outreach Programs. Retrieved May 31, 2019, from www.fsa.usda.gov/programs-and-services/outreach-and-education/outreach-programs/index

Weston, K. (1995). Get thee to a big city: sexual imaginary and the great gay migration. *GLQ: A Journal of Lesbian and Gay Studies*, *2*(3), 253–277.

White, M. M. (2018). *Freedom Farmers: Agricultural resistance and the black freedom movement*. Chapel Hill, NC: University of North Carolina Press.

Wildseed. (n.d.). About. Wildseed Community Farm & Healing Village. www.wildseedcommunity.org/

Williams, R. (1973). *The Country and the City*. New York: Oxford University Press.

Woods, M. (2008). Social movements and rural politics. *Journal of Rural Studies*, *24*(2), 129–137.

Wypler, J. (2018). Farmer or queer? Researching the herstory, challenges & triumphs surrounding lesbian & queer farmers. Retrieved May 31, 2019, from https://invisiblefarmer.net.au/blog/2018/2/5/invisfarmer/queerfarmer.

6 Masculinity and Environment

Kathryn Gregory Anderson

Over the holidays a few years ago, I met up with an old friend from high school at a local pub. He was wearing a t-shirt with Earth First! printed in large bold letters over a drawing of the globe. I thought he was supporting the radical environmental advocacy group until he turned around and showed me the message on the back: "We'll log the other planets later." Over the previous few years, he had close-shorn his curly locks, lowered his voice, and traded in his city-boy SUV for a troublesome old pickup truck and a gun. He was learning how to be a man in Montana.

This chapter explores how, in cultures where male power is hegemonic, claiming and mobilizing a masculine identity and avoiding the feminine can become problematic for environmental sustainability. I begin with basic tenets of critical gender theory and masculinities studies that help explain the relationship between masculinity and environment. I then examine recent research suggesting that anti-environmental attitudes and behavior as well as indifference or skepticism about environmental science and risk can be explained in part as performances enacted to signal a masculine identity (Brough, Wilkie, Ma, Isaac, & Gal, 2016; Kahan, Braman, Gastil, Slovic, & Mertz, 2007). After discussing these direct anti-environmental masculine performances, I explore how enacting core masculine-coded performances *not* directly related to the environment can also obstruct environmental protection. Privileging the rational, technical, and competitive, and avoiding feminine-coded emotion and cooperation can result in excluding social justice arguments for environmental action (Caniglia, Brulle, & Szasz, 2015) and favoring technological and business-friendly solutions that may be untested and dangerous – such as climate geoengineering – over vital regulations and multi-lateral cooperation (Anshelm & Hultman, 2014; Fleming, 2017; Hultman, 2017; Swim, Vescio, Dahl, & Zawadzki, 2018).

1 Hegemony of Male Power Incentivizes a Gender Binary and Hyper-Masculine Performances

Research on gender and environment has mainly focused on how women are disproportionately affected by and try harder to resist environmental destruction, with scant attention paid to how different masculinities influence environmental problems and resistance (Hultman, 2017). Ecofeminist scholars have long proclaimed the connections between social oppression (with particular attention to the oppression of women) and environmental destruction (Plumwood, 1993; Warren, 2000). Yet there has not been a deep and sustained investigation into how different masculinities (as an unmarked category) are connected to environmental phenomena, a lacuna that gender and environment scholar Sherilyn MacGregor calls "curious, given the role that hegemonic forms of masculine power – in institutions of the state such as the military and scientific agencies, as well as in corporations and environmental movement organizations – have played in shaping both environmental problems and how they have (not) been addressed" (2017, p. 5). Men tend to be portrayed as the villains without much sustained analysis of why they come to think and act as they do (Connell & Pearse, 2015, cited in Hultman, 2017).

Recent environmental sociology scholarship has begun to understand masculinity through a critical gender theory lens (e.g., Bell & York, 2010), incorporating insights from the field of masculinities studies. This approach understands gender as a social structure and a power relation rather than simply as learned characteristics that individuals internalize into their core identities (Connell, 2016). Gender is something one does, actively, to achieve valuable outcomes, rather than something one naturally is (Fenstermaker & West, 2002) or something stable that one performs consistently (Butler, 1993). Multiple different ways of doing masculinity and femininity arise in different cultures, historical time periods, social categories of men and women (Connell & Messerschmidt, 2005), and even within the same individual at different moments. While all performances are active, they vary from subconscious choices not felt to be choices at all, to various degrees of self-reflective intentionality.

The prominent gender theorist Raewyn Connell conceived of masculine *practices* as meso-level patterns of behavior performed by situated agents grappling with the situations they face, as they perceive them (Ferree, 2018). This helps us see how masculinity is both a social structure and is also embedded in individuals, who enact different masculine performances depending on their perception of what the specific context demands. For example, one of the practices explored here is adopting an anti-environmental stance. Rather than taking environmental concern at face value – as one's actual motivation to protect natural resources and ecosystems – a critical gender theory lens explores how people instrumentally use their stance towards the environment as a signaling device to convey a masculine identity and to influence their social status.

To understand why individuals choose certain practices and why certain masculinities arise where they do, it is critical to recognize that in modern industrialized nations male power is largely hegemonic. This means that men have power over women and this power is largely unquestioned, unseen, and accepted as natural. Enacting overtly masculine performances is often vital to accessing material rewards, social status, esteem from others, connections, jobs, and autonomy (Connell, 2005). Furthermore, because of the structural subordination of women, the core elements of many masculine performances serve to construct or reinforce a gender binary that dichotomizes gender into two non-overlapping categories and to distance oneself from the feminine (Bosson & Michniewicz, 2013).

Patriarchal structures and hegemonic male power don't only spur hyper-masculine performances, including anti-environmentalism, in men. In an androcentric society, where the masculine is valued above the feminine, all genders, including women and transgender people, may at times enact masculine performances to access power and authority. While men more than women learn that they are entitled to domination (Johnson, 2017) and the rewards that accrue for enacting masculine-coded performances are often limited to men, an intersectional analysis suggests that certain rewards will be available to certain categories of women who perform masculinity in certain ways. Consider Sarah Palin's chant, "Drill baby drill" or Anne Gorsuch, who gutted the EPA under Reagan.

Over time and place, there has been tremendous variation in both the degree of gender dichotomization and in the specific behaviors that count as idealized or even acceptable masculine performances (Brandth & Haugen, 2006; Schrock & Schwalbe, 2009). Researchers have explored the phenomena of cowboy rural masculinities (Campbell & Bell, 2000), hipster urban masculinities (Bridges & Pascoe, 2018), managerial businessman masculinities (Sinclair, 1995), the strength and endurance of manual laborer masculinities (Paap, 2006), and positive masculinities that legitimize egalitarian relations (Messerschmidt & Messner, 2018), among many others. During the masculinity crises of the 1920s (Dubbert, 1979; Rome, 2006) and the 1980s (Kimmel, 1987), men's insecurity about their manhood and the devaluing of the feminine lead to compensatory hyper-masculine norms that hamstrung the environmental movement (Rome, 2006). Both in the progressive era around the 1920s and again since the 1980s, an anti-green stance or environmental indifference became a masculine performance/characteristic in some sub-cultures (Brough et al., 2016; Rome, 2003). The anti-green policing of masculinity in the 1920s is evident in newspaper cartoons depicting prominent environmentalist men in women's clothing, engaged in women's activities, like sweeping (Rome, 2006).

Performances of masculinity not only enforce a hierarchy between men and women but also a hierarchy among different categories of men. Critical gender scholars often use the term hegemonic masculinity to describe an idealized type of masculine performance that has especially strong access to power, that naturalizes gender inequality, and that subordinates other categories of men (Connell & Messerschmidt, 2005). The characteristics of hegemonic masculinity vary across groups and contexts (Schrock & Schwalbe, 2009). For example, the stereotype applied to an environmental economist in Washington D.C. emphasizes the rationality, emotional control, and technical proficiency idealized in that sub-culture, while physical domination, toughness for its own sake, and overt sexism would be disparaged. In contrast, the stereotype of an alumnus of certain football-centered college fraternities would still value toughness and physical strength, virility or even sexual aggression, substantial gender dichotomization or even overt sexism, and certain emotions like anger and passionate loyalty. These hegemonic ideals of masculinity have power in the individual's relevant sub-culture, even if not always lauded more broadly.

Individuals are often unselfconsciously sophisticated students of the variation in hegemonic masculinity across the social worlds within which they circulate. Thus, the environmental economist who is an alumnus of a football-centered fraternity might drive the family Prius to a professional meeting about speculation in carbon markets and later drive his SUV to a football game. Perhaps without thinking, he makes the calculation that while he'll spend more money on gas and emit more carbon in his SUV (about which he might feel remorse), he'll feel more "like a man" tailgating from his SUV. Not being held accountable for proving their masculinity is of considerable value to many men.

2 Anti-Green Stance Signals Valuable Masculine Identity

While most men don't broadcast anti-environmentalism through slogans on t-shirts, there is a robust literature showing the tendency for men to report fewer pro-environmental values, beliefs, and attitudes and fewer private environmental behaviors than do women (Kennedy & Dzialo, 2015; McCright & Xiao, 2014; Yates, Luo, Mobley, & Shealy, 2015). For example, McCright (2010) found that 37 percent of women but only 28 percent of men believe global warming will threaten their way of life. Xiao and McCright (2014) found that 48 percent of women but only 39 percent of men report that they "always or often reduce the energy or fuel you use at home for environmental reasons." While this "environmental gender gap" has been the subject of much discussion and debate, it is most often explained by gendered socialization in values and personality (e.g., women as more empathetic and altruistic; men as more detached and competitive) and/or to different social roles (women in mothering/caretaking roles and men in bread-winner/market/public roles) (Arnocky & Stroink, 2011; Dietz, Kalof, & Stern, 2002; McCright, 2010; Zelezny, Chua, & Aldrich, 2000).

This chapter suggests that rather than taking environmental concern or resistance at face value or as an expression of one's altruism and empathy more generally, an environmental stance, pro or anti, can also be understood as a signaling device to convey identity and to achieve particular social consequences. This approach does not dispute gender socialization explanations, but it relies less on them, recognizing that learned characteristics of gender are neither entirely stable nor performed in a consistent way in people's daily lives (Butler, 1993). Rather, gender is strategic in the sense that people perform different culturally shaped skills and habits depending on the situation they face. Different gender identities can be signaled at different times to advance specific goals. Furthermore, gender identities and practices are not random assortments of characteristics, but rather are continually shaped by a gender order in which male power is hegemonic. Thus, it's not that men simply learn at an early age to repress empathy or to be indifferent about the environment, but rather these are strategies to signal a valued masculinity in androcentric cultures where masculinity confers material and symbolic rewards.

2.1 Anti-Environmental Masculine Overcompensation

Evidence that environmental stance is used instrumentally specifically to signal masculinity comes from Brough et al. (2016), who found that when their masculinity was threatened in an experimental setting, men in their sample reacted with anti-environmental behavior. Based on past research suggesting that "going green" is considered more feminine by a majority of Americans (Bennett & Williams, 2011), Brough et al. (2016) test the hypothesis that men's unwillingness to engage in pro-environmental behaviors stems from a cultural association of green and feminine. Their experiments first confirmed the association between green and feminine, both for green products and for green behaviors, and both among men and women. They then demonstrated that even subjects' own gender self-perceptions can be manipulated: subjects told to write a few sentences about an action that was environmentally friendly reported feeling more feminine compared with those describing an action that harmed the environment. Next, they administered a previously verified masculinity threat to an all-male sample, instructing all participants to imagine receiving a $150 gift card from coworkers with a note saying, "we thought this card was perfect for you – happy birthday!" and varying whether participants got a pink floral card with Happy Birthday written in frilly font or an age threat control group card. Compared with the control group, respondents receiving the feminine, masculinity-threatening card were significantly more likely to choose conventional products over environmentally friendly options. In contrast, in a separate masculine affirmation experiment, male participants whose masculinity was affirmed by feedback that they have a very masculine writing style were significantly more likely to choose an environmentally friendly product over a conventional option compared to men in the control condition.

These results suggest that environmental concern is, to a significant degree, flexible and plastic, performed inconsistently and used instrumentally to signal gender identity based on immediate situational cues interacting with embodied gender structures. The random assignment to experimental condition essentially controlled for stable differences in environmental concern or socialized characteristics like empathy, altruism, hierarchism, individualism, social class, or partisan identity. Among men, the manipulated desire to signal masculinity had an independent and important influence on their environmental behavior.

Brough et al.'s (2016) study is an example of a larger experimental literature on *masculine overcompensation*, the phenomenon wherein men react to insecurity about their masculinity with hyper-masculine behavior to recover their masculine status (Willer, Rogalin, Conlon, & Wojnowicz, 2013). In numerous experiments, men have responded to masculinity threat with more negative views of homosexuality, more support for war, less support for gender equality, more derogation of women, greater probability of subsequently sexually harassing a woman, and more risk-taking in the stock market (Bosson, Vandello, Burnaford, Weaver, & Wasti, 2009; Kosakowska-Berezecka, Besta, & Vandello, 2016; Maass, Cadinu, Guarnieri, & Grasselli, 2003; Willer et al., 2013). In a review of this literature, Willer et al. (2013) concluded that "the core aspect of masculinity that men enacted in the face of threats was dominance, a fundamental basis of hierarchy and status differentiation among men and boys" (p. 1013). This corresponds with research findings that a heightened orientation to social dominance mediates the gender-environmentalism link (Milfont & Sibley, 2016).

2.2 Why Is Eco-Friendly Feminine in the First Place?

Some have conjectured that the green-feminine stereotype may have arisen from marketers using feminine branding to sell eco-friendly products to the majority female market for domestic goods (Brough et al., 2016; Stafford & Hartman, 2012). Critical gender theory would suggest several important supplementary explanations. Most basically, environmental concern signals feminine-coded empathy and caring for other people and even non-human beings. It signals an appreciation of interconnection and inter-dependence over the hegemonic masculine values of independence and individualism. In patriarchal hierarchies, competition and dominance are legitimized (Schrock & Schwalbe, 2009), whereas being green requires cooperating with instead of dominating over nature. Residents of a natural gas fracking boomtown community, for example, describe the industry as "very male ... enthrallment with their trucks, their tools ... making noise ... it's a rape of the land ... the aggressiveness, and the takeover aspect of it" (respondent interview, Filteau, 2016, p. 535). On the other hand, practicing sustainable agriculture requires respecting ecological processes and engaging in a back-and-forth dialogue with nature, in contrast with a desire to control nature through chemicals and big machines (Peter, Bell, Jarnagin, & Bauer, 2000). Alternative and organic farmers historically have subscribed to more egalitarian gender norms and have had more egalitarian relationships in general, including with the earth (Peter et al., 2000; Trauger, 2004).

Additionally, green became feminine, in part, due to the historical contingency that women have always constituted a substantial presence among environmentalists (Rome, 2003, 2006). During times when women have been extremely constrained in what they are allowed to do publicly, "municipal housekeeping" and preserving nature were arguably their most legitimate public activities (Rome, 2003, 2006). Women have been indispensable in environmental activism long before Rachel Carson published *Silent Spring*.

Finally, the green-feminine stereotype has been strategically enhanced by marketing and political campaigns that have explicitly connected extractive industries to idealized male identities as a way to reduce community resistance to ecologically destructive activities. The coal industry in Appalachia, for example, has run commercials associating coal with the traditional male identities of protector, defender, hero, and bread-winner, while equating environmentalism with anti-man attitudes and loss of male privilege (Bell & Braun, 2010; Bell & York, 2010).

2.3 The "White Male" Effect – Hierarchism, Individualism, and Identity-Protective Cognition

Another line of research supporting the idea of anti-environmental stance as identity-signaling device is the Cultural Cognition Project (CCP). The CCP has found that people arrive at their environmental views through a largely pre-conscious process of *identity-protective cognition*, whereby selective interpretation of environmental information preserves valued identities and social relationships, often at the expense of logical and scientific analysis of information (Kahan et al., 2007).

The CCP divides individuals into groups along the "cultural worldview" axes of hierarchical vs. egalitarian, and individualistic vs. communitarian. People who share hierarchical and individualistic cultural worldviews with important social networks will tend to downplay environmental risks relative to people with egalitarian and communitarian cultural worldviews (Kahan et al., 2012).

Kahan et al. (2007) explicitly link the gender gap in environmental concern to hierarchism and individualism, two core performances of prevalent hegemonic masculinities. They investigate the "white male effect" – a widely accepted finding that women and non-white men are significantly more concerned about environmental risk than are white men (Finucane, Slovic, Mertz, Flynn, & Satterfield, 2000; Flynn, Slovic, & Mertz, 1994). They find that while it appears that gender and race are driving differences in climate change skepticism and other environmental risk perceptions, the more influential variables are hierarchism and (to a lesser extent) individualism, and white males tend to be more hierarchical and more individualistic.

While this empirical evidence is extremely useful, their interpretation of the data could be enriched by an intersectional analysis that understands the multiplicative impact of multiple social locations on what identities are available to different people (Collins, 2005). The rewards that accrue for performances of hierarchical and individualistic worldviews may be circumscribed for many women and for many non-white men, and the penalties for attempting those performances may be higher. In their review of the literature on African American masculinity, Nickleberry and Coleman (2012) point out that African American expressions of masculinity represent the skills, beliefs, and values necessary to cope with racial discrimination. Understanding race as a social construction highlights the ways that hierarchical and individualistic worldviews may be part of whiteness intersectionally performed with masculinity. Kahan et al. (2007) reduce gender too readily to sex category when they argue that hierarchism and individualism – two core performances of gender that are often only safely performed by white men – are more influential in environmental concern than gender. Re-framed in these ways, the Kahan et al. (2007) results help us understand the pathways through which race and gender influence environmental values.

3 Masculine Performances of the "Rational" and Technological Can Hurt the Environment

While lack of environmental concern may be performed to signal valued masculinity in a range of cultures and contexts, enthusiastically embracing environmentalism can be consistent with some varieties of hegemonic masculinity as well. Even among self-described environmentalists, however, performing certain masculinities has been found to handicap effective environmental action. Because compassion, cooperation, risk-aversion, self-restraint, and even justice ethics are frequently coded as feminine and weak, masculine performances that serve to distance the actor or the action from femininity avoid such associations (Rome, 2006). When debating climate change policies, Swim et al. (2018) found that the men in their sample tended to use arguments that focus on science and business, while the women cited policy arguments that focus on ethics and justice. The men even attributed feminine traits to other men who used ethics and environmental justice arguments.

What looks like "rationality" is often a ruse, covering up efforts to maintain current power structures and obscuring strong underlying emotions and interests. Even though many men learn to stifle emotion (Gardiner, 2000), it's not necessarily the case that they succeed or that women end up actually being more emotional, less "rational," or less objective. In her classic ethnography of nuclear defense intellectuals, Cohn (1987) illustrated how technological language and claims to the high ground of rationality hid their irrational logic and highly emotional motivations. Anshelm & Hultman (2014) found that among older white men apparently rational climate skepticism stems from sentimentality for a bygone era of unquestioned deference to male authority.

The fiction of masculine rationality, however, doesn't mean that it doesn't have powerfully real environmental consequences. In the progressive era, environmental reformers responded to a macro-cultural masculinity crisis by hiding their humanistic motivations under the guise of economic and scientific arguments, resulting in many environmental issues being excluded from the reform agenda because they could not be described in masculine terms (Rome, 2006). In contrast, the opening of masculinity in the 1960s, to include pacifism, emotion display, and caringness, contributed to the passage of the United States' cornerstone environmental laws, justified not primarily on economic and technical arguments, but largely on ethical and moral grounds (Rome, 2003).

The calculating, competitive, self-interest maximizing "rational actor model" at the heart of economic theory reflects hegemonic masculine patterns of valuing autonomy and detachment over dependence and connection (Nelson, 1995). Coinciding with the macro-cultural masculinity crisis of the 1980s, economics became the social science of environmental policy and of government more broadly. Since 1989, economists have been referenced more than twenty times more than sociologists or anthropologists in both the Congressional Record and the *New York Times* (Wolfers, 2015). The president does not have a Council of Sociological Advisers, nor does the USDA have a Sociological Research Service. Yet decades of empirical research have challenged rational actor assumptions (Kahneman, 2011; Thaler & Sunstein, 2009; Weber & Dawes, 2005) and have demonstrated the importance of cooperative norms, altruism, and reciprocity for designing and maintaining sustainable environmental management (Ostrom, Burger, Field, Norgaard, & Policansky, 1999). While economics has much to offer, its narrow prescriptions have been found empirically to limit the effectiveness of environmental governance (Brueckner, 2007), and its logic has been used to justify rolling back environmental regulations and gutting the EPA. Market logic pushes an agenda that sociologists have described as the individualization and privatization of environmental responsibility (Kennedy & Kmec, 2018; Sandilands, 1993), causing a narrowing of our "environmental imagination" in ways that defer vital public actions and institutional solutions (Maniates, 2001).

Hultman (2013) coins the term "ecomodern masculinity" to characterize current prevalent environmental performances that favor technological expansion and business justifications that serve to maintain the power structures of climate-destroying systems at the expense of real environmental protection. Climate geoengineering, for example, despite potentially being shortsighted, untestable, and dangerous (Pierrehumbert, 2015), appeals to certain "alpha male" environmental masculinities (Fleming, 2017) and can even generate more trust in climate science among skeptical (typically white male) hierarchical individualists (Kahan, Jenkins-Smith, Tarantola, Silva, & Braman, 2015).

While as a society we can justify certain specific environmental measures "rationally," for example in terms of avoided health costs, and while green technology is critical to a sustainable future, broadly effective environmental action requires mobilizing concern about ecosystems and vulnerable communities of people. Privileging rational and technical frames translates into environmental policies evaluated in terms of technical efficiency and economic growth rather than social welfare and social equity (NAACP, 2012; Salleh, 2009). Ultimately, favoring top-down technological fixes and increased production results in less broad ecological protection than solutions carrying the taint of feminine-coded strategies like conservation, sharing, restraint, and cooperation (Kronsell, 2013).

Conclusion

This chapter reviews evidence that everyday masculine performances that reinforce a gender dichotomy and create distance from the feminine routinely obstruct environmental protection. Anti-environmental attitudes and behaviors as well as indifference or skepticism about environmental science and risk are often used to signal core masculine characteristics like hierarchism and individualism and even masculinity itself. Even among environmental advocates, masculine-coded performances of valuing the rational, top-down, and technical over the sentimental and cooperative result in excluding vital justifications, metrics, and solutions from the environmental agenda. While academic scholarship is coming to understand these gendered performances as means to safeguard a threatened identity, it has not been sufficiently emphasized that these environmentally destructive performances are incentivized by a hegemonic gender order in which men have privileged access to material and social rewards, in which masculinity is inherently relative and precarious, and in which men are strongly policed for gender deviance.

Masculine branding of environmental products and behaviors, along the lines of Pepsi Max (the first diet cola for men) or "broga," or the macho and highly successful "Don't Mess with Texas" anti-littering campaign (Stafford & Hartman, 2012), has been proposed to make environmentalism more appealing to men. However, making green more macho is insufficient for addressing the environmental risks that underpin our political economy as a whole. As long as androcentric dominance hierarchies incentivize gender dichotomization and hyper-masculinity, green cannot be adequately coded as masculine. Truly adequate solutions – opposing coal and tar sands, restricting commerce that produces greenhouse gasses and other toxic emissions, and supporting massive government expenditures on green infrastructure – will require leveling gender power hierarchies and reconstructing masculinity. We need to raise consciousness about the virulence of male power, the gender binary, and toxic masculinity. We need to raise children to believe that men are not entitled to privilege, and to valorize cooperation and caring for each other and the earth.

References

Anshelm, J., & Hultman, M. (2014). A green fatwā? Climate change as a threat to the masculinity of industrial modernity. *NORMA: International Journal for Masculinity Studies*, *9*(2), 84–96.

Arnocky, S., & Stroink, M. (2011). Gender differences in environmentalism: The mediating role of emotional empathy. *Current Research in Social Psychology*, *16*, 1–14.

Bell, S., & Braun, Y. (2010). Coal, identity, and the gendering of environmental justice activism in central Appalachia. *Gender and Society*, *24*(6), 794–813. https://doi.org/10.1177/0891243210387277

Bell, S., & York, R. (2010). Community economic identity: The coal industry and ideology construction in West Virginia. *Rural Sociology*, *75*(1), 111–143.

Bennett, G., & Williams, F. (2011). *Mainstream Green: Moving Sustainability from Niche to Normal.* Retrieved from www.goodlifer.com/2011/04/mainstream-green-moving-sustainability-from-niche-to-normal/

Bosson, J. K., & Michniewicz, K. S. (2013). Gender dichotomization at the level of ingroup identity: What it is, and why men use it more than women. *Journal of Personality and Social Psychology*, *105*(3), 425–442. https://doi.org/10.1037/a0033126

Bosson, J. K., Vandello, J. A., Burnaford, R. M., Weaver, J. R., & Wasti, S. A. (2009). Precarious manhood and displays of physical aggression. *Personality and Social Psychology Bulletin*, *35*(5), 623–634. https://doi.org/10.1177/0146167208331161

Brandth, B., & Haugen, M. (2006). Changing Masculinity in a Changing Rural Industry: Representations in the Forestry Press. In H. Campbell, M. M. Bell, & M. Finney (eds.), *Country Boys: Masculinity and Rural LIfe* (pp. 217–234). University Park, PA: Pennsylvania State University Press.

Bridges, T., & Pascoe, C. (2018). On the elasticity of gender hegemony: Why hybrid masculinities fail to undermine gender and sexual inequality. In J. W. Messerschmidt, M. A. Messner, R. Connell and P. Yancey Martin (eds.), *Gender Reckonings* (pp. 254–274). New York: New York University Press.

Brough, A. R., Wilkie, J. E. B., Ma, J., Isaac, M. S., & Gal, D. (2016). Is eco-friendly unmanly? The green-feminine stereotype and its effect on sustainable consumption. *Journal of Consumer Research*, *43*(4), 567–582. https://doi.org/10.1093/jcr/ucw044

Brueckner, M. (2007). The Western Australian Regional Forest Agreement: Economic rationalism and the normalisation of political closure. *Australian Journal of Public Administration*, *66*(2), 148–158. https://doi.org/10.1111/j.1467–8500.2007.00513.x

Butler, J. (1993). *Bodies that Matter: On the Discursive Limits of "Sex."* New York: Psychology Press.

Campbell, H., & Bell, M. (2000). The question of rural masculinities. *Rural Sociology, 65*(4), 532–546.

Caniglia, B. S., Brulle, R. J., & Szasz, A. (2015). Civil society, social movements, and climate change. In R. E. Dunlap and R. J. Brulle (eds.), *Climate Change and Society: Sociological Perspectives*. New York: Oxford University Press, pp. 235–268.

Cohn, C. (1987). Sex and death in the rational world of defense intellectuals. *Signs, 12*(4), 687–718.

Collins, P. (2005). *Black Sexual Politics: African Americans, Gender, and the New Racism*. New York: Routledge.

Connell, R. (2005). *Masculinities*. Los Angeles, CA: University of California Press.

Connell, R. (2016). Afterword. In E. Enarson & B. Pease (eds.), *Men, Masculinities and Disaster* (p. 234). New York: Routledge.

Connell, R., & Messerschmidt, J. (2005). Hegemonic masculinity: Rethinking the concept. *Gender and Society, 19*(6), 829–859. https://doi.org/10.1177/0891243205278639

Connell, R., & Pearse, R. (2015). *Gender in World Perspective*. Malden, MA: Polity Press.

Dietz, T., Kalof, L., & Stern, P. (2002). Gender, values, and environmentalism. *Social Science Quarterly, 83*(1), 353–364.

Dubbert, J. (1979). *A Man's Place: Masculinity in Transition*. Englewood Cliffs, NJ: Prentice-Hall.

Fenstermaker, S., & West, C. (2002). Introduction. In *Doing Gender, Doing Difference: Inequality, Power and Institutional Change*. New York: Routledge, pp. xiii–xviii.

Ferree, M. (2018). Theories don't grow on trees: Contextualizing gender knowledge. In J. Messerschmidt, P. Martin, M. Messner, & R. Connell (eds.), *Gender Reckonings*. New York: New York University Press, pp. 13–34.

Filteau, M. (2016). "If you talk badly about drilling, you're a pariah": Challenging a capitalist patriarchy in Pennsylvania's Marcellus Shale region. *Rural Sociology, 81*(4), 519–544. https://doi.org/10.1111/ruso.12107

Finucane, M., Slovic, P., Mertz, C., Flynn, J., & Satterfield, T. (2000). Gender, race, and perceived risk: The "white male" effect. *Health, Risk, and Society, 2*(2). https://doi.org/10.1080/713670162

Fleming, J. (2017). Excuse us, while we fix the sky: WEIRD supermen and climate engineering. In Sherilyn MacGregor & N. Seymour (eds.), *Men and Nature: Hegemonic Masculinities and Environmental Change* (Vol. 4, pp. 23–28). RCC Perspectives: Transformations in Environment and Society. https://doi.org/doi.org/10.5282/rcc/7979

Flynn, J., Slovic, P., & Mertz, C. (1994). Gender, race, and perception of environmental health risks. *Risk Analysis, 14*(6), 1101–1108.

Gardiner, J. K. (2000). Masculinity, the teening of America, and empathic targeting. *Signs, 25*(4), 1257–1261.

Hultman, M. (2013). The making of an environmental hero: A history of ecomodern masculinity, fuel cells and Arnold Schwarzenegger. *Environmental Humanities, 2*, 79–99.

Hultman, M. (2017). Exploring industrial, ecomodern, and ecological masculinities. In Sherilyn MacGregor (ed.), *Routledge Handbook of Gender and Environment* (pp. 261–274). London: Routledge. https://doi.org/10.4324/9781315886572-28

Johnson, A. (2017). Every day like today: Learning how to be a man in love. In Sherilyn MacGregor & N. Seymour (eds.), *Men and Nature: Hegemonic Masculinities and Environmental Change* (pp. 45–50). RCC Perspectives: Transformations in Environment and Society.

Kahan, D., Braman, D., Gastil, J., Slovic, P., & Mertz, C. (2007). Culture and identity-protective cognition: Explaining the white male effect in risk perception. *Journal of Empirical Law Studies*, *4*(3), 465–505. https://doi.org/10.1111/j.1740-1461.2007.00097.x

Kahan, D., Jenkins-Smith, H., Tarantola, T., Silva, C., & Braman, D. (2015). Geoengineering and climate change polarization: Testing a two-channel model of science communication. *Annals of the American Academy of Political and Social Science*, *658*(1), 192–222. https://doi.org/10.1177/0002716214559002

Kahan, D., Peters, E., Wittlin, M. et al. (2012). The polarizing impact of science literacy and numeracy on perceived climate change risks. *Nature Climate Change*, *2*(10), 732–735. https://doi.org/10.1038/nclimate1547

Kahneman, D. (2011). *Thinking, Fast and Slow*. New York: Farrar, Straus and Giroux.

Kennedy, E. H., & Dzialo, L. (2015). Locating gender in environmental sociology. *Sociology Compass*, *9*(10), 920–929.

Kennedy, E. H., & Kmec, J. (2018). Reinterpreting the gender gap in household pro-environmental behaviour. *Environmental Sociology*, 1–12. https://doi.org/10.1080/23251042.2018.1436891

Kimmel, M. (1987). The contemporary "crisis" of masculinity in historical perspective. In H. Brod (ed.), *The Making of Masculinities: The New Men's Studies*. London: Allen and Unwin.

Kosakowska-berezecka, N., Besta, T., & Vandello, J. (2016). If my masculinity is threatened I won't support gender equality? The role of agentic self-stereotyping in restoration of manhood and perception of gender relations. *Psychology of Men & Masculinity*, *17*(3), 274–284.

Kronsell, A. (2013). Gender and transition in climate governance. *Environmental Innovation and Societal Transitions*, *7*, 1–15. https://doi.org/10.1016/j.eist.2012.12.003

Maass, A., Cadinu, M., Guarnieri, G., & Grasselli, A. (2003). Sexual harassment under social identity threat: The computer harassment paradigm. *Interpersonal Relations and Group Processes*, *85*(5), 853–870. https://doi.org/10.1037/0022-3514.85.5.853

MacGregor, S. (2017). Gender and environment: an introduction. In S MacGregor (ed.), *Routledge Handbook of Gender and Environment* (pp. 1–24). New York: Routledge.

Maniates, M. F. (2001). Individualization: Plant a tree, buy a bike, save the world? *Global Environmental Politics*, *1*(3), 31–52.

McCright, A. M. (2010). The effects of gender on climate change knowledge and concern in the American public. *Population and Environment*, *32*(1), 66–87. https://doi.org/10.1007/s11111-010-0113-1

McCright, A. M., & Xiao, C. (2014). Gender and environmental concern: Insights from recent work and for future research. *Society and Natural Resources*, *27*(10), 1109–1113. https://doi.org/10.1080/08941920.2014.918235

Messerschmidt, J., & Messner, M. (2018). Hegemonic, nonhegemonic, and "new" masculinities. In J. Messerschmidt, P. Martin, M. Messner, & R. Connell (eds.), *Gender Reckonings*. New York: New York University Press, pp. 35–56.

Milfont, T. L., & Sibley, C. G. (2016). Empathic and social dominance orientations help explain gender differences in environmentalism: A one-year Bayesian mediation analysis. *Personality and Individual Differences*, *90*, 85–88. https://doi.org/10.1016/j.paid.2015.10.044

NAACP. (2012). *Coal Blooded Putting Profits Before People*. Retrieved from www.naacp.org/wp-content/uploads/2016/04/CoalBlooded.pdf

Nelson, J. A. (1995). Feminism and economics. *Journal of Economic Perspectives*, *9*(2), 131–148.

Nickleberry, L., & Coleman, M. (2012). Exploring African American masculinities: An integrative model. *Sociology Compass*, *6*(11), 897–907.

Ostrom, E., Burger, J., Field, C. B., Norgaard, R. B., & Policansky, D. (1999). Revisiting the commons: Local lessons, global challenges. *Science*, *284*(5412), 278–282. https://doi.org/10.1126/science.284.5412.278

Paap, K. (2006). *Working Construction*. Ithaca: Cornell University Press.

Peter, G., Bell, M., Jarnagin, S., & Bauer, D. (2000). Coming back across the fence: Masculinity and the transition to sustainable agriculture. *Rural Sociology*, *65*, 215–233.

Pierrehumbert, R. T. (2015). Climate Hacking Is Barking Mad. *Slate*. Retrieved from www.slate.com/articles/health_and_science/science/2015/02/nrc_geoengineering_report_climate_hacking_is_dangerous_and_barking_mad.single.html

Plumwood, V. (1993). *Feminism and the Mastery of Nature*. New York: Routledge.

Rome, A. (2003). Give Earth a chance: The environmental movement and the sixties. *Journal of American History*, *90*(2), 525–554.

Rome, A. (2006). Political hermaphrodites: Gender and environmental reform in progressive America. *Environmental History*, *11*, 440–463.

Salleh, A. (ed.). (2009). *Eco-Sufficiency & Global Justice. Women Write Political Ecology*. New York: Pluto.

Sandilands, C. (1993). On 'green consumerism': Environmental privatization and 'family values.' *Canadian Women's Studies/ Les Cahiers de La Femme*, *13*(3), 45–47.

Schrock, D., & Schwalbe, M. (2009). Men, masculinity, and manhood acts. *Annual Review of Sociology*, *35*, 277–295. https://doi.org/10.1146/annurev-soc-070308-1

Sinclair, A. (1995). Sex and the MBA. *Organization*, *2*, 295–317.

Stafford, E. R., & Hartman, C. (2012). Making green more macho. *The Solutions Journal (Rocky Mountain Institute)*, *3*(4), 25–29.

Swim, J. K., Vescio, T. K., Dahl, J. L., & Zawadzki, S. J. (2018). Gendered discourse about climate change policies. *Global Environmental Change*, *48*(January), 216–225. https://doi.org/10.1016/j.gloenvcha.2017.12.005

Thaler, R. H., & Sunstein, C. R. (2009). *Nudge: Improving Decisions About Health, Wealth, and Happiness*. London: Penguin Books.

Trauger, A. (2004). 'Because they can do the work': Women farmers in sustainable agriculture in Pennsylvania, USA. *Gender, Place and Culture*, *11*(2), 289–307. https://doi.org/10.1080/0966369042000218491

Warren, K. J. (2000). *Ecofeminist philosophy: A Western Perspective on What It Is and Why It Matters*. Lanham, MD: Rowman and Littlefield.

Weber, R., & Dawes, R. (2005). Behavioral Economics. In N. J. Smelser & R. Swedberg (eds.), *The Handbook of Economic Sociology* (Second Edition). New York: Princeton University Press.

Willer, R., Rogalin, C. L., Conlon, B., & Wojnowicz, M. T. (2013). Overdoing Gender: A test of the masculine overcompensation thesis. *American Journal of Sociology*, *118*(4), 980–1022. https://doi.org/10.1086/668417

Wolfers, J. (2015, January 23). How economists came to dominate the conversation. *New York Times*, www.nytimes.com/2015/01/24/upshot/how-economists-came-to-dominate-the-conversation.html.

Xiao, C., & McCright, A. M. (2014). A Test of the biographical availability argument for gender differences in environmental behaviors. *Environment and Behavior*, *46*(2), 241–263. https://doi.org/10.1177/0013916512453991

Yates, A., Luo, Y., Mobley, C., & Shealy, E. (2015). Changes in public and private environmentally responsible behaviors by gender: Findings from the 1994 and 2010 general social survey. *Sociological Inquiry, 85*(4), 503–531. https://doi.org/10.1111/soin.12089

Zelezny, L. C., Chua, P.-P., & Aldrich, C. (2000). Elaborating on gender differences in environmentalism. *Journal of Social Issues, 56*(3), 443–457. https://doi.org/10.1111/0022-4537.00177

7 Toxicity, Health, and Environment

Jennifer S. Carrera and Phil Brown

Introduction

Many diseases are caused by toxic substances in our varied environments – the places we live, work, go to school, and play. Factory workers become sick from chemicals and air particles in their work settings; farmworkers become sterile from pesticides they apply to crops; women suffer miscarriages from toxic wastes long-buried in their neighborhood; melanoma increases as ozone-depleting gases (that include toxic pesticides like methyl bromide) reduce the ozone layer and allow in ultraviolet rays; reproductive abnormalities occur from exposure to phthalates in plastics and other sources; and various cancers are linked to PFOA, the chemical used to make Teflon. These are a few of the endless examples of environmentally induced illnesses that are the subject of growing public awareness and activism. Our understanding of the relationship between our health and our environments throughout the modern era has progressed slowly while the number of new contaminants to which we are exposed rapidly grows.

We can take insight from John Snow's classic study, which led to the discovery of Vibrio cholerae (cholera) and established the field of epidemiology, as an example of an observational study that demonstrates linkages between disease, social structures, and social conditions. While prior to 1853, it was generally held that illness was caused by bad odors, "miasmas," Snow hypothesized a connection with the local drinking water. He conducted a systematic health survey in the area of highest illness and measured proximity to nearby wells. He found a strong correspondence between a specific well, the Broad Street pump, and mortality, where people living further from the well became ill only when they made a point of using that well. Mortality rates in areas further from the well were equal to pre-epidemic rates. Among those who could have drunk contaminated water but did not show high rates of cholera, Snow tracked down the sources of water intake and found that workhouse residents had their own water supply and that brewery workers were often drinking beer instead of water (Goldstein and Goldstein 1986). His work showed a clear connection between water source and cholera incidence and spurred a fundamental shift in the understanding of cholera, specifically, and disease, broadly.

Snow did not confront an intentional pollution episode but located a seemingly natural environmental situation that caused disease, and he made an historic analysis of the cause. Most notably, Snow had a very modern public health approach to the problem, as seen in his recommendations: "The communicability of cholera ought not to be disguised from the people, under the idea that the knowledge of it would cause a panic, or occasion the sick to be deserted" (Snow 1853). His public health and sociological prescriptions were equally important: sanitation, cleanliness, better living conditions, and even work breaks for miners to go home for lunch (Goldstein and Goldstein 1986). Snow's approach, often called "barefoot epidemiology," is quite similar to Brown's (1992) notion of "popular epidemiology," which explains what affected residents do in what Edelstein (1988) terms "contaminated communities."

In Snow's founding of epidemiology, we realize the social embeddedness of disease and the environment; we learn the importance of group, rather than individual, clinical data; we learn the need to use both survey methods and intense firsthand observation to attain adequate information; and we learn about corporate, medical, and governmental inability and reluctance to face environmentally induced disease. Environmental sociologists who study environmental health are examining, and at times *are,* modern-day John Snows – committed professionals and laypeople who have shown astounding capacity to identify, research, and act on environmental factors in health.

For this discussion, we focus on the health effects caused by toxic substances in people's immediate or proximate surroundings: chemical-related, air pollution-related, and radiation-related diseases and symptoms that affect groups of people in neighborhoods and communities. There are many good reasons for focusing on toxic substances. It is an area that has engendered an enormous amount of conflict, policymaking, legislation, public awareness, media attention, and social movement activity. It puts into sharp relief a variety of disputes between laypeople and professionals, citizens and governments, and among professionals, because environmentally induced diseases (EIDs) are among the most prominent types of "contested illnesses" (Brown 2007). Precisely because environmental diseases are so centered in daily life and all aspects of the economy, these diseases have become highly politicized and have spurred a significant social movement.

Legacy and Scope of the Anti-Toxics Movement

The publication of Rachel Carson's *Silent Spring* created a watershed moment for mass public attention to environmental health effects of toxics. Carson (1962) synthesized an enormous amount of observational and experimental data to demonstrate how pesticides – especially DDT, which was ubiquitous at that time – were serious hazards, causing morbidity and mortality in animals and humans. Like many other pioneers in public health, Carson was sharply criticized by many for being unscientific and for attacking major economic sectors. Carson's impact led to significant regulation of pesticides and other chemicals, a new wave of the environmental movement, and eventually to the passage of the National Environmental Protection Act and the establishment of the Environmental Protection Agency

(EPA). Most importantly, Carson made the first link to breast cancer and the role of endocrine-disrupting compounds (EDCs), which would later be shown as central to so many diseases and conditions.

In the context of widespread attention to and fear around exposure to toxics, the discovery of hazardous waste under a school in Niagara Falls, New York in 1978 became a key event around which the budding anti-toxics movement coalesced. This was the first time that human health was raised as a central concern in an environmental crisis and the outcome of Love Canal prompted the creation of the Superfund Program by the EPA. During the Love Canal crisis, residents were warned by state health officials that toxic chemicals permeated the Love Canal neighborhood. This knowledge stemmed from the International Joint Commission on the Great Lakes, which traced the pesticide Mirex to the dumpsite that had been sold for one dollar by Hooker Chemical Company to the city in order to build a school. The revelation meshed with residents' awareness of having seen noxious substances oozing from the site and experiencing unusual health effects. As residents organized to learn more, they discovered high rates of miscarriages, birth defects, cancer, and chromosome damage (Levine 1982). Rather than rely on writing and education, activists took direct action by community organizing, demonstrating, organizing health studies, and demanding action by state and federal governments. That approach to toxic crisis stimulated a new anti-toxic waste movement (Brown and Masterson-Allen 1994) and eventually led to thousands of communities around the country taking similar action when faced with toxic contamination.

A number of key episodes furthered public awareness and government response, including the dioxin explosion in Seveso, Italy; the Union Carbide explosion of methyl isocyanate (MIC) in Bhopal, India; Michigan feed contamination with polybrominated biphenyls (PBBs); and the leukemia cluster at Woburn, Massachusetts. In some instances, there was a major contamination episode with immediate health effects, such as widespread animal (e.g. bird and horse) deaths due to the spreading of dioxin-laced fuel oil to contain dust in Times Beach, Missouri. Elsewhere, a contamination episode might be without immediate health effects, but such effects were expected to follow, as at Three Mile Island where radioactive gases were released in a nuclear reactor meltdown. In other cases, people discovered prior contamination episodes, such as both accidental and deliberate (to examine exposure pathways) releases of radioactivity from the Hanford nuclear facility in Washington State.

Much of our knowledge about environmental health effects has come from actually or potentially affected people. Lay participation in research began in earnest in the early 1980s with the Woburn leukemia cluster, where a major health survey was conducted by an alliance of Harvard School of Public Health biostatisticians and Woburn residents. This study was a massive source of information on adverse pregnancy outcomes and childhood disorders from 5,010 interviews, covering 57 percent of Woburn residences which owned telephones. The researchers trained 235 volunteers to conduct the survey, taking precautions to avoid bias. Other data included information on twenty cases of childhood leukemia (ages nineteen and under) that were diagnosed between 1964 and 1983 and the state environmental agency's water model of the two contaminated wells.

Childhood leukemia was found to be significantly associated with exposure to water from two hydrologically low-lying wells. Children with leukemia received an average of 21.2 percent of their yearly water supply from the wells, compared to 9.5 percent for children without leukemia. Controlling for risk factors in pregnancy, the investigators found that access to contaminated water was associated with perinatal deaths since 1970, eye/ear anomalies, and CNS/chromosomal/oral cleft anomalies. With regard to childhood disorders, water exposure was associated with kidney/urinary tract and lung/respiratory diseases (Lagakos et al. 1986). Due to lack of resources, this study would not have been possible without community involvement. Yet precisely this lay involvement led professional and governmental groups – the Department of Public Health, the Centers for Disease Control, the American Cancer Society, and the EPA – to charge that the study was biased (Brown and Mikkelsen 1990). Such charges erroneously assumed that laypeople must have biases that will interfere with research and that scientists automatically have no biases.

Environmental justice activists have also raised concerns about toxics, where the environmental justice movement has placed special emphasis on the targeting of politically vulnerable communities. These communities tend to have high proportions of residents of color and low-income. Blacks, Latinos, and Native Americans discovered more frequent placement of landfills, incinerators, and toxic waste sites in neighborhoods with majority populations of color or low-income people, when compared with predominately white communities. Superfund cleanups take longer to arrange and are less thorough in communities that are predominately of color. There is also disproportionate exposure to lead poisoning, which became a national scandal with the Flint crisis in 2014. Just as the anti-toxics movement pushed the broader environmental movement to take on the issues of human health, the environmental justice movement pushed both the anti-toxics movement and the broader environmental movement to link environmental health with race and class stratification (Brown 1995; Bullard 1993; Hofrichter 1993; Mohai and Bryant 1992).

Finally, the dramatic growth in women's involvement as leaders and members of anti-toxic waste activist groups has heightened the importance of gender issues in environmental health (Blocker and Eckberg 1989; Brown and Ferguson 1995; Cable 1992; Krauss 1993; Perkins 2012). These activists challenged the dominant perspectives of environmental agencies and researchers who downplayed women's embodied concerns, and in doing so advanced relational approaches to contamination and the importance of subjectivity (Brown et al. 2004). Breast cancer activists emphasized environmental causes of cancer while showing medical and governmental slowness in investigating that link. This led to increasing attention to the gendered dimensions of environmental health. Beginning in the 1990s the environmental breast cancer movement (EBCM) challenged the mainstream breast cancer movement's biomedical focus on treatment rather than prevention and overemphasis on genetic factors and individual lifestyle factors such as diet, exercise, and alcohol consumption rather than pushing for structural change. EBCM activists also pushed for the inclusion of lay perspectives in the design and peer review of scientific research. Because of their focus, the EBCM allied with anti-toxic activists and environmental justice activists more than with mainstream breast cancer organizations (Zavestoski et al. 2004).

Environmental reproductive justice emphasizes the need for research into toxic factors involved in reproductive health, and efforts to remove them through consumer activism and government regulation. Scholars and activists in this area are centrally concerned with environmental justice, focusing on women of color who face much greater exposures (Hoover 2018). Such work advances intersectional analyses of environmental health problems through considering how multiple social locations intersect multiplicatively to generate unique experiences as synergies among linked social categories, not otherwise reducible to the simple cumulation of those effects as broadly shared experiences within categories (Crenshaw 1991).

Women have led in organizing around environmental reproductive justice. This growing movement stems in part from the larger reproductive justice movement that focuses on women's freedom to choose in all aspects of reproductive life, but to this it adds consideration of environmental causation of reproductive harms (Gurr 2011; Hoover et al. 2012; Zavella 2016). This movement deals with threats to reproductive health from many endocrine disrupting compounds that affect fertility, pregnancy, fetal health, and the birth process, as well as affecting the growth of the child after birth. State and federal attacks on reproductive health coverage, combined with general cuts in health care, contribute to decreased health care access for the most vulnerable people.

In all of the given examples, the role of laypeople, through their incredible self-education, self-organization, and scientific success in identifying environmentally induced diseases and their likely causes, is one of the most significant phenomena in shaping the field of environmental health and reshaping environmental activism more broadly. Along the way, sociologists have worked to study the formation and organization of anti-toxics mobilization, partnered with community members and other scientists to conduct formal scientific studies, and supported residents in their community-driven efforts to collect data outside of established scientific channels.

Types of Environmental Health Effects

Physical Health Effects

Health effects of toxic substances are an increasingly common part of our landscape. Colborne et al. (1996), formulating the originally controversial but now broadly accepted "endocrine disrupter hypothesis," amassed a wealth of evidence that many chemically based products, ubiquitous in our daily surroundings, act as endocrine disruptors, affecting female and male reproductive development and capacities, including miscarriages and lowered sperm count. The "hormone havoc" of these chemicals also causes thyroid deficiency and neurological deficits, such as weaker reflexes and lowered IQ scores. Pervasive polychlorinated biphenyls (PCBs) and dioxins are of particular concern.

Toxic substances have been implicated in diseases of all organ systems. Although some toxics have "signature" diseases (e.g. asbestosis and mesothelioma), most do not. Hence, neither sufferers nor clinicians are likely to attribute individual cases to a toxic exposure. People usually notice environmentally induced diseases when they observe clusters, especially if there is a known source (e.g. abandoned toxic waste sites, operating incinerators and deep injection toxic disposal wells, nearby chemical and other factories). At other times, people learn about excess cancer rates from annual cancer registry reports. Sometimes people notice health effects on animals and become concerned that humans too will be affected. The die-off of a farm family's entire cattle herd in the Mid-Ohio Valley due to contamination from their leasing to DuPont Chemical, supposedly a non-toxic dumpsite, is one such case. The Tennant family sued the DuPont and eventually won a major class action lawsuit. Support for the case included a 69,000-person epidemiological study that linked PFOA to six diseases and conditions and raised national attention on the entire class of per- and polyfluorinated compounds (PFAS), making it one of the most prominent contaminants today (Frisbee et al. 2009).

Biomarkers, which are physical changes that may be tied to disease or indicative of toxics exposure, are the subject of intense debate among health experts. Examples of biomarkers connected to exposure events include chromosome damage among residents of Love Canal and changes in eye blink reflexes and trigeminal nerve damage attributable to trichloroethylene exposure among Woburn residents (Brown and Mikkelsen 1990). Advocates argue that such markers are important evidence that warrants immediate health care, site remediation, and regulatory action while typically adversarial parties (e.g. regulators, municipalities, industry) argue that biomarkers are not observations of actual health harm that affect daily life, so no action is needed. The case of lead illustrates this dynamic well. Actionable lead exposure is now defined at blood levels which do not necessary show up as clinical morbidity (CDC 2018), but are likely to become clinically significant based on extensive experience. Elevated blood lead levels do not ensure symptoms of lead toxicity but there is considered to be no safe level of exposure to lead, especially for children (WHO 2010). Children in Flint who have expressed fears that their potential has been destroyed by lead exposure add nuance to this debate.

Mental Health Effects

Researchers have paid much attention to mental health effects because toxic contamination episodes leave a residue of fear and anxiety. Depression is often found among residents of toxic neighborhoods. Depression among the Woburn families was so deep that after getting reports prepared by a psychiatrist examining the families on behalf of the companies' defense, the attorneys chose not to call him as an expert witness. Only an order from the judge enabled the plaintiffs to obtain copies of his notes, which referred to widespread depression (Brown and Mikkelsen 1990). At Love Canal, too, depression was common. Many residents threatened suicide and several attempted it (Gibbs 1982). Three Mile Island residents also experienced high levels of depression. During the year after the accident, mothers experienced new episodes of affective disorder at a rate three times higher than a non-exposed control group in another location (Bromet et al. 1982)

Psychological effects are amplified by lack of broad social validation and support. Unlike with natural disasters which are typically sudden and highly visual, people living in contaminated communities have to fight to gain public recognition of pollution, its impacts, and its human causes. Litigation, so common to toxic sites, also contributes to ongoing depression by forcing parents to relive the horrors of a child's illness through a seemingly endless process of depositions, testimony, and numerous consultations with independent experts for both defendants and litigants. This points to one of the most disturbing aspects of community action on toxics: remedies and justice are available only at the cost of intense reliving and public display of the suffering. People are robbed of privacy for their sorrow and are forced to grieve in public. People often lose perspective on what are their private feelings and what is the public record. They are compelled to justify their suffering by legal proof, lest it be diminished and discredited.

Hypervigilance is common in toxic waste sites, as residents become preoccupied with searching for causes of and responsibility for contamination and with protecting their future health (Vyner 1988). Post-traumatic stress disorder (PTSD) has been found in many toxic waste sites. People living in these communities exhibit most characteristics of what Lifton (1968) calls the "general constellation of the survivor," a phenomenon he studied extensively in relation to atomic bomb survivors. They retain a "death imprint" of images and memories of the disaster and feel a strong "death anxiety." Many family members experienced "death guilt," based on feeling that they could have done something to prevent a loved one's death. "Psychic numbing" was also common, with its diminished capacity for feelings, particularly involving apathy, withdrawal, depression, and overall constriction in living. People also suffer "impaired human relationships," sometimes manifested as an inability to comfort one another, even when family members recognized the need for mutual support (Lifton and Olson 1976).

The generalized distress that people who live near toxic waste sites experience can be understood as "demoralization," a worried feeling indicating a considerable degree of being upset and out of control (Dohrenwend 1981). People in contaminated communities often start to view the environment as a malevolent force, and even the home, typically a place of refuge and safety, becomes threatening. Residents lose faith that the world is just and may even think that they deserve what happened to them (Edelstein 1988). Ultimately, as Brown and Mikkelsen (1990) found, people come to believe that there is "no safe place," since contamination is so widespread. Such attitudes even occur in the midst of highly organized communities that are fighting corporations and government.

Community-Wide Effects

Rather than only affecting individuals, contamination episodes harm entire communities, disrupting ordinary social relations as well as damaging health (Erikson

1976). Sociologists have added importantly to this area through in-depth case study analyses. Adeline Levine's (1982) study of Love Canal offered the first sociological case study that explained the conflict between lay discoverers of disease on the one hand, and government and corporate parties on the other. Michael Edelstein (1988) introduced the term "contaminated communities" as a standard for research on toxic-affected communities. Couch and Kroll-Smith's (1990) study of an underground mine fire in Centralia, PA emphasized how internal divisions concerning health effects and corporate/governmental responsibility pitted groups of residents against each other, further undermining the binds of community. This problem of "corrosive communities" (Freudenburg and Jones 1991) leads to fundamental rifts in the social bonds of community, including rancorous disputes between individuals and between groups.

Communities facing contamination episodes are often in the midst of "chronic technological disasters" (Couch and Kroll-Smith 1985), long-term, unfolding problems that typically had no visible starting point, but which corrode the social fabric. Vyner (1988) pursued this idea with the concept of "invisible environmental contaminants," in his case mostly concerning ionizing radiation. Because they were unseen, Vyner termed them "occult," which were mysterious factors that people had a hard time seeing or identifying.

Social Responsibility of Government for Protecting Public Health

Residents in contaminated communities often expect local, state, and federal government support for their claims, at least in earlier years. The failure of many government bodies to support victims' claims and to properly enforce regulations has led to increased mistrust of government (Szasz 1994). Residents in many contaminated communities have become politicized in response to governmental failure to respond to citizen complaints concerning corporate polluters (Krauss 1993).

The ongoing Flint Water Crisis, which began in 2014 with the switch from the Detroit water system to the Flint River, offers a recent case in which community members all across Flint, Michigan expressed their outrage at the local, state, and federal government for failing to prioritize the community's health needs above financial austerity goals. Neoliberal tendencies of privatizing public services are called out by many as the underlying driver of the water crisis (Fasenfest and Pride 2016), while some scholars point specifically to austerity urbanism within neoliberalism as the core of the community's devastation (Peck 2012). Still, many argue, and we would agree, that such policies to strip social services and protections from certain populations (specifically communities of color) are only possible under a racialized capitalist system that determines which populations are appropriate for abandonment (Pulido 2016).

Government agencies have been reluctant to affirm most relationships between contaminants and disease. One reason is the fear that business will be harmed.

Corporations, especially producers, have fought against recognition of environmental health effects because of the financial cost involved in settlements, fines, production restructuring, and alternative forms of disposal. Some major retailers, though, have been responsive to consumer demands and pressed suppliers to provide them with products free of certain chemicals that activists have targeted, e.g. flame retardants, per- and polyfluorinated compounds, parabens, and phthalates.

Residents in Love Canal (Levine 1982), Woburn (Brown and Mikkelsen 1990), and many other sites have emphasized how environmentally induced disease sufferers and their allies target the corporations they hold responsible for contamination episodes, as well as various levels of government for failing to do their job in recognition and remediation. Sociologist McGoey (2012) argues that in some cases corporations and regulators work strategically to know as little as possible about potential health harms to avoid accountability. Corporate unwillingness to take responsibility, combined with government inaction, delays the appropriate understanding of environmental health effects, their causes, and their treatment.

Lack of government attention to environmental causes of disease is self-reinforcing, with fear over being criticized for regulatory shortcomings leading to failures to regulate chemicals and other toxic substances that are a source of public anger. Much of the data collected about toxicity occurs through regulatory programs, e.g. compliance with the Clean Air Act, Clean Water Act, and Safe Drinking Water Act. But regulatory agencies often lack the resources or a clear mandate to review the data and there is a long history of lack of enforcement. While environmental law includes provisions to empower citizens–such as publication of hazards data through "right-to-know" provisions of the Superfund Reauthorization Act of 1986 Title III, known as the Emergency Planning and Community Right-to-Know Act (EPCRA) (Dillon et al. 2017) – the data itself is far from perfect and the average person has little awareness of their rights to access the data. Even if they are aware, they are unlikely to have resources to analyze it and even fewer resources for doing their own testing to independently assess or fill any gaps in the existing data. It is significant that the burden of proof is on the contaminated communities, rather than the parties who contaminated it or the public health and regulatory agencies.

These problems are even more pronounced in the Trump administration with recently resigned EPA Administrator Scott Pruitt leading an agency he previously said he wanted to destroy. The United States has withdrawn from the Paris climate accord, cut back on enforcement, approved the pesticide chlorpyrifos that EPA scientists had recommended against, reversed Obama's halting of the Dakota Access Pipeline, attacked the Clean Power Plan, opened up oil drilling on coastlines and delicate Arctic land, dismissed the Science Advisory Board, prohibited EPA-funded scientists from being on any advisory boards, removed climate change material from websites, and opposed alternate energy programs (Bomberg 2017; Dillon et al. 2017).

Since environmental health effects are so widespread, it might seem surprising that we do not have more research, medical education, government effort, and social recognition. But given how environmental health effects are tied to fundamental societal structure, it is not surprising that we have seen so much resistance to dealing with environmental health. The next section examines difficulties and disputes in dealing with environmental health; these include methodological issues, disputes over whose voice counts, data availability, and reluctance of engagement by the medical community.

Barriers in the Production of Environmental Health Knowledge

Standards of Proof

Neutra (1990) points out that environmental epidemiology studies usually have such small numbers of exposed persons that observed rates must often be twenty times that of expected rates in order to attain statistical significance. This presents many problems for people trying to determine if there is a disease cluster, and that difficulty is even prior to the difficulty of linking the cluster to a purported cause (such as a local polluter). Some scientists argue that these problems make it unlikely to determine specific causes of environmental disease.

Others, including the National Research Council's (NRC) Committee on Environmental Epidemiology, argue that traditional approaches are not necessarily appropriate. The NRC argues that the same flaws in data can be understood in the opposite fashion: we cannot reject claims of environmental causation precisely because we do not have the right data. Further, small effect sizes should not be viewed as obstacles – even low relative risks are very powerful if large numbers of people are affected. Along the same line, consistency across many studies should be acceptable as evidence of etiological linkages, even if not all of those studies meet statistical significance (National Research Council 1991).

Although it is not always easy to evaluate the effects of removal of a hazard, researchers have shown such findings: no new cases of leukemia developed in Woburn once contaminated wells were shut, and birth weights returned to normal in Love Canal and the Lipari (NJ) landfill after exposure declined (National Research Council 1991). Ozonoff and Boden (1987) distinguish statistical significance from public health significance, since an increased disease rate may be of great public health significance even if statistical probabilities are not met. They believe that epidemiology should mirror clinical medicine more than laboratory science, by erring on the safe side of false positives.

A sociological perspective on health and environment research pushes us to reconceptualize our level of analysis and to take on a more community-oriented approach. Traditional epidemiology likes to differentiate itself from clinical medicine in that it takes into account *populations*. Hence it can see patterns, and even causes, that would be invisible to health professionals who look at individual cases. Such a population-based approach is clearly necessary but is only partly sufficient. In fact, much of what epidemiology does is to examine *populations* while using *individual* level data, for instance, studying personal risk factors and heart disease in a state. In some cases, this approach is insufficient. For example, if you want to study race or class difference in toxic hazards, you could easily be stymied by small numbers. One alternative is to selectively study the toxic exposures of large numbers of minority populations, to characterize their risks, and to compare this with known populations that are predominately white.

There are often problems in studying environmental health, such as inadequate history of the site, lack of clarity regarding the route of contaminants, determining appropriate sampling locations, small numbers of cases, bias in self-reporting of symptoms, getting appropriate control groups, people's movement in and out of exposed areas, lack of knowledge about characteristics and effects of certain chemicals, and unknown or varying latency periods for carcinogens and other disease-causing agents. Given the many forms of cancer reported for all municipalities or counties in a state (e.g. on annual state cancer registry report), it is likely that some elevations in cancer prevalence will be attributable to chance (National Research Council 1991).

Increasingly, many in the environmental health field follow the "precautionary principle," which argues for "1) taking preventive action in the face of uncertainty; 2) shifting the burden of proof to the proponents of an activity; 3) exploring a wide range of alternatives to possibly harmful actions; and 4) increasing public participation in decision-making" (Kriebel et al. 2001). Even without statistical significance we may find a clear association based on strength of association; consistency across persons, places, circumstances, and time; specificity of the exposure site and population; temporality of the exposure and effect; biological plausibility of the effect; coherence with known facts of the agent and disease; and analogy to past experience with related substances. The latter is especially true as we move to considering classes of chemicals, rather than one chemical at a time.

Data Availability

In the last two decades, there has been considerable development in methods and surveillance for environmental health, including expansion of the Centers for Disease Control's (CDC) National Health and Nutrition Examination Survey (NHANES) to include biomonitoring, complemented by increasing precision of laboratory instruments to measure chemicals and their toxicity. Lay input has been central to this process, not only in lay-involved studies, but in the pressure on government and science to do more. However, even when data is available often multiple barriers are in place to discourage its use. For example, extracting data from the EPA is cumbersome, sometimes requiring tedious Freedom of Information Act requests and lengthy waits.

While such datasets are invaluable, they are scarce. To better understand and document exposure and its impacts we need more routine monitoring for toxics. Advocacy for the inclusion of questions on environmentally related illnesses to regular national health surveys is one way to address this need. Such data would provide the opportunity to consider impacts related to morbidity in addition to mortality. Toxic disease registries are also needed. While the Agency for Toxic Substances and Disease Registry has such a registry, it is only for Superfund sites and it requires individuals to specifically give permission to be included. Monitoring of toxic diseases like reportable infectious diseases would provide similar opportunities for early warning for cluster impacts and the possibility to run analyses to identify possible sources.

In order for data to be meaningful, it must be available at appropriate geographical scales for analysis. While some studies currently collect data at the census tract, zip code, city, or metropolitan statistical area (MSA) level, examining impacts on populations can be challenging when merging multiple data sources (e.g. census with health exposure data). As observed in Flint, incongruent geographic scales can lead to false conclusions with potentially devastating consequences (Sadler 2016).

Reluctance of Medical Professionals

Environmental health has been inadequately studied both by epidemiology and by medicine. In particular, epidemiologists have found it hard to gain federal and private grant support for environmental research. In medical training, occupational and environmental medicine have been relegated to a minor position, and the environmental dimension typically gets less attention than the occupational (Castorina and Rosenstock 1990). Extensive evidence from clinician surveys and case reports of environmental health trainings shows that health professionals are not sufficiently literate in environmental health (Trasande et al. 2010; Brown et al. 2018). In one survey of all Massachusetts physicians whose practices are located within communities that have EPA National Priority List sites (Superfund sites) within their borders, these physicians were no more knowledgeable about environmental health hazards than those physicians whose practices are not located in communities with NPL sites (Brown and Kelley 1996). Even in the wake of a major phenomenon like widespread exposure to lead in the water supply in Flint, Michigan in 2014, local resident and faculty physicians at Hurley Medical Center felt ill prepared to handle the questions arising from exposure and the potential health outcomes (Taylor et al. 2017). Medical centers and research institutions are generally not necessarily interested in such work since environmental health has low prestige, little funding, and can put medical facilities in conflict with local centers of power.

Collaborations between Academics and Lay Activist-Researchers

Community-Based Participatory Research

In the context of these various obstacles, citizens have had to be central actors in discovering and dealing with toxic contamination. As a result, a long tradition of public science has developed. The National Institute of Environmental Health Sciences (NIEHS), part of the National Institutes of Health (NIH), has made tremendous strides in environmental health research by incorporating social scientists. By 1995, NIEHS became the first NIH institute to create a community-based participatory research (CBPR) grant initiative. New programs focused on environmental justice and the ethical, legal, and social implications of scientific research. This provided primary relevance and an infrastructure for social scientists and community groups to enter the NIEHS scientific research space. Annual meetings brought together grantees, creating a network in which environmental health and social science researchers learned from one another and developed additional collaborations. Eventually, social science research became a requirement for some NIEHS programs and projects, an essential step for promoting interdisciplinary environmental health research (Baron et al. 2009). NIEHS inaugurated its Partnerships for Environmental Public Health (PEPH) in 2008, providing an umbrella for community engagement and research translation across its center programs.

The Household Exposure Study (HES), a CBPR project to evaluate exposures to pollutants from legacy contaminants, consumer products, and local emissions is one of these transdisciplinary research partnerships (Brody et al. 2009). Silent Spring Institute, an independent research center, collaborated with academics (including many sociologists), and the EJ organization Communities for a Better Environment to collect data in multiple communities using biomonitoring, a tool used by environmental health scientists to explore the body burden of exposure (Brody et al. 2009). Community members were engaged at every level, as participants rather than subjects, about their report-backs and their scientific understanding (Adams et al. 2011; Brown et al. 2012). The integration of social science in the HES has facilitated the development of new theories such as the "research right-to-know" (Morello-Frosch et al. 2009), "exposure experience" (Altman et al. 2008), and "politicized collective illness identity" (Brown 2007) that have redefined and restructured exposure studies as a whole, while also increasing public understanding, environmental health literacy, community empowerment, and mutual trust and respect between researchers and study communities

Advocacy Biomonitoring

Advocacy biomonitoring involves laypeople, working through activist organizations to produce important environmental health science. These projects are often initiated by non-scientists, usually NGOs, who contract outside laboratories to conduct the chemical analyses. Sample sizes tend to be small, typically ranging from three to thirty people, so results are not intended to be analyzed statistically but rather to illustrate the chemicals in ordinary people. In many of these projects, individuals publicly share their chemical exposure data along with photographs and biographies. Laypeople use these studies to focus on public narratives and social change, even though they also work on personal solutions. Projects typically target chemicals that are less-studied and poorly regulated, and for which health implications and exposure sources often uncertain. In examining these studies, sociologists emphasize the importance of going beyond individual solutions to press for regulatory and corporate reform in order to reduce exposures (Washburn 2013; Morello-Frosch and Brown 2014; MacKendrick 2018). A new variant, conducted by Silent Spring Institute, uses crowd-sourced biomonitoring using the DetoxMe Action Kit, in which people pay to participate in urine biomonitoring for ten emerging contaminants, as part of a national collaborative of participants.

Civic Science

Civic science (also called citizen science) has been a growing mechanism for affected communities to gather data. While citizen participation in scientific data collection has a long history (e.g. public ornithology), citizen partnerships in scientific knowledge production beyond merely serving as an instrument of data collection are more recent. One significant example of such work can be seen in the efforts of the Louisiana Bucket Brigade to collect air quality samples in fenceline communities (Ottinger 2010; 'fenceline' is a commonly used term in US environmental circles to refer to communities directly adjacent to polluting facilities). Another community used a catcher to track drift from pesticide applications (Harrison 2011). The Public Lab for Open Science has been a pioneer in developing tools for public monitoring of environmental quality, with a range of techniques such as helium balloons equipped with digital cameras to detect oil spill effects from the BP oil spill, hydrogen sulfide detectors using photographic paper to see the toxic hazards from fracking, and thermal bobs to detect water temperature increases from thermal pollution. These tools and other similar approaches enable communities to report toxic releases that are often unknown or overlooked by regulatory agencies (Wylie 2018).

Critical Epidemiology

Lay discovery has prompted many scientists to reevaluate their traditional approaches. In addition, the expansion of the nation's toxic waste crisis has shaped the awareness of epidemiologists and other scientists in a more critical direction, leading to the development of "oppositional professionals" (Brown 2007). Scientists who find themselves as oppositional professionals tend not to come from an activist background, but find a large kinship with the activists and seek to reform their professions and institutions as a consequence.

Some scientists begin with a traditional perspective and are simply won over to supporting contaminated communities. Beverly Paigen (1982), who worked with laypeople in Love Canal, spoke of her new awareness:

> Before Love Canal, I also needed a 95 percent certainty before I was convinced of a result. But seeing this rigorously applied in a situation where the consequences of an error meant that pregnancies were resulting in miscarriages, stillbirths, and children with medical problems, I realized I was making a value judgment ... whether to make errors on the side of protecting human health or on the side of conserving state resources.

John Till, who headed the Dose Reconstruction Study at the Hanford plutonium site in Washington State, which is reconstructing historical dosages of radioactive iodine releases, recounted how he had been transformed by the process of being part of what he terms a "public study." He came to grasp the importance of openness to the public, and of access to classified information. He found a great empathy for the Hanford downwinders and a special understanding of the concerns of Native Americans in the area who viewed Hanford as an assault on their entire heritage (Kaplan 2000).

As the number of what we term "critical epidemiologists" grows, we may see a greater number of well-designed health studies in which laypeople play a central role. While such oppositional professionals still draw from the basic scientific methods of their disciplines, they find themselves working towards creating a new praxis of engaged research that questions the default assumptions of those methods. Steve Wing (1994) puts forth an alternative conceptualization of epidemiology that taps the concerns of both popular epidemiology activists and critical epidemiologists:

1) Ask not what is good or bad for health overall, but for what sectors of the population.
2) Look for connections between many diseases and exposures, rather than looking at merely single exposure-disease pairs.
3) Examine unintended consequences of interventions.
4) Utilize people's personal illness narratives.
5) Include in research reporting the explicit discussion of assumptions, values, and the social construction of scientific knowledge.
6) Recognize that the problem of controlling confounding factors comes from a reductionist approach that looks only for individual relations rather than a larger set of social relations. Hence, what are nuisance factors in traditional epidemiology become essential context in a new ecological epidemiology.
7) Involve humility about scientific research, combined with a commitment to supporting broad efforts to reform society and health.

This critical epidemiology approach has been taken up in recent years by social scientists studying complex exposure pathways and syndromes, such as in the case of those experiencing multiple chemical sensitivities (Phillips 2010). Scholars push to rethink epidemiology through a lens of social theory that is sensitive to institutional and structural forces such as the role of racism in shaping exposures and health outcomes (Senier et. al 2017; Wemrell et al. 2016). An important dimension of this work draws attention to power in the ways that classical epidemiological data and scientific knowledge about health and healthcare are produced, with particular consideration to the role of experts in maintaining control over this knowledge production process in exclusion of publics (Joyce and Senier 2017; Cordner 2015).

Future Directions of Social Science Environmental Health Scholarship

In its most recent turn, environmental sociology (and other social science work on environmental contamination) engages in boundary crossing efforts that integrate social science with environmental health practice. This new approach moves beyond pure research to intervention, reflecting increasing collaboration between social scientists and environmental health scientists to measure exposures, press for cumulative exposure to be addressed, and prepare research data to affect health and environmental policy. Such efforts variously focus on mitigating primary (direct hazards) and secondary (individual, community, and societal) impacts of past exposures and preventing new exposures from occurring. Further, these efforts typically involve collaboration with community-based organizations. In such participatory research, the social scientist becomes an actor in events rather than a mere observer or analyst.

Transdisciplinary environmental health research has increased awareness of effects beyond the physical and health consequences of environmental disaster and contamination to include community empowerment, ethical practices of sharing data, and policy implications. This approach to environmental health research has created the opportunity and need for involvement of social scientists and environmental sociologists have been instrumental to these collaboratives.

As sociologists, we need to spend more time thinking about how to make clear connections between community contexts, environmental variables, and health outcomes. Through community engagement, we can consider ideal scenarios for improving data access and plausible connections to environmental quality. Ongoing, data access should be considered a key dimension of environmental justice struggles because, either through intent or benign neglect, the necessary data to show harm to vulnerable communities due to toxics exposure is lacking.

Beyond just data access, which is severely threatened under the Trump administration, sociologists must join others in confronting the White House, EPA, OSHA, Department of Energy, and other agencies that have halted action to stem climate change; supported unfettered coal, oil, and gas industries; cut back on enforcement; approved harmful pesticides and other chemicals; and reduced clean energy approaches.

Sociologists offer important insight into the structural conditions that lead to community exposure and environmental health effects. We need to develop and advance creative models that study health outcomes in relation to a variety of inputs, including political structure, public participation, access to health services, economic factors, population density, racial/ethnic proportions, environmental quality, and access to the natural environment. This calls in particular for qualitative work on the experience of illness of people living in toxic areas. As elaborated in this chapter, to better engage with communities to understand and convey environmentally compromised contexts, more environmental sociologists will need to develop partnerships with environmental health scientists to advance a necessary transdisciplinary approach.

References

Adams, C., Brown, P., Morello-Frosch, R., Brody, J. G., Rudel, R. et al. (2011). Disentangling the Exposure Experience: The Roles of Community Context and Report-Back of Environmental Exposure Data. *Journal of Health and Social Behavior*, 52(2), 180–196.

Altman, R., Brody, J., Rudel, R. et al. (2008). Pollution Comes Home and Pollution Gets Personal: Women's Experience of Household Toxic Exposure. *Journal of Health and Social Behavior*, 49, 417–435.

Baron, S., Sinclair, R., Payne-Sturges, D. et al. (2009). Partnerships for Environmental and Occupational Justice: Contributions to Research, Capacity and Public Health. *American Journal of Public Health*, 99, S517–S525.

Blocker, J., & Eckberg, D. L. (1989). Environmental Issues as Women's Issues: General Concerns and Local Hazards. *Social Science Quarterly*, 70(3), 586–593.

Bomberg, E. (2017). Environmental Politics in the Trump Era: An Early Assessment. *Environmental Politics*, 26(5), 956–963.

Bromet, E. J., Parkinson, D. K., Schulberg, H. C., Dunn, L. O., & Gondek, P. C. (1982). Mental Health of Residents Near the Three Mile Island Reactor: A Comparative Study of Selected Groups. *Journal of Preventive Psychiatry*, 1, 225–276.

Brody, J. G., Morello-Frosch, R., Zota, A. et al. (2009). Linking Exposure Assessment Science with Policy Objectives for Environmental Justice and Breast Cancer Advocacy: The Northern California Household Exposure Study. *American Journal of Public Health*, 99, S600–S609.

Brown, P. (1992). Popular Epidemiology and Toxic Waste Contamination: Lay and Professional Ways of Knowing. *Journal of Health and Social Behavior*, 33, 267–281.

Brown, P. (1995). Race, Class, And Environmental Health: A Review and Systematization of The Literature. *Environmental Research*, 69, 15–30.

Brown, P. (2007). *Toxic Exposures: Contested Illnesses and the Environmental Health Movement*. Columbia University Press.

Brown, P., Brody, J. G., Morello-Frosch, R. et al. (2012). Measuring the Success of Community Science: The Northern California Household Exposure Study. *Environmental Health Perspectives*, 120, 326–331.

Brown, P. & Ferguson, F. (1995). "Making a Big Stink": Women's Work, Women's Relationships, and Toxic Waste Activism. *Gender & Society*, 9, 145–172.

Brown, P. & Mikkelsen, E. J. (1990/1997). *No Safe Place: Toxic Waste, Leukemia, and Community Action*. University of California Press.

Brown, P. & Kelley, J. (1996). Physicians' Knowledge of and Actions Concerning Environmental Health Hazards: Analysis of a Survey of Massachusetts Physicians. *Industrial and Environmental Crisis Quarterly*, 9, 512–542.

Brown, P. & Masterson-Allen, S. (1994). The Toxic Waste Movement: A New Kind of Activism. *Society and Natural Resources*, 7, 269–286.

Brown, P., Zavestoski, S., McCormick, S. et al. (2004). Embodied Health Movements: Uncharted Territory in Social Movement Research. *Sociology of Health and Illness*, 26, 1–31.

Brown, P., Clark, S., Zimmerman, E., Miller, M., & Valenti, M. (2018). Health Professionals' Environmental Health Literacy. In Symma Finn and Liam O'Fallon, eds. *Environmental Health Literacy*. Springer.

Bullard, R. ed. (1993). *Confronting Environmental Racism: Voices from the Grassroots*. South End Press.

Cable, S. (1992). Women's Social Movement Involvement: The Role of Structural Availability in Recruitment and Participation Processes. *Sociological Quarterly*, 33, 35–47.

Carson, R. (1962). *Silent Spring*. Harcourt.

Castorina, J. & Rosenstock, L. (1990). Physician Shortage in Occupational and Environmental Medicine. *Annals of Internal Medicine*, 113, 983–986.

Caufield, C. (1989). *Multiple Exposures: Chronicles of the Radiation Age*. Perennial Books.

Centers for Disease Control and Prevention (CDC). (2018). Lead. Retrieved from www.cdc.gov/nceh/lead/default.htm. Accessed August 21, 2018.

Colborne, T., Dumanoski, D., & Meyer, J. (1996). *Our Stolen Future: Are We Threatening Our Fertility, Intelligence, and Survival? A Scientific Detective Story.* Dutton.

Cordner, A. (2015). Strategic Science Translation and Environmental Controversies. *Science, Technology, & Human Values*, 40(6), 915–938.

Couch, S. R. & Kroll-Smith, J. S. (1990). *The Real Disaster Is Above Ground: A Mine Fire and Social Conflict.* University Press of Kentucky.

Couch, S. R. & Kroll-Smith, J. S. (1985). The Chronic Technical Disaster: Towards a Social Scientific Perspective. *Social Science Quarterly*, 66, 564–575.

Crenshaw, K. W. (1991). Mapping the Margins: Intersectionality, Identity Politics, and Violence against Women of Color. *Stanford Law Review*, 43(6), 1241–1299.

Dohrenwend, B. (1981). Stress in the Community: A Report to the President's Commission on the Accident at Three Mile Island. *Annals of the New York Academy of Sciences*, 365, 159–174.

Dillon, L., Walker, D., Wylie, S., Shapiro, N., Lave, R. et al. (2017). Environmental Data Justice and the Trump Administration: Reflections on Forming EDGI. *Environmental Justice*, 10(6), 186–192.

Edelstein, M. (1988). *Contaminated Communities: The Social and Psychological Impacts of Residential Toxic Exposure.* Westview.

Erikson, K. (1976). *Everything in Its Path: The Destruction of Community in the Buffalo Creek Flood.* Simon and Schuster.

Fasenfest, D., & Pride, T. (2016). Emergency Management in Michigan: Race, Class and the Limits of Liberal Democracy. *Critical Sociology*, 42(3), 331–34.

Freudenburg, W. & Jones, T. (1991). Attitudes and Stress in the Presence of Technological Risk: A Test of the Supreme Court Hypothesis. *Social Forces*, 69, 1143–1168.

Frisbee, S. J., Brooks, A. P. Jr., Maher, A., Flensborg, P., Arnold, S. et al. (2009). The C8 Health Project: Design, Methods, and Participants. *Environmental Health Perspectives*, 117(12), 1873.

Gibbs, L. (1982). Community Response to an Emergency Situation: Psychological Destruction and the Love Canal. Paper presented at the American Psychological Association, August 24.

Gill, D. A. and Picou, J. S. (1995). Environmental Disaster and Community Stress. Paper presented at Third International Conference on Emergency Planning and Disaster Management, Lancaster, England.

Goldstein, I. & Goldstein, M. (1986). The Broad Street Pump. pp. 37–48 in John R. Goldsmith, ed., *Environmental Epidemiology.* CRC Press.

Gurr, B. (2011). "Complex Intersections: Reproductive Justice and Native American Women." *Sociology Compass* 5(8), 21–735.

Harrison, J. (2011). *Pesticide Drift and the Pursuit of Environmental Justice.* MIT Press.

Hofrichter, R. ed. (1993). *Toxic Struggles: The Theory and Practice of Environmental Justice.* New Society Publishers.

Hoover, E. (2018). Environmental reproductive justice: Intersections in an American Indian community impacted by environmental contamination. *Environmental Sociology*, 4 (1), 8–21.

Hoover, E., K. Cook, R. Plain, K. Sanchez, V. Waghiyi, P. et al. (2012). Indigenous Peoples of North America: Environmental Exposures and Reproductive Justice. *Environmental Health Perspectives* 120(12), 1645–1649.

Joyce, K. & Senier, L. (2017). Why Environmental Exposures? *Environmental Sociology*, 3 (2), 101–106.
Kaplan, L. 2000. Public Participation in Nuclear Facility Decisions: Lessons from Hanford. In Daniel Kleinman, ed., *Science, Technology, and Democracy*. SUNY Press, pp. 67–86.
Kriebel, D., Tickner, J., Epstein, P., Lemons, J., Levins, R. et al. (2001). The Precautionary Principle in Environmental Science. *Environmental Health Perspectives*, 109(9), 871–876.
Krieg, E. (1995). A Socio-Historical Interpretation of Toxic Waste Sites: The Case of Greater Boston. *American Journal of Economics and Sociology*, 54, 1–14.
Krauss, C. (1993). Women and Toxic Waste Protests: Race, Class and Gender as Resources of Resistance. *Qualitative Sociology*, 16, 247–262.
Lagakos, S.W., Wessen, B.J., & Zelen, M. (1986). An Analysis of Contaminated Well Water and Health Effects in Woburn, Massachusetts. *Journal of the American Statistical Association*, 81, 583–596.
Levine, A. (1982). *Love Canal: Science, Politics, and People*. Heath.
Lifton, R. J. (1968). *Death in Life: Survivors of Hiroshima*. Random House.
Lifton, R. & Olson, E. (1976). The Human Meaning of Total Disaster: The Buffalo Creek Experience. *Psychiatry*, 39, 1–18.
MacKendrick, N. (2018). *Better Safe Than Sorry: How Consumers Navigate Exposure to Everyday Toxics*. University of California Press.
Maxwell, N. (1996). Land Use, Demographics, and Cancer Incidence in Massachusetts Communities. Sc.D. dissertation, Boston University School of Public Health.
McGoey, L. (2012). The Logic of Strategic Ignorance. *The British Journal of Sociology*, 63 (3), 553–576.
Mohai, P. & Bryant, B. eds. (1992). *Race and the Incidence of Environmental Hazards*. Westview.
Morello-Frosch, R., Brody, J. G., Brown, P. et al. (2009). 'Toxic Ignorance' and the Right-to-Know: Assessing Strategies for Biomonitoring Results Communication in a Survey of Scientists and Study Participants. *Environmental Health*, 8, 6.
Morello-Frosch, R. & Brown, P. (2014). Science, Social Justice, and Post-Belmont Research Ethics: Implications for Regulation and Environmental Health Science, In Daniel Kleinman and Kelly Moore (eds.) *Handbook of Science, Technology, and Society*. Routledge, pp. 479–491.
National Research Council. (1991). *Environmental Epidemiology, Volume 1, Public Health and Hazardous Wastes*. National Academy Press.
Neutra, R. R. (1990). Counterpoint from a Cluster Buster. *American Journal of Epidemiology*, 132: 1–8.
Ottinger, G. (2010). Buckets of Resistance: Standards and the Effectiveness of Citizen Science. *Science, Technology, & Human Values*, 35(2), 244–270.
Ozonoff, D. & Boden, L.I. (1987). Truth and Consequences: Health Agency Responses to Environmental Health Problems. *Science, Technology, and Human Values*, 12, 70–77.
Paigen, B. (1982). Controversy at Love Canal. *Hastings Center Reports*, 12(3), 29–37.
Peck, J. (2012). Austerity Urbanism. *City*, 16(6), 626–655.
Perkins, T. E. (2012). Women's Pathways into Activism: Rethinking the Women's Environmental Justice Narrative in California's San Joaquin Valley. *Organization and Environment*, 25, 76–94.

Pulido, L. (2016). Flint, Environmental Racism, and Racial Capitalism. *Capitalism Nature Socialism*, 27(3), 1–16.

Phillips, T. (2010). Debating the Legitimacy of a Contested Environmental Illness: A Case Study of Multiple Chemical Sensitivities (MCS). *Sociology of Health and Illness*, 32(7), 1026–1040.

Sadler, R. (2016). How ZIP codes nearly masked the lead problem in Flint. *The Conversation*. Available at theconversation.com/how-zip-codes-nearly-masked-the-lead-problem-in-flint-65626. Accessed February 14, 2018.

Senier, L., Brown, P., Shostak, S., & Hanna, B. (2017). The Socio-Exposome: Advancing Exposure Science and Environmental Justice in a Postgenomic Era. *Environmental Sociology*, 3(2), 107–121.

Snow, J. (1853). On the Prevention of Cholera. *The Medical Times and Gazette*, 7, 367–369. Retrieved from www.ph.ucla.edu/epi/snow/onpreventioncholera.html. Accessed August 21, 2018.

Szasz, A. (1994). *Ecopopulism: Toxic Waste and the Movement for Environmental Justice*. University of Minnesota Press.

Taylor, D. K., Lepisto, B.L., Lecea, N., Ghamrawi, R., Bachuwa, G. et al. (2017). Surveying Resident and Faculty Physician Knowledge, Attitudes, and Experiences in Response to Public Lead Contamination. *Academic Medicine*, Mar, 92(3), 308–331.

Trasande L., Newman, N., Long, L., Howe, G., Kerwin, B.J. et al. (2010). Translating Knowledge about Environmental Health to Practitioners: Are We Doing Enough? *Mt Sinai Journal of Medicine*, 77, 114–123.

Vyner, H. M. (1988). *Invisible Trauma: The Psychosocial Effects of the Invisible Environmental Contaminants*. Lexington Books.

Washburn, R. (2013). The Social Significance of Human Biomonitoring. *Sociology Compass*, 24, 162–179.

Wemrell, M., Merlo, J., Mulinari, S., & Hornborg, A.-C. (2016). Contemporary Epidemiology: A Review of Critical Discussions Within the Discipline and A Call for Further Dialogue with Social Theory. *Sociology Compass*, 10(2), 153–171.

Wing, S. 1994. Limits of Epidemiology. *Medicine and Global Survival*, 1, 74–86.

World Health Organization (WHO). (2010). Childhood Lead Poisoning. Geneva. Retrieved from www.who.int/ceh/publications/leadguidance.pdf. Accessed August 21, 2018.

Wylie, S. A. (2018). *Fractivism: Corporate Bodies and Chemical Bonds*. Duke University Press.

Zavella, P. (2016). Contesting Structural Vulnerability through Reproductive Justice Activism with Latina Immigrants in California. *North American Dialogue*, 19(1), 36–45.

Zavestoski, S., McCormick, S., & Brown, P. (2004). Gender, Embodiment, and Disease: Environmental Breast Cancer Activists' Challenges to Science, the Biomedical Model and Policy. *Science as Culture*, 13(4), 563–586.

8 The Environment's Absence in Medicine: Mainstream Medical Coverage of Leukemia

Manuel Vallée

Introduction

Human health is inextricably intertwined with ecological health. This point has been repeatedly underscored by environmental health research that elucidates the close relationship between environmental pollution and disease. Whether we are discussing cancers, reproductive issues, birth defects, developmental disorders, musculo-skeletal problems, metabolic disorders, immunological problems or practically any other disease category, environmental health research shows that disease is invariably related to environmental pollution, in often subtle but intimate ways (CHE, 2018a; Schettler et al., 2000; Steingraber, 2009).

Curiously, however, environmental pollution's role is usually obscured or downplayed in mainstream medical information. For example, Brown and colleagues (2001) showed print media's coverage of breast cancer consistently downplays the role of toxicants, in favor of an individualizing frame that emphasizes genes and lifestyle choices. A similar pattern is found with information provided by the medical profession, as demonstrated by Steingraber's (2009) analysis of cancer educational materials distributed in clinics, hospitals, and waiting rooms. Vallée (2013) found the same to be true with the 2011 clinical practice guidelines for Attention Deficit/Hyperactivity Disorder, which failed to mention lead or other toxicants associated with the condition. Such information discrepancies matter because they conceal the significant role environmental pollution plays in producing disease, thereby making it more difficult for patients and families to protect themselves, as well as to advocate for stronger policy and regulations.

Although social scientists have extensively analyzed the medical information provided in mass media (Atkin et al., 2008; Brown et al., 2001; Lewison et al., 2008), the same cannot be said for online medical publishing websites, such as WebMD.com and Healthline.com. This lacuna is significant for three reasons. First, their information is more accessible than conventional print sources because most content is free, can be instantaneously accessed, and can be accessed at a distance. Second, lay audiences are likely to give more credibility to the websites, due to features suggesting a close association with the medical profession: 1) including "health," "medicine," "MD," or "Dr" in their website names; 2) clearly identifying medical doctors as content reviewers; and 3) including medical doctors on the governing boards. Third, the medium has grown significantly in recent years, as indicated by the proliferation of websites, the followings many have developed, and that it has become a multi-billion dollar industry (Bray, 2017). The medium's growth means the websites now exert a greater capacity to influence how people understand and address disease.

To shed light on how medical issues are discussed on such websites, I examine WebMD's leukemia coverage. Leukemia is a bone marrow cancer that affects over 385,000 people in the United States and is the most common cancer in children under 15 (NCI, n.d.; NIH, n.d.). Although environmental health researchers have identified twenty-two toxicants associated with leukemia, WebMD's coverage systematically obscures the environmental causation perspective by failing to identify most toxicants and by emphasizing a genetic and lifestyle causation frame. Building on Brown et al.'s work (2001), I also illuminate rhetorical mechanisms through which dominant sources downplay the environmental causation perspective, including placing toxicant information in subordinate locations, surrounding the information with negating statements, and treatment discussions that fail to address toxicants. I also discuss how obscuring toxicant information places humans at greater risk and makes it more difficult for patients and families to protect themselves. Finally, to unearth the problem's sources, I consider WebMD's reliance on the medical profession, as well as the profession's financial motivations and ideological orientation.

Background: Environmental Pollution and Disease Framing

While industrialization accelerated environmental pollution during the nineteenth and early twentieth centuries, John Bellamy Foster (2009) argues the problem has worsened since World War II because societies have become increasingly reliant on plastics, chemical pesticides, and other non-biodegradable chemicals. The rampant use of these products, he argues, undermines the very life-support systems that make life possible on Earth. Moreover, the resulting pollution has impacted human bodies. Because we rely on the environment for the food we eat, the water we drink and the air we breathe, pollution has been building up in human tissues, which has been linked to cancer and numerous other medical conditions (CHE, 2018a; Schettler et al., 2000; Sexton et al., 2004).

However, the relationship between toxicants and disease is often difficult to discern in mainstream medical information, which tends to favor reductionist disease framings that emphasize genetics and lifestyle choices (such as smoking and alcohol consumption). For instance, Brown et al. (2001) found print media coverage of breast cancer seldom references environmental causation (i.e. chemicals, pollutants, and radiation), and focuses instead on genetics and personal lifestyle factors. Similarly, a follow-up study by Atkin et al. (2008) revealed that toxicant risk factors were only mentioned in 4 percent of breast cancer stories in print and television media. Additionally, Lewison et al. (2008) found that breast cancer stories on the BBC website typically emphasized an individualizing frame, where the cause was implicitly attributed to genetic factors and/or personal lifestyle choices. In short, the research has repeatedly demonstrated that mainstream disease coverage invariably obscures the environmental causation perspective.

Consequences of Reductionist Disease Framings

Disease frames matter for several reasons. First, an individualizing frame encourages the public to adopt a disease understanding that ignores the causal role played by environmental pollution, focusing instead on genes and harmful personal choices (Brown et al., 2001; MacKendrik, 2010). In turn, this determines how we seek to address the symptoms. For example, if the dominant asthma framing emphasizes genetic causes, then doctors will steer patients towards symptom suppression, which will invariably be pursued, in twenty-first-century industrialized countries, through pharmaceutical medications. Moreover, they will ignore environmental factors (such as a moldy living environment or exposure to chromium, latex, plastic fumes, or the many other toxicants associated with the condition [CHE, 2018a]). Consequently, this ensures patients will continue to live in symptom-causing environments that undermine their ability to eliminate or, at least, minimize the symptoms. On the other hand, if asthma symptoms are attributed to external factors, then doctors will steer patients to environment-changing interventions, such as diet alteration, mold remediation, and/or cleaning up toxic contamination in their home and workplace.

Disease framings also have important socio-economic implications. First, they significantly impact healthcare expenditures. If the prevailing frame obscures underlying causes of disease, society is far less likely to take steps to eliminate them. In turn, this maintains disease incidence rates, which drives up healthcare expenditures and further stresses our already overburdened healthcare systems. Second, because they steer patients towards using medical treatments, individualizing disease frames are a boon for treatment manufacturers (including pharmaceutical manufacturers) and the medical sub-disciplines who rely on those treatments for their medical authority.

Disease frames also impact who is viewed as being responsible for the problem, and whether individual troubles come to be seen as public issues. As Brown et al. (2001) emphasize:

> If the media focus blame and responsibility on the individual, it is likely that the problem will not be considered a social problem that merits public or governmental attention. If, however, the problem is framed so that structural or institutional causes receive the blame, it becomes a social problem of concern to all members of a community. (p. 752)

In turn, disease frames can have major consequences for social policy. If the disease is seen as being caused by industry and/or government failure to regulate industry, the framing can galvanize public will towards pressuring political representatives to tighten regulations. If, on the other hand, the disease continues to be seen as a personal issue, little will be done towards generating a collective solution.

Given their social impact, dominant disease frames should not be taken for granted but rather should be interrogated, which includes identifying their social implications and illuminating the actors and processes that enable those frames to become dominant.

Analyzing Online Medical Publishing Websites

Beyond print media, Brown et al. (2001) identify other entities that adjudicate which disease frames become dominant, including other private sector entities (such as think tanks, disease groups, activists, and social movements), government (including agencies and politicians), as well as the scientific field (including individual researchers, professional organizations, as well as journals and their editors).

Although much attention has been directed towards mass media, online medical publishing sites have been relatively underanalyzed. This matters for several reasons. First, the field of online medical publishing has grown significantly over the last two decades, and is populated by numerous competing websites, including WedMD.com, Medicine.Net, Healthline.com, Doctoroz.com, Mercola.com, and DrWeil.com, many of which have attained significant followings (Comscore, 2016). The medium's significant growth means it now has the capacity to significantly influence public perceptions about disease causes and attribution of responsibility.

To address this gap, I examine WebMD's leukemia coverage. WebMD is a strategic case because it is the largest online medical publisher: in 2015 its network of websites were visited by more unique visitors than any other private or government website dedicated to health matters (Comscore, 2016). In turn, this enabled the company to generate $705 million in 2016, which led it to be purchased the next year for $2 billion (Bray, 2017). WebMD is also closely affiliated with the medical profession, as indicated by the medical doctors who review the articles and sit on its board of directors. While it targets physicians through a professional portal, it also targets consumers through its webMD.com website and *WebMD The Magazine*, a patient-directed publication typically on display in physician waiting rooms (*The Write News*, 2005).

Environmental Health Scholarship on Leukemia

The environmental health community has three authoritative bodies that track the relationship between toxicant exposure and disease. The first is the International Agency for Research on Cancer (IARC), which is the World Health Organization's specialized cancer agency. It was founded in France in 1965 to "lighten humanity's ever-growing burden of cancer" (IARC, 2018) by promoting international collaboration in cancer research. It focuses on identifying how lifestyle and environmental risk factors interact with genetics to produce cancer. This focus implicitly recognizes that "most cancers are, directly or indirectly, linked to environmental factors and thus are preventable" (IARC, 2018). A key contribution has been the IARC Monograph Programme, where international working groups evaluate the carcinogenicity of toxicants and publicly disseminate their findings.

The second authoritative body is California's Office of Environmental Health Hazard Assessment (OEHHA), one of six agencies within the California Environmental Protection Agency (CalEPA). OEHHA's mission is to "protect and enhance public health and the environment by scientific evaluation of risks posed by hazardous substances" (OEHHA, 2018a), which it pursues by evaluating the health and environmental risks posed by hazardous substances, including pesticides, air pollutants, carcinogens, reproductive toxins, chemical exposures in the workplace, and chemical contaminants in food and water. It also "implements the Safe Drinking Water and Toxic Enforcement Act of 1986, commonly known as Proposition 65, and compiles the state's list of substances that cause cancer or reproductive harm" (OEHHA, 2018b). Additionally, it establishes exposure limits for air, water, and soil contaminants, which guides the general public, NGOs, all boards and departments within CalEPA, as well as federal agencies, including the Department of Justice and Department of Public Health (OEHHA, 2018a; OEHHA, 2018b).

The third authoritative organization is The Collaborative on Health and the Environment (CHE), which was founded in 2002 to create an interdisciplinary research network on environmental health issues (CHE, 2018b). A significant contribution has been the "Toxicants and Disease" database (CHE, 2018a), which enables users to identify all toxicants associated with a disease and, conversely, all diseases associated with a particular toxicant. Additionally, if a toxicant is associated with a disease, the database will signal the evidence strength. CHE is the only authoritative body that addresses leukemia directly, as OEHHA and IARC only identify whether toxicants are carcinogenic. For this reason, the CHE database was my main reference point.

Over the last four decades, environmental health researchers have identified numerous toxicants related to leukemia's development. As seen in TABLE 8.1, the CHE database lists twenty-two toxicants or toxicant classes associated with the disease, including formaldehyde, benzene, ionizing radiation, ethylene oxide, and 1,3-butadiene, which are each classified as having a "strong" level of evidence. A "strong" classification means "a causal association with the disease has been verified" (CHE, 2018c) and the toxicity of the chemical is well accepted by the scientific research community. Additionally, the seventeen others are all considered to have a "good" level of evidence, which is the classification given to "toxicants associated with a disease through epidemiological studies (cross-sectional, case-series, or case-control studies) or for toxicants with some human evidence and strong corroborating animal evidence" (CHE, 2018c).

IARC has also flagged many of those chemicals as being harmful or potentially harmful to human health, with eleven listed as recognized carcinogens, including formaldehyde, benzene, ethylene oxide, ionizing radiation, TCDD, and 1,3-butadiene (see TABLE 8.1). Additionally, the evidence for two others (DDT/DDE, chlorinated solvents) was deemed sufficiently strong to warrant a "probable carcinogen" designation, with three others (carbon tetrachloride and 1,2-dichloroethane) classified as "possible carcinogens." OEHHA has also recognized the majority of the toxicants as carcinogens (see TABLE 8.1). While being classified as a carcinogen

Table 8.1 *Toxicants associated with adult-onset leukemias*

Class of Toxicants	Toxicant	CHE Database Evidence Rating *	IARC Classification **	OEHHA – Health Ailment it associated with the toxicant (date added to Prop 65's List) ***	WebMD pages mentioning the toxicant
	1,3-Butadiene	strong	Category 1****	Cancer (04/01/1988)	0
	Ethylene Oxide	strong	Category 1	Cancer (07/01/1987)	0
	Formaldehyde	strong	Category 1	Cancer (01/01/1988)	1
	Ionizing Radiation	strong	Category 1	NA	7
	Benzene	strong	Category 1	Cancer (02/27/1987)	38
	1,2-Dichloroethane	good	Category 2B ††	Cancer (10/01/1987)	0
	Alachlor	good	NA	Cancer (01/01/1989)	0
Aromatic Amines	2-Naphthylamine	good	Category 1	Cancer (02/27/1987)	0
Aromatic Amines	4-Aminobiphenyl	good	Category 1	Cancer (02/27/1987)	0
Aromatic Amines	4,4'methyleneibis	good	Category 1	Cancer (01/01/1987)	0
Aromatic Amines	Auramine	good	Category 2B	Cancer (07/01/1987)	0
Aromatic Amines	Benzidine	good	Category 1	Cancer (02/27/1987)	0
	Arsenic	good	Category 1	Cancer (02/27/1987)	12
	Carbon Disulfide	good	NA	Reproduction & Developmental Problems (07/01/1989)	8
	Carbon Tetrachloride	good	Category 2B	Cancer (10/01/1987)	0
Chlorinated solvents		good	Category 2A†††	NA	1
	DDT/DDE	good	Category 2A	Cancer (10/01/1987)	0
Pesticides		good	NA	NA	20
Phenoxyacetic herbicides	2,4-Dichloro phenoxyacetic Acid	good	Category 2B	NA	0

Table 8.1 (*cont.*)

Class of Toxicants	Toxicant	CHE Database Evidence Rating *	IARC Classification **	OEHHA – Health Ailment it associated with the toxicant (date added to Prop 65's List) ***	WebMD pages mentioning the toxicant
Phenoxyacetic herbicides	Agent Orange †	good	Contains TCDD dioxin, which IARC classifies as Category 1 toxicant	NA	43
Dioxins	TCDD	good	Category 1	Cancer (01/01/1988)	0
Air Pollution	tobacco smoke (active smoking)	good	NA	Reproduction & Developmental Problems (04/01/1988)	72

*: all CHE data taken from their "Toxicant & Disease" database (CHE, 2018a)
**: all IARC data was obtained from the "IARC Monographs" website (IARC, 2018)
***: all OEHHA data was obtained from the "Chemicals" page on the OEHHA website (OEHHA, 2018c)
****: Category 1= Carcinogenic to humans
†: Agent Orange is a combination of 2,4-D and 2,4,5-T
††: Category 2B: Possibly carcinogenic to humans
†††: Category 2A: Probably carcinogenic to humans

does not necessarily link a toxicant to leukemia, it does underscore its capacity to contribute to cancer-producing processes.

WebMD's Coverage of Toxicants

Although environmental health researchers have identified many toxicants associated with leukemia, WebMD's leukemia coverage fails to cover most of them. Moreover, while some are mentioned, WebMD downplays their importance through several rhetorical strategies.

Obscuring Environmental Causation by Tokenistic Coverage of Toxicants

An October 2018 search for "leukemia" on WebMD's website yielded 1,213 results and the first step was identifying the coverage provided for each toxicant. TABLE 8.1 shows WebMD's coverage failed to mention fourteen of twenty-two toxicants linked to leukemia. Besides omitting most toxicants, the website ignored two of five strongly linked to leukemia (1,3-Butadiene and ethylene oxide). Additionally, toxicants that were mentioned appeared on few pages. For instance, only five toxicants (benzene, pesticides, arsenic, agent orange and tobacco smoke) appeared on ten or more of the 1,213 pages addressing leukemia.[1] Moreover, even the best covered toxicants only appeared on a small fraction of leukemia pages. For example, agent orange was the chemical with the most coverage and it appeared on less than 3.5 percent of leukemia pages. Such spotty coverage effectively obscures the links between toxicant exposure and leukemia.

Tokenistic Coverage on Key Webpages

Another important issue is the coverage on pages viewers are more likely to see, such as initial search results. Initial results are particularly important because viewers are unlikely to read all 1,200 results and are likely to stop after reading results from the initial pages. The first page provided links to thirteen readings, but only two (i.e. "What is leukemia? What causes it?" and "Slideshow: Guide to Leukemia") mentioned toxicants (see TABLE 8.2). The second page was a bit better as five out of ten links mentioned toxicants. While toxicants were mentioned on seven of the first twenty-three results, this was woefully limited. Moreover, the problem was compounded by the fact none covered all toxicants linked to leukemia.

The "What is leukemia? What causes it?" (WebMD, 2017a) page is particularly important because it provides an overview of leukemia, which will strongly interest those who do not have prior knowledge about the condition. As well, its focus on "causes" makes it the most likely page to discuss environmental toxicants. However, its coverage was poor as it only mentioned tobacco smoke, high doses of radiation, and "some chemicals" (WebMD, 2017a). Although this covers two toxicants (smoking and radiation) from the CHE list, it leaves out twenty others, including four that have "strong" evidence (benzene, formaldehyde, ethylene oxide, and 1,3-butadiene). Even though the page mentions "some chemicals," this vague statement fails to alert readers to the specific chemicals that can prove harmful, how they might be exposed to them, or what they can do to protect themselves.

The "What is Acute Myeloid Leukemia?" page is also very important (WebMD, 2017b). Although it is the last result on the second page of search results, Acute Myeloid Leukemia (AML) is leukemia's most prevalent form and will therefore be sought out by many viewers. The coverage on this page was a bit better as it mentioned benzene, chemotherapy drugs, cleaning products, detergents, and paint strippers. Identifying these chemicals alerts readers to the harmfulness of chemicals in their environment, which supports the environmental causation perspective. However, the coverage suffers from vagueness, failing to identify chemicals that are either known culprits or for which there is growing evidence of harm. Moreover, while the page added a chemical with "strong" evidence (i.e. benzene), it failed to mention the four others with "strong" evidence (ethylene oxide, formaldehyde, ionizing radiation, and 1,3-Butadiene), as well as the seventeen others associated with leukemia.

---

1 For each toxicant, I conducted a search pairing the name of the toxicant and "leukemia." Then, each page was analyzed to verify that the mention related to the toxicant's disease-causing potential. Pages that did not meet that criteria were eliminated from the count.

Table 8.2 *WebMD articles on leukemia*

Position in the search	Page Location	Title of Result	Name of WebMD Reviewer	Toxicants Mentioned
1	First	Non-Hodgkin's Lymphoma	NA	none
2	First	Slideshow: Chronic Myelogenous Leukemia Phases and Treatment	Laura J. Martin, MD (Jan. 2018)	none
3	First	Video on the Stages of Multiple Myeloma	Neha Pathak, MD (May 2017)	none
4	First	Vitamin and Supplement results for Leukemia	NA	none
5	First	What is leukemia? What causes it?	Melinda Ratini, DO, MS (Sept. 2017)	radiation, tobacco smoke, and certain chemicals
6	First	Leukemia (Directory)	NA	none
7	First	Blood Cancer (Directory)	NA	none
8	First	What is leukemia?	William Blahd, MD (Dec. 2016)	none
9	First	What is bone marrow cancer?	Louise Chang, MD (Feb. 2017)	none
10	First	What is leukemia?	William Blahd, MD (Dec. 2016)	none
11	First	Childhood Leukemia: Symptoms, Treatments, Risk Factors, Tests	Neha Pathak, MD (Sept. 2017)	none
12	First	Childhood Leukemia Directory	NA	none
13	First	Slideshow: Guide to Leukemia	Laura J. Martin, MD (April 2018)	smoking, tobacco, and very high levels of radiation
14	Second	How is leukemia grouped?	Melinda Ratini, DO, MS (Sept. 2017)	none
15	Second	What causes leukemia?	Melinda Ratini, DO, MS (Sept. 2017)	smoking, high radiation exposure, and certain chemicals
16	Second	How is leukemia treated?	Melinda Ratini, DO, MS (Sept. 2017)	none

Table 8.2 (cont.)

Position in the search	Page Location	Title of Result	Name of WebMD Reviewer	Toxicants Mentioned
17	Second	How is bone marrow diagnosed?	Louise Chang, MD (Feb. 2017)	none
18	Second	What is childhood leukemia?	Louise Chang, MD (Feb. 2017)	none
19	Second	What are treatments for bone marrow cancer	Louise Chang, MD (Feb. 2017)	none
20	Second	What is childhood leukemia?	Neha Pathak, MD (Sept. 2017)	"exposure to chemotherapy or chemicals such as benzene (a solvent)"
21	Second	Juvenile Myelomonocytic leukemia	William Blahd, MD (March 2017)	"some theories are that having a virus or being around toxic chemicals or radiation can cause this to happen"
22	Second	Chronic Myeloproliferative Neoplasms Treatment (PDQ): Treatment-Health Professional Information [NCI]-Chronic Neutrophilic Leukemia	William Blahd, MD (March 2017)	"some theories are that having a virus or being around toxic chemicals or radiation can cause this to happen"
23	Second	What is Acute Myeloid Leukemia? What Causes it?	Laura J. Martin, MD (Nov. 2017)	smoking, benzene, certain cleaning products, detergents, paint strippers, high doses of radiation, chemotherapy drugs

Emphasizing a Reductionist Framework

Even when toxicants are mentioned, the environmental causation perspective can be downplayed by the surrounding context. As argued by Brown et al. (2001), "the context within which environmental causation is mentioned says much about the way it is legitimized or delegitimized" (p. 764). In particular, they found it can be undermined by an article's focus on other causal factors, such as genes, medical treatments and lifestyle factors (such as smoking).

Similarly, WebMD's leukemia coverage also emphasizes genes, as underscored by the second paragraph of the "What is leukemia? What causes it?" page:

> There's really nothing you can do to prevent leukemia. It's cancer of your blood cells caused by a rise in the number of white blood cells in your body. They crowd out the red blood cells and platelets your body needs to be healthy. All those extra white blood cells don't work right, and that causes problems.

This statement clearly situates leukemia in the body and obscures the role of environmental pollutants. The first sentence is particularly problematic as it suggests that even if we know certain toxicants contribute to leukemia, such knowledge can not stop leukemia's development. It is a subtle nod to a gene-based explanation, where genetic programming runs its course regardless of environmental context.

This causal framework is reinforced by the next sub-section ("How does it happen?"), which provides additional information about the physiological process through which leukemia happens, with no information about how environmental factors mediates that process. Reductionist disease framings are deficient because they fly in the face of environmental health research, which shows that while individuals may have genetic susceptibilities to developing disease, those susceptibilities are invariably triggered by the environmental context, not the genes (Steingraber, 2009). Moreover, Steingraber (2009) argues "shining the spotlight on inheritance focuses us on the one piece of the puzzle we can do absolutely nothing about" (p. 291).

The genetic framework is further emphasized at the section's end, which points to family history as a leukemia risk factor: "if an identical twin gets a certain type of leukemia, there is a 20% chance the other twin will have it within a year." This support for the genetic frame assumes that any similarities between twins will be due to genetics. However, the assumption ignores that fetuses can be significantly exposed to toxicants in the womb, and this is particularly true in underprivileged communities. For instance, Goodman (2009) found an average of 200 chemicals in newborn umbilical cord blood. Thus, if disease similarities are found between twins, toxicant exposures have to be considered as a potential contributing factor. At the very least, it is a factor twin studies should control for.

Contextual Factors that Undermine the Environmental Causation Perspective

Besides emphasizing a reductionist framework, there are three other ways WebMD's coverage undermines the environmental causation perspective: 1) placing toxicant information in subordinate positions; 2) surrounding the toxicant information with negating statements; and 3) providing treatment discussions that ignore environmental remediation.

Regarding the first, when toxicants are mentioned, it tends to be deep within the text, after readers get substantial exposure to the reductionist paradigm. For instance, "What is leukemia? What causes it?" does not address toxicants until the article's fourth section ("Causes"), where the authors finally state:

> It may be possible that certain things in your environment could trigger the development of it. For example, if you are a tobacco smoker, you are more prone to some types of leukemia than a nonsmoker. It's also associated with a high amount of radiation exposure and certain chemicals.

That statement is preceded by three sections, with the first telling readers: 1) who is most likely to get leukemia (adult men); 2) "there's nothing you can do to prevent leukemia"; and 3) the key role played by defective white blood cells. The second section describes how blood cells work and what happens when white blood cells act abnormally, while the third discusses different types of leukemia. By the time the reader reads the statement about toxicants, they have been repeatedly exposed to statements that individualize leukemia. Moreover, when toxicants are finally addressed, it is a statement that suffers from the sins of omission and vagueness (i.e. "certain chemicals").

Toxicant information can also be downplayed by surrounding it with negating statements. For instance, the "Causes" section opens with "No one knows exactly what causes leukemia," which erroneously suggests there is a dearth of solid research linking the condition to environmental factors. The subsequent sentence reinforces the reductionist frame by stating "people who have it have certain abnormal chromosomes," while failing to identify the environmental factors that can alter those chromosomes. And the third sentence begins with "You can't really prevent leukemia," which reinforces the notion that knowledge about toxicants will not help people avoid the condition.

Third, the environmental causation perspective can be undermined by treatment discussions that completely ignore the importance of assessing and remediating, if necessary, the patient's living and working environments. On this point, the "What is leukemia? What causes it?" article discusses numerous treatments (including chemotherapy, radiation therapy, stem cell therapy, and even surgery) without mentioning the benefits to be gained from ensuring that the patient's environments are not re-exposing them to harmful toxicants.

Social Consequences

WebMD's leukemia coverage has important public health implications. First, obscuring the toxicants shields chemical manufacturers from blame, thereby reducing their likelihood of: 1) being penalized for their pollution; 2) being held responsible for cleaning it up; 3) having to face tougher regulations; and/or 4) risk profit-harming consumer boycotts. In turn, this means many polluted environments will remain unremediated, manufacturers will continue to pollute, and more humans will be exposed to harmful substances.

Second, the coverage shields politicians from the political repercussions of weak and ineffective toxicant regulations, which weakens pressure to enact tougher policy and regulations. In turn, this also helps maintain a situation where more and more people will be exposed to harmful substances.

Third, the disease framing makes it harder for citizens to protect themselves and their families. WebMD's poor toxicant coverage maintains peoples' ignorance about carcinogenic substances in their living and work sites, which decreases their chances of addressing the problem. This is particularly important for those routinely exposed to toxicants, such as farm workers, families living near farms, and the surprisingly large numbers exposed to carcinogens in the workplace (Fritschi & Driscoll, 2006; Harrison, 2011). The problem is also vitally important for recovering patients. If they survive, their ignorance about toxicants will return them to potentially polluted and disease-exacerbating homes and workplaces. Sandra Steingraber (2009) argues all cancers have "ecological roots" and we have a *human right* to knowledge that will help us uncover those roots. However, that task is made much more difficult when medical information fails to identify known toxicant culprits.

Fourth, obscuring toxicants perpetuates an individualizing understanding of disease, which leads patients to pursue symptom-suppressing treatments, which are themselves toxic and laden with side effects that require further medical attention and medical expenditures (Lazarou et al., 1996). In leukemia's case, it is estimated 5 to 20 percent of AML cases, which are leukemia's most prevalent form, can be attributed to previous cancer treatments. Moreover, the figure is even higher for those treated for breast cancer, gynecologic cancers, and lymphomas, which tend to be treated with particularly toxic medications (O'Donnell et al., 2012).

Accounting for WebMD's Coverage

In trying to account for mainstream disease framings one should consider the surrounding political economy. Brown et al. (2001) argued that print media's individualization of breast cancer is related to the fact "it is easier to press individual responsibility than corporate and/or governmental responsibility" (p. 771). Their statement underscores that disease framings have significant economic and political consequences, and that they need to be related to the dominant political economy. As this pertains to WebMD's leukemia coverage, chemical manufacturers benefit significantly because the individualizing disease framing shields them from blame. Politicians are also protected from the political consequences of weak and ineffective regulations, which could: 1) damage to their reputation; 2) weaken their reelection campaigns; and 3) force them to pass legislation that could sever their relations with industry. The latter is significant as many, if not most, politicians rely on industry election contributions. This is particularly true in the United States, where it is so costly to run for office (Scherer, Rebala & Wilson, 2014). Beyond campaign funding, many politicians benefit from the revolving door with industry, whereby they pass industry-friendly legislation while in office and get rewarded with lucrative industry appointments when they leave office (Faber, 2008).

While a political economy approach provides important context, it does not provide a sufficient explanation. To shed more light on WebMD's coverage we also need an institutional analysis that considers the organization's primary objectives and that relates its knowledge production to its primary social relationships. Although WebMD provides medical information and presents itself as an extension of the medical profession, it is in fact a for-profit entity, whose primary objective is profit accumulation. Their provision of medical information is a means to the end of drawing viewers to their website, in order to generate advertising revenue.

This business model makes them particularly dependent on advertising revenue. As others have shown, such dependence inevitably leads to editorial content being altered to suit the advertisers' interests (Campbell, 2009; Steinem, 2011). In WebMD's case, pharmaceutical companies are major advertisers and they benefit significantly from an individualizing disease framing. Specifically, maintaining the social ignorance about toxicants and disease decreases the likelihood that environmental pollution will be effectively addressed, which means people will continue to be exposed to toxicants, become sick, and create demand for pharmaceutical products.

Although advertiser influence is an important consideration, WebMD's business model makes it even more reliant upon the medical profession. Not only does medical research provide the basis for website content, the development of website articles is itself overseen by medical professionals, as exemplified by the fact its "What is leukemia?" page was reviewed by William Blahd, MD (WebMD, 2017a). WebMD's content is a reflection of mainstream medicine's tendency to reproduce individualizing disease frames that obscure the role of toxicants. Consequently, it behooves us to better understand the social dynamics that contribute to the production of such medical knowledge, including financial motivations and ideological tendencies.

Prioritizing Financial Interests Over Public Health

Even though the medical profession is often portrayed as nobly fighting disease and helping patients overcome illness, research suggests the profession has consistently prioritized financial interests. Paul Starr (1982), in particular, argues the American Medical Association (AMA) has, since its 1847 inception, consistently prioritized protecting and expanding the physicians' financial interests, which it has done by increasing its professional standards, embracing biomedical approaches, and working to undermine its healthcare competitors. An example of the latter is the profession's steadfast opposition to public health's prevention initiatives. For instance, in the 1920s the AMA and its lobbyists thwarted initiatives to establish neighborhood public health centers, due to their fear the centers would provide free care to people who would otherwise pay for medical care (Brandt & Gardner, 2000). Similarly, in 1921 the profession derailed public health's initiative to provide pre and postnatal care for infants and their mothers (Brandt & Gardner, 2000). Moreover, while public health professionals have consistently supported proposals to provide universal healthcare to all United States citizens, such initiatives have been consistently opposed by the American medical profession, including in the late 1950s, when they used red-scare tactics to reduce public support from 75 percent to 25 percent (Quadagno, 2004).

The profession's financial orientation is also manifested by the "medical politicking" (Conrad & Schneider, 1994) the AMA has pursued vis-a-vis other medical practitioners, where they have sought to weaken their competition by stigmatizing and delegitimizing their services. An example is the AMA's nineteenth-century moral crusade to stigmatize abortion, which led to abortion being criminalized in many states in 1866 and eventually the rest of the country. Although physicians did not provide abortions at that time, many competing disciplines did, with many practitioners having lucrative practices that suffered significantly with abortion's criminalization (Conrad & Schneider, 1994). Similarly, during the 1960s and 1970s American medicine used similar tactics against chiropractic, which had emerged as an economic threat (Winnick, 2009). Such tactics have also been used against midwives, acupuncturists and, in more recent decades, against naturopathy (Baer, 2001; Winnick, 2009).

These examples underscore that the medical profession prioritizes financial considerations even when it threatens the public's health. In turn, this provides a frame through which to understand why the profession would produce medical knowledge that obscures the role of toxicants. Producing such knowledge financially benefits the profession in three ways. First, it helps maintain public ignorance about the dangers of toxicants, which reduces public pressure on politicians to enact tougher regulations, guarantees people will continue to be exposed to dangerous products, and maintains a steady flow of new patients. Second, it preserves the profession's market share. While mainstream medicine offers little to undo the health effects of toxicant exposure, there are other practitioners (including osteopaths, naturopaths, doctors of Chinese medicine, and doctors of environmental medicine) who claim that ability. However, patients are less likely to seek them out if they are unaware of the relationship between toxicants and disease.

Third, obscuring the role of toxicants enables medical organizations (like the AMA and other professional societies) to maintain lucrative partnerships with industry, such as pharmaceutical manufacturers, which benefit significantly from concealing the environmental sources of disease. However, medical organizations also form partnerships with environmental polluters, who benefit mightily from obscuring the relationship between toxicants and disease. For example, the National Comprehensive Cancer Network (NCCN) (i.e. the producer of the clinical practice guidelines for leukemia) website lists General Electric (GE) as one of their sponsors, in addition to numerous pharmaceutical companies. GE has a lengthy track record of polluting the environment, including with PCBs and other carcinogens, as well as deliberately exposing citizens to nuclear radiation (Multinational Monitor, 2001). By 2001 they were deemed wholly or partially liable for at least 78 federal Superfund sites, had paid hundreds of thousands of dollars in fines, and was forced to pay a $200 million settlement for its pollution of the Housatonic River in

Massachusetts (Multinational Monitor, 2001). Sponsorship from such companies undoubtedly comes with an implicit, if not explicit, understanding that disease coverage will downplay the role of toxicants.

Ideological Opposition to Preventive Approaches

Beyond material interests, the omission of environmental pollutants can also be attributed to a worldview that is hostile to public health and its prevention initiatives. Although Brandt & Gardner (2000) identify many ways organized medicine has opposed public health over the twentieth century, they caution against solely attributing this to financial self-interest, arguing it can also be traced to medicine's deep ideological adherence to the reductionist biomedical paradigm.

When the AMA was founded the profession suffered from a poor reputation, low scientific credibility and low moral authority. To some extent, this was due to the profession's low level of professionalization and standardization (Starr, 1982). However, it was also due its reliance on "heroic medicine," which relied heavily on bloodletting, blistering, vomiting, and purging (Starr, 1982). The potential harmfulness of such practices stoked public opposition and contempt for the profession. The AMA addressed the issue by steering physicians away from such practices and towards bio-medicine, which included a growing reliance on diagnostic technologies (including x-rays and stethoscopes), a firm adherence to bacteriology and the germ theory of disease, which located disease in the individual (Starr, 1982).

The reductionist paradigm gave physicians an understanding of disease and approach that enabled them to decouple disease from its social roots, thereby making public health's broad social and environmental agenda appear unnecessary. In particular, physicians were attracted to the paradigm's "science-based objectivity and technique," which "contrasted with the tumultuous world of public health" (Brandt & Gardner, 2000, p. 711). Physicians were suspicious of public health endeavors, as they believed it was difficult to address disease-causing social conditions and such efforts were rendered even more difficult because they were tainted by "politics, advocacy, individual noncompliance, and social diversity" (ibid). The physician worldview dictated that "medicine could not solve the problems of poverty, illiteracy, and inequity–but it could, at least potentially, cure the diseases that these social forces produced" (ibid). As well, many medical advocates argued that improving health and life expectancy through medical interventions would eventually lead to reductions in poverty and social inequities (ibid).

In turn, this worldview strongly influenced medical training and medical practice, orienting both toward disease-focused reductionism and away from a concern about contextual sources of disease (Brandt & Gardner, 2000). In the 1920s there were many physicians who recognized the importance of prevention-oriented research and teaching. However, this objective was overshadowed by the medical schools' focus on intensive scientific and clinical training (Brandt & Gardner, 2000). This was also true in the 1930s and 1940s, where attempts to introduce preventive medicine failed to alter the dominance of the reductionist paradigm (Brandt & Gardner, 2000).

In the 1990s the problem was still present as 25 percent of medical schools offered no instruction in environmental medicine and those that did averaged less than ten hours of instruction over four years (Schenk et al., 1996). Relatedly, two thirds of medical school deans reported their schools offered minimal coverage of environmental issues (Graber et al., 1995), with similar results being reported for residency training (Musham et al., 1996; Lees, 1996). While the issue has garnered attention in recent decades, the problem continues to persist. For instance, in 2010 35 percent of graduating medical students reported being under-educated in environmental health (AAMC, 2010). Moreover, surveys find that while physicians acknowledge the importance of environmental factors, few receive the training to conduct a proper environmental history. For instance, Kilpatrick et al. (2002) found only 21 percent of Georgia pediatricians had received such training, while Zachek et al. (2015) found the same was true for only 7 percent of pediatric hematologists and oncologists.

These problems significantly impact medical practice as physicians lack the competence and confidence to engage with environmental issues, and invariably ignore these issues in the clinical encounter (Kilpatrick, 2002; Trasande et al., 2010; Zachek et al., 2015). For example, 73 percent of pediatric hematologists and oncologists reported rarely or never seeing a case they suspected was related to the patient's environment (Zachek et al., 2015). Moreover, 44 percent reported discomfort with having conversations with patients and their families about potential environmental causes of disease (Zachek et al., 2015). Additionally, Trasande et al. (2010) found that while Michigan pediatricians voiced high self-efficacy with addressing problems related to lead and second-hand smoke, they were far less confidant when dealing with pesticides, air pollution, PCBs, mercury, and mold exposures. Moreover, while pediatricians routinely refer patients to lead/toxicology clinics, they typically do not refer patients to regional pediatric environmental health specialty units, which could help patients address exposures to other toxicants (Trasande et al., 2010).

Thus, while medicine has financial interests for producing medical information that obfuscates the role of environmental pollutants, such knowledge production has deep ideological roots, which have created an education system that encourages doctors to ignore environmental pollution and deprives them of the tools to address it. If environmental medicine was more emphasized in medical school, we would have physicians who demand and produce medical information that better acknowledges the relationship between environmental pollution and disease.

Conclusion

Environmental pollution is an important source of disease. However, this information tends to be obscured by mainstream medical information, which fails to identify most toxicants related to disease and presents toxicant information in ways that systematically downplays its importance. This reinforces the dominance of the reductionist medical paradigm, which attributes disease to genes and personal choices. In turn, this makes it harder for individuals to protect themselves and their families from polluted contexts and the industries who produce them. Moreover, while polluting industries and politicians are shielded from public scorn, pharmaceutical manufacturers and the medical profession benefit from a steadily growing flow of patients.

Although mainstream sources obscure the environmental causation frame, many are working to change the situation. Besides the environmental health researchers who document and publicize the links between disease and pollution, there are scientific organizations (such as Silent Spring Institute and The Collaborative on Health and the Environment) that work to educate the public (Brown et al., 2009). Additionally, environmental justice activists are publicizing the links between environmental pollution and disease in local communities, as well as organizing the communities to eliminate those problems (Brown et al., 2009). In the medical field, individual healthcare practitioners, such as functional medicine doctors, have pursued training in environmental health and have incorporated the knowledge into their clinical practice. Moreover, some practitioners (such as Dr. Mercola, Dr. Weil, and Dr. Oz) have sought to educate the public by integrating toxicant information into their website content.

While these efforts are important, their capacity to bring about change is limited because they are not tackling the problem's roots. One root cause is medical curricula that obscure or marginalize the role of toxicants. We need research that illuminates how such deficiencies are socially constructed and reproduced, which includes identifying: 1) the process through which school curricula are revised; 2) the people who make key decisions; 3) the values in their calculus; and 4) the social forces that shape that value system.

Another root cause is the public's general ignorance about toxicant harmfulness to humans and ecosystems, which can also be traced to educational shortcomings. In particular, Woodhouse & Howard (2009) argue North American universities are failing to educate most, let alone all, students about the ecological and human harms associated with the production, use, and disposal of common toxicants. This is another area that should be tackled by future research. Key in this regard would be to analyze places in academia were such information would be most expected (such as chemistry courses) and to explore why those sites are failing to provide it. In turn, this knowledge would enable us to correct these major educational failings and, in the process, create a populace that is more aware of toxicant harmfulness, better able to protect themselves and their families, as well as more demanding of tighter regulations and medical information that accurately conveys the disease-causing effects of environmental pollution.

References

Association of American Medical Colleges (AAMC). (2010). *Medical School Graduation Questionnaire: All Schools Summary Report: Final*. Retrieved from http://docplayer.net/7369062-Gq-medical-school-graduation-questionnaire-all-schools-summary-report-final.html.

Atkin, C., Smith, S., McFeters, C., & Ferguson, V. (2008). A Comprehensive Analysis of Breast Cancer News Coverage in Leading Media Outlets Focusing on Environmental Risks and Prevention. *Journal of Health Communication*, 13(1), 3–9.

Baer, H. (2001). The Sociopolitical Status of U.S. Naturopathy at the Dawn of the 21st Century. *Medical Anthropology Quarterly*, 15(3), 339–346.

Brandt, A., & Gardner, M. (2000). Antagonism and Accommodation: Interpreting the Relationship Between Public Health and Medicine in the United States During the 20th Century. *American Journal of Public Health*, 90, 707–715.

Bray, C. (2017, July 24). K.K.R. to Buy WebMD and Take Majority Stake in Nature's Bounty. *The New York Times*, p. B7. Retrieved from www.nytimes.com/2017/07/24/business/dealbook/kkr-webmd-natures-bounty.html

Brown, P., Zavestoski, S., McCormick, S., Mandelbaum, J., & Luebke, T. (2001). Print Media Coverage of Environmental Causation of Breast Cancer. *Sociology of Health & Illness*, 23(6), 747–775.

Brown, P., Zavestoski, S., McCormick, S. et al. (2009). Embodied Health Movements: New Approaches to Social Movements in Health. In Peter Conrad (ed.), *The Sociology of Health & Illness: Critical Perspectives* (pp. 592–604). New York: Worth Publishers.

Campbell, E. (2009). Corporate Power: The Role of the Global Media in Shaping What We Know About the Environment. In Kenneth Gould and Tammy Lewis (eds.), *Twenty Lessons in Environmental Sociology* (pp. 68–84). New York: Oxford University Press.

Collaborative on Health and the Environment (CHE). (2018a). Toxicant and Disease Database. *Collaborative on Health and the Environment*. Retrieved from www.healthandenvironment.org/our-work/toxicant-and-disease-database/

Collaborative on Health and the Environment (CHE). (2018b). A Brief History of CHE. *Collaborative on Health and the Environment*. Retrieved from www.healthandenvironment.org/about/a-brief-history

Collaborative on Health and the Environment (CHE). (2018c). About the Toxicant and Disease Database. *Collaborative on Health and the Environment*. Retrieved from www.healthandenvironment.org/our-work/toxicant-and-disease-database/about-the-toxicant-and-disease-database

Comscore. (2016, January 21). Comscore Ranks the Top 50 U.S. Digital Media Properties for December 2015. Retrieved from www.comscore.com/Insights/Market-Rankings/comscore-Ranks-the-Top-50-US-Digital-Media-Properties-for-December-2015/

Conrad, P., & Schneider, J. (1994). Professionalization, Monopoly, and the Structure of Medical Practice. In Peter Conrad and Rochelle Kern (eds.), *The Sociology of Health & Illness: Critical Perspectives* (4th ed.) (pp. 167–173). New York: St. Martin's Press.

Faber, D. (2008). *Capitalizing on Environmental Injustice: The Polluter-Industrial Complex in the Age of Globalization*. Lanham: Rowman & Littlefield.

Foster, John Bellamy. (1999). *The Vulnerable Planet: A Short Economic History of the Environment*. New York: Monthly Review Press.

Fritschi, L., & Driscoll, T. (2006). Cancer Due to Occupation in Australia. *Australian and New Zealand Journal of Public Health*, 30(3), 213–219.

Goodman, S. (2009, December 02). Tests Find More than 200 Chemicals in Newborn Umbilical Cord Blood. *Scientific American*. Retrieved from www.scientificamerican.com/article/newborn-babies-chemicals-exposure-bpa/

Graber, D., Musham, C., Bellack, J., & Holmes, D. (1995). Environmental Health in Medical School Curricula: Views of Academic Deans. *Journal of Occupational and Environmental Medicine*, 37(7), 807–811.

Harrison, J. (2011). *Pesticide Drift and the Pursuit of Environmental Justice.* Cambridge, MA: MIT Press.

IARC. (2018). IARC Monographs on the Evaluation of Carcinogenic Risk to Humans. Retrieved from http://monographs.iarc.fr/ENG/Monographs/vol77/index.php

Kilpatrick, N., Frumkin, H., Trowbridge, J. et al. (2002). The Environmental History in Pediatric Practice: A Study of Pediatricians' Attitudes and Practices. *Environmental Health Perspectives*, 110, 823–827.

Lazarou, J., Pomeranz, B., & Corey, P. (1996). Incidence of Adverse Drug Reactions in Hospitalized Patients. *Journal of the American Medical Association*, 279(15), 1200–1205.

Lees, R. (1996). Occupational and Environmental Health: Preparing Residents to Treat Related Illnesses. *Canadian Family Physician*, 42, 594–596.

Lewison, G., Tootell, S., Roe, P., & Sullivan, R. (2008). How Do the Media Report Cancer Research? A Study of the UK's BBC Website. *British Journal of Cancer*, 99, 569–576.

MacKendrik, N. (2010). Media Framing of Body Burdens: Precautionary Consumption and the Individualization of Risk. *Sociological Inquiry*, 80(1), 126–149.

Multinational Monitor. (2001, July/August). GE: Decades of Misdeeds and Wrongdoing. *The Multinational Monitor*, 22(7–8). Retrieved from www.multinationalmonitor.org/mm2001/01july-august/julyaug01corp4.html

Musham, C., Bellack, J., Graber, D., & Holmes, D. (1996). Environmental Health Training: A Survey of Family Practice Residence Program Directors. *Family Medicine*, 28(1), 29–32.

National Cancer Institute (NCI). (n.d). Leukemia – Patient Version. National Cancer Institute. Retrieved from www.cancer.gov/types/leukemia

National Institute of Health (NIH). (n.d). "Cancer Stat Facts: Leukemia. National Institute of Health. Retrieved from https://seer.cancer.gov/statfacts/html/leuks.html

O'Donnell, M., Abboud, C., Altman, J., Applebaum, R., Arber D. et al., (2012). Acute Myeloid Leukemia: Clinical Practice Guidelines in Oncology. *Journal of the National Comprehensive Cancer Network*, 10(8), 984–1021.

OEHHA. (2018a). About. Office of Environmental Health Hazard Assessment. Retrieved from https://oehha.ca.gov/about

OEHHA. (2018b). Risk Assessment. Office of Environmental Health Hazard Assessment. Retrieved from https://oehha.ca.gov/risk-assessment

OEHHA. (2018c). Chemicals. Office of Environmental Health Hazard Assessment. Retrieved from https://oehha.ca.gov/chemicals

Quadagno, J. (2004). Why the United States Has No National Health Insurance: Stakeholder Mobilization Against the Welfare State, 1945–1996. *Journal of Health and Social Behavior*, 45(extra issue), 25–44.

Schettler, T., Stein, J., Reich, F., Valenti, M. & Wallinga, D. (2000). *In Harm's Way: Toxic Threats to Child Development.* Greater Boston Physicians for Social Responsibility. Boston, MA: Red Sun Press.

Schenk, M., Popp, S., Neale, A.V., & Demers, R. (1996). Environmental Medicine Content in Medical School Curricula. *Academic Medicine*, 71(5), 499–501

Scherer, Michael, Pratheek Rebala, and Chris Wilson. (2014, October 23). The Incredible Rise in Campaign Spending; *Time.* Retrieved from http://time.com/3534117/the-incredible-rise-in-campaign-spending/

Sexton, K., Needham, L., & Pirkle, J. (2004). Human Biomonitoring of Environmental Chemicals: Measuring Chemicals in Human Tissues Is the "Gold" Standard for Assessing People's Exposure to Pollution. *American Scientist*, 92, 38–45.

Starr, P. (1982). *The Social Transformation of American Medicine*. New York: Basic Books.

Steinem, G. (2011). Sex, Lies, and Advertising. In Gail Dines and Jean Hurney (eds.), *Gender, Race, and Class in Media: A Critical Reader* (3rd edition) (pp. 235–242). Thousand Oaks, CA: Sage Press.

Steingraber, S. (2009). The Social Construction of Cancer: A Walk Upstream. In Leslie King and Deborah McCarthy (eds.), *Environmental Sociology: From Analysis to Action* (pp. 287–299). Lanham, MD: Rowman & Littlefield.

The Write News. (2005, April 22). WebMD Corporation Launches Print Magazine. Retrieved from www.writenews.com/webmd-corporation-launches-print-magazine-42220055

Trasande, L., Newman, N., Long, L., Howe, G., Kerwin, B. et al. (2010). Translating Knowledge About Environmental Health to Practitioners: Are We Doing Enough? *Mount Sinai Journal of Medicine*, 77, 114–123.

Vallée, M. (2013). Perpetuating a Reductionist Medical Worldview: The Absence of Environmental Medicine in the ADHD Clinical Practice Guidelines. *Advances in Medical Sociology*, 15, 241–264.

WebMD. (2017a). What Is Leukemia? Retrieved from www.webmd.com/cancer/lymphoma/understanding-leukemia-basics#1

WebMD. (2017b). What Is Acute Myeloid Leukemia (AML)? Retrieved from www.webmd.com/cancer/lymphoma/qa/what-is-acute-myeloid-leukemia-aml

Winnick, T. (2009). From Quackery to "Complementary" Medicine: The American Medical Profession Confronts Alternative Therapies. In Peter Conrad (ed.), *The Sociology of Health & Illness: Critical Perspectives* (pp. 261–275). New York: Worth Publishers.

Woodhouse, E., & Howard, J. (2009). Stealthy Killers and Governing Mentalities: Chemicals in Consumer Products. In Merrill Singer and Hans Baer (eds.), *Killer Commodities: Public Health and the Corporate Production of Harm* (pp. 35–66). Lanham, MD: Altamira Press.

Zachek, C., Miller, M., Hsu, C. et al. (2015). Children's Cancer and Environmental Exposures: Professional Attitudes and Practices. *Journal of Pediatric Hematology/Oncology*, 37(7), 491–497.

PART III

Beyond the Human

9 Interventions Offered by Actor-Network Theory, Assemblage Theory, and New Materialisms for Environmental Sociology

Katharine Legun and Abbi Virens

Introducing the Post-Human

Central to work in environmental sociology is a philosophical question about how we should understand the interrelationships between people and their environments. It was a concern with the ways that the environment had been omitted from social analyses that inspired some of the foundational work introducing environmental sociology to mainstream sociology, with Catton and Dunlap (1978) suggesting we abandon a Human Exceptionalism Paradigm, later updated to the Human Exemptionalism Paradigm (Catton and Dunlap 1980; Dunlap and Catton 1994), that had dominated sociology for a New Ecological Paradigm that embraced the centrality of the environment in social life. If we pursue a better society without considering how embedded we are in an ecological world, as Catton and Dunlap suggested, sociologists are just participating in degradation for the pursuit of progress.

While the call to position the environment more centrally into social research was clear, we still have a broad range of ideas about *how* we can better incorporate ecology into our understanding of social dynamics. Indeed, in his review of changes in the field of environmental sociology, Dunlap (2010: 15) suggests, "societal–environmental interactions remain the most challenging issue, and divergent approaches to them, the source of our most fundamental cleavages." These cleavages can be linked to early debates about the degree to which the environment can be directly known and understood as having universal, material properties, or whether it is always a product of our discursive, social interpretations.

These challenges emerge partly along divisions between modern or post-modern scholarly approaches that exist in many realms of academic work. Yet, these divisions may be more profound and paradoxical in relation to the environment. Modernity has been grounded in a commitment to science and technology as a driver of social improvement and progress. Those same commitments have been seen to lead to environmental degradation, and in some cases, environmental catastrophes, due to their use for environmental exploitation and their seeming levels of objective certainty, paired with the inherent limitations of knowledge (Saurin 1993; Beck 1992; Giddens 1990). This is compounded by ambivalence towards those experts who seemingly hold a monopoly on environmental insights (see Beck 1992), particularly when that science is being used to support policies and projects that may undermine the values and interests of populations who may have alternative sources of knowledge (see, Béné 2005; Agrawal 2002).

As a result, there has been a resistance to taking the insights of natural science at face value, and more post-modern approaches have considered the social construction of scientific knowledge as a culturally and politically laden activity. This extends an underlying philosophical commitment to the possibility that the same reality can be experienced and interpreted differently by individual people. However, this can present a conundrum for those who are concerned about the environment, where a reliance on the natural sciences are necessary for a better understanding of ecological phenomena, which are often imperceptible to bare human senses. As environmentally concerned scholars with genuine interests in the health of our planet, we need to both engage in a critical approach that attends to the social nature of science, while also taking seriously the alarming trends being identified by scientists (see Latour 2004). Environmental sociologists are somewhat trapped between modernism and post-modernism, and perhaps this is why so many have embraced new ways to be both realist and ambivalent about that realism.

In this chapter, we will propose that new post-human or more-than-human approaches provide one method for thinking deeply about how society is produced and reproduced through the environment, but without relying on essentialist understandings of the environment and its effects, typical of more modernist approaches. It also helps us avoid being paralyzed by the inability to accept concrete environmental effects on humans, often associated with post-modernism (for a thorough discussion, see Latour 2012). We do this by considering how *things*, be they plants, technologies, or toxins, participate materially in shaping our practices, subjectivities, and social lives, while recognizing that our knowledge of them is defined by that particular material participation. We also recognize that the effects of things, while meaningful, are dependent on the complex social domains in which they are situated. We see some of these themes emerging through actor-network theory or assemblage theory, which are particular approaches that have grown in popularity in sociology and geography, and part of an overall engagement with "new materialisms" present across social sciences and humanities.

From Materialism to New Materialisms

More-than-human approaches can be seen to fall under the umbrella of "new materialism." While the role of materials in society is not a new idea, new materialist approaches aim to embrace the real qualities of things in our environments and take seriously their effects, while also resisting a more determinist or modernist approach to materiality that sees those qualities as stable and those effects as inherent. If we immediately dismiss the real features of the things around us, the world becomes primarily constituted of subjective understandings and socially constructed realities, and materials are inconsequential to understanding our world and its dynamics. If we take materials as stable, and their effects inherent, they once again become less meaningful in understanding our world, as they become flat objects, only interesting in how they are influenced by humans or distributed amongst us. As Coole and Frost (2010) suggest, new materialisms aim to operate outside of more post-structural and structural approaches by taking materials seriously while embracing their multiplicities.

Materiality has been an undercurrent in a great deal of sociological work. After all, class, one of the central concepts to sociology, is grounded in the unequal distribution of goods, and the unequal power that some have over the material conditions under which we live our lives. Historical materialism is a perspective describing the social effects of increased technological development and concentration of capital. In this work, goods and material conditions are extra-human factors in social life that have effects on how people live their everyday lives, and also how they interpret how they can and should live their everyday lives. Yet, in most aspects of class analysis, materials themselves are not a focal point and have little meaning independently of the meaning attributed by society. In other words, materials do not create inequality, but rather property rights that give some exclusive and enduring access to certain materials create inequalities.

Materials still play a central role in the specific dynamics of change described by Marx and Engels. In *the German Ideology*, the fact that weaving required machinery made it a catalyst for the emergence of industrial social relations; bringing people to towns, generating more extensive trade and capital accumulation, and reducing the power of guilds. In other words, we could see the dependence of weaving on the loom as having practical consequences with social and political effects. As new productive technology develops for post-industrial purposes, we could anticipate some new social and political effects, even within the basic dynamics of capitalism. Attending to those differences can offer some new insights that help us understand the current shape of our economy and our society. Work focusing on the more-than-human aims to hone that focus by considering what exactly our material lives encourage us to do, how, why, and with what effects. By focusing on these questions, we can look intricately at why particular practices have emerged and continue, without implying that they are natural, inevitable, or wholly intentional.

We can see both the reproduction of basic capitalist systems, and their evolution and change, with respect to new technologies. For example, new computer technologies and the Internet have been linked with the increased magnitude of accumulation and financialization (Ma and McGroarty 2017; Roberts 2014), new forms of exploitation in the so-called "sharing economy" and "gig economy" (Kenney and Zysman 2016) as well as the explosion of creative self-exploitation through online entrepreneurialism (Van Doorn 2017). The material characteristics of these technologies that network people and their everyday lives into an online world governed by algorithms, is important. As James Bridle (2018) suggests in his book, *New Dark Age: Technology and The End of the Future*, our understandings of these technologies and how they interact with each other is dwarfed by their significance, capacities, and effects. In short, we are at the mercy of material forces of change that we struggle to see and understand, let alone steer. Humans are being redefined by technologies in ways that illuminate the force of technology and the limitations of individual human agency.

Bridle's work echoes critiques of modernization generally, in the ways that he describes people becoming enrolled in a world being driven by technologies and their profiteers, without the knowledge, power, or language for democratic control. In *Tasting Food, Tasting Freedom*, Sidney Mintz opens the text by drawing on a narrative by Nigerian author Amos Tutuola from *The Palm-Wine Drunkard*. Tutuola's story is of a child who develops rapidly and exercises insatiable hunger. His consumption is unstoppable, as those around him are unable to keep up with him or protect their own consumption in his presence. His name is Zurrjir, "which means a son who would change himself into another thing very soon." Mintz uses this as a metaphor for modernity. He remarks that the story was released on the eve of Nigerian independence, making the metaphor even more meaningful. There is a desire for transformation, so fixating and demanding that it cannot be stopped, but it is a burdensome and destructive process, as much as a promissory one. He goes on to use this to foreground a broader review of his work on the historical and political crafting of food traditions, one aspect of which seems to be this separation of pleasure from consumption. Sugar and coffee became utilitarian as they helped the delivery of calories to the labour force and the extension of the work day. Tutola's character well encapsulates these dynamics, as a child who has hunger so demanding that food becomes utilitarian and almost anti-relational, or anti-sensory. For the human body, it becomes both necessary and qualitatively, experientially, unknowable.

In both the work of James Bridle and Sidney Mintz, there are features of technologies that are knowable in an abstract, applied, and utilitarian sense, and these are driving their incorporation into the economy. But these technologies also have effects beyond their purported use. They become necessary to our everyday lives and powerful, shaping the rational calculus from which we build our lives.

While these features of technology – the power of their compelled integration into the functioning of our everyday lives – may become visible during periods of transition, post-human scholars would claim that it is true of all materials, all the time. Materials seem stable and controllable and knowable when a whole bunch of other features of that world are held constant. This is because materials are, at once, real (or natural), social (or human), and semiotic (discursive). While we understand real features of our world to make informed decisions, we have never pursued progress with a totalizing commitment to a singular reality. Instead progress is always partially about the exercise of power due to the social and discursive elements of our reality. This basic approach is part of Bruno Latour's claim that *We Have Never Been Modern*: the objects around us are never fully understood and their purpose is never singular or inherent. When we know things, like atoms or rivers or air, it is due to some kind of consistent conditions under which those materials are experienced.

In other words, in order for us to know things and get things to do what we want them to do, we have to do a whole lot of work as well to essentially set them up. So are the behaviours of objects, as we know them, inherent in themselves? Or a reflection of us – our intentions and our work? Or something in between: a relational effect? Those taking a more-than-human approach would suggest the latter. We use things, we depend on them, we incorporate them into our activities in ways that seem sensible and reliable, but we are always, to some extent, at their mercy and orienting our own activities to making them work. And why do we do this? Because we want the outcome of that collaboration. A post-human approach considers how we have incorporated things into our lives, and considers what capacities for action have been enabled through their incorporation. For environmental sociology, a post-human approach allows us to place features of the environment at the centre of our social analyses by considering how they affect our capacities, and our very ability to be human. The environment in this framework is not simply an object of our attention as we extract resources or learn botany or frolic in the wild. We are subjects of our environment, in that every action that we undertake, and every action that we imagine we can undertake, relies on the cooperation of environmental agents. What's more, those relationships could easily change, shifting our own capacities and the capacities of our non-human comrades as well.

Working with Post-Humans

There are a wide range of ideas that are generally categorized as post-human or more-than-human, but a central feature is their emphasis on the ways that agencies, practices, or performances of a person are contingent on the agencies, behaviours, practices, or performances of other, non-human things. For actor-network theory (ANT), the focus tends to be on the kinds of agency created through collective, heterogeneous collaboration, and the meaning-making that happens through that collective process. For assemblage theory, the focus tends to be slightly more on the historical and contextual significance of a particular constellation of things that seem to be operating together. Others have embraced a broad, fluid approach, situating their work within the "new materialisms," which aims to turn away from more modernist and humanist approaches that take the individual body as an assumed unit of analysis, and the capacities of humans as a given. New materialism within this frame has been a particularly meaningful intervention in feminist and queer studies around destabilizing, deconstructing, and critiquing the material-discursive realities that shape things like gender and sexuality. Below, we'll elaborate on each of these approaches, and how they might inform environmental sociology.

Actor-Network Theory

Actor-network theory (ANT) emerged in the 1980s, largely as a critique of science and through empirical research of scientists and laboratories. Michel

Callon's work on scallops was perhaps one of the clearest early examples of ANT in action (Callon 1984). In his work, he described the process through which scientists learned about the behaviours and reproduction rates of the local scallops – the results of which would influence policy around fisheries. The tanks housing the scallops would have a particular water temperature, a particular salinity and oxidation, and food would be introduced to the tanks in a carefully measured and consistent way. Only under very controlled conditions could the scientists define the true behaviour of the scallop. The laboratory conditions produce behaviours that should be replicable under the same conditions, and any changes in those conditions were shown to have different results. The implication is that, in order to make the real, wild scallops into coherent scientific objects, there were a range of other agents that also needed to be part of the network. The process through which the wild "real" scallop becomes the social scallop, and the discursive scallop, is called translation, and while we may deploy that discursive scallop in a range of different contexts, its meaning is the product of a specific material network and its actions or behaviours are generated through that network. The relationship between specific material networks, agency, and meaning is often referred to as *material semiotics* (for more discussion of material semiotics see Law 2008).

We could expand material semiotics to think more carefully about human agency. For example, when we drive our cars, we often think that we have the agency to drive, and are personally engaging in that activity. And yet, our ability to drive is contingent on a wide array of things that are outside of our personal capacities. We actually do very little in the activity of driving – we merely turn a key, place our feet on the pedals, and guide the vehicle as we go. A lot of the action of driving is done by the car, and the car depends on fuel, roads, and also other humans. We could see all of these components together producing the action of driving, and should one of these components be absent, driving would not happen, and at least not in the way that we are accustomed. ANT scholars approach the world as being constituted by such networks, so that there are no inherently meaningful objects that exist in the world independently of the networks in which they are enrolled (for a review of ANT, see Latour 1996).

For thinking about society and the environment, actor-network theory can help us attend to the ways that; 1) our human actions, and capacities for action, rely on an extensive cast of ecological characters who collaborate with us in our various endeavours 2) technology influences how the characteristics of ecological agents figure into our activities, making some kinds of human–ecological relationships possible 3) our knowledge and use of the environment reflects just one iteration of many possible ways that we could network together the world around us and 4) what we know of ecological processes and capacities could change, and these changes could also enable us to do different things as well and 5) there is a relationship between how we materially experience the world and how it is defined discursively, but these materialities and discursive understandings are only one of many possibilities, and tenuously held together.

Some argue that actor-network approaches ignore power, or are hostile to it. This point comes from the ways that an ANT approach stresses the ways that agency is a product of relationships between things that do not do anything on their own. No single entity in the network is individually producing the action across the network, leading to the claim that there is a kind of ontological flattening of all things along a similar agency plane. On the other hand, we can use power to explain why some networks become successful and others do not. Latour uses the concept of "immutable mobiles" to explain how, for example, scientific definitions of things in nature spread outside of the networks in which they were created (Latour 1987). We could also say that some understandings of ecological life align with some kind of moral or ideological preferences, or have some unequal and strategic advantages in terms of the kinds of actions afforded to particular network players, so they catch on. This kind of application of ANT will be discussed later in the context of apples.

Other researchers tend to draw on assemblage theory to talk more explicitly about power, but in a way that echoes ANT in its post-modern articulations. It is postmodern because it rejects any claim that materials have singular, inherent characteristics; or that knowledge of reality follows a progressive linear trajectory over time; or that power is passively experienced in coherent social structures. Because those aligning with a more-than human approach consider our material experiences of reality to be fundamentally unstable and not predetermined by those materials, there is an embrace of multiple possible futures, a rejection of linear progress, and a reading of power which relies on action and practice.

Assemblage Theory

Assemblage theory emerged initially from the work of Gilles Deleuze and Felix Guattari, where seeds of the ideas were embedded in their text *A Thousand Plateaus, Capitalism and Schizophrenia*. Like ANT, assemblage theory refers to heterogeneous things hanging together, or forming symbiotic relationships. The idea was not particularly well developed by Deleuze or Guattari, but it featured enough to be picked up later. Manuel DeLanda has been at the forefront of this later elaboration, describing *agencement* as a foundation to the conceptualization of assemblage (DeLanda 2016). *Agencement* refers to both the process of fitting together components, and the resulting assemblage of components that fit together. It is perhaps because of the ways that *agencement* has stressed the processes of bringing things together that it has been used more widely to focus on how these particular heterogeneous configurations came together, with some of the most forceful applications being used by scholars to explain neoliberal processes (Ong 2007; Brady 2014), or the local geographical experiences of globalization (McFarlane 2009).

The assemblage approach has been particularly forceful because it allows for the analysis of highly diffuse micro-power, following the work of Foucault. Foucault argued that governmentality was happening in new ways, where governing forces would not be exerted top-down from the state, but instead through seemingly disparate entities and institutions that would tell people how they should behave to be good citizens. People willingly participate and subject themselves to these forces. We would expect that neoliberalism, or the reduction of regulation for the expansion of market forces would increase the kinds of governmentality processes Foucault identified. As Ong (2007) argues, if neoliberalism is the unraveling of structure, we need analytic tools that can better capture coherent moments of governance that are being cobbled together across space and with more diverse resources. Assemblage theory has been particularly useful in thinking about how heterogeneous forces are brought together in ways that exert a form of governing force.

The focus on governance is perhaps due to ways that assemblage theory has often come to focus more on the process of bringing things together, rather than the relationships between them. While the term assemblage may be used as a "descriptor for some form of provisional unity across difference" (Andersen and McFarlane 2011), its analytic application is often more action oriented. From an assemblage approach, we can see power being exercised in the act of assembling, or assemblying – the latter is sometimes used to denote a form of gathering by actors, specific to the intentional formation of a particular kind of assemblage. As Tania Murray Li (2007: 264) emphasizes in her work, "assemblage links directly to practice, to assemble." She identifies four processes that can be part of assembling and that create governing forces: 1) *forging alignments*, or bringing heterogeneous objects together 2) *rendering technical*, or representing messy processes as a set of causal relations 3) *authorizing knowledge*, or defining the arena of relevant knowledge and 4) *Managing failures and contradictions*. Li also suggests that *anti-politics* are created to redefine political questions as technical questions and *reassembling* is undertaken to constantly tweak the assemblage. All of these ways of assembling as an action emphasize the ways that assemblages are constantly in flux.

While the analytic focus in assemblage theory is often on processes of governance, materials, such as environmental or landscape elements, play a significant role in assemblages and assemblying as a factor that shapes how those processes occur and develop. In her work on creating land as an object for investment, for example, Li (2014) describes how in Sulawesi, Indonesia, there was no native word for land apart from the social understanding of what the land meant for the community. She explains how material features of landscape would relate to the social uses of it (2014: 590):

> Primary forest, for example, means forest in which no one has ever taken axes to trees. Since highlanders consider the investment of labour to create individual property, when they note that a patch of forest is *do 'at*, they are not just commenting on the enormous size of the trees. They are noting that no one owns it yet, and hinting towards its potential for use, and future status as individual property when labour is applied.

Li describes how the introduction of cacao required a new word for land as an investible commodity. This new crop created a new way of thinking about land, and a new land assemblage. The crop, in a sense, played a significant part in shaping the formation of a kind of assemblage.

As described later in the chapter with respect to apples, plants and their characteristics can shape what kinds of assemblages are possible and their governing dynamics. The implication is that governance processes can emerge from or with particular material conditions. For environmental sociology, this approach helps us explore how particular ecologies are mobilized and supported by governance relationships, like commodity crops and monocultures supporting particular types of market-based governance systems. Some crops make better commodities, due to their yield, the ease of mechanization in production, their storability and ability to travel, and their possible standardization. A more-than-human approach also allows us to see how governance systems rely on and require those ecologies in their operation, and may shift when ecological conditions change (see, for example, Dwairtama 2017).

New Materialisms

New materialisms is a term that captures the sentiments of actor-network theory and assemblage theory, and may be considered an umbrella term encapsulating both approaches. However, it includes a broader range of work, approaches, and topics, and is more often aligned with anthropology and cultural studies and takes aim at more ontological questions about the nature of being. It is considered both a rejection of anthropocentrism – by suggesting that non-humans are vibrant, self-organizing, have a kind of willful capacity – and a philosophical critique of dualistic and modernist ways of thinking that posit people as subjects, non-humans as objects, implying a clear distinction between the human mind and the corporeal world. Instead, for those taking a new materialist approach, what it means to be human and human subjectivity is a function of an expansive cast of non-human characters. Another way that these more-than-human subjectivities can be framed is like a distributed cognition.

If we went back to our driving example, we could consider this kind of distributed cognition to be evident in the ways that we bring motor vehicles into our mental processes, so that those vehicles become extensions of our body. When we think about turning a car at an intersection, we rarely process it as our bodies doing something to the vehicle to get the response that we want, but instead just engage with the vehicle as if it were a limb. Those who ride a bicycle may have a similar experience – the function of the bicycle almost becomes part of our bodies, and part of our cognition. In this way, the materials and technologies can be understood to influence our subjectivity: that is, our sense of personhood, our consciousness, and our sense of agency.

One focus of new materialisms has been on the ways that human subjectivity is created by extra-human actors and forces, and perhaps most profoundly, technology. This was part of Donna Haraway's thesis in the canonical "Cyborg Manifesto," (1991) which became a foundation of post-human feminism. In the Manifesto, she describes how the human subject is a hybridized configuration including human and post-human elements, where we are "cyborgs" or fused with the technoscientific world around us. In other words, our experiences and knowledge of ourselves is produced and mediated through technologies and science. Such an approach rejects forms of feminism that suggest that femininity or female experience exists internally, bounded within the human body as a subjective position, and instead considers how subjectivities are created through relations external to the body with discourses, technologies, and materialities. In taking this approach, Haraway can better account for the vast diversity of female experience, and better reject forms of feminism that centre more on identity or physical features.

Along these lines, we could also consider how ecological elements are composite pieces of our very experience of being: our understanding of ourselves as persons, with certain kinds of bodies, and particular types of capacities. They influence us in ways that are outside of our control. Jane Bennett (2010) has described how elements of the non-human world all have a vibrancy. They have an indeterminant quality, and a kind of willfulness that can be seen to both challenge our control over the world, and also outline the limits of our human agency. In doing so, they create a context in which we work with non-humans and come to rely on them and their perceived stable behaviour, while also being confronted with their externality.

The ecological materials we interact with also work their way into our extended ontological being. We form symbiotic relationships with animals, and have evolved with them, as Haraway elaborates in *When Species Meet* (2013). As a result, we may know where they are without looking for them, and seamlessly coordinate ourselves with them. We have reliable personal experiences of breathing air and drinking water, or running on grass or wading through a stream, and we experience these sensations subconsciously, as though we already knew their textures and material characteristics, while our connections to those materials tell us about our embodied humanity. For a thoughtful meditation on this, consider Annemarie Mol's (2008) paper "I Eat an Apple," where she describes how eating an apple tells us about our own subjectivities, as something outside of us becomes part of our body: "an eater has semi-permeable boundaries" (30). In a sense, we are what we are because of the things outside of ourselves that we bring in, and many of those things are material, and many are ecological.

Plants and More-Than-Human Agency

Actor-network theory, assemblage theory, and new materialisms are all interconnected approaches that expand our analytic lens beyond traditional, human-bound contours. They consider action, governance, and even the very essence of being, to be related to materials beyond the human. For those interested in societal–environment intersections, this can involve including a diverse range of actors with greater analytic depth into our understanding of the social world. For example, Andrew Barry (2013) has looked at how metals influence the politics around oil drilling. John Liu (2017) has discussed how the materiality of carbon has shaped the development, or lack of development, of carbon markets. Anna Tsing (2015) has considered how the matusake mushroom, with its growth in old logging sites, participates in what she calls a *ruin economy*. Paul Robbins (2012) demonstrated how the physical features of grass have propelled its maintenance into a massive industry, with suburban homeowners compelled to participate in burdensome and sometimes toxic lawn care. In these works, we can see how materials and their qualities participate in shaping economic and political systems that govern industries. By attending to the ways that materials participate in these systems, we can better understand how they are reproduced, how they change, and whether there may be opportunities to redefine our relationships to these materials in ways that may steer us towards an alternative arrangement.

We will explore the use of a more-than-human approach by looking at plants. Plants operate an important intersection between society and the broader, material world. It would be impossible to overstate how important plants are for society – plants provide our food, our air, and much of the materials for our shelter and clothing. They also differ from other materials in how we understand and orient to them. Michael Marder (2013; see also Marder's chapter in this volume) has discussed how they operate in-between the living and inert, and so philosophically play a unique role in our phenomenological understanding of the world. Plants are certainly alive, but they differ from animals or other species with an individual kind of sentience, and so our interaction with them and relationship to them differs. They can be quite a force in our industries, but we do not often think of them as forceful.

It is perhaps for this reason that many plant characteristics have featured in commodity studies in agriculture. The botany of the banana plant is a player in the history of the banana industry in Soluri's *Banana Cultures* (2005), beans are a character in Freidberg's *French Beans and Food Scares* (2004), and the climatic needs of the coffee plant relates to the politics around its development in Paige's *Coffee and Power* (1998). In these case studies, plants and their botany has been found to play a significant role in the political economic systems that emerge around them, and contribute to the burgeoning discipline of political ecology.

Using apples as a case, Legun's research has traced how the botany of apple trees has shaped the various formal and informal rules in the industry, and in doing so, has influenced the logics, norms, and culture of the apple industry (2015a; 2015b). For example, many apples are bi-coloured, meaning that they contain both red and yellow, or red and green, both on a ripe apple. Some climatic features shape the colouring of the apple, so that hot days and cold nights typically tend to produce apples with a slightly greater proportion of red on their surface area. By the 1930s, apple grading standards had become more formalized and differentiated prices of the apple based on the amount of redness. As a result, there was a desire to produce redder apples (see Legun 2015).

Apple trees are normally reproduced by taking a small branch from an existing tree, called a scion, and grafting it onto a rootstock. This makes a new tree exactly the same as the previous one. If you plant the seed of an apple, it will grow a completely unknown variety of apple, and one that may not taste good at all. All apple trees of a variety are an exact replica of each other, unless the scion that was used in the grafting process had a mutation. One possible mutation is that a branch of an apple tree may produce apples that may be redder. When these are successfully grafted onto rootstock, and consistently produce apples that are redder, they are called a "red sport." Some growers will scan their orchards regularly, looking for these mutated branches, and hoping to find a red sport (Legun 2016). The use of colour in apple standards, paired with the botany of the tree – grafting to make varieties and the existence of mutations – has had a huge influence on the apple industry and how growers can strategize in the market (see also Legun 2017).

The need to graft trees to produce a variety, paired with the grading standards and red mutations, also has created the backdrop against which new modes of managing apple varieties has emerged. "Club apples" or "managed apple varieties" are apple varieties that are managed by a group of growers or a corporation who hold property rights to a variety (Legun 2015a; 2015b). The prevalence of new varieties in the supermarket is a result of the proliferation of these new clubs, which also reflects the unique botany of apple trees – their propensity for variation and the required human intervention necessary for maintaining a variety through grafting (i.e., cloning). Apple trees, in this case, are a plant that has had a significant influence on the economic agency of apple producers.

We can see this influence both in the ways they are true collaborators in the action of apple producing, or true *actants* in the actor-network, and in the ways that they shape the kind of assembling that can happen and the politics that result. Trees and humans are collaborators in the production of the apple, and both are necessary for the apple to be produced and enable and constrain the capacities of the other in the process. Apple trees, in order to produce an edible variety, require grafting, and grafting allows the tree to generate something predictable. The tree has the botanical capacity to be reproduced in this way, enabling the human to express the capacity to graft. This collaborative enterprise has also set the stage for redder apples to be produced. The tree mutates, but for that mutation to become a meaningful change to a variety, it requires human recognition and grafting.

These activities also do not happen in a vacuum, but this simple dyad of agencies between human and trees also allows for other features of the apple industry to be assembled together, with significant organizational, and political implications. The patent enables new varieties, the material outcome of the human-plant dyad, to first be used for rents through royalties, and then later for production control through cooperative patent ownership and license control. We can now see a heterogeneous network operating within the apple industry: trees and their botanical properties enable grafting, and varieties have been salient ways of differentiating apples both on the farm and in the market; farmers and their management of trees enable them to use varieties in highly strategic ways; patents and their ability to restrict ownership, planting, and marketing enables apple varieties to become the foundation of "clubs"; the supermarket form of food distribution creates a particularly functional space for apple varieties to become brands. The resulting is a new kind of assemblage with new political potentialities for growers, who can participate in a number of clubs, which have new kinds of power to avoid overproduction, negotiate with retailers over price, and send apples strategically into different markets to reduce competition (see Legun 2015a; 2015b)

By highlighting the role that plants play in shaping human agency, we may be better able to understand dynamics of change. We also may be better able to recognize opportunities for change that may be more in line with the types of human–plant relationships and cooperative agencies we value. New apple assemblages tell us about challenges in the industry historically that created the incentives to build new assemblages. Overproduction, waste, and cycles of price instability in food systems are a significant problem for farmers, and a byproduct of capitalist ideals of free market competition. These new assemblages exemplify one way that actors are cobbling together the relations that enable them to do things differently, albeit in a way that reproduces some elements of capitalism – exclusion from variety production, inequality among producers, new corporate forms of variety management – and ushers in a different set of challenges and relational complexities.

Conclusion

Interest in new materialisms has been bubbling away in sociology, geography, philosophy, and anthropology since the 80s when actor-network theory and assemblage theory first started to emerge. Its elaboration in an environmental context has been forcefully articulated within political ecology circles, through the work of scholars like Jane Bennett (2009), Tania Murray Li (2007, 2014) and Paul Robbins (2012). In these works, we see how different kinds of things are actors in the economic, political, and social worlds that develop around them. Taking them seriously allows us to consider, with great depth, how we are engaging with them and how we might do so otherwise. It also allows us to place ecology more centrally in our social analyses.

These new approaches take things seriously, in that they recognize that things in our world are not static and thus a constant in our social lives, nor completely fluid and malleable so that they are entirely social. They also recognize that the current qualities we see expressed by the things around us relate to how we use them. Taking this approach allows for an analysis that can draw from structuralism and post-structuralism, or modernism and post-modernism, by allowing us to recognize multiple material realities and a vast possibility for alternative realities, yet to be experienced. Traditional modernity may see progress along a linear continuum, with knowledge of the world always growing or becoming more accurate. Instead we can consider what modernities are being pursued, what ways of knowing the world are becoming better developed, and how what other kinds of networks or assemblages may be available to us and enable us to pursue another path.

References

Agrawal, A. (2002). Indigenous knowledge and the politics of classification. *International Social Science Journal, 54*(173), 287–297.
Anderson, B., & McFarlane, C. (2011). Assemblage and geography. *Area, 43*(2), 124–127.
Barry, A. (2013). *Material politics: Disputes along the pipeline.* John Wiley & Sons.
Beck, U. (1992). *Risk society: Towards a new modernity.* Sage.
Béné, C. (2005). The good, the bad and the ugly: Discourse, policy controversies and the role of science in the politics of shrimp farming development. *Development Policy Review, 23*(5), 585–614.

Bennett, J. (2010). *Vibrant matter: A political ecology of things*. Duke University Press.
Brady, M. (2014). Ethnographies of neoliberal governmentalities: From the neoliberal apparatus to neoliberalism and governmental assemblages. *Foucault Studies, 18*, 11–33.
Bridle, J. (2018). *New dark age: Technology and the end of the future*. Verso Books.
Callon, M. (1984). Some elements of a sociology of translation: domestication of the scallops and the fishermen of St Brieuc Bay. *Sociological Review*, 32(1_suppl), 196–233.
Catton, W. R., & Dunlap, R. E. (1980). A new ecological paradigm for post-exuberant sociology. *American behavioral scientist, 24*(1), 15–47.
Catton, W. R., & Dunlap, R. E. (1978). Environmental sociology: A new paradigm. *The American sociologist, 13*(1), 41–49.
Coole, D., & Frost, S. (2010). Introducing the new materialisms. In *New materialisms: Ontology, agency, and politics*, eds D. Coole & S. Frost. Duke University Press, pp. 1–43.
DeLanda, M. (2016). *Assemblage theory*. Edinburgh University Press.
Deleuze, G., & Guattari, F. (1988). *A thousand plateaus: Capitalism and schizophrenia*. Bloomsbury Publishing.
Dunlap, R. E. (2010). The maturation and diversification of environmental sociology: from constructivism and realism to agnosticism and pragmatism. In *The international handbook of environmental sociology*, eds. M. R. Redclift and G. Woodgate. Edward Elgar, pp. 15–32.
Dunlap, R. E., & Catton, W. R. (1994). Struggling with human exemptionalism: The rise, decline and revitalization of environmental sociology. *The American Sociologist, 25*(1), 5–30.
Dwiartama, A. (2017). Resilience and transformation of the New Zealand kiwifruit industry in the face of Psa-V disease. *Journal of rural studies, 52*, 118–126.
Freidberg, S. (2004). *French beans and food scares: Culture and commerce in an anxious age*. Oxford University Press on Demand.
Giddens, A. (1990). *The consequences of modernity*. John Wiley & Sons.
Haraway, D. J. (1991). A cyborg manifesto: Simians, cyborgs, and women. In *Simians, cyborgs and women: The reinvention of nature*. New York: Routledge, 149–181.
Haraway, D. J. (2013). *When species meet*. University of Minnesota Press.
Kenney, M., & Zysman, J. (2016). The rise of the platform economy. *Issues in Science and Technology, 32*(3), 61.
Latour, B. (1987) *Science in action*, Milton Keynes: Open University Press.
Latour, B. (1996). On actor-network theory: A few clarifications. *Soziale welt, 47*, 369–381.
Latour, B. (2004). Why has critique run out of steam? From matters of fact to matters of concern. *Critical inquiry, 30*(2), 225–248.
Latour, B. (2012). *We have never been modern*. Harvard University Press.
Law, J. (2008). Actor network theory and material semiotics. In *The new Blackwell companion to social theory*, ed. Bryan S. Turner. John Wiley and Sons, pp. 141–158.
Legun, K. A. (2015a). Club apples: a biology of markets built on the social life of variety. *Economy and Society, 44*(2), 293–315.
Legun, K. (2015b). Tiny trees for trendy produce: Dwarfing technologies as assemblage actors in orchard economies. *Geoforum, 65*, 314–322.
Legun, K. (2016). Ever-redder apples: How aesthetics shape the biology of markets. In *Biological Economies*, eds Ri. Le Heron, H. Campbell, N. Lewis, and M. Carolan. Routledge, pp. 139–152.
Legun, K. (2017). Desires, sorted: Massive modern packing lines in an era of affective food markets. *Journal of Rural Studies, 52*, 110–117.

Li, T. M. (2007). Practices of assemblage and community forest management. *Economy and Society, 36*(2), 263–293.

Li, T. M. (2014). What is land? Assembling a resource for global investment. *Transactions of the Institute of British Geographers, 39*(4), 589–602.

Liu, J. C. E. (2017). Pacifying uncooperative carbon: Examining the materiality of the carbon market. *Economy and Society, 46*(3–4), 522–544.

Ma, T., & McGroarty, F. (2017). Social machines: How recent technological advances have aided financialisation. *Journal of Information Technology, 32*(3), 234–250.

Marder, M. (2013). *Plant-thinking: A philosophy of vegetal life*. Columbia University Press.

Marx, K., & Engels, F. (1970 [1845]). *The German Ideology*. International Publishers Co.

McFarlane, C. (2009). Translocal assemblages: Space, power and social movements. *Geoforum, 40*(4), 561–567.

Mintz, S. W. (1996). *Tasting food, tasting freedom: Excursions into eating, culture, and the past*. Beacon Press.

Mol, A. (2008). I eat an apple. On theorizing subjectivities. *Subjectivity, 22*(1), 28–37.

Ong, A. (2007). Neoliberalism as a mobile technology. *Transactions of the Institute of British Geographers, 32*(1), 3–8.

Paige, J. M. (1998). *Coffee and power: Revolution and the rise of democracy in Central America*. Harvard University Press.

Robbins, P. (2012). *Lawn people: How grasses, weeds, and chemicals make us who we are*. Temple University Press.

Roberts, J. M. (2014). *Digital Publics: Cultural Political Economy, Financialisation and Creative Organisational Politics*. Routledge.

Saurin, J. (1993). Global environmental degradation, modernity and environmental knowledge. *Environmental Politics, 2*(4), 46–64.

Soluri, J. (2005). *Banana cultures: Agriculture, consumption, and environmental change in Honduras and the United States*. University of Texas Press.

Tsing, A. (2015). *The mushroom at the end of the world: On the possibility of life in capitalist ruins*. Princeton: Princeton University Press

Tutuola, A. (1994). *The palm-wine drinkard: And, my life in the bush of ghosts*. Grove Press.

Van Doorn, N. (2017). Platform labor: on the gendered and racialized exploitation of low-income service work in the 'on-demand' economy. *Information, Communication & Society, 20*(6), 898–914.

10 Plants *and* Philosophy, Plants *or* Philosophy

Michael Marder

1

When, just over ten years ago, I started working on the nexus of philosophy and plant life,[1] very few studies of the theme were in existence. A subterranean presence in the intellectual history of the West, the vegetal world was still not a legitimate theme in the repertoire of self-respecting philosophical research in 2008. A remarkable exception to the general rule was Elaine Miller's meticulous study *The Vegetative Soul: From Philosophy of Nature to Subjectivity in the Feminine*.[2] I much appreciated Miller's explorations of how plant subjectivity was construed in nineteenth-century, particularly Hegelian, thought and how its traits were then ascribed to human femininity.

My own worry was with the place of plants in the metaphysical tradition of the West, the tradition that, beginning with Plato, had valorized immutable being, immune to changes in the "empirical" world: ideas, the unmoved mover, substance, God, transcendental subjectivity, and so forth. I concluded that not only were plants defined by their capacity for metamorphosis, growth, and decay at the antipodes of these metaphysical daydreams but also that metaphysics drew its *raison d'être* from the overturning and negation of vegetal being.

The marginalization of plants in the intellectual canon of the West is not a mere oversight; it is a symptom of unease, if not of repression, at the sight of metaphysical philosophy's disavowed source, which presents it with an inverted mirror image. Rather than the conjunction of "plants *and* philosophy," the tendency of metaphysics has been to produce a disjunction of "plants *or* philosophy," while conveniently

---

1 Thus far, besides dozens of articles and book chapters on the subject, I have published five books dedicated to philosophy and plants: *Plant-Thinking: A Philosophy of Vegetal Life* (New York: Columbia University Press, 2013); *The Philosopher's Plant: An Intellectual Herbarium*, with drawings by Mathilde Roussel (New York: Columbia University Press, 2014); *The Chernobyl Herbarium: Fragments of an Exploded Consciousness*, with artworks by Anaïs Tondeur (London: Open Humanities Press, 2016); (with Luce Irigaray), *Through Vegetal Being: Two Philosophical Perspectives* (New York: Columbia University Press, 2016); and *Grafts: Writings on Plants* (Minneapolis, MN: University of Minnesota Press, 2016).
2 Elaine Miller, *The Vegetative Soul: From Philosophy of Nature to Subjectivity in the Feminine* (Albany, NY: SUNY, 2002).

omitting the fact that metaphysical thought had consolidated itself by virtue of its self-understanding as *not-a-plant*.

The approach to vegetal life I developed aimed to revolutionize metaphysics from within: to turn it around and to compel it to confront its disavowed reflection in a plant. The implications of such a gesture were bound to exceed the scope of theoretical philosophy *proper*, extending into the realms of ethics, politics, aesthetics, and ecology, among others. For me, these were and are some of the positive ramifications of the critique of metaphysics, whence my thinking had received its initial impetus.

In the meantime, a group of plant scientists committed to elaborating the problem of "plant intelligence" – which they later on rebranded "plant signaling and behavior"[3] – sought to upend the way the subject matter of their discipline had been framed as an object steeped in nonconscious torpor. Their experiments in plant learning, decision-making, and other cognitive processes required daring interpretations imputing agency and activity to plants. They did not, however, dislodge the coordinates for a dominant interpretation of the world and left the notion of agency unscathed, despite rendering it potentially more inclusive of entities previously deemed passive.

While I am deeply sympathetic to the heterodox scientific project, I believe that a thinking encounter with plants calls for a more radical change: to uncouple the active, autonomous, sovereign, and, at bottom, domineering features of subjectivity from the very idea of a subject and to interrogate the rationale behind this philosophical category. The choice between passive objects to act upon and active-productive subjects is a false one; indeed, it is a chunk of the metaphysical legacy that stubbornly endures today, even while the subject–object split is undergoing a massive dismantling or deconstruction. This is the problem with Matthew Hall's otherwise well-intentioned book *Plants as Persons*.[4] In order to justify the need for an ethical treatment of plants, Hall feels he must prove that they are subjects with their autonomous standing in the world, pursuit of determinate ends, and personhood. It escapes him that these are, by and large, the impositions of a metaphysical mindset, which is (here comes a bitterly ironic twist!) at the root of the unethical treatment of plants, of animals, and countless humans, to boot. The liberation of plants from *their* objectification is equally an opportunity for human emancipation from the manufactured bonds of our identity as subjects that separates us from the outside world.

Lest you think that I am indulging in conceptual hairsplitting, the issue at stake is not purely academic. Scientific discussions of what plants know,[5] of vegetal intelligence,[6] or of plant signaling and communication[7] give off the appearance of

---

3 Referring to the association and scientific journal by the same name.
4 Matthew Hall, *Plants as Persons: A Philosophical Botany* (Albany, NY: SUNY Press, 2011).
5 Daniel Chamovitz, *What a Plant Knows: A Field Guide to the Senses* (New York: Scientific American / Farrar, Staus & Giroux, 2012).
6 Stefano Mancuso & Alessandra Viola, *Brilliant Green: The Surprising History and Science of Plant Intelligence*, translated by Joan Benham (Washington, DC: Island Press, 2015).
7 František Baluška & Stefano Mancuso (eds.), *Communication in Plants: Neuronal Aspects of Plant Life* (Berlin & Heidelberg: Springer, 2007); Anthony Trevawas, *Plant Behavior and Intelligence* (Oxford: Oxford University Press, 2014).

a break with disrespect toward (if not the abuse of) our "green cousins" now assigned their rightful place as subjects. We should harbor no illusions, however: plant knowledges are not spared the fate reserved for all other modes and systems of knowing under capitalism that extracts them from the knowers as a profitable form of value. To count as a nonhuman subject, or a nonhuman person,[8] is not a panacea from politico-economic exploitation; on the contrary, it is subjects and persons who are the temporary placeholders of economic value in "knowledge economies." The unconscious danger lurking in the shadows of granting subjectivity to plants, animals, and entire ecosystems is not just that global capitalism may cunningly coopt challenges to anthropocentrism but that the newfangled status of other-than-human lives may actually be the next logical step in the extension of immaterial, subjective, cognitively mediated commodities. The enlargement of the subjective sphere is conducive to the growth not of plants but of capital. After all, the dominant form of the commodity today is not a consumable object; it is the subject itself, in all its multicolored, pluralist splendor.

So, how should philosophy engage with plants in order to avoid falling into the traps of metaphysics and capitalism, the latter a contemporary avatar of the former? Addressing this question, which has been bothering me for some time now, will make up the bulk of the text below. Before I do so, a few words are in order on a unique contribution environmental sociology can make to the rethinking of vegetal life.

A veritable explosion of "critical plant studies" in the last three years has seen humanities scholars from a variety of backgrounds (primarily from literary, cultural, and film studies) take up the threads of scientific and philosophical investigations. In many cases, critical endeavors are content with applying the emergent theories of vegetal life to their own disciplinary concerns.[9] Keeping to this method, they risk replicating the exploitative dynamics of capitalism that extracts value from plant knowledges. To applied scholarship, philosophy and science provide a sort of surplus-value, which contains *in nuce* the insights the humanistic disciplines and social sciences subsequently transfer onto a particular film, piece of literature, or socio-political process at hand. A myriad of empirical cases may be subsumed under the generality of philosophical and scientific theses and conclusions. A publishing industry gets off the ground virtually overnight.

Environmental sociology, for its part, has a chance to go well beyond the trodden path of applying readymade theory. To begin with, the name of the discipline befits the being of plants. More faithfully than botany, psychology, and philosophy, socio-logy reflects vegetal subjectivity. First, each plant is a socium, a society of semi-independent growths – a "collective being" in the words of nineteenth century French botanist Brisseau-Mirbel[10] – and of fungi, bacteria, and other microorganisms

---

8 Recently, the term "nonhuman people" was used in Timothy Morton, *Humankind: Solidarity with Nonhuman People* (New York & London: Verso, 2017).

9 One notable exception here is the collection edited by Patricia Vieira, John Ryan, and Monica Gagliano, *The Language of Plants: Science, Philosophy, Literature, and Cinema* (Minneapolis, MN: University of Minnesota Press, 2017).

10 Quoted in Georges Canguilhem, *Knowledge of Life*, translated by Stephanos Geroulantos and Daniela Ginsburg (New York: Fordham University Press, 2008), p. 41.

inhabiting transition zones around the roots or of insects and animals hovering around its aboveground portions. Second, rather than a random pile, the assemblage of vegetal and nonvegetal multiplicities is an articulation, or, to put it in Greek, the *logos* of life and of lives. It follows from these theses that a robust theory of plant being must be socio-logical. The qualifier "environmental" is also far from purely decorative; vegetal subjectivity is not an interiority withdrawn from the world (in the case of humans, as well, such an interiority may be no more than a metaphysical fiction), but a mode of existence turned outwards, a coexistence with its milieu.

Interpreted along these lines, environmental sociology overcomes the polarization of human society and nonhuman environment. A society in its own right, replete with its own *logoi*-articulations, the latter can no longer denote the inconspicuous context for our cultural, political, and economic existence. Especially, if it has to do with plant life.

2

Let's return to the question of philosophy's engagement with plants, the question we will be orbiting throughout these reflections. To say that the issue is purely theoretical, unconnected to the social and political processes, to which it may be subsequently applied, is to miss the point from the get-go. As I argue in my most recent, still unpublished works, vegetality, or the essence of plants, is inseparable from historical vicissitudes, be they a part of "natural" or "cultural" histories:

> Whatever we are collectively inflicting on plants at the current and at any other intersection of human and vegetal histories meddles directly with their essence. The commercial production of sterile seeds, for instance, robs plants of their reproductive potential and consolidates vegetality itself as something sterile, finite, nonreproducible in itself yet marshalled toward infinite growth, corresponding to the infinite, nonsatisfiable demands of capital, to which vegetality is forced monotonously to say its *yes*.[11]

Differently stated, our interactions with plants write what Michel Foucault termed "a history of the present," overwritten by the ahistorical (plantless) concept of essence.

I want to highlight two moments of the vegetal history of the present, along with their multifaceted consequences, the moments roughly standing for plant ontology and epistemology.

On the one hand, a focus on the collective constitution of plants – the boundaries between individual specimens and entire communities blurred – defies conventional evolutionary theory, caught up in a vicious feedback loop with the historically and culturally situated human self-conceptions. A typical representation of the organism–environment interaction is that of an opportunistic maximization of energy and other resources in the service of individual survival, whether of the phenotype or of the genotype's "selfish gene" that quasitranscendentally regulates behavioral patterns behind the specimen's back, as it were. Works such as Eduardo Kohn's *How*

---

[11] Michael Marder, "Vegetality," unpublished manuscript.

Forests Think[12] and Peter Wohlleben's *The Hidden Life of Trees*[13] offer a vegetal counternarrative, according to which the subject of evolution is not a naturalized version of bourgeois utility-maximizing individuals but a plural agency of altruistic sharing and communal resource allocation. Each of these authors composes a history of the present with regard to plants as a site of an ontological and political contestation of the dominant ideology that reigns over the supposedly disinterested scientific research.

On the other hand, the promise of the decentralized mode of intelligence characteristic of plants (say, the swarm-like thinking of roots[14]) may be greatly exaggerated. While it is true that the nineteenth- and twentieth-century theory of the state as an organic totality, modeled after an animal body, is outdated, there are no guarantees that vegetal (or any other kind of) decentering is the magic key to emancipation. We already live in a world of networks lacking a single command and control center, that is, in a social and political reality that is vegetal, even though we have not quite become cognizant of this transformation. It is not easy to shake off millennia-old definitions of the human as a political *animal* (Aristotle) and to consider ourselves as political *plants*. Nonetheless, the pliability of capitalism means that it can accommodate this ground-shift, if not profit from it as value production transitions from the industrial to the postindustrial mode. When dispersed desires, pleasures, and knowledges are harnessed for the purpose of value creation and extraction, "the plant in us" becomes a locus of meta-surplus-value, the blind spot of the entire disseminated system that bestows meaning on that system and enables the more or less imperceptible continuation of exploitation. Critiques of the subjectivized commodity form in consumer capitalism are yet to contend with the vegetal shape of decentralized subjectivity, the "inner network" (now including the structure of the brain[15]) that predicates the smooth functioning of the "outer" social, economic, and political network. Or, more fundamentally still, with the interface of the two kinds of network that seamlessly pass into one another in the manner of plant existence, eschewing the hard-and-fast barriers between interiority and exteriority.

The vegetal challenge to the egoism ingrained in evolutionary theory may indeed interfere with and disrupt the subjective substratum of the capitalist variation on metaphysics, namely a utility-maximizing, possessive individual. When the hyperbolic separation between a living entity and otherness (other living entities, ecosystems, the inorganic world of the elements, and so forth) is mitigated, when solid partitions metamorphose into breathable membranes, then cutthroat competition over finite resources, which both evolutionary thought and capitalist ideology take for granted, sheds its veneer of the inevitable.

---

12 Eduardo Kohn, *How Forests Think: Toward an Anthropology Beyond the Human* (Berkeley and Los Angeles, CA: University of California Press, 2013).
13 Peter Wohlleben, *The Hidden Life of Trees: What They Feel, How They Communicate – Discoveries from a Secret World*, translated by Jane Billinghurst (Vancouver: Greystone Books, 2016).
14 Marzena Ciszak, Diego Comparini, Barbara Mazzolai, Frantisek Baluska et al., "Swarming Behavior in Plant Roots," *PLoS One* 7(1), January 2012, e29759.
15 Martijn van den Heuvel & Hilleke Hulshoff Pol, "Exploring the Brain Network: A Review on Resting-State fMRI Functional Connectivity," *European Neuropsychopharmacology* 20(8), August 2010, pp. 519–534.

Furthermore, since plants do not hoard energy but channel it vertically (the earth–atmosphere axis), horizontally (amongst themselves), and laterally (between themselves and nonvegetal life forms), the accumulation of un-restituted value that defines capital is foreign to them. What is of vital significance to them circulates on the surface: the leaf unfurled and receptive to solar energy, the moisture and minerals the roots imbibe through osmosis ... Plants do not need to resort to the operations of extraction, of destroying the outer shells of things, of enucleating what is of the essence. In this, too, vegetal processes diverge from capitalist operations that are inherently extractive insofar as they wrench the immaterial exchange-value out of the materiality of use-value. If the extraction of oil, natural gas, and coal from the body of the earth has much in common with the extraction of labor and knowledge from human and nonhuman bodies and minds, that is because both are forms of value extractivism prevalent in metaphysics and perfected in its capitalist incarnation.

In the tradition of German philosophy, Marx calls a series of hyperbolic separations between each individual human and all the others, as well as between humanity and nonhuman nature, *alienation*. Various types of alienation instigate one another: the more alienated we are from fellow human beings, the more distance opens between each individual and the nonhuman world; the more detached humanity (our species-being) is from nonhuman life, the more psychically and socially fragmented the specimens of *Homo sapiens*. Plant being, conversely, holds out the prospect of a direct coincidence of individuality and collectivity, seeing that each vegetal individual is inherently collective. On its terrain, the good of one is the good of all, "private" interest and "universal" wellbeing are immediately one and the same. What is plant being if not communism?

That said, overcoming alienation by claiming that the sundered elements are actually undifferentiated and immediately identical is a recipe for disaster. Besides the strong urge to turn vegetal life into a utopia, as opposed to its representation as a "jungle" with the raging war of all against all, the problem with this solution is that it regards this life as a *model*, as though one mode of existence were directly translatable into another. It is true that the gap between the individual and the universal levels of human life has widened into an abyss. It is also true that, even in the human, "individuality" is plural (dividuality) and that its consolidation into a single-minded pursuit of private interest is a by-product of repressing the wildly proliferating vegetal heritage of our subjectivity. But, no matter how successful we are in narrowing it, a minimal gap between the one and the many in human existence is irreducible, which makes the dialectical work of mediation all the more necessary. To believe otherwise is to sweep an actual, practical contradiction under a theoretical rug, bearing a pretty floral design.

Like ontological plurality, the dispersal of intelligence into a multiplicity of minds, distributed across the sentient extension of plants, is not an assured escape route from metaphysical and capitalist domination. In our "knowledge economies," intelligence is the commodity that produces and reproduces itself with the excess of surplus-value, over and above what is strictly required for its self-reproduction. Why would plant intelligence be any different? Capitalism and metaphysics coax knowledges out, extract, attribute value to, and traffic in them. The surplus over the knowing and the known is the *capacity* to know, a potentiality prior to its actualization. Doesn't the surge of interest in plant intelligence zero in (and capitalize) on this capacity of plants, which it then converts into the principles of vegetal robotics, environment sensing, or biochemical signaling? There is nothing inherently wrong with learning these things from plants in a cross-species or cross-kingdoms pedagogy that is not limited to capitalism. The troublesome bit is the *form* such learning and its objective outcomes assume: a commodity. Of course, nothing and no one is ensured against the far-reaching power of commodification, insinuating itself into the previously non-economic domains of life (in the discourse of economics: "externalities"). If, however, plant intelligence is also under the spell of the commodity form, then we cannot assert that it maintains and fosters an innately redemptive potential in the midst of the current capitalist-metaphysical onslaught.

It is for this reason that I much prefer *plant-thinking*, an expression I coined with the inspiration of Plotinus' *phutiké noesis* ("vegetal mind"), to *plant intelligence*. In a nutshell, intelligence is instrumental; thinking is not. Intelligence is meant to solve problems and achieve determinate goals; thinking problematizes things and makes them indeterminate. Intelligence is the triumphant, algorithmically verifiable[16] application of the mind to matter (or to the environment), forced to do the mind's bidding. Thinking happens when the instrumental approach fails; it is a positive sign of failure, of disquiet, of an unending albeit finite search. Intelligence is the tool of evolutionary success, enabling the survival of the fittest; thinking is a mark of in-adaptation, without which, nevertheless, adaptation is not possible. Finally, intelligence enables commodification, while thinking may resist this sprawling phenomenon.

Plant-thinking, then, is not plant intelligence; it could very well be that the two are mutually incompatible. The extension of plant parts – leaves, roots, shoots – to the other, their tending in all directions at once, is a dynamic, material, living image of *in*tentionality.[17] Vegetal extended and extending intentionality goes beyond the use of that to which it strives, first, because it does not represent sunlight and other vital "resources" as objects and, second, because the solar target of its striving is unreachable. Unfolding between earth and sky ever since the initial bifurcation of the germinating seed, the spatialized thinking of a plant is the unrest of and in the middle.[18] Plant-thinking is a growing in-between, the middle place or the milieu, later on formalized into *environment*. In a certain sense, plant-thinking is environment-thinking.

More often than not, our unrest is downright nonvegetal: it expresses itself in dissatisfaction with the physical or symbolic place where one finds oneself. Building upon existential turbulence, the disquietude of self-reproducing surplus-value (i.e., capital) percolates to our lives, which it uproots, for instance, by citing the need to make labor "flexible," that is, to leave all personal attachments behind and undergo displacement in tandem with the shifting job market conditions. Thinking may

---

16 Francisco Calvo Garzón, "The Quest for Cognition in Plant Neurobiology," *Plant Signaling & Behavior* 2(4), July/August 2007, pp. 208–211.
17 See the chapter titled "The Wisdom of Plants" in my *Plant-Thinking*.
18 Dov Koller, *The Restless Plant* (Cambridge, MA: Harvard University Press, 2011).

follow suit, its genetic incompletion presenting every place as a point of passage to another, on the way to nowhere. The plant's in-between state indicates its belonging to more than one element; it inhabits at the same time the day and the night, basking in sunlight aboveground and navigating the labyrinths of water, minerals, fungi, and bacteria belowground. The in-betweenness of the human is, by contrast, a badge of our nonbelonging, our exclusion from every objective milieu. Accentuated by metaphysics and capitalism, this difference turns the noninstrumentality of thinking into a force of pure negativity, deadly in its repudiation of every determinate form and place of finite existence.

3

To repeat: What is the task of a philosophy that encounters plant life? (Will it be still recognizable as philosophy?) So formulated, the question braids together ontology and epistemology, ethics and politics. I put aside the truism that a sounder understanding of who or what plants are should ground a more sensitive way of treating them. If thinking unseats knowing, then understanding, whereby the subject interiorizes its object, is no longer the be-all and end-all of a properly philosophical venture. Thus, I ask: How to think with plants, which amounts to being with them in the in-between? And how to do so while remaining vigilant to the time lags, gaps, and divergences between human and vegetal modes of thinking-being?

Much rides on the meaning of thinking and being unrefracted through the prism of modern philosophy, notably through the triangulation of the subject, the object, and their relation. It is utterly uncontroversial to announce that outside these categories, the staples of epistemology and ontology become essentially environmental (and, hence, vegetal). Less obvious and less consensual is the premonition that our emergency exit from modern philosophical discourses and practices may further escalate capitalist expansionism.

Consumer capitalism and knowledge economies have subjectivized the commodity form, drawing value and surplus-value from the knower or consumer subject. Today, something else is afoot. All around us, "subject–object binaries" break down: in object-oriented ontology (OOO), the environmental humanities, posthumanism, participatory action research ... The collapse of the subject–object relation undeniably affects the status of plants and, in light of advances in quantum physics, what used to be considered inanimate matter. Descartes is buried, time and again, to loud self-congratulatory applause of his gravediggers. But what if the commodity form has mutated once more? What if, neither objective nor subjective, it stems from the breakdown of the subject–object coupling?

Consider the following. The boundaries of the subject have been distended to the extent that nothing is a (manipulable) object any more, resulting in the end of subjectivity as a determinate concept. That everything *is* an object – from global warming to jetlag, from a tree to a unicorn – is the reverse side of the same coin, which some currents of thought, such as OOO, favor. The totalizing assertion about the nature of "everything" robs objecthood of *its* meaning. The articulations of the extremes, each of them claiming to have gained an upper hand over the other, malfunction beyond repair: we are way past subject–object correlations, correspondences, or dialectical syntheses. As a result, we survive amidst the ruins of the dyad, surreptitiously receiving negative energy from its ongoing disintegration.

The initial thrust of fragmentation gives way to the haphazard piling up of multiplicities. It is not by chance that *logos*, translated not as "study" (as in the usual renditions of socio-logy, bio-logy, etc.) but as "articulation," is now a dirty word, and *ecology* has grown incomprehensible.[19] In response to Timothy Morton's cheers for ecology without nature,[20] I would say that ours is the age of the environment without ecology – our environs, our surroundings, void of integral connections. Just as our metaphysics is the downfall and bankruptcy of metaphysics.

As far as the logic of self-valorizing value, or capital, is concerned, the amassing ruins of the subject–object split (and we should never forget that "we" are a part of these ruins) serve a double purpose: they hide the relations of exploitation and imperceptibly supplant the hylomorphic (from the Greek "formed matter" or "form-matter") organization of beings with an alien form, which is none other than that of the commodity. The subjectivization of the commodity has already fulfilled, to a significant extent, the first purpose of obfuscation. With the innermost potentialities of the intelligent and consuming subject commodified, exploitation went into hiding, tying its routines to dreams, pleasures, and desires, including those of knowing – Sigmund Freud's and Melanie Klein's "epistemophilic drive." The nearly psychotic breakdown of the subject–object paradigm further occludes relations of exploitation, in that it precludes the articulation of any relations whatsoever. The subject–object split is itself split, resulting in a conceptual explosion akin to the one nuclear fission instigates. The infamous "butterfly effect," according to which everything is interconnected and any action can cause any reaction, remote as it may be from its source, is the highest stage of this disarticulation, analogous to statements "Everything is an object" or "Everything is a subject."

And this leads us to the second purpose in replacing formed matter with the commodity form. If, paradoxically, global interconnectedness coincides with the absence of external relations, not to mention of a theory of relationality, this is attributable to the dearth of inner relations (or self-relations, if you will, whether harmonious or clashing and self-contradictory) between the form and the content of existences. The ensuing disintegration by far outstrips the nonorganismic, open structure of vegetal life: not cohering together, lacking any sort of *logos*, the fragments of the subject–object paradigm receive their sense from the outside, from the commodity form imposed in lieu of the phosphorescent glow of their intimate meaning.

---

19 Some of the legitimate translations of *logos* in eco-logy are mutually incompatible. So, the "study of the dwelling" expels one from the practical "articulation of the dwelling" but increasing the distance between the dwellers and their abode.

20 Timothy Morton, *Ecology without Nature: Rethinking Environmental Aesthetics* (Cambridge, MA: Harvard University Press, 2009).

When it comes to plants, they (along with all other entities and processes with which experimental science occupies itself) are both physically and conceptually analyzed into hormonal networks and biochemical elements, electrical signals and genomes, calcium transduction pathways and transmembrane proteins. The plant as plant disappears. On the one hand, what comes to the fore, in its stead, are the vegetal tendencies that the phenomenologically accessible shape of plants has been hitherto obscuring. On the other hand, its disjointed component parts supply easily manipulable, formless materials. The promise with which the dynamic view of the plant beckons us, stressing its tendencies (I resist the scientifically correct "functions") as opposed to fixed structures, grows dim as soon as science, technology, and the capital that animates them recruit its capacities.

Needless to say, the logic I am describing is not new. Its prototype is the system of metaphysics that spirits away the sense of entities it explains on the grounds of a single, omnipotent and omnipresent, concept or Being. Two of the salient novel aspects of this logic's current instantiation are: 1) its remarkable capacity to construct reality not only at the ideational-ideological but also at the physical-material level; and 2) its derivation of philosophical, socio-cultural, and economic values from the disintegration of previous conceptual unities and corporealities.

For centuries, metaphysics has been obliquely influencing human and nonhuman lives by generating the blueprints for our approaches to what is and to ourselves. Now, in its anti-metaphysical, perverse shape, it is directly molding the world, lending the nightmarish fantasy an actual body.

Take the patenting of plants' genetic sequences[21] or of their medicinal properties known to traditional healers well before the advent of capitalism. Relying on the apparatus of applied science and starting from the decimation and evisceration of the plant, biomedical and biogenetic research offers an ideal basis – the code: life translated into information – for a possible reconstruction of what has been destroyed. The code is private property, and the realization of the possibility of putting flesh on the abstract skeleton all over again hinges on paying for it. Half the passage from the ideal to the real is folded into capital, while the other half, moving away from the real, ends up in an ecological disaster, the irrecoverable derealization of biodiversity and mass extinction. Gene banks and seed vaults, such as the one in Svalbard, Norway,[22] artificially bridge the dialectical thesis and its antithesis: they are the repositories of information regarding plants reduced to mere genetic materials (germplasm) and deemed useful, or potentially useful, to human beings who may resort to them in the event of their extinction.

In response to the impasses of objectifying and subjectivizing plants, as well as the extremely profitable collapse of the subject–object split, premodern metaphysics is not a solution. The ideality of the genome, its primacy over and indifference towards past, present, and future phenotypes, is still that of Platonic ideas, their cold faces discernable behind the death-masks of scientific and technological "neutrality" and

---

21 Osmat A. Jefferson, Deniz Köllhofer, Thomas H. Ehrich, and Richard A. Jefferson "The Ownership Question of Plant Gene and Genome Intellectual Properties," *Nature Biotechnology* 33, 2015, 1138–1143.
22 Karl Gruber, "Agrobiodiversity: The Living Library," *Nature* 544, April 2017, S8–S10.

antimetaphysics. If anything, Plato was a philosopher of intelligence, of polar opposites, uncomfortable with what lay between them. Socrates, for his part, was a thinker, conscious of his place in the middle (as "the midwife of ideas"), of nonaccomplishment, dedicated not to extracting insights but to moving them along, letting them circulate in conversations with others, seeing to it that they grow, often without a final destination in sight. Although he did not believe he could learn anything from plants,[23] he was a plant-thinker *par excellence*, who had to be physically eliminated so that the metaphysical tradition could take root (hence, Plato's tragic complicity in the verdict he bemoaned). Knowing that he knew nothing save for his own not-knowing, Socrates carved out a niche for thinking that achieved no determinate goals and solved no problems but created them aplenty, including for the thinker himself.

Are we to go back to Socrates without Plato, with the view to resisting commodification? Or accompany Heidegger on his way to the pre-Socratics, below and behind metaphysics, in order to hasten the other beginning of thinking after the end of metaphysics? And is there an "after" at all coming on its heels?

It has become clear that twentieth-century intellectual wrangling about the meaning of the end with regard to the metaphysical tradition is a debate about the end of capitalism in disguise, provided that capitalism is *our* avatar for metaphysics. The nuanced view is that, rather than an abrupt cut, the end of metaphysics (and of history, which is invariably *of* metaphysics) is itself endless, an ostensibly infinite repetition of exhausted past possibilities and a protracted, deepening, all-embracing nihilism. (In 1888, Nietzsche thought that the future history of nihilism would last for another two hundred years.[24]) This means that the end of capitalism and of total commodification *à la* György Lukács would be equally endless, protracted and ever-deepening. Trouble is, metaphysics and capitalism have endangered a livable planet, so that their theoretical endlessness may be cut short by the end of the material substratum they have silently presupposed and despoiled. The end of the world would be a contingent "externality" on the logical plane of the history of metaphysics, ideally indifferent to the actuality or inactuality of existence.

Even here, in the practical effects of the latest metaphysical program, we detect the inversion of vegetal influence: the latter makes the world livable, the former – unlivable. We face a choice between plants and philosophy *qua* metaphysics, thinking and philosophy, the world and philosophy. Moreover, the moment of the inversion is not a singular, onetime occurrence; it is incessantly repeated so long as vegetal life, among other kinds of vitality, tenaciously persists despite all odds. And the decisive choice we are to make is similarly replayed over and over again: plants or philosophy, thinking or philosophy, the world or philosophy. That is how a Nietzsche-inspired revaluation of all values happens, or ought to happen, today. Far from a return to any given historical figure or period, it is necessary to circle back (and to go forward, seeing that the movement is continuing) to the revolutionary,

---

[23] "You see, I am fond of learning. Now the country places and the trees won't teach me anything, and the people in the city do" (Plato, *Phaedrus* 230d).

[24] Michael Allen Gillespie, *Nietzsche's Final Teaching* (Chicago, IL & London: University of Chicago Press, 2017), p. 35.

violently foundational instances scattered throughout the history of metaphysics and capitalism that claims to be ending, endlessly. Hitting the *pause* button in our historical replay at those precise moments, we will linger on the edge, at the border separating philosophy from plants, philosophy from thinking, philosophy from the world. There, in the no man's, no woman's, and no plant's land, we will have experienced the powerless power of thinking.

11 Animals and Society: An Island in Japan

Margo DeMello

We are surrounded by animals. Not only are we ourselves animals, but our lives, as humans, are intimately connected with the lives of non-human animals. Animals share our homes as companions whom often we treat as members of the family; we even may buy clothing for them, celebrate their birthdays, and take them with us when we go on vacation. We can view animals on the "Animal Planet" network or television shows such as "My Cat from Hell" or "Pit Bulls and Parolees" and subscribe to magazines such as *Bark* or *Rabbits USA*. We eat animals, or their products, for most every meal, and much of our clothing, and most of our shoes, are made up of animal skins, fur, hair, or wool. We wash our hair with products that have been tested on animals and use drugs that were created using animal models. We visit zoos, marine mammal parks and rodeos in order to be entertained by performing animals, and we share our yards – often unwillingly – with wild animals whose habitats are being eroded by our presence. We refer to animals when we speak of someone's being "pig-headed" or call someone a "bitch" or "cow." We include them in our religious practices and feature them in our art, poetry, and literature. Political protest ignites because of disagreement over the status and treatment of animals. In these and myriad other ways, the human and nonhuman worlds are inexorably bound.

For thousands of years, animals of all kinds have figured prominently in both the material foundations and the ideological underpinnings of human societies. Human–animal studies (HAS) – sometimes known as anthrozoology or animal studies – is an interdisciplinary field that explores the spaces that animals occupy in human social and cultural worlds and the interactions humans have with them. Central to this field is an exploration of the ways in which animal lives intersect with human societies.

Much of human society is structured through interactions with non-human animals or through interactions with other humans about animals. Indeed, much of human society is based upon the exploitation of animals to serve human needs. Yet, until very recently, academia has largely ignored these types of interaction. In fact, the invisibility in scholarly inquiry of animals was perhaps as great as their presence in our daily lives.

Their presence, however, becomes difficult to ignore when we consider the magnitude of animal representations, symbols, stories, and their actual physical presence in human societies and cultures. Animals have long served as objects of study – in biology, zoology, medical science, anthropology, and the like – but were rarely considered to be more than that, and were even more rarely considered to be "subjects of a life" rather than simply objects of study. One possible reason has to do with the human use of non-human animals: when we grant that animals have subjectivity, including their own interests, wants, and desires, then it becomes more difficult to justify many of the practices that humans engage in on animals, such as meat consumption or medical experimentation.

Human–animal studies is one of the newest scholarly disciplines, only emerging in the last twenty years in academia. But unlike fields like political science, anthropology, English, and geology, HAS is both multidisciplinary and interdisciplinary. That is, it is a field of study that crosses disciplinary boundaries and it is itself composed of several disciplines. In other words, HAS scholars are drawn from a wide variety of distinct disciplines (interdisciplinary) and HAS research uses data, theories, and scholarship from a variety of disciplines (multidisciplinary).

Human–animal studies is the only scholarly discipline to take seriously the relationships between human and non-human, whether real or virtual. Like feminist scholars in the 1970s, who brought issues of gender into scholarly inquiry, HAS scholars have been inserting "the animal" into the humanities, social sciences, and natural sciences. As humans' dependence on non-human animals increases and as our relationship with them changes in the twenty-first century, not examining this relationship within the context of academia seems bizarre – especially given the increased presence of animal rights activism in the world around us.

Using Sociology

Of all of the disciplines which are a part of human–animal studies, sociology is perhaps the discipline that is most logically associated with the field, both because of its theoretical approaches as well as its methods. Sociology's key theoretical paradigms – symbolic interactionism, functionalism, and conflict theory – are all applicable to HAS, and are all seen in some classic HAS research in the field. As sociologist Cheryl Joseph writes,

> Given sociology's premise that human beings are social animals whose behaviors are shaped by the individuals, groups, social structures, and environments of which we are part, it seemed both logical and timely to enjoin the discipline with the study of other animals in the context of human society. (Joseph 2010: 299)

For example, functionalism, which focuses on social stability and the function of the social institutions, can be used to analyze the roles that animals play in human society and the basis for human attitudes towards those animals. Conflict theory, another macro approach widely used in sociology, derives from the work of Karl Marx and focuses on conflict and power struggles within society. It too can be used within HAS, and is especially valuable when looking at the exploitation of animals for human economic gain.

But perhaps the most important approach found in sociology in terms of its relevance to HAS is symbolic interactionism. Symbolic interactionism is a micro-level theory – that is, it focuses on person-to-person interactions rather than on large social forces – and is a perfect approach to study the interactions and relationships between humans and non-human animals. In addition, symbolic interactionism looks at how humans construct the social world and create meaning within it via interaction and the use of symbols. This approach allows sociologists to not only study the interaction between humans and animals but to analyze the meanings given to those interactions. In addition, sociology's use of participant observation, combined with interviews, surveys, and other tools, provides a way for us to observe and analyze humans' interactions with, and attitudes toward, animals. For instance, the now-classic HAS text, *Regarding Animals* (Arluke and Sanders 1996), used a sociological lens and participant observation to examine, among other things, the relationships between people, their animals, and veterinarians, the work done by animal lab technicians with laboratory animals, and the lives of animal shelter workers. In addition, in another classic study, Alger and Alger (2003) used participant observation to study cats in a cat shelter, and found that the cats can communicate with humans via some of the same symbols used by humans.

In terms of methodology, human–animal studies suffers from a problem of how to understand not just the human interactant, but the animal interactant as well. Understanding the human side of that relationship is one thing – sociologists and anthropologists, for example, can draw on classic methods like participant observation or surveys in order to understand human attitudes towards animals, while literary critics or art historians can analyze the role played by animals within literature or art. But how can we ever understand the feelings, attitudes, and perceptions of the animals themselves?

What does the worldview of a dog look like? Can we imagine what it is to see through dog eyes? Or to smell through a dog nose? Whether or not we can truly answer these questions (we certainly cannot use surveys or other traditional methods to answer them), HAS scholars need to remind ourselves that other ways of knowing and being in the world exist.

Anthropologist Talal Asad, in a discussion about the problematics of cultural translation within anthropology, discusses how the translation of other cultures can be highly subjective and problematic due in part to the "inequality of languages" (1986: 156). The ethnographer is both the translator and the author of that which is being translated, because it is he or she who has final authority in determining the meaning of the behavior being studied. Cultural translation, thus, is inevitably enmeshed in conditions of power, with the anthropologist inevitably holding the power in the relationship.

This same problem exists, arguably to a much greater extent, when trying to understand, and put into human words, the minds of non-human animals. Studies of animal behavior, for example, still rely primarily on objective accounts by scientists who attempt to suppress their own subjectivity. The suppression of the author from the text (through the use of technical, reductionist language and passive voice) contributes to the objectification of the animal, and, in some ways, makes it even harder to see the animal within the account. Still, these studies, especially those conducted in recent years which move away from the starkly reductionist studies of the past, can be extremely valuable in terms of the insight that they can give us into animal behavior, and, hopefully, animal emotions and consciousness.

Here, too, sociology is a particularly apt field from which to draw. One of sociology's key methods, shared with cultural anthropology, is participant observation, or ethnography. This method of both observing and interacting with the subjects of one's research can be used with non-human animals as well as with humans.

Drawing off of the practices of ethnographers, human–animal studies scholars are developing new methods in an attempt to bring the animal's side of the relationship into focus (see Kirksey and Helmreich 2010, Hamilton and Taylor 2012, and Smart 2014). Multispecies ethnography is one of those new methods.

Multispecies ethnography is a way in which we might try to understand not just the behaviors and, whenever possible, inner states of both animals and humans, but this method allows us to try to understand the complicated ways in which our lives, as members of different species, are wrapped up together, influencing and shaping each other. In addition, we try to understand how our lives and theirs are simultaneously produced through larger cultural, economic, and political forces. We know now, for example, that non-human lives are not exempt from these forces. Through a multispecies ethnographical lens, we are able to move animals from objects to subjects, with their own histories, interests, and agendas.

Sociology is useful to human–animal studies scholars in other ways as well. Sociologists are well known for a way of seeing the social world known as the sociological imagination. Coined by C. Wright Mills in the 1950s, the sociological imagination provides us with a way of seeing our lives in social context; it allows us to see the ways in which social forces shape our lives. As I mentioned before, we now understand that animal lives, like human, are shaped by these same social forces.

But in addition, human–animal studies is, like sociology itself, a field of study, as well as a way of seeing. HAS is defined by its subject matter – human–animal relationships and interactions – but also in part by the various ways in which we understand animals themselves. HAS is not simply about understanding animal behavior; we do want to understand animals *in the context of* human society and culture. We explore the literary and artistic usage of animals in works of literature or art, the relationship between companion animals and humans within the family, the use of animals as symbols in religion and language, the use of animals in agriculture or biomedical research, and people who work with animals as a part of their job. Our focus then is to look at animals wherever they exist within the human world, and to try to see them as they really are – and as they are constructed by us.

While animals exist *as animals*, in the world around us, once they are incorporated into human social worlds, they take on human categories – often based on their use to humans – and it is these categories (for instance, laboratory animal, pet, livestock) that shape not only how the animals are seen, but how they are used and treated. As sociologist Keith Tester wrote, "A fish is only a fish if you classify it as one." (1991: 46). Moreover, these classifications are not neutral – they are politically charged in that they serve to benefit some (humans, some animals) at the expense of others (other animals).

Ultimately, human–animal studies tries to get to the core of our representations of animals and to understand what it means when we invest animals with meanings. What do animals mean to us, then, and how does our meaning shape their lives?

Animal Tourism

In order to illuminate how sociological approaches can be used to understand the lives of other animals, their relationships to humans, and even the impact of human–animal relations on the environment, I want to discuss an island with a unique population.

Okunoshima, an island located in the Hiroshima Prefecture in southern Japan, has been host to a large population of feral rabbits since at least the 1970s, if not before. The rabbits of the island have limited access to food and water, and thus rely for their survival on the tourists who visit the island to feed them. These tourists, who are largely drawn to the island in order to see, touch, and spend time with the rabbits, have altered the rabbits' lives in ways that have been complicated and unexpected.

Animal tourism, a type of ecotourism (see de Lima and Green 2017), is an increasingly popular, albeit increasingly problematic, way in which animal lovers incorporate their love of watching, and interacting with, wild animals into their holiday travels. Rather than visiting zoos, which most of us know are not good for animals, instead we research tours and sites where we might find wild animals whom we can watch in a satisfyingly "wild" state – but not too wild, however. We do, after all, want to get close to animals, which is almost impossible to do in any truly wild setting. So we spend our money on a range of animal encounter sites, from visiting an orangutan sanctuary in Borneo to going on a safari in Kenya to going to Seal Bay in Australia to watch seals or the Antarctic to see penguins. Some of the really problematic forms of animal tourism include riding elephants in India, swimming with dolphins in the Florida Keys, or holding turtles in the Cayman Islands. These are the activities which go far beyond just watching wild animals.

We know from the research (see Bulbeck 2012 and Fennell 2014) that people's attitudes towards the animals they encounter at these sites differ wildly based on level of education (those with a higher education enjoy viewing animals in the wild more than in enclosures, wanted an educational component to their visit, and overall showed a more conservationist approach), gender (women felt more of an emotional or humanistic connection to the animals than did men, while men adopted a more conservationist approach), the authenticity of the encounter (zoo-goers looked for entertainment over education while the opposite was true for more natural sites), and geography (rural visitors showed a more pragmatic approach towards animal use and conservation).

What drives animal tourists in all of these cases seems to be a quest to get close to something that is wild, "authentic," or even primitive. Ultimately, however, the quest for wildness that drives the tourists and provides the income to support so many of these sites is itself a fabrication: in order to allow visitors access to wild animals, the animals must be, in fact, contained in some way.

And while most educated animal tourists claim that they want a truly wild experience, for many visitors the less authentic, the more pleasurable the experience: who could deny the bodily pleasure from cuddling a baby koala or swimming with a dolphin? In fact, the more wild the site, the less the animals' movements and behaviors are controlled but the more that the visitors' activities are constrained, increasing the animals' freedom (including their freedom to not be present) but for many, decreasing the visitors' pleasure.

Other visitors, on the other hand, experienced some guilt about visiting these sites, knowing that the presence of humans is not good for the animals. Ultimately, though, self-interest (the desire to see or touch the animals) wins out, even for the more conservation-minded of the tourists. We know by now of the extremely high costs in water, food, and energy to the host countries of ecotourism and tourism in general, because of westerners' much higher consumption of resources than indigenous populations, as well as the negative impacts of tourism on many of the animals themselves.

Chilla Bulbeck, in her work on Australian ecotourism (2012), notes that visitors continue to view the animals anthropomorphically in the more managed sites, and want to connect with the animals via touching or feeding them. The question then becomes: if visitors do not come to hear the conservation messages conveyed at these sites, preferring instead to have a physical or emotional connection with the animals (often based on a Disneyfied understanding of the animals), how will we ever manage to enact positive change for wildlife?

One solution might be a different form of animal tourism entirely. What about having tourists visit sites in which the animals are not wild, and thus do not suffer from human contact the way so many wild animals do? And what if, at these sites, you could actually feed, touch, and interact with the animals, and what if doing that were actually GOOD for the animals?

Okunoshima

In March of 2015, I took a group of students on a research trip to Okunoshima, known to many as "Rabbit Island" (*Usagi Shima* in Japanese) because of the large number of rabbits who live there. We wanted to use multispecies ethnography to explore how the rabbits' lives are affected by their contact with the tourists who visit the island, as well as why the tourists are drawn there in the first place. We also wanted to look at the impact that the rabbits, and the people, had on the ecology of the island, and finally, what impact that larger cultural forces may play in shaping all of these elements.

Because of its relative isolation, after World War I, the Japanese government chose Okunoshima, which was (and is) uninhabited by people, as the location of a major poisonous gas factory. In total, 6,616 tons of mustard gas were produced on the island from 1929 until Japan's defeat in 1945. Japanese White rabbits served as test subjects for the gas; some people think these rabbits were set free on the island to act as the proverbial canaries in a coal mine, alerting observers (through their deaths) to dangerous levels of gas on the island, while others think they were simply used as laboratory test subjects.

From at least the 1970s, Japanese visitors began visiting Okunoshima in order to learn about the island's grim history. Today, visitors tour the ruins of the factory and munitions, and visit the island's Poison Gas Museum. Tourists can also visit the island's hot springs, and can camp or stay in a large hotel on the island. (Before the facilities were destroyed by the rabbits, visitors could also play golf and tennis.) There are no permanent human residents on the island; even the workers take the ferry to the island to work each day.

The current rabbits are most likely eightieth-generation descendants of a group of eight rabbits who were abandoned on the island by a group of school children in the 1970s; it is unknown whether any of the rabbits who were gas test subjects survived the end of the war.

My students and I think that there are approximately 1,000 rabbits on the island, who live in thirty-nine separate colonies. The colonies appear to be kinship-based, at least in part, with young rabbits of both sexes most likely staying in their natal communities.

The rabbits, like their wild European cousins, are committed to their territories, and mark them by creating large feces piles, which are usually found in the middle of the daytime territory. While in the two most high-traffic areas, near the hotel and the ferry, there was some overlap between colonies, for the most part the rabbits do not stray far from their home territory. Clearly, the rabbits know not only their territories, but also which members make up their communities.

In 2013, a video was uploaded to YouTube (www.youtube.com/watch?v=RdeX4NqvDZw) which shows a young tourist from Hong Kong being chased by a large herd of island rabbits, who were demanding food from her. This video quickly went viral, with more than two million people viewing it, and, according to our interviews, leading to countless people deciding to visit Okunoshima to experience the rabbits first-hand. This video's popularity followed on the heels of an earlier bump in Okunoshima's popularity during 2011, the Year of the Rabbit, when the island also experienced a great deal of media exposure. These two media events – the Year of the Rabbit and the YouTube video – are directly responsible for the conditions in which the rabbits now find themselves.

The media exposure led to an explosion in visitors to Okunoshima, which itself caused an explosion in the rabbit population – from about 400 ten years ago to about 1,000 today, based on our interviews with island staff and long-term visitors. Because the island (which lacks water and only has a 4 km circumference) cannot comfortably support that many rabbits (and there are no natural predators), the rabbits have now resorted to begging for a living. Tourists today, which include large numbers of foreign visitors, come less frequently for the military history and more frequently for the rabbits – to feed and photograph them. Tourists journey to the island on a ferry from the mainland, weighed down with vegetables, fruits, and rabbit pellets, as well as bottled water for the dozens of plastic and metal bowls scattered around the island.

The rabbits of Okunoshima, whose ancestors were abandoned on the island at least forty years ago (and perhaps as long as seventy years ago), have had to adapt to a remarkable existence – one that is equally constrained by the physical environment in which they find themselves, as well as by the humans who interact with them on almost a daily basis. And at the same time, they have carved out a life for themselves by both altering their environment, and manipulating humans into caring for them.

These rabbits are not especially well-suited to living on the island – they are, after all, a domesticated species, and their wild relatives come from a very different environment (the Iberian Peninsula). Nor are they especially well-suited to 'making their living' by interacting with humans. They are a prey species and the relatively short amount of time that they have been domesticated has not eliminated their flight response. In fact, during most of rabbits' history with humans, they lived not with them, as dogs and cats do, but outside of human homes, being kept, for the most part, for food and fur. It was only in the last 150 years that they have been kept, and bred, as pets.

The rabbits who live on Okunoshima are liminal animals – they live in an uneasy space between the domestic and the wild. They are not native to Japan and are not even wild. But they have had to utilize some of the behaviors that their wild European cousins use to survive – like digging elaborate burrows – as well as the species' famously prolific reproductive abilities which now serve them well. These rabbits build homes, create communities, defend territorial boundaries, find friends, mate, and rear babies – all without any human intervention.

But at the same time, because there is not enough food or water on the island to support all the rabbits, they have shed the cautious aspects of their nature in favor of an almost shocking level of aggressive friendliness – they are more friendly, in fact, than many pet rabbits. This is how they feed themselves – by begging.

The rabbits of Okunoshima live lives that are both wild and non-wild. Like, They dig their burrows, as European wild rabbits do, in the mountain in the center of the island. The perimeter of the island, which is surrounded by a walking trail, is where the rabbits hustle for food during the daytime. At night, when the visitors are gone, most (but not all) rabbits retreat to their burrows, where the mothers also rear their young. Some mothers, on the other hand, build their burrows in small hills along the well-trafficked walkways. These burrows are much more vulnerable to human disturbance, but also make it easier for the mothers to beg for food, while still protecting their young. Both wild and domesticated rabbits are crepuscular, which means they are most active at dawn and dusk. Island rabbits, on the other hand, have become largely diurnal, building their daily activities around the ferry schedule.

But the rabbits are not equally friendly, nor do they all behave the same. The rabbits have exploited the different ecological zones on the island and have created a number of different socio-zones which overlap those eco-zones. For instance, the rabbits on the north and east sides of the island, as well as in the central mountain region, which are overgrown with tropical trees and shrubs and are less accessible to humans, are much shyer than those on the more accessible (and arid) sides of the island. Those living outside of the hotel and at the pier are the most friendly and aggressive, because of the easy access to guests.

European rabbits are a highly social species, preferring to spend the majority of their time in the company of others. They are non-vocalizing animals, so they use their bodies – the movements of their ears, tails, legs, and bodies, their tongues and teeth, and their scent glands – to communicate their needs and desires to others. They box, grunt, pee on each other and on their surroundings, climb, dig, and stand on two legs. All of these gestures both express their emotions and have an impact on others. Even the ways in which they sit or lie, often lightly touching the body of a friend, express their emotions. But on the island, many of these gestures are extended to people as well. What my students and I witnessed is a mutual understanding on the part of the human and lapine interactants: Both are aware, at least in part, of the "other's" motivations, and act accordingly. The rabbits know, for example that the tourists are not here to hurt them; they are here to feed them, while island workers mostly ignore them. Unfortunately, while the tourists are drawn to the island by a love of animals, their interactions with the rabbits may not, after all is said and done, be in the rabbits' best interests.

Animals, Humans and "Nature"

Agustin Fuentes has written about what he calls "domesticatory practices" (Fuentes 2007), to refer to, for example, temple monkeys in Bali and the large amount of interaction that they have with humans there. He notes that monkeys who live in long-term zones of sympatry with humans end up sharing social and cultural relationships with those humans.

In the case of Okunoshima, the rabbits were already domesticated prior to their arrival on the island. I would suggest that these rabbits have experienced both a re-wilding, in which they re-learn the behaviors of their wild cousins, but also a re-domestication, in which they adapt themselves to their human visitors. But unlike Fuentes' monkeys, these rabbits are not living in a human space. Rather they are living in a wild space which they have shaped to their own needs – with burrows carved into the mountain, feces piles to mark territory, urine marking of their space and their comrades, and competition over who controls the most desirable territories – i.e. those with the most access to humans. They don't live at the borders of human society; they have created a new society to which humans have been drawn, through the consumption of media about Okunoshima. In the process, these rabbits, and the humans who inadvertently created the situation in which they live, have co-constituted a unique landscape and animal-scape. This space – inhabited primarily by animals, with humans as visitors – is a landscape occupied by rabbits who act autonomously to demand what they need from humans who travel across the world to give it to them. Unfortunately, what the tourists give them has led to the current unsustainable numbers on the island, as well as an extremely short life expectancy.

Yet the rabbits are also aliens. They did not evolve on this island and have destroyed much of the natural vegetation there. Because of this, there are those who argue that the rabbits should be destroyed (or at the very least heavily controlled), in order to let the island go back to its "natural" state (which includes the ruins of the gas plant, a hotel, a museum, a visitor center, a golf course, tennis courts, a campground, and paved roads and trails). These aliens are not living in a fully human space; but because this space has been claimed by Japan as a site of national remembrance, those needs are now in direct conflict with the needs of the rabbits – and the tourists who love them.

This is far from the first time that European rabbits, both prized and hated for their reproductive abilities, have been seen as an "invasive alien species," with all of the governmental and conservationist worrying that this label carries. Whether the concerns are about the impact of the species on biodiversity, native species, or the economic goals of the administrators, rabbits often pay the price after they have been introduced to a new area.

The most well-known example of "rabbit as invasive species" is the case of the European rabbits who were introduced to Australia in 1788 as food animals, and again in 1859, when they were released into the wild for hunting (see Jernelöv 2017). As most people now know, the rabbits thrived in this new environment, and, as their population exploded, so did Australian farmers' concerns, as the rabbits ate millions of dollars-worth of crops. The result has been the reclassification of rabbits from game to pest and innumerable strategies – all violent – aimed at curtailing the animals' numbers, including poisoning, trapping, shooting, warren ripping, and biological warfare. Luckily for the rabbits of Okunoshima, the Japanese government has not yet moved to take similar lethal actions against them.

This represents another way that the Okunoshima rabbits embody liminality. There are two distinct and competing views of the rabbits: the government and their agents see them as wildlife, but do not consider them, as they do the other animals on the island, as "nature." Instead, these agents see the rabbits as something set apart from nature, and clearly not belonging in nature. They are, in anthropologist Mary Douglas' term, "matter out of place" (Douglas 2003). Because of this, the official government policy is that feeding the rabbits is discouraged and the hotel can no longer sell rabbit pellets for them. On the other hand, the tourists see them through the lens of *kawaii* (a particular rendering of "cute" in Japanese). Part of what makes these rabbits kawaii, besides their obvious cuteness, is the fact that there are so many rabbits in one place. Unfortunately, these huge numbers are unsustainable on such a tiny island. And finally, the hotel management, tasked by the government with overseeing the island, is dismayed by the damage done to the island, but also understands the economic value of the rabbits, and uses pictures of the rabbits in all of their promotional materials, which draws in yet more tourists, and ultimately, drives up the number of rabbits. As we can see, not only are there many different social constructions of the rabbit; these competing constructions each shape, in a different way, the lives of the rabbits themselves.

But blaming the rabbits for the damage to the island obscures the human activities that both caused the problem and allow it to continue. It is tourism, after all, that caused the rabbit population to explode to such unsustainable levels. And, ironically, if the rabbits are considered to be an alien species because they did not live on the island until they were first brought to it, and because of the environmental damage they have caused, they do not fulfill the final criteria usually mentioned by those concerned about alien species: economic damage. The work that the hotel staff must do to try to clean up their damage pales in comparison to the money that the tourists now bring in, largely because of the rabbits.

But how are the rabbits themselves doing? On the one hand, the rabbits act as if they are happy. They run, play, dig nap holes in the sand, and are, especially in the high tourist areas, not only remarkably easy-going around people, but seem to want to spend time with them, even when not being fed. To someone who has lived with domestic rabbits for almost thirty years, they look well-adjusted indeed.

On the other hand, my students and I concluded, based on body condition and behavior, that the rabbits are overwhelmingly young, with the vast majority being under two years old, with a large number of babies and pregnant females. This means that birth rates are very high, but so are mortality and morbidity, resulting in an overwhelmingly young population. According to a hotel employee, about fifteen rabbits die of various causes per day, and their bodies are picked up and disposed of by hotel staff so that visitors cannot see them. In addition, the rabbits were suffering from a variety of illnesses and injuries. Of the rabbits that we counted, 26 percent had visible injuries or illnesses, with the rabbits living near the hotel being both the sickest and the most injured, thanks to competition over food – approximately half of those rabbits were clearly unhealthy or hurt.

While the rabbits are dependent, at least in part, on the food brought by humans, this food, ironically, contributes to the short life spans of the rabbits. Much of the food is not species-appropriate, and because it comes in large quantities at some times (Saturdays during the summer, for example), and rarely at other times (winter, rainy days), the rabbits lack the consistency that is needed for their delicate digestive systems. Thus, one common cause of death is gastrointestinal stasis, because of the overload in unhealthy and inconsistent foods.

Still, the rabbits, most likely unaware of the tenuous nature of their existence, continue to do whatever they need to do to survive: building their homes, mating, raising their young, and teaching those young how to find food and water in a rabbit-centered, but human-defined, world.

So, given all of these issues, what, if anything, should be done? Should tourists stop visiting Okunoshima? Should the Japanese government step in and regulate the tourists? Should someone just come in and "rescue" all of the rabbits?

As an animal lover, I feel that there is something to be said about encouraging the emotional connection to animals that so many animal tourists seek as well as, when carefully and respectfully done, tactile experiences like bunny cuddling, in addition to more educational endeavors. To the question of how to impact positive change for animals in the face of a public unwilling to educate themselves about conservation, Chilla Bulbeck suggests that animal encounter sites should exploit, rather than discourage, the public's emotional and physical responses to animals, turning them towards a new kind of conservation, which she calls "respectful stewardship" of a "hybrid nature."

This involves taking the desire for authentic animal experiences and combining it with an obligation to manage such experiences and the wilderness sites in which they are found. Because nature is no longer "nature," and there is no longer a pristine wilderness in which wild animals live and humans can visit, people who care about animals and the environment need to manage the natural world as respectful stewards, even if that means limiting human desires in order to save what's left. A respectful stewardship would take into account a more complex understanding of the natural/artificial world, and would also take into account its multiple stakeholders – animals and human alike. Ultimately, a respectful stewardship would combine the scientific and conservationist approach long favored by environmentalists and combine it with the humanistic, emotional approach taken by animal encounter visitors.

In the case of Okunoshima, the rabbits there have not only come to depend on their human visitors, but have created a brand new hybrid culture which I think is worth celebrating. I would, however, love to see some education about rabbits' dietary needs, perhaps on the hotel website and on the ferry on the way over to the island, combined with some stricter regulations about what kinds of foods can be taken on the island. Ultimately, if we restrict humans from visiting the island altogether, the rabbits will suffer, although in time, the population may shrink to a level that would be sustainable without human intervention. But I think that would be a shame, for humans and rabbits alike.

Multi-Species Ethnography, Sociology, and the Future of Human–Animal Relations

In some ways, humans live lives that are more separated from those of animals and the "natural" world than ever before. The majority of the world's citizens now live in urban areas, and in the next decade, that number will continue to rise. Most people spend very little time in nature, and for many people, the only relationship that they have with other species is through their pets, and through the food – much of which came from animal bodies – on their plates.

Yet at the same time, we are absolutely dependent on animals. Besides providing most people much of their food, they provide much of our clothing, they are used to test many of our household products and all of our drugs, medical devices, and medical procedures, and animals continue to be an important source for much of the world's entertainment. They continue to proliferate in our religions, art, and literature, and they provide labor – as beasts of burden, military working animals, or assistance animals – for millions of people worldwide. And of course, a small number of species have carved out increasingly important roles in our families as pets.

How we make sense of these important, and complicated, relationships will become ever more critical as our planet undergoes the most severe degradation of its habitats and experiences the greatest loss of animal (and plant) species the world has ever known.

Sociology, like all social and natural sciences, needs to respond to these changes, and our students need to be provided with the tools to help them to negotiate this new world. Multi-species ethnography, which is simply a tool which can allow us to bring "the animal's perspective" into our work, is just one method by which we might do this.

In our work on Okunoshima, my students and I used multispecies ethnography in order to investigate an interesting set of relationships that has developed between a large population of domesticated, but feral, rabbits, and a growing population of tourists. By investigating both the humans' *and* the animals' behaviors, as well as the underlying motivations and attitudes of both groups, we began to understand that what at first seemed like a cute story – visit Okunoshima to take selfies with the rabbits! – is in fact much more complicated than it appeared. Like so many of our relationships with other species, or even our relationships with people from another cultural or social group, our actions and interactions have consequences, which may not always be visible. And more importantly, the consequences of our actions often have far-reaching, and negative, repercussions – repercussions which typically impact those with the least power.

As we continue to grapple with some of the irreversible changes that our actions have caused to this planet, and to the people and animals who live on it, we may want to start using some of the tools that we have to understand the world to shed light on those changes – so that we may, perhaps, begin to fix the problems that we have caused.

References

Alger, Janet, and Steve Alger. (2003). *Cat Culture: The Social World of a Cat Shelter*. Temple University Press.
Arluke, Arnold, and Clinton Sanders. (1996). *Regarding Animals*. Temple University Press.
Asad, T. (1986). The concept of cultural translation in British social anthropology. *Writing Culture: The Poetics and Politics of Ethnography*, *1*, 141–164.
Bulbeck, C. (2012). *Facing the Wild: Ecotourism, Conservation and Animal Encounters*. Earthscan.
de Lima, I. B., and R. J. Green (eds.). (2017). *Wildlife Tourism, Environmental Learning and Ethical Encounters: Ecological and Conservation Aspects*. Springer.
Douglas, M. (2003). *Purity and Danger: An Analysis of Concepts of Pollution and Taboo*. Routledge.
Fennell, D. A. (2014). *Ecotourism*. Routledge.
Fuentes, A. (2007). Monkey and Human Interconnections: The Wild, the Captive, and the In-Between. In *Where the Wild Things Are Now: Domestication Reconsidered* edited by R. Cassidy and M. Mullin. Berg, pp. 123–146.
Hamilton, L., & Taylor, N. (2012). Ethnography in evolution: Adapting to the animal "other" in organizations. *Journal of Organizational Ethnography*, *1*(1), 43–51.
Jernelöv, A. (2017). Rabbits in Australia. In *The Long-Term Fate of Invasive Species* (pp. 73–89). Springer.
Joseph, C. (2010. Teaching Human–Animal Studies in Sociology. In *Teaching the Animal: Human Animal Studies across the Disciplines* edited by Margo DeMello. Lantern Press, pp. 299–339.

Kirksey, S., & Helmreich, S. (2010). The emergence of multispecies ethnography. *Cultural Anthropology, 25*(4), 545–576.
Mills, C. W. (2000). *The Sociological Imagination*. Oxford University Press.
Smart, A. (2014). Critical perspectives on multispecies ethnography. *Critique of Anthropology, 34*(1), 3–7.
Tester, K. (1991). *Animals and Society: The Humanity of Animal Rights*. Routledge.

PART IV

Sustainability and Climate Change

12 Possibilities and Politics in Imagining Degrowth

Valérie Fournier

Introduction

Few would now deny the urgency to develop economic models that would enable us to live within the planet's ecological limits. The fact that the earth has finite capacities to fuel growth and absorb pollution has been well recognised since the publication of Meadows et al's (1972) *Limits to growth* and become a central motif of international climate agreements and policies (Steffen et al, 2015).

The most common response to this ecological challenge is to find 'sustainable' ways to grow the economy through a combination of market, technological and regulatory solutions that would all ensure that growth stays within the earth's capacities. Thus, whilst environmental concerns have moved centre stage of government and sometimes corporate agenda, the founding principles of Western patterns of consumption and production remain non-negotiable. Ecological sustainability, as framed by national governments or international institutions (for example from the Kyoto Protocol in 1997 to the Paris Agreement in 2016) remains firmly subordinated to economic growth. Whilst there are different approaches to Ecological Modernisation or Sustainable Development (e.g. Barry, 2005), the central motif is to reconcile the tensions between technology and ecology, economic growth and ecology, and competitive market and ecology (Blühdorn and Welsh, 2007). In this convenient rendering, economic growth and environmental protection are intimately linked through win-win strategies, suggesting we can have everything we want without facing any risk (Milne et al, 2006).

Against this idea of sustainable development, degrowth offers a more radical vision that doesn't try to make growth greener but rather takes aim at the very idea of growth, or the 'tyranny' it exercises on modern imagination and practices. From this perspective, if there is to be any hope of a sustainable future, it is precisely economic growth that needs to be called into question (e.g. Milne et al, 2006; Scott-Cato, 2006). Discourse and policies of sustainable development and ecological modernisation only serve to 'sustain the unsustainable' (Blühdorn, 2007). In the context of recent financial, economic, social and environmental crises, degrowth offers a promising alternative as it seeks to decouple well-being (of people and the planet) from growth; it emerges as a radical call for voluntary and equitable downscaling of the economy towards a more just, sustainable and participatory society (Weiss and Cattaneo, 2017).

In what follows, I first trace briefly the development of the degrowth movement. I will then outline the main ideas of degrowth, starting from its critique of growth, to the emphasis on social sustainability, the re-politicisation of the economy, and the sort of cultural, economic and institutional measures envisaged. Finally, I will turn my attention to the work of re-imagination or re-conceptualisation central to building a degrowth society.

The Development of the Degrowth Movement

Following the first Degrowth Conference in Paris in 2008, degrowth can be defined as a 'voluntary transition towards a just, participatory and ecologically sustainable society' (Weiss and Cattaneo, 2017). It involves a downscaling of production and consumption in such a way that increases human well-being and enhances ecological conditions (Schneider et al, 2010).

The idea of degrowth and the debates it has generated are inscribed within a critical tradition that has challenged neo-liberal understanding of economic development and modernity for some time ; thus the movement identifies Gandhi, Gorz (1975), Arendt (1958), Illich (1973) and Schumacher (1973) (to mention only a few) as its precursors. It also has roots in ecological economics and the critique of productivism for its destructive effects on the environment.

The term degrowth was first used by Georgescu-Roegen's (1971) seminal work *The Entropy Law and the Economic Process*. Georgescu-Roegen introduced the notion of degrowth in response to what he regarded as the irreversible damage inflicted by the politics of endless growth preached by neo-liberal economics. He argued that classical economics was based on a mechanistic vision that ignores the principle of entropy, the second principle of thermodynamics. According to the entropy law, whilst it may be the case that energy is conserved ('nothing gets lost, everything is transformed'), it is nevertheless degraded or transformed by its use, and therefore cannot be returned to its original state and used again in the same way. For example, the energy that goes into the making of a computer can never be returned to its original state and be used to make another computer. Thus endless growth, supported by an ever-growing use of natural resources, will lead to the exhaustion and despoiling of the earth capacity and is a physical aberration.

At a theoretical level, degrowth draws upon a wide range of intellectual traditions in ecological economics, social ecology or economic anthropology (Martinez-Alier et al, 2010). But degrowth is not only an academic debate; it also embraces social and environmental movements (Martinez-Alier et al, 2010). These two levels of academic debates and activist movements have been closely connected through international networks, and in particular a series of international degrowth conferences that started in Paris in 2008 (Weiss and Cattaneo, 2017).

Whilst the ideas associated with degrowth had been in circulation in academic and activist circles before, it is in France in the 1990s and 2000s that these ideas became crystallised under the broad term of 'decroissance' (Weiss and Cattaneo, 2017). Since, degrowth has become an umbrella term that embraces and connects various ideas, policies and social movements. The movement soon spread from France to Italy and Spain, and then other countries mostly in Europe and Latin America (D'Alisa et al, 2015). In just over a decade, degrowth has evolved into a broad international network that connects social and environmental grassroots movements as well as multi-disciplinary academic research, connections that have been facilitated in part by the six biannual international conferences on degrowth organised at the time of writing (from Paris in 2008, to Barcelona in 2010, Montreal in 2011, Venice in 2012, Leipzig in 2014, Budapest in 2016). There have also been various special issues on degrowth in academic journals such as the *Journal of Cleaner Production* and *Ecological Economics*, together with numerous articles elsewhere (Weiss and Cattaneo, 2017).

Whilst degrowth has mostly developed in the Global North, it is closely aligned with movements in the Global South such as Buen Vivir in South America, Ubuntu in South Africa or the Gandhian Economy of permanence in India (D'Alisa et al, 2015).

What Does Degrowth Stand For?

The degrowth movement suggests that we need to find ways to reorganise economic relations so as to bring them within ecological limits, and do so in a way that is socially fair. Degrowth dos not refer to a specific programme, but to a broad political vision for socially and ecologically transformative changes, or a framework that connects different ideas, policies and citizens' initiatives (Kallis, 2011).

The critical and transformative intent of the degrowth project is well captured by the 'de' prefix. Whilst this could be seen as casting a negative shadow on degrowth, such a reading would only be valid to the extent that 'de' was understood in terms of 'negative growth'. However, as many have insisted, the 'de' prefix may best be understood in terms of a liberation process of getting rid of the imperative of growth institutionally and mentally (e.g. Ariès, 2005; Latouche, 2014; Schneider et al, 2010). So degrowth does not stand for negative growth, or recession, but rather for a decolonisation of the imaginary away from the 'tyranny of growth' (Kallis, 2011; Latouche, 2014). In this sense, degrowth is a symbolic weapon, or 'missile concept' (Ariès, 2005) that may enable us to imagine and practice a prosperous society without growth, a 'prosperous way down' (Odum and Odum, 2006), or a 'society of frugal abundance' (Latouche, 2014). It aims to do so firstly by breaking the deeply entrenched connection between prosperity and growth, and this is its critical function as we'll see below with the critique of growth. And secondly, it encourages us to imagine prosperity or well-being in different ways, for example by connecting prosperity with 'down', or abundance with frugality; this work of re-imagination will be the focus on the next section. Together, the work of critique and re-imagination central to degrowth suggests that it can serve the same subversive function as utopia (Kallis and March, 2015); thus degrowth is not meant to be a blueprint for an ideal society, but rather to help us think about the limitations of growth, and imagine alternative futures; it refers to the process of imagining, experimenting with, shaping a society without growth, rather than to a fixed destination. And this process of imagination and experimentation starts with a critique of growth.

Critique of Growth

Drawing on the work of Georgescu-Roegen, proponents of degrowth denounce economic thinking and systems that see growth as a taken for granted and ultimate good, or an axiomatic economic necessity (Kallis, 2011). Growth in GDP is considered as the holy grail of economic policy by most national governments and international institutions from the Word Bank to the European Central Bank (Harvie et al, 2009; Wright, 2010). Growth has become the fetish of capitalism and is supposed not only to deliver increased profit for capitalist firms, but also jobs, prosperity and better lives for all (Hamilton, 2003).

Against this blind faith in growth, the degrowth movement, alongside many others in ecological economics, insists that growth is ecologically and socially unsustainable. Or to use Daly's (1996) terms, growth is 'uneconomic' in that it makes us worse off by using up our natural capital and resources, whilst providing no benefits to human well-being and welfare. Notably, the degrowth movement insists not just on the ecological limits to growth, but also, and maybe more fundamentally, on its social limits (Cosme et al, 2017; Kallis et al, 2012). In line with many recent studies, it points to the growing inequalities that have been brought about by growth as wealth is increasingly concentrated in the hands of a few (e.g. Jackson, 2009; Pickett and Wilkinson, 2009; Picketty, 2014). Another argument underlying the social limits of growth is that above a certain level of material wealth, growth does not increase happiness or well-being (e.g. Kallis et al, 2012; Norgard, 2013). Instead, it may produce rising rates of depression and anxiety (Wilkinson and Pickett, 2010). The notion of Affluenza suggests that increasing consumption and wealth leave us with feelings of emptiness (e.g. James, 2007). The competitive drive for money, status and material things leads to anomie, alienation and addiction, whilst undermining our capacity to build connections with others.

Beside pointing out that growth and ever-increasing consumption are unsustainable socially and ecologically, degrowth proponents also question measures of growth such as the Gross National Product (e.g. Ariès, 2005; O'Neill, 2015). Such measures only take into account the production and sale of commodified goods and services, ignoring the damaging effects these have on other 'goods': justice, equality, democracy, human and ecosystems' health, quality of life, social relations. They point to the absurdity of an economic system based on 'growth' when what is meant to 'grow' remains arbitrary. Thus increasing cancers, road accidents, obesity, ecological disasters, wars all contribute to economic growth through the consumption of insurance, medical products and services, the cleaning industry, weapons and so on, a point made by many in green politics (e.g. Scott-Cato, 2006). Not only do measures of growth fail to take into account the damaging effects of production and consumption (e.g. on the environment, our health or quality of life), but they also count as positive contributions the whole industries that have emerged to deal with these effects (e.g. pharmaceutical industry, waste management).

The degrowth movement is not only critical of growth, or the growth fetish, but also of any attempt to reconcile growth and environmental concerns. In particular, the concept of sustainable development is seen as an 'oxymoron' (Latouche, 2004), for growth, be it sustainable, green or however qualified, cannot be sustainable. Indeed, the sustainable development discourse that has gained broad appeal following the publication of the Brutland report in 1987 (WCED, 1987) seems to have done little to keep resource use and consumption within ecological limits, and this casts serious doubts on its ability to keep growth in check (Martinez-Alier et al, 2010). At the core of this argument is a critique of the reliance on eco-technologies to fuel 'greener' growth. The main point made here relates to what has become known as the 'rebound effect' (e.g. Binswanger, 2001; Schneider, 2003): any gain in energy derived from the use of more efficient technology is usually cancelled out by an increase in consumption. Eco-efficient technologies only make us consume more, a finding confirmed by others (e.g. Princen, 2003; Herring, 2002); for example, fuel efficient cars enable us to travel more; the use of solar energy enables us to heat our house or water more, and so on. The problem is not with these eco-efficient technologies themselves – potentially they can be useful – but with their inscription within a paradigm of growth; their deployment towards increased consumption and production. Thus the much heralded decoupling of growth from resource use is denounced as an illusion because energy efficiency improvements tend to rebound (Kallis, 2011).

It is within this context of mounting ecological and social wreckage, of the absurd and arbitrary definition of growth, and the pursuit of growth (albeit 'sustainable') at all cost, that degrowth is offered first and foremost as a conceptual or ideological weapon. Proponents of degrowth seek to challenge the 'naturalness', the supposed inevitability and desirability of growth; they oppose the ideology of growth (more than growth in itself which is no more than an arbitrary calculation), and offer degrowth as a political weapon to decolonise the collective imagination and free it from the tyranny of growth (Kallis and March, 2015; Latouche, 2005a).

Voluntary and Democratic Transition

But if the degrowth movement is intent on getting us out of the economic imperative to grow, it is equally suspicious of political or ecological imperatives. Thus whilst ecological limits may make degrowth inevitable, the most fundamental question is a social and political one: that is, how can degrowth be made socially sustainable and provide for a prosperous future, or, to refer back to some of the terms used in the definition provided above, how can it increase well-being in a way that is just and democratic.

Thus degrowth proponents are wary of the possible danger of emerging authoritarian responses to environmental crisis. There is a long-lived tension within environmentalism between a commitment to democracy and grassroots participation on the one hand, and a concern for immediate action and results in the light of rapid ecological degradation on the other (e.g. Doherty and de Geus, 1996; Latta, 2007). Indeed, the sense of an environmental crisis could become another means (beside for example invoking the threat of terrorism) of reinforcing state authority (Blühdorn and Welsh, 2007). The degrowth movement recognises the danger of 'ends-oriented' thinking where the perception of a crisis produces a 'political imperative' that pushes aside democratic debate and urges us all to act in a concerted manner.

Whilst recognising the threat of environmental degradation, proponents of degrowth are not prepared to sacrifice democracy to some 'ecological imperative' any more than to some economic one. The fact that there are limits to growth or that growth may come to an end is to be embraced as an opportunity to redefine a better social order, one where we would have greater autonomy in collectively deciding on our own future. Thus degrowth is not just for the environment; it is also put forward as a way of developing a more participative, democratic and just society, for example, by bringing decisions closer to users (D'Alisa et al, 2015). Indeed it is this emphasis on democracy that has attracted much interest and maybe demarcates the degrowth movement within environmental politics (Weiss and Cattaneo, 2017).

Unlike some radical ecologists and in particular deep ecologists (e.g. Næss, 1989; Sessions, 1995), degrowth proponents are as keen to escape from the 'force' of nature as they are from the force of capitalism or the market. Degrowth is not presented as an ecological imperative (although it may be that too), but as an opportunity to initiate debates and reclaim decisions about the organisation of economic and social activities. Thus Ariès (2005) insists that degrowth is not a 'forced option' in the face of catastrophic environmental crisis; he is keen to move away from apocalyptic visions that could legitimise imposed solutions, and insists that degrowth is a choice that defenders would make without the oncoming ecological crisis, 'simply to be human'. Degrowth is not defended as a necessity but as a choice, one that has to be made democratically and openly. Similarly, for Latouche (2006) whilst degrowth may impose itself through natural limits, it is an opportunity to democratically reclaim and rethink the way we live, or 'to make a virtue out of necessity'. Thus the idea is to turn some 'inevitable' economic degrowth into a window of opportunities for designing socially sustainable systems (Kallis, 2011). The material conditions defined by limited ecological space and its current over-use may create an imperative for radical change in the ways we organise ourselves, but it does not in itself dictate how this should be done.

This emphasis on democratic choice over 'imperative' is accompanied by a privileging of human and social values above ecological ones. Whilst degrowth may have to operate within ecological limits, it is strongly anchored in humanist values; and various proponents are at pain to show that their concerns are primarily with human values and social justice rather than ecological values. Thus as noted above, Ariès (2005) would stand for degrowth even without the oncoming ecological crisis, 'simply to be human'. Similarly, for Latouche (2006), degrowth is not just about protecting the environment, it is a question of social justice. More generally, it is argued that the ecological crisis is just one consequence of an ideology of growth that destroys the social fabric (by creating inequalities and poverty) and democracy as well as the environment. In sum, degrowth is not defended as a necessity but as a political choice: we are not condemned to degrowth, rather it is an open path we can choose and shape.

Thus degrowth is not envisaged in terms of recession, sacrifice, austerity or scarcity (as it is often accused of), but as an opportunity to reconsider collectively what constitutes the good life, to define ways of life that are both ecologically and socially more sustainable (Kallis, 2011). And this places collective democratic choices, rather than the 'natural' forces of the market, competition or growth, at the centre of economic relations. It is in foregrounding these collective debates and choices that degrowth is an eminently political project; as discussed below, it seeks to re-politicise the economy.

The Re-Politisation of the Economy

For proponents of degrowth, in order to challenge neo-liberal economics of growth, we need to start with value and politics, we need to oppose economic determinism or 'economism' by going back to the terrain of the political. Thus one of the starting points of the degrowth movement is to politicise the economy, to reveal it as an abstract idea, a self-referential system of representations (Latouche, 2005b) rather than an objective reality, a set of 'given' facts and forces as it is commonly presented. Of course, on this point, degrowth advocates draw upon a whole tradition that since Polanyi (1944) has sought to deconstruct the naturalness of the economy (see for example Caillé, 2005; Callon, 1998; Gibson-Graham, 1996, 2006) and has called into question the 'hardness', 'fact-ness' or supposed inevitability of economic 'realities' such as the market, work or value. From this perspective, the main culprit is not growth itself but the ideology of growth, a system of representation that translates everything into a reified and autonomous economic reality inhabited by self-interested consumers, or *homo economicus*. It follows that to challenge the 'tyranny of growth', it is not sufficient to call for lesser, slower or greener growth for this would leave us trapped within the same economic logic; rather we need to escape from the economy as a system of representation (Fournier, 2008). This means re-imagining economic relations, identities, activities in different terms, for example by privileging the figure of the citizen over that of the consumer (e.g. Ariès, 2005; Dobson, 2003; Fournier, 2008), or as will be discussed in the next section, by foregrounding abundance rather than scarcity, or amateur work rather than paid work.

This re-politicisation of the economy involves placing economic decisions firmly within the hands of people (rather than the market, or other supposedly natural forces). Thus decisions about economic relations and activities should be subject to deliberative democracy and public debate rather than left to experts or the markets. For example, we could decide that we have enough material wealth, or that we want to work less – at least in paid employment, as I'll go on to discuss below. Or we could have collective debates about selective degrowth. Not everything can or should 'degrow', we could maybe do with fewer high-speed trains, or factories producing unnecessary gadgets, but more renewable energy infrastructure, education, caring systems, local food production systems (Kallis, 2011). The choice as to which activity should grow or degrow should not be left to the whims of the market but be placed in the hand of citizens through open political debates.

From Politics to Practice

So far, I have articulated the main ideas underpinning degrowth in terms of its ecological and social critique of growth, the emphasis on a just and democratic society, and the re-politicisation of the economy. But moving on from these broad ideas in order to consider the sort of practical measures that could take us to a degrowth society is essential if degrowth is to become an enabling framework rather than remain at the level of critique. If some ecologically inevitable degrowth is to be turned into a window of opportunities for some socially desirable arrangements, we need to learn to manage without growth. Currently capitalist societies are built on growth; without radical change in institutional arrangements, imagination and social practices, degrowth in capitalist societies could have dire social consequences and be unstable (for example, high level of unemployment, fiscal crisis of the state, as indeed we saw in Greece recently). So the question is what sort of practices and institutional arrangements could facilitate the transition to a stable and prosperous degrowth society (Kallis et al, 2012).

Cosme et al's (2017) review of the literature on degrowth suggests that proposals for action are articulated around three broad goals: an ecological one involving the reduction of the environmental impacts of human activities; a social one related to the redistribution of income and wealth both within and between countries; and one related to values and the promotion of a transition from a materialistic to a convivial and participatory society. This threefold framework is useful to get a sense of the sort of concrete practices put forward by the degrowth movement in that each level connects theoretical ideas with practical arrangements; it also gives a sense of the range of institutional, cultural and economic reforms that would be necessary to build a degrowth society.

The ecological objective to reduce the environmental impact of human activities could broadly take the form of reducing consumption, encouraging conservation and recycling. Concretely, this could be achieved through measures such as ad bans or pollution taxes, a ban on harmful activities such as nuclear or coal industry, or the provision of incentives for local production (Cosme et al, 2017).

The second objective regarding social justice and a more equal distribution of wealth and resources could be achieved through concrete measures such as new forms of ownerships (e.g. cooperative organisations), redistributive taxation, reduction in working hours (e.g. the twenty-one-hour working week), work-sharing, salary caps, or the introduction of a basic or citizen income (Cosme et al, 2017). There have been several experiments throughout the world recently (e.g. in Finland, in Ontario in Canada, in the city of Stockton in California, in Kenya, to mention a few examples) aiming to provide a guaranteed minimum income to people on low income. For example, the city of Stockton in California has recently launched a basic income scheme that provides 500$ a month to some families on low income in order to tackle inequality and poverty (Goodman, 2018). This income is paid out without any requirements or conditions, and ensures that people can meet basic needs, whilst giving them recognition for the 'free work' they may do in the community. Another emerging practice that can contribute to social equality is the 'pay-what-you-can' scheme implemented in some establishments such as Grace Café in Kentucky (www.gracecafeky.org/). The idea here is that customers who can afford it in effect subsidise those on low income by paying more for the food they consume.

The third objective involves redefining the good life and encouraging a transition away from a materialistic society to one emphasising conviviality and participation. Concretely, this could be achieved through community building, education, downshifting, co-housing and various projects encouraging resource sharing. Particular emphasis is also placed on recognising the value of voluntary work, or promoting the decommodification and deprofessionalisation of work (D'Alisa et al, 2015), something that will be discussed below.

The range of measures envisaged spans grassroots initiatives as well as institutional reforms, and involves multiple strategies from multiple actors. It also suggests that moving away from a society built on growth requires wide ranging cultural, political and economic changes (Joutsenvirta, 2016; Kallis, 2011).

What runs through the three axes identified by Cosme et al (2017) is a reorganisation of society away from growth and material wealth, towards frugality, sharing, conviviality and care. And this in turns calls for a re-articulation of values, and the way they govern the economy. We need to redefine social and economic relations away from the logic of economic growth and its emphasis on calculative rationality, consumerism and productivism. Thus one important pillar of degrowth is the work of re-imagination that needs to be done to free ourselves from the compulsion of growth and to be able to envisage life with degrowth. As many have insisted degrowth is about the decolonisation of the imagination and the re-imagination of what it means to live a 'good life', for example in terms of conviviality, frugality, sharing, abundance to use some of the values often put forward by degrowth advocates (e.g. D'Alisa et al, 2015; Latouche, 2014). It is to this work of re-imagination that we now turn our attention.

Re-Imagining the Economy

As already suggested, degrowth is not about 'less of the same' (as in recession, or declining GDP) but calls for 'an altogether new, qualitatively different world that will evolve with confrontation with the existing one' (Kallis and March, 2015: 362). So degrowth is not a quantitative question of doing less, or the opposite of 'growth', but instead refers to a fundamental change of our key references, a paradigmatic change that aims to take us out of the domain of economic rationality, and asks fundamental questions about the nature of wealth, its distribution, its use, and misuse (Chertkovskaya et al, 2017; D'Alisa et al, 2015; Martinez-Alier et al, 2010). This paradigmatic change means breaking away from the key figure of consumer capitalism, the *homo economicus* always seeking to maximise utility, be it as a consumer always wanting more goods, or as a worker wanting more pay. This radical break means re-imagining economic relations and identities in different terms, and below I suggest two examples which illustrate how this work of re-imagination can work in the domain of consumption and work.

But before we move on to these specific examples, it is worth saying a few words about the approach taken for re-imagining the economy. Maybe a useful starting point is the pioneering work of Gibson-Graham (1996, 2006) in developing a vocabulary that would enable us to imagine economic difference (from capitalism). Gibson-Graham (1996, 2006) have called for a re-conceptualisation of economic relations and identities away from 'capitalocentric' thinking and have sought to reframe economic activities in terms of the co-existence of different forms of transactions, labour, and ways of producing and distributing surplus. In their involvement in various community regeneration programmes, Gibson-Graham have invited people to re-imagine their economic activities in terms other than those made available by capitalism. For example, they have sought to re-present different forms of transactions beside commodity market (e.g. local trading schemes, gifts, mutual exchange between households), different forms of labour beside wage labour (e.g. self-employment, volunteering, domestic work), and different forms of surplus distribution beside capital accumulation and profit 'imperative' (e.g. governed by social or environmental ethic). This broadening of the 'economy' has opened up possibilities for people to re-imagine their 'economic activities' in terms of the voluntary contributions, mutual help and provision of 'free services' in which they routinely engage in the home, the neighbourhood, or the broader community.

The degrowth movement shares much of Gibson-Graham's critical intent, and also insists that the economy is open to choices and multiple possibilities; both approaches contribute to freeing the imagination and conceptualisation of material practices from the grip of neo-liberal capitalism, or the growth imperative (Chertkovskaya et al, 2017). In the same way as Gibson-Graham proposed a new vocabulary for thinking about the diverse economy beyond capitalism, D'Alesia et al. (2015) provide a new vocabulary to think past growth, to articulate alternatives to growth not in terms of their difference from a growth-based economy (e.g. more or less growth, consumption, production ...), but in their own terms. The aim of this sort of lexicon for degrowth is to convey the diversity of approaches, concepts, ideas that can be mobilised to achieve a paradigm shift in the way we think about and practice the economy (D'Alisa et al, 2015); for example, it foregrounds ideas such as sharing, simplicity, conviviality, commons or care. As explained by the editors in the Introduction to *Degrowth: A Vocabulary for a New Era*, the objective of degrowth is not to make an elephant leaner but to turn an elephant into a snail, and this requires new points of reference and perspective. Below, I illustrate this work of re-conceptualisation with two examples, the first one is about the way we relate to scarcity and abundance, the second one is about the way we value work.

'Having Enough' and Abundance

One prominent term in the degrowth literature is that of abundance. With what may appear like the paradoxical idea of 'frugal abundance' (Latouche, 2014), degrowth advocates offer a way of turning the scarcity central to capitalist accumulation and growth into abundance. Latouche argues that the idea of scarcity is closely intertwined with that of growth. Capitalism relies on endless, unlimited needs and wants to nourish its growth, but in the process, it produces relative scarcity, through the enclosure of resources but also through positional inequalities and the promise of unlimited choice. The limitless nature of wants and needs means that there is always more to be had, so there is always scarcity. Growth will never provide 'enough' for everyone (Skidelsky and Skidelsky, 2012) because it will always lead to more needs to be satisfied with positional goods, hence it creates scarcity. And scarcity in turn calls for economising, accumulating and generating more growth.

Degrowth is about imagining and creating a different society that accepts that it has had enough (D'Alisa et al, 2015). Indeed, most civilisations apart from capitalism had a sense of 'enough'; and it is only if we accept that we have enough that we can escape scarcity and get a sense of abundance (Chertkovskaya et al, 2017). From this perspective, scarcity emerges as the product of (unequal) social relations of production and distribution rather than the product of ecological limits, and hence is avoidable. Indeed, the degrowth movement, like Illich (1973) before, stands against the over use of natural resources not so much because of their scarcity or 'objective limits', but because accepting or defining limits is conducive to a more democratic and egalitarian society. Only a society that 'has had enough' can liberate itself from scarcity (Kallis and March, 2015); in this sense self-limitation is liberating in that it is an exercise of choice to live within some (limited) resources. The question then becomes one of how a society can limit itself. And sharing is one avenue for self-limitation: sharing work, housing, cars, bikes, resources; sharing resources means they are not available for private accumulation, but it also means equal access and hence eliminates the sense of scarcity and unmet wants produced by inequality.

Developing a sense of 'having enough' also generates relative abundance as there will always be surplus resources. Through the notion of '*dépense*', the degrowth movement calls for the 'dispensing' or wasting of these surplus (in for example big feasts or ostentatious projects) rather than their investment in productive activities that would fuel growth and generate environmental destruction as well as social inequalities; surplus has to be 'dispensed' if we are to move away from growth (Romano, 2015).

Thus a key concern for degrowth is to recognise that many of us (at least in the Global North) have enough, and to develop institutions as well as social and economic relations that will allow us to live with enough (Kallis and March, 2015).

Amateur Work

Another key tenet of the degrowth movement concerns the re-conceptualiation of work, and in particular its decommodification and deprofessionalisation (D'Alisa et al, 2015). In a degrowth society characterised by lower output, hours of work must also fall, or labour productivity must decline (Norgard, 2013; Schor, 2015); both options are seen as connected. To avoid degrowth leading to unemployment for some, worksharing and a reduction in working hours have been suggested; for example the idea of the twenty-one-hour week (NEF, 2010) has attracted much interest as it would enable work to be spread out more widely and equally (Norgard, 2013).

But as suggested earlier, degrowth is not just about doing less of the same, doing less work in this case, but calls for re-conceptualising work, for example thinking about what it is for and what it produces; thus work can provide satisfaction, meaning, dignity, care, beauty, besides products or income (Chertkovskaya et al, 2017; Nierling, 2012). And indeed, much voluntary work is by definition productive of personal satisfaction (Norgard, 2013; Stebbins, 2017). As Norgard (2013) argues, whilst the professional economy is driven by money, the voluntary economy, or what he calls the amateur economy, is driven by love (from the Latin *amare* – to love) or affective motivation. Privileging the amateur economy would be a way of reclaiming control over, and satisfaction in, one's work. In addition, the transfer of some activities from the capital intensive professional economy to the labour intensive amateur economy would reduce overall labour productivity, and hence resource throughput, but would be productive in other terms: of satisfaction, meaning, or creativity (Norgard, 2013). So if we take into account not just outputs in terms of products or services but all benefits, including satisfaction from the production process, amateur work is not so 'unproductive'. Moreover, much voluntary work is oriented towards the reproductive sphere, such as cooking, gardening, caring, or community activities; and as such enables more sustainable lifestyles by contributing to subsistence activities and political participation (Nierling, 2012). Privileging volunteer work would in this respect be a way of valuing care work which has long been discarded by the growth economy as it tends not to be monetised (or gets only poorly monetised). Indeed, many feminist economists have long insisted on the need to re-value care work to promote gender equality and environmental justice (e.g. Dengler and Strunk, 2018; Gibson-Graham, 2006).

Overall, degrowth involves a re-conceptualisation of work, of its relation to income, and of what it is productive of. The decommodification of work is central to degrowth and means its (at least partial) decoupling from income as well as conventional understanding of productivity (Gerber and Gerber, 2017). Voluntary or amateur work may be less productive of goods and services, but it is more productive of benefits that cannot be easily quantifiable such as satisfaction, sense of recognition, care, or dignity (Nierling, 2012). The Worldwide Opportunities on Organic Farms movement (WWOOF) provides a good illustration of work that takes place outside the monetised economy and draws upon, as well as contributes to, the values of cooperation, care, conviviality and environmental sustainability (wwoof.net/). WWOOF is a worldwide movement that links volunteers with organic farmers; volunteers help with everyday tasks on an organic farm in exchange of free board and lodging, but also skills and knowledge related to organic farming and different cultures. The idea is to promote cultural and educational exchange, an exchange that is not based on money but trust, care, and the sharing of skills, knowledge, and sustainable lifestyles (Lans, 2016).

However re-valuing voluntary work requires institutional reforms that would ensure that everyone can afford to work without pay, in the forms of a basic income or worksharing. Hence the relationship between voluntary and paid work needs to be thought about holistically (Nierling, 2012)

Conclusion: Nonscalability Rather than Descaling

Throughout this chapter, I have suggested that degrowth refers to a more radical break than just scaling down, or descaling. Thinking of degrowth simply in terms of descaling, or negative growth, would be falling victim to capitalocentrism and reducing degrowth to the opposite of growth. Instead, degrowth involves the imaginative challenge of thinking about what it would be like to live without growth, one of the main assumptions that has underpinned much of the twentieth and twenty-first centuries' economic thinking and practice. What is required is not to think against growth but in different terms, thus as suggested earlier, the prefix 'de' is not to be read as less growth but thinking without growth. This radical break is signalled firstly by foregrounding democratic choices and debates in the shaping of the economy, and secondly by re-imagining economic relations and identities in different terms, for example thinking in terms of abundance, of having 'enough', or thinking about work in terms of satisfaction or social relations; both moves can be captured under a core concern of the degrowth movement: escaping from the economy (e.g. Fournier, 2008; Latouche, 2005b).

But let's be clear about what this escape means, or what it is we are escaping from. Insisting on the need to escape from the 'economy' is not to deny the importance of the various material practices (e.g. work, exchange, distribution) that go towards meeting our needs. What we need to escape from is a particular understanding of the economy as dictated by certain imperatives we have no choice over: growth, competition, market and so on. Thus for proponents of the degrowth movement, economic practices need to be re-embedded within the social and the political rather than be seen as belonging to an autonomous, reified field of 'the economy'.

This escape from the economy is at least as much a question of decolonising the imagination as one of enacting new practices, it calls for rethinking the economy (or as Caillé, 2005, puts it 'de-thinking the economic'), or rethinking ourselves outside the sort of economic relations envisaged by growth based capitalism, for example by fighting against the reduction of human beings to their economic function, as workers and consumers (Ariès, 2005). This re-thinking outside the economic positions made available by capitalism was illustrated here with various conceptual moves within degrowth debates, all seeking to affirm different values to that of 'economic rationality': for example, replacing the discourse of economic imperatives by that of choice and democracy, replacing the discourse of scarcity with that of abundance, of endless growth with that of 'having enough', or re-conceptualising work in terms of its capacity to produce satisfaction or conviviality rather than incomes or goods.

Maybe much of this work of re-conceptualisation can be captured in terms of learning to see the nonscalable (Tsing, 2015). Tsing suggests that scalability, that is the ability of projects (be it production or research) to change scale smoothly without any change in the project frame has been the hallmark of modernity and industrialisation; it underpins the dream of progress. She argues that in order to live within the ruins of capitalism, we need to turn our attention to the nonscalable, that which exists and lives in transformative relations. The nonscalable cannot be reproduced or scaled up without being qualitatively transformed; it is not geared for expansion. Maybe the sort of amateur work discussed earlier provides a good example of nonscalable quality; for example the amateur gardener's work and the satisfaction it produces is contingent on certain relations with plants, with soil, maybe with a certain sense of time and season, all of which would be undermined by demands to produce on a larger scale. Similarly, the idea of 'having enough' relies on the sense that past a certain point, more or scaling up (consumption, production) would produce qualitatively different relations, relations to work that may become more instrumental, relations to others that may become more unequal, and relations to 'things' that may become more reliant on exchange value and accumulation than use value and 'depense'. Thus it could be argued that degrowth is not about descaling but recognising nonscalability, that which cannot be scaled up without losing its relational quality.

References

Arendt, H. (1958) *The Human Condition*. Chicago: University of Chicago Press.
Ariès, P. (2005) *Décroissance ou Barbarie*. Lyon: Golias.
Barry, J. (2005) Ecological modernisation. In J. Dryzek & D. Schlosberg (eds.), *Debating the Earth*, pp. 303–21. Oxford: Oxford University Press.
Binswanger, M. (2001) Technological progress and sustainable development: what about the Rebound Effect? *Ecological Economics*, 36: 119–32.
Blühdorn, I. (2007) Sustaining the unsustainable: symbolic politics and the politics of simulation. *Environmental Politics*, 16(2): 251–75.
Blühdorn, I., and Welsh, I. (2007) Eco-politics beyond the paradigm of sustainability: a conceptual framework and research agenda. *Environmental Politics*, 16(2): 185–205.
Caillé, A. (2005) *Dé-penser l'Economique: Contre la Fatalité*. Paris: La Découverte.
Callon, M. (ed.) (1998) *The Laws of the Markets*. London: Blackwell.
Chertkovskaya, E., Paulsson, A., Kallis, G., Barca, S. and D'Alisa, G. (2017) The vocabulary of degrowth: a roundtable debate. *Ephemera*, 17(1): 189–208
Cosme, I., Santos, R., and O'Neill, D. (2017) Assessing the degrowth discourse: a review and analysis of academic degrowth policy proposals. *Journal of Cleaner Production*, 149: 321–34.
D'Alisa, G., Demaria, F., and Kallis, G. (eds.) (2015) *Degrowth: A Vocabulary for a New Era*. Abingdon: Routledge.
Daly, H. (1996) *Beyond Growth: The Economics of Sustainable Development*. Boston, MA: Beacon Press.

Dengler, C., and Strunk, B. (2018) The monetized economy versus care and the environment: degrowth perspectives on reconciling an antagonism. *Feminist Economics*, 24(3): 160–83.

Dobson, A. (2003) *Citizenship and the Environment*. Oxford: Oxford University Press.

Doherty, B., and de Geus, M. (eds.) (1996) *Democracy and Green Political Thought: Sustainability, Rights and Citizenship*. London: Routledge.

Fournier, V. (2008) Escaping from the economy: the politics of degrowth. *International Journal of Sociology and Social Policy*, 28 (11/12):528–45.

Georgescu-Roegen, N. (1971) *The Entropy Law and the Economic Process*. Harvard University Press: Cambridge, MA.

Gerber, J-D. and Gerber, J-F. (2017) Decommodification as foundation for ecological economics. *Ecological Economics*, 131: 551–6.

Gibson-Graham J. K. (1996) *The End of Capitalism (As We Knew It): A Feminist Critique of Political Economy*. Oxford: Blackwell.

Gibson-Graham, J. K. (2006) *A Postcapitalist Politics*. Minneapolis, MN: University of Minnesota Press.

Goodman, P. (2018) Free cash to fight income inequality? California city is first in US to try. *The New York Times*, May 30. Available at www.nytimes.com/2018/05/30/business/stockton-basic-income.html

Gorz, A. (1975) *Ecologie et Politique*. Paris: Galilée.

Hamilton, C. (2003) *Growth Fetish*. Sydney: Allen and Unwin.

Harvie, D., Slater, G., Philip, B. and Wheatley, D. (2009) Economic well-being and British regions: the problem with GDP per capita. *Review of Social Economy*, 67(4): 483–505.

Herring, H. (2002) Is energy efficiency environmentally friendly? *Energy & Environment*, 11: 313–25.

Illich, I. (1973) *Tools for Conviviality*. London: Calder and Boyars.

Jackson, T. (2009) *Prosperity without Growth: Economics for a Finite Planet*. London: Earthscan.

James, O. (2007) *Affluenza: How to Be Successful and Stay Sane*. London: Vermillon.

Joutsenvirta, M. (2016) A practice approach to the institutionalization of economic degrowth. *Ecological Economics*, 128: 23–32.

Kallis, G. (2011) In defence of degrowth. *Ecological Economics*, 70: 873–80.

Kallis, G., and March, H. (2015) Imaginaries of hope: the utopianism of degrowth. *Annals of the Association of American Geographers*, 105 (2): 360–8

Kallis, G., Kerschner, C., and Martinez-Alier, J. (2012) The economics of degrowth. *Ecological Economics*, 84: 172–80.

Lans, C. (2016) Worldwide Opportunities on Organic Farms (WWOOF) as part of the existing care economy in Canada. *Geoforum*, 75: 16–19.

Latouche, S. (2004) Degrowth Economics: why less should be much more, *Le Monde Diplomatique*, November (available at http://mondediplo.com/2004/11/14latouche)

Latouche, S. (2005a) *Décoloniser l'imaginaire*. Lyon: Parangon

Latouche, S. (2005b) *L'invention de l'Economie*. Paris: Albin Michel.

Latouche, S. (2006) The globe downshifted: how do we learn to want less? *Le Monde Diplomatique*, English edition, January (available at http://mondediplo.com/2006/01/13degrowth).

Latouche, S. (2014) *Essays on Frugal Abundance*. Paris: Simplicity Institute. Available at http://simplicityinstitute.org/wp-content/uploads/2011/04/FrugalAbundance1SimplicityInstitute.pdf

Latta, P. (2007) Locating democratic politics in ecological citizenship, *Environmental Politics*, 16(3): 377–93.

Martinez-Alier, J., Pascual, U., Vivien, F-D. and Zaccai, E. (2010) Sustainable de-growth: mapping the context, criticisms and future prospects of an emergent paradigm. *Ecological Economics*, 69: 1741–7.

Meadows, D., Meadows D. and Randes, J. (1972) *Limits to Growth*. New York: Universe Books.

Milne, M., Kearins, K. and Walton, S. (2006) Creating adventures in wonderland: the journey metaphor and environmental sustainability. *Organization*, 13 (6): 801–39.

Næss, A. (1989) *Ecology, Community and Lifestyle: Outline of an Ecosophy*. Translated by D. Rothenberg. Cambridge: Cambridge University Press

NEF (2010) *21 Hours: Why a Shorter Working Week Can Help Us All to Flourish in the 21th Century*. London: New Economic Foundation.

Nierling, L. (2012) "This is a bit of the good life": recognition of unpaid work from the perspective of degrowth. *Ecological Economics*, 84: 240–6.

Norgard, J. (2013) Happy degrowth through more amateur economy. *Journal of Cleaner Production*, 38: 61–70.

O'Neill, D. (2015) Gross National product. In D'Alisa, G., Demaria, F, and Kallis, G. (eds.) *Degrowth: A Vocabulary for a New Era*: 103–8. London: Routledge.

Odum, H.T., Odum, E.C. (2006) The prosperous way down. *Energy*, 31: 21–32.

Picketty, T. (2014) *Capital in the 21st Century*. Cambridge, MA: Harvard University Press.

Polanyi, K. (1944). *The Great Transformation*. New York: Rinehart.

Princen, T. (2003) Principles for sustainability: from cooperation and efficiency to sufficiency. *Global Environmental Politics*, 3(1): 33–50.

Romano, O. (2015) Dépense. In D'Alisa, G., Demaria, F. and Kallis, G. (eds). *Degrowth: A Vocabulary for a New Era*: pp. 86–89. London: Routledge.

Schneider, F. (2003) L'effet Rebond. *l'Ecologiste*, 4(3): 45.

Schneider, F., Kallis, G., Martinez-Alier, J. (2010) Crisis or opportunity? Economic degrowth for social equity and ecological sustainability. *Journal of Cleaner Production*, 18: 511–18.

Schor, J. (2015) Work sharing. In D'Alisa, G., Demaria, F. and Kallis,. G. (eds). *Degrowth: A Vocabulary for a New Era:* pp. 195–8. London: Routledge.

Schumacher, E.F. (1973) *Small Is Beautiful: Economics as if People Mattered*. London: Blong and Briggs.

Scott-Cato, M. (2006) *Market, Schmarket: Building the Post-Capitalist Economy*. Cheltenham: New Clarion Press.

Sessions, G. (ed) (1995) *Deep Ecology for the Twenty-first Century*. Boston: Shambhala.

Skidelsky, R. and Skidelsky, E. (2012) *How Much Is Enough?* New York: Other Press.

Stebbins, R. (2017) *Between Work and Leisure: The Common Grounds of Two Separate Worlds*. New York: Routledge.

Steffen, W., Richardson, K., Rockstrom, J. et al. (2015) Planetary boundaries: guiding human development on a changing planet. *Science*, 347 (6223): 1–15.

Tsing, A.L. (2015) *The Mushroom at the End of the World: On the Possibility of Life in Capitalist Ruins*. Princeton, NJ: Princeton University Press.

WCED (1987) *Our Common Future*. Oxford University Press, Oxford.

Weiss, M. and Cattaneo, C. (2017) Degrowth: taking stock and reviewing an emerging academic paradigm. *Ecological Economics*, 137: 220–30.

Wilkinson, R. and Pickett, K. (2010) *The Spirit Level*. London: Bloomsbury Press.

Wright, E. (2010) *Envisioning Real Utopias*. London: Verso.

13 Sustainable Consumption

Emily Huddart Kennedy

Introduction

During a commercial break in the 2017 Super Bowl, viewers were treated to a comical advertisement for a hybrid car (Kia, 2017). The commercial portrays actor Melissa McCarthy placing herself in danger in order to save whales, forests, icebergs, and rhinoceroses. In each of these moments of direct action, McCarthy is physically injured; between these moments of activism, she is calmly driving in a comfortable (hybrid) car, so we are unsurprised when the closing narration states, "It's hard to be an eco-warrior. But it's easy to drive like one." This commercial humorously captures a dominant narrative around sustainable consumption: that acting on one's environmental values in the marketplace (through so-called "green consumption") is far more straightforward than being involved in direct action, while being just as effective as these more traditional forms of engagement (Maniates, 2001). What is more, in a political climate that sees climate skeptics and fossil fuel industry insiders in charge of federal environmental policy, as is currently the case in the United States, making the switch to a more eco-friendly car may well feel like a more effective course of action. In addition to its views during the Super Bowl, at time of print, this ad has also been viewed at least 25 million times on YouTube, speaking to its resonance with the general population. Green consumption, in this case purchasing a hybrid vehicle, is as accepted (if not more so) a path to sustainable consumption as engagement in traditional social movement activities.

Sustainable consumption is a trendy and heterogeneous topic – at once a normative goal, a social practice, a topic of scientific inquiry, and a theoretical framework (Liu et al., 2017; Reisch et al., 2016). As a goal, it is fairly straightforward to define: a level of consumption of goods and services that can be sustained across time and space indefinitely (Griggs et al., 2013). In more detail, ecological economist Herman Daly distills the pursuit of sustainability to three goals that have clear points of connection to sustainable consumption. First, sustainable use of renewable resources means that the pace at which we use resources should not exceed the rate at which they are replaced. Second, sustainable use of non-renewable resources means that the pace at which we use them should not be faster than the rate at which we develop substitutes. And third, sustainable rates of pollution and waste means that emissions should not exceed the capacity of natural systems to absorb these effluents (Daly, 1990).

Yet as a social practice the conceptual clarity that is apparent when it comes to the scope of the problem is quickly lost as we consider questions like,: *How do we evaluate whether the consumption of something is sustainable? How do we evaluate whether a substitute is acceptable? Should consumption be equitable? Do we each bear equal responsibility for consuming sustainably?* Questions like these demand inquiry from the social sciences in general, and from environmental sociology in particular.

As a topic, sustainable consumption requires that learners become familiar with measures from the natural sciences (e.g., capturing information on stores of natural resources and lifecycle assessments of the environmental impact of a product) *and* concepts from sociology (e.g., agency, structure, power, and culture). Not surprisingly, given the multidisciplinary interests in the topic (see for instance Hertwich, 2005; Jackson, 2005; Middlemiss, 2018; Sanne, 2002), sustainable consumption scholars lack a coherent theoretical framework, though some promising avenues exist. Given the disciplinary breadth of study on the subject, and the focus of this handbook on environmental sociology, the following chapter approaches and reviews sustainable consumption quite narrowly. In particular, I argue that the topic of sustainable consumption can provide us with insight into human–environment relationships, and that sociological explanations of the increasing ubiquity of the "political consumer" (Stolle and Micheletti, 2013) can help us understand how human actions and interactions affect the environment.

The chapter proceeds as follows: I first situate the topic of sustainable consumption within broader environmental sociological questions about the causes of, impacts of, and solutions to environmental issues. This part of the review presents a paradox – while evidence is showing us that issues like climate change and biodiversity loss are complex because they result from various levels of social organization (from national-level government policy and legislation on corporate ethics, to urban design and rituals around gift-giving), at the same time, individual consumption choices face almost unparalleled levels of scrutiny in the public sphere (Maniates, 2001; Middlemiss, 2018; Szasz, 2007). The second section reviews literature that helps to explain this paradox by calling attention to ideologies that implicate the individual as an appropriate agent of change to bring about sustainability goals. In this section I describe the social practices theoretical framework as a promising avenue to explain sustainable (and unsustainable) consumption patterns. Next, I outline two related problems with mainstream, contemporary sustainable consumption. The material problem is that individual sustainable consumption activities like buying a hybrid car are largely unrelated to individuals' environmental impact. This is closely tied to the second, symbolic, challenge – like any social practice, sustainable consumption is built upon class, gender, and race inequality. Inattention to these axioms of inequality constitutes a threat to the pursuit of a just sustainability (Agyeman and Evans, 2004). The chapter concludes with a discussion of future research directions in this rich and timely area of environmental sociological scholarship.

1 Environmental Issues: Causes, Impacts and Solutions

Current rates of consumption of resources and pollution are unsustainable. Perhaps one of the most compelling heuristics to gauge the rate of human impact on the environment – or the extent to which we are (not) meeting Daly's three goals – is Earth Overshoot Day. Earth Overshoot Day estimates the date in the calendar year when humanity has used up its "budget" of natural resources. After that day, we are borrowing against future generations. In 2018, this date is August 1. That is, after August 1, resources are being consumed on the planet at a rate faster than they can be replenished (Earth Overshoot Day, 2018).

It is clear that we are currently consuming renewable and non-renewable resources more quickly than they can be regenerated or substituted, and that we are exceeding the Earth's carrying capacity to absorb pollutants (Rockström et al., 2009). However, seeking to understand the causes, impacts, and solutions associated with this clear statement of the problem raises many thorny and complicated sociological questions: what are the elements of social order that gave rise to humanity's impact on the non-human world (causes)? How are these impacts distributed across individuals and groups around the world (impacts)? Who bears the responsibility to ameliorate such issues? And how are we to act on that responsibility (solutions)? These questions animate many prominent, social scientific questions on the topic of sustainable consumption (e.g., Cohen, Brown, and Vergragt, 2013; Kennedy, Cohen and Krogman, 2016; Reisch et al., 2016).

Within environmental sociology, two competing frameworks tackle the question "what elements of social order gave rise to our impact on the non-human world?" (see Volume I, Chapters 10 and 11 on treadmill of production theory and ecological modernization theory, respectively). The next question, "how are these impacts distributed" is addressed under the mantle of environmental justice (see Volume I, Chapters 27 through 31), a topic that is intimately tied to the ecological impacts witnessed as part of climate change as well as to the disproportionate burdens held by low-income people of color (see Volume II, Chapters 14–17). The last two questions, concerning the responsibility to protect the environment and the social action this responsibility entails, fall within the domain of sustainable consumption (though see also Chapters 9, 11, 12 and 13 in Volume II as well as sections V, VI and VII in Volume II).

A brief tour through the themes of causes and impacts demonstrates first the inseparability of consumption from production and second, the disproportionate impact of production and waste on those living at the margins of society (Anantharaman, 2017). Though there is a division among those who feel environmental problems stem from the incompatibility of capitalist modes of economic growth and environmental protection (treadmill theory) or from autocratic governance structures and unharnessed scientific and technological advances (ecological modernization theory), I set that debate aside to focus on the shared emphasis on consumption as inextricable from production (Carolan, 2004; Gould, Pellow and Schnaiberg, 2004; Mol and Spaargaren, 2004). The take-away message here is that trying to achieve sustainable consumption without tackling production (and vice versa) will never result in truly sustainable human–environment relationships (Vergragt, Akenji, and Dewick, 2014). Yet the common-sense logic of seeing production and consumption as part of the same cycle is obfuscated when popular wisdom sees the individual *consumer* as both capable of and responsible for upholding the common good (Gabriel and Lang, 2015; Szasz, 2007). Or, as stated by Soneryd and Uggla (2015, p. 913), "Paradoxically, 'simple solutions', such as changing light bulbs, having meat-free days, and choosing public transport, are being highlighted at a time when the global, transboundary, and complex character of environmental problems is being acknowledged." That is, individual consumption choices are being ever more scrutinized while the multifaceted, global, and complex nature of environmental problems is being ever better documented. I address this paradox through a discussion of green governmentality.

2 Sustainable Consumption in Green Governance Models: The Role of the Individual

Traditional models of governance (including environmental governance) were based on the related concepts of duties, solidarity, and citizenship. In contrast, current models of governance are increasingly linked with the themes of consumption, responsible choice, and lifestyles (Rose, 2000). It is not just that the individual is of paramount importance in contemporary green governance models, but that it is in their roles as consumers and in the shape of their daily lives that individuals are presumed to effect pro-environmental change (Middlemiss, 2014). Such arguments rest on the assumption that individual actors exercise free choices and that, concomitantly, greening society requires prompting individuals to buy eco-friendly products (Shove, 2003; 2010; Warde, 2017). As Melissa McCarthy's representation of the eco-warrior in the Super Bowl ad makes clear, the comfort of this market-based sustainable consumption model is that it upholds a common belief that individuals act with free will – that they are autonomous, uninfluenced by a collective entity such as civil society or the state (see Middlemiss, 2014 for more detail). Within this theory of action, realizing sustainable consumption need not entail a wholesale redesign of our political, sociotechnical, and cultural systems but rather rests on offering consumers the choice of a greener product. The role of the state and the market, then, is to communicate to consumers the benefits of these sustainable products, and to label them as eco-friendly so consumers can differentiate among the panoply of options available in the modern marketplace (Johnston, 2008; Southerton, Warde, and Hand, 2004). While this individualized responsibility is not the only mode of governing human–environment relationships, it is currently central to the pursuit of sustainable consumption in many jurisdictions (Soneryd and Uggla, 2015).

Sociologists have provided important insights to aid our understanding of why it is that consumer behavior is subject to such a high degree of public scrutiny – and why so many theories of consumer behavior assume actors make their shopping choices voluntarily (e.g., Johnston, 2008; Soneryd and Uggla, 2015). What these authors argue is that the tendency for individuals to feel responsible for their environmental impact is partly a result of an ideological stance that the consumer is an autonomous and powerful entity capable of (and responsible for) making social change through choices in the marketplace (Soneryd and Uggla, 2015). Scholars see this same assumption of individuals as autonomous actors in social scientific theories of pro-environmental behaviors. For example, Shove (2010) critiques theories of action that rest on the premise that attitudes give rise to behaviors that can be evidenced in individual choices (Shove terms this the ABC model where "A" stands for attitude, "B" stands for behavior, and "C" stands for choice). These ABC theories have an extensive history within environmental sociology; as evidenced by frequent and influential accounts of consumption behavior from the perspective of social psychological theories like value-belief-norm theory and variations of the theory of planned behavior (Stern et al., 1999). In general, the study of environmental concern and environmental behavior has largely adopted an ABC approach, as can be seen in models that seek to identify the socio-demographic correlates of concern and the attitudinal correlates of pro-environmental behaviors (e.g. Clark, Kotchen and Moore, 2003; Wiidegren, 1998). Policy recommendations resulting from this approach tend to align quite well with the green governance model that Soneryd and Uggla (2015) argue implicates individual consumption practices in resolving complex sustainability challenges.

In contrast to an ABC model of behavior, many sociologists envision quite different models of consumption behavior, sustainable and otherwise. A prominent alternative to these ABC theories is the so-called "social practice" approach (e.g., Halkier, 2009; Kennedy, Cohen and Krogman, 2016a; Shove, 2010; Warde, 2005; 2017). Social practice theories shift the analytic emphasis from individual behaviors to actual practices enacted across time and space. By decentering the individual, scholars can focus on "broader, more complex patterns of activities and 'how efforts at governing consumption engage creatively with people's existing ethical disposition'" (Barnett et al., 2011 as cited in Soneryd and Uggla, 2015, p. 916). Social practice theories seek to make sense of the myriad consumption practices that are highly routinized and analytically distinct from actions where consumption is the focal point of activity (e.g., shopping as a leisure activity) (Shove and Warde, 2002). This "inconspicuous consumption" is relevant to sustainable consumption since it takes place every day and is implicated in a substantial portion of the residential sector's consumption of resources (i.e., laundering, showering). Clearly, these practices are not solely the direct result of the launderers' values and attitudes but relate to cultural norms of cleanliness and comfort (Shove, 2003), economic structures and classed lifestyles (Soneryd and Uggla, 2015), and technological developments in plumbing and heating (Southerton et al., 2004).

Social practice theories draw from earlier sociological efforts to hybridize agency-heavy (subjective) accounts of social action with structural (objective) accounts (e.g., Bourdieu, 1977; Giddens, 1984). In this way, voluntaristic theories of sustainable consumption (e.g., Kollmuss and Agyeman, 2002; Stern et al., 1999) are shown to be relatively inattentive to power and social structures as important influences on the development of individual values and the actor's capacity to manifest values into actions (Swidler, 1986). Further, overly structural accounts of sustainable consumption (e.g., Baudrillard, 2016; Schor, 2008) are seen to be inattentive to culture and identity and thus unable to explain the emergence of subcultures of intentionally sustainable lifestyles or even variation within the lifestyles of socioeconomically homogenous groups (e.g., Lorenzen, 2012; Schelly, 2017).

An emerging focus on alternative lifestyles and the place of consumption in defining alterity examines anti-consumer identities, such as voluntary simplicity and freeganism, groups that use alternative consumption practices as a tool to challenge socially constructed definitions of waste, sufficiency, and the good life. Voluntary simplifiers seek to reduce their engagement in the labor force (made possible by reducing their level of consumption) in order to regain satisfaction and wellbeing (Lorenzen, 2012). Freegans consume from the end of the production cycle – as products are relegated to the trash (Barnard, 2016). Voluntary simplifiers and freegans reject green consumption as a viable path to a just sustainability, instead seeking to work on socio-ecological change outside of capitalist modes of production. In short, this is sustainable consumption of a different sort – voluntary simplicity and freegan practices require significant reconfigurations of work, home, and education (Kennedy, 2011) – that is, these are not sustainable consumption activities that can easily be integrated into mainstream lifestyles and nor do these activities rest solely on the purchase of high-cost eco-friendly products.

What this review should make clear is that individualizing environmental protection is both ubiquitous and multifaceted. We can see traces of individualization in our theories of sustainable consumption when we assume that behaviors arise from individually held attitudes. We see individualization in our social worlds when we feel we are being spoken to as responsible consumers as exemplified in the commercial featuring Melissa McCarthy. However, sociologists have challenged both the theory and practice of individualized responsibility. *Theoretically*, social practice theories draw attention to practices in a way that brings into focus norms, policies, social difference, and technical infrastructure as shaping consumption patterns, rather than individual attitudes. And *topically*, sociologists are studying groups that use alternative (and lower-impact) consumption practices to resist dominant, resource-intensive lifestyles.

3 Two Related Problems with Sustainable Consumption

The shortcomings of relying on individualized sustainable consumption practices as a solution to environmental issues are both material and symbolic. Looking first to the material problem, the primary issue here is that the way sustainable consumption is commonly operationalized (e.g., recycling, buying eco-friendly goods) does not actually add up to a substantially lower environmental impact (Csutora, 2012; Geiger et al., 2017; Kennedy, Krahn and Krogman, 2014). That is, what looks like a sustainable lifestyle in our popular imagination – think of cloth bags, farmers' markets, and hybrid cars – is unlikely to actually be sustained by current and projected stores of natural resources. The symbolic problem is more multifaceted.

Symbolically, a taste for sustainable consumption seems to have become part of the repertoire of high cultural capital consumers – those who already enjoy a degree of privilege in society (see Volume I, Chapter 17). In what seems to be the most systematic analysis of this topic to date, Carfagna et al. (2014, p. 158) report that ethical consumers – those reporting purchasing goods claiming to offer environmental and / or social benefits – are "overwhelmingly high cultural capital consumers." That is, high status taste appears to have developed an ecological orientation in that "good" consumers are expected to display an ecological consciousness. This taste has less to do with material impact than with the purchase of relatively costly goods (e.g., hybrid cars, organic food) and the avoidance of brands deemed to employ unjust or unsustainable practices. In addition to these so-called "buycotts" and "boycotts," ethical consumers are also likely to engage in relatively time-consuming activities (e.g., gardening, DIY workshops) (Curtis, 2016). These two problems – of material and symbolic drawbacks to individualized sustainable consumption – intersect to create a challenge to sustainability. This challenge can be described as follows: we lack a social value for truly low-impact lifestyles while we accord social status to those who can afford to engage in expensive and time-consuming activities that may do little to reduce their overall environmental impact. Meanwhile, these choices are made in a context shaped by a tendency to individualize the responsibility to protect the environment while leaving corporate and state power unchecked (Freudenburg, 2006; Szasz, 2007).

Not only is sustainable consumption classed in complex ways, it is also a site of gendered engagement (Cairns, Johnston, and MacKendrick, 2013; Middlemiss, 2014). Empirical studies have demonstrated that there is a gender gap in engagement in pro-environmental behaviors, with women undertaking these actions more frequently than men. This observation is the starting point for some scholars to ask what women's disproportionately greater engagement in environmental protection might mean for women's lives, and women's status more specifically (Kennedy and Kmec, 2018). Within the context of the purchase of organic products, scholars have found that women feel social pressure to protect the purity of their children by purchasing healthy foods (Cairns et al., 2013). The authors describe this as a "gendered burden" (p. 97) for women since these ethical food choices are often out of sync with the pragmatic ideals of food shopping and preparation. In short, sustainable consumption activities are imbued with undertones of caring; in a neoliberal setting, one's ability to demonstrate goodness is coupled with their ability to engage in sustainable consumption (Johnston and Cairns, 2013). As discussed elsewhere in this volume (see Volume II, Chapter 25), health is an important motivator for engagement in sustainable consumption, and one that is experienced differently by men and women.

An under-examined area of study concerns the relationship between sustainable consumption and race. Yet, rather than assume this is because the practice of sustainable consumption is somehow race-neutral, an important area of research highlights ways that discourses of sustainability can exacerbate racial discourses and how environmental protection policies can further marginalize people of color. For example, Alkon (2012) argues that the so-called "green economy," which promises a widespread gains to environmental and individual health, in fact largely fails to deliver such benefits to low-income people of color. Only when the needs of these populations are considered in the design and operations of green enterprises, such as farmers' markets, do green practices appeal to and benefit people beyond the white middle classes (Alkon, 2012). Scholars of race and sustainability encourage us to consider the myriad ways that the past influences our current, inequitable distribution of wealth and opportunity. For example, those studying the environmental injustices suffered by Indigenous peoples force us to pay attention to the ways that genocide and forced cultural assimilation shape contemporary patterns of hunger and environmental activism (Norgaard, Reed, and Van Horn, 2011).

The scholarship on class, gender, and race in the context of sustainable consumption opens up much-needed space to interrogate the relationships among environmentalism and these axioms of inequality. While sustainable consumption may well have a patina of virtue due to the underlying promise of improving the quality of the natural environment for all, it is clear that the benefits of these efforts, the work to bring about sustainability, and the rewards for doing so, are distributed in ways that will continue to privilege dominant actors in society unless inequality is embedded in the discourse and practice of sustainability.

Some years ago, Hawken, Lovins and Lovins (2013) argued that if we want to achieve sustainability goals then we must design societies such that the sustainable option is the default option. That is, we should not be required to pay more if we want recycled paper or eat meat that has been treated humanely, or buy bottled water if we want to feel protected from contamination. Consumption is one point along a continuum of extraction, processing, shipment, use and disposal. Given that the consumer is only (tangentially) involved at the points of use and disposal, making sustainable choices the default option requires considerably more attention be placed "upstream" – on the extraction, processing and shipment of products and the policies and laws that govern choices in each of these domains.

4 Selected Future Research Directions

To date, the majority of academic scholarship and policy briefs on sustainable consumption assume first, that more engagement in sustainable activities is unquestionably positive and second, that individuals have considerable agency in the arena of consumption. Within environmental sociology, the literature on this topic has evolved from, and remains closely tied to, research on "pro-environmental behaviors." The focus on pro-environmental behaviors, as already discussed, relies on individual-level variables, but additionally, this scholarship often presumes that individual-level change is one of the more tractable pathways to sustainability (Middlemiss, 2018). Yet, as reviewed above, material and ideological problems have emerged, complicating the assumption that more is always better in the case of sustainable consumption. As a result, future sociological inquiry should delve into more detailed exploration of how sustainable consumption practices can be used as a strategy to exclude. Two related developments I argue would strengthen the study of sustainable consumption are, first, to study the businesses, policy makers, and elected representatives at various levels of government to understand how the landscape for sustainable consumption is structured, and second, to examine sustainable consumption in the context of information and communication technologies (ICT). In sum, these future research trajectories include: examining the social and symbolic boundaries associated with sustainable consumption practices, studying actors in the state and market realms whose decisions impact on sustainable consumption, and theorizing how sustainable consumption practices change (or not) in the digital context.

The first recommended avenue of future research is to document the use of sustainable consumption as a tool for social exclusion. To do so, scholars might delve into the lived experiences of those who do not participate in commonly measured "sustainable consumption" practices to understand how non-elite actors engage with the non-human environment and how these actors perceive mainstream sustainable consumption practices and their place in this field. Materially, the positive association between sustainable consumption activities and carbon dioxide emissions (e.g., Kennedy, Krahn and Krogman, 2015) has cast doubt on whether individualized solutions like hybrid vehicles and organic food constitute a robust response to global climate change and other environmental crises. And ideologically, the possibility that sustainable consumption is a gendered raced, and classed performance (e.g., Cairns et al., 2013; Carfagna et al., 2014) creates unexpected pathways between engaging in sustainable activities and bolstering, or even exacerbating, social inequality. Consequently, promising theoretical approaches to understanding sustainable consumption decenter the individual, as evidenced by the social practices approach (Shove, 2010; Warde, 2017). However, there exist few efforts to examine how sustainable consumption discourse is perceived and experienced by those who lack the cultural and economic capital to engage in low environmental impact practices (Middlemiss, 2018). Following this mandate demands interrogating the topic of sustainable consumption using theories more sensitive to power and cultural exclusion. Promising pathways include cultural sociological theories of class boundaries and class distinction (e.g., Bourdieu, 1984; Lamont, 1992; Lamont et al., 2016b) and theories of power that decenter the individual (e.g., Foucault's notion of technologies of the self [1988]).

A second area of future research could examine the organizational logics driving sustainable consumption. A great deal of sociological research into individual consumption patterns focuses on the household; even social practice theories (which decenter the individual) focus largely on private-sphere practices. In other words, although social practice theories shift emphasis from individuals to *practices*, the empirical focus remains on how the residential sector's consumption habits and routines are organized (Kennedy, Cohen, and Krogman, 2016b). We need to better understand the governance of our cities, countries, and globe, of our public institutions, and of the green marketplace in order to fully understand how the continued reproduction of unsustainable systems of provisions occurs. Promising pathways here include critical theories of governance and communities, particularly within a neoliberal context wherein much responsibility to uphold or advance the public good is downloaded onto civil society (Aiken et al., 2017). While little research exists within environmental sociology explicating the role of the market in promoting unsustainable consumption products and practices, one promising example is Carolan's (2015) study of "affective barriers." Carolan examines food tastes by interviewing industry experts whose job is to induce people to eat a particular product. In this way, his research illuminates how tastes that feel "natural" to the consumer are in large part engineered by personnel in corporations whose aims are unrelated to ecological sustainability. Similar efforts to understand vehicle choice, housing choices, and other large domains of consumption would enrich the study of sustainable consumption.

A final area for future research into sustainable consumption in environmental sociology is to consider the influence of information and communication technology (ICT) on the consumption habits of individuals, firms, and governments. The digital environment creates the potential for consumers to produce what they consume more readily (as captured in the increasingly common term, the "prosumer," see Ritzer and Jurgenson [2010]). It also offers the possibility of catalyzing groups of interest not bound in geographic space. At the same time, social media seems to amplify social pressure to consume profligately, the ubiquity of smartphones means we can check into work anytime-anywhere thus extending the work domain further into our personal lives, and our ready access to a digital interface both exposes us to frequent and targeted advertising and allows our data to be collated by marketing firms and others. Given the rapidly growing breadth and depth of ICT, as evidenced by rising numbers of internet-enabled devices in our homes and workplaces and the tendency for ever-younger audiences to be connected to online environments, it will be increasingly important to consider how sustainable consumption is contested, produced and reproduced in the digital context.

Conclusion

Those of us who feel for Melissa McCarthy's rendition of the modern-day eco-warrior likely can relate to the desire to do something to save the planet and the feeling that there are myriad barriers to doing so. Current research indicates that traditional political activity is often mistrusted as a strategy for social change (see Bennett et al., 2013) and even those involved in grassroots organizations struggle to imagine or make use of a rich, diverse tool kit of engagement tactics to tackle socio-ecological issues (Kennedy, Parkins, and Johnston, 2016; Perrin, 2006). The current political climate in the United States represents such a significant threat to the pursuit of sustainable consumption, that a retreat from pursuing pro-environmental reforms at the federal level seems eminently reasonable. One of President Trump's first decisions in office was to back out of the Paris Agreement (the international agreement to reduce greenhouse gas emissions). And, as this chapter goes to press, the head of the Environmental Protection Agency is Andrew Wheeler – who arrived at this post after working as a lobbyist for the coal industry. In contrast, for those individuals who can afford the time and cost burdens of channeling concern for the environment through green products and practices without incurring any sacrifice to taste, style, or quality (see Soper [2004] on alternative hedonism), the pursuit of sustainable consumption through individual, consumer-based choices likely feels both more satisfying and more effective. In short, traditional engagement is not only seen as time-consuming, but also as relatively ineffective, while consumption-focused solutions are felt to be pleasurable and significant (Kennedy, Johnston and Parkins, 2017).

I have argued in this chapter that an environmental governance model that attributes undue responsibility to individuals structures the contemporary practice of sustainable consumption. Research from a social practices standpoint shows us that individual consumption patterns are strongly influenced by policy, sociotechnical systems, and norms – observations that cast doubt on the degree of agency and autonomy individual actors have when as they move throughout daily life. Sustainable consumption as commonly conjured up in our contemporary milieu, involves an array of activities and consumer choices that are typical of high cultural capital lifestyles. These two themes intersect: in a neoliberal governance model in which individuals are believed to have as much power and responsibility to act to protect the natural environment as those in state and market spheres, there are few challenges to the notion of individualized environmental responsibility. Further, with a decline in trust in traditional politics, channeling individual concerns in the marketplace is often experienced as the most feasible and efficacious course of action for many. However, this stratifies ecological citizenship such that those with the time, money, and taste required to "go green" appear to be the most environmentally concerned among us, even while their carbon footprints grow ever larger. Finally, these patterns are exacerbated by evidence that sustainable consumption is also deeply gendered. In sum, the structures that organize sustainable consumption practices and the resultant patterns of sustainable consumption implicate the individual as a responsible entity in a way that fails to consider the role of more powerful actors and that may sharpen the boundaries between social strata.

I concluded the chapter with suggested directions for future research. In particular, I called for research looking past the typical, privileged sustainable consumer to understand how working class groups experience the consumer arena, sustainability-focused or otherwise. I also argued for balancing the considerable emphasis on household-level consumption with greater emphases on the state and market-based actors whose decisions impact the structure of household-level consumption and also entail environmental impact of their own through production, disposal, military activities, and the development and maintenance of public institutions. Finally, I argued that information and communication technology (ICT) has both possible benefits for making consumption more sustainable (through pro-sumption and by catalyzing groups dispersed in geographic space into communities of interest) and the potential to amplify consumption patterns – two materially and symbolically important pathways worthy of environmental sociological inquiry.

As a topic, sustainable consumption will likely continue to grow in importance to those interested in the theory and practice of human–environment interactions. As our population climbs past 7.5 billion people, our economies continue to grow, and promised decoupling of environmental impacts and economic growth fail to transpire at the global level, we require theoretically informed, empirical research to catalyze debates and discussions about how we can live comfortably within ecological limits in a way that does not depend on continued marginalization of those with less power and privilege.

References

Alkon, A. H. (2012). *Black, White, and Green: Farmers' Markets, Race, and the Green Economy*. Athens, GA: University of Georgia Press.

Agyeman, J., and Evans, B. (2004). "Just sustainability": The emerging discourse of environmental justice in Britain? *The Geographical Journal, 170*(2), 155–164.

Aiken, G. T., Middlemiss, L., Sallu, S. and Hauxwell-Baldwin, R. (2017) Researching climate change and community in neoliberal contexts: An emerging critical approach. *Wiley Interdisciplinary Reviews: Climate Change, 8*(4), e463.

Anantharaman, M. (2017). Elite and ethical: The defensive distinctions of middle-class bicycling in Bangalore, India. *Journal of Consumer Culture, 17*(3), 864–886.

Barnard, A. V. (2016). *Freegans: Diving into The Wealth of Food Waste in America*. Minneapolis, MN: University of Minnesota Press.

Baudrillard, J. (2016). *The Consumer Society: Myths and Structures*. Thousand Oaks, CA: Sage Publications.

Bennett, E. A., Cordner, A., Klein, P. T., Savell, S., and Baiocchi, G. (2013). Disavowing politics: Civic engagement in an era of political skepticism. *American Journal of Sociology, 119*(2), 518–548.

Bourdieu, P. (1984). *Distinction: A Social Critique of the Judgement of Taste*. Oxon, UK: Routledge.

Cairns, K., Johnston, J., and MacKendrick, N. (2013). Feeding the "organic child": Mothering through ethical consumption. *Journal of Consumer Culture, 13*(2), 97–118.

Carfagna, L. B., Dubois, E. A., Fitzmaurice, C. et al. (2014). An emerging eco-habitus: the reconfiguration of high cultural capital practices among ethical consumers. *Journal of Consumer Culture, 14*(2), 158–178.

Carolan, M. S. (2004). Ecological modernization theory: What about consumption? *Society and Natural Resources, 17*(3), 247–260.

Carolan, M. (2015). Affective sustainable landscapes and care ecologies: Getting a real feel for alternative food communities. *Sustainability Science, 10*(2), 317–329.

Clark, C. F., Kotchen, M. J., and Moore, M. R. (2003). Internal and external influences on pro-environmental behavior: Participation in a green electricity program. *Journal of Environmental Psychology, 23*(3), 237–246.

Cohen, M. J., Brown, H. S., and Vergragt, P. (eds.). (2013). *Innovations in Sustainable Consumption: New Economics, Socio-Technical Transitions and Social Practices*. Northampton, MA: Edward Elgar Publishing.

Csutora, M. (2012). One more awareness gap? The behaviour–impact gap problem. *Journal of Consumer Policy, 35*(1), 145–163.

Curtis, R. B. (2016). Ethical markets in the artisan economy: Portland DIY. *International Journal of Consumer Studies, 40*(2), 235–241.

Daly, H. E. (1990). Toward some operational principles of sustainable development. *Ecological Economics*, 2(1), 1–6.

Earth Overshoot Day. (2018) Retrieved from www.overshootday.org/. Accessed on July 9, 2018.

Elliott, R. (2013). The taste for green: The possibilities and dynamics of status differentiation through "green" consumption. *Poetics*, 41(3), 294–322.

Foucault, M. (1988). *Technologies of The Self: A Seminar with Michel Foucault*. Boston, MA: University of Massachusetts Press.

Freudenburg, W. R. (2006). Environmental degradation, disproportionality, and the double diversion: Reaching out, reaching ahead, and reaching beyond. *Rural Sociology*, 71(1), 3–32.

Gabriel, Y., and Lang, T. (2015). *The Unmanageable Consumer* (20th Anniversary edition). London: Sage.

Geiger, S. M., Fischer, D., and Schrader, U. (2017). Measuring what matters in sustainable consumption: An integrative framework for the selection of relevant behaviors. *Sustainable Development*, 26(1), 18–33.

Giddens, A. (1984). *The Constitution of Society: Outline of the Theory of Structuration*. Berkeley, CA: University of California Press.

Griggs, D., M. Stafford-Smith, O. Gaffney, et al. (2013). Policy: Sustainable development goals for people and planet. *Nature*, 495(7441), 305.

Gould, K. A., Pellow, D. N., and Schnaiberg, A. (2004). Interrogating the treadmill of production: Everything you wanted to know about the treadmill but were afraid to ask. *Organization and Environment*, 17(3), 296–316.

Halkier, B. (2009). A practice theoretical perspective on everyday dealings with environmental challenges of food consumption. *Anthropology of Food*, (S5) http://aof.revues.org/index6405.html, accessed July 9, 2018.

Hawken, P., Lovins, A. B., and Lovins, L. H. (2013). *Natural Capitalism: The Next Industrial Revolution*. New York: Routledge.

Hertwich, E. G. (2005). Life cycle approaches to sustainable consumption: A critical review. *Environmental Science and Technology*, 39(13), 4673–4684.

Huddart Kennedy, E., Parkins, J. R., & Johnston, J. (2016). Food activists, consumer strategies, and the democratic imagination: Insights from eat-local movements. *Journal of Consumer Culture*, 18(1), 149–168.

Jackson, T. (2005). Motivating sustainable consumption: A review of evidence on consumer behaviour and behavioural change. A report to the Sustainable Development Research Network. Centre for Environmental Strategies, University of Surrey.

Johnston, J. (2008). The citizen-consumer hybrid: Ideological tensions and the case of Whole Foods Market. *Theory and Society*, 37(3), 229–270.

Johnston, J., and Cairns, K. (2013). Searching for the "alternative," caring, reflexive consumer. *International Journal of Sociology of Agriculture and Food*, 20(3), 403–408.

Kennedy, E. H. (2011). Rethinking ecological citizenship: The role of neighbourhood networks in cultural change. *Environmental Politics*, 20(6), 843–860.

Kennedy, E. H., Krahn, H., and Krogman, N. T. (2014). Egregious emitters: Disproportionality in household carbon footprints. *Environment and Behavior*, 46(5), 535–555.

Kennedy, E. H., Krahn, H., and Krogman, N. T. (2015). Are we counting what counts? A closer look at environmental concern, pro-environmental behaviour, and carbon footprint. *Local Environment*, 20(2), 220–236.

Kennedy, E. H., Cohen, M. J., & Krogman, N. (eds.). (2016a). *Putting Sustainability into Practice: Applications and Advances in Research on Sustainable Consumption*. Northampton, MA: Edward Elgar Publishing.

Kennedy, E. H., Cohen, M. J., & Krogman, N. T. (2016b). Social practice theories and research on sustainable consumption. In *Putting Sustainability into Practice: Applications and Advances in Research on Sustainable Consumption* (3–22). Northampton, MA: Edward Elgar Publishing.

Kennedy, E. H., Johnston, J., and J. R. Parkins. (2017). Small-p politics: How pleasurable, convivial and pragmatic political ideals influence engagement in eat-local initiatives. *The British Journal of Sociology*. DOI:0.1111/1468-4446.12298. [Epub ahead of print].

Kennedy, E. H. and J. A. Kmec. 2018. (2018). Reinterpreting the gender gap in household pro-environmental behaviour. *Environmental Sociology*, 4(3), 299–310.

Kia. (2017). *Hero's Journey*. Retrieved from http://adage.com/article/special-report-super-bowl/melissa-mccarthy-kia-s-super-bowl-ad/307782/.

Kollmuss, A., and J. Agyeman. (2002). Mind the gap: Why do people act environmentally and what are the barriers to pro-environmental behavior? *Environmental Education Research*, 8(3), 239–260.

Lamont, M., Silva, G. M., Welburn, J. et al. (2016b). *Getting Respect: Responding to Stigma and Discrimination in The United States, Brazil, and Israel*. Princeton, NJ: Princeton University Press.

Lamont, M. (1992). *Money, Morals, And Manners: The Culture of The French and the American Upper-Middle Class*. Chicago, IL: University of Chicago Press.

Liu, Y., Y. Qu, Z. Lei, and H. Jia. (2017). Understanding the evolution of sustainable consumption research. *Sustainable Development*, 25(5), 414–430.

Lorenzen, J. A. (2012). Going green: The process of lifestyle change. *Sociological Forum* 27 (1), 94–116.

Maniates, M. F. (2001). Individualization: Plant a tree, buy a bike, save the world? *Global Environmental Politics*, 1(3), 31–52.

Middlemiss, L. (2018). *Sustainable Consumption: Key Issues*. New York: Routledge.

Middlemiss, L. (2014). Individualised or participatory? Exploring late-modern identity and sustainable development. *Environmental politics*, 23(6), 929–946.

Mol, A. P., and Spaargaren, G. (2004). Ecological modernization and consumption: A reply. *Society and Natural Resources*, 17(3), 261–265.

Norgaard, K. M., Reed, R., and Van Horn, C. (2011). A continuing legacy: Institutional racism, hunger, and nutritional justice on the Klamath. In *Cultivating Food Justice: Race, Class, and Sustainability*, eds. A. H. Alkon and J. Agyeman. Cambridge, MA: MIT Press, pp. 23–45.

Perrin, A. J. (2009). *Citizen Speak: The Democratic Imagination in American Life*. Chicago, IL: University of Chicago Press.

Reisch, L. A., Cohen, M. J., Thøgersen, J. B., and Tukker, A. (2016). Frontiers in sustainable consumption research. *GAIA-Ecological Perspectives for Science and Society*, 25 (4), 234–240.

Ritzer, G., & Jurgenson, N. (2010). Production, consumption, prosumption: The nature of capitalism in the age of the digital "prosumer." *Journal of Consumer Culture*, 10(1), 13–36.

Rockström, J., W. Steffen, K. Noone, Å. Persson, F. S. Chapin et al. (2009). A safe operating space for humanity. *Nature* 461(7263): 472–475. doi:10.1038/461472a.

Rose, N. (2000). Community, citizenship, and the third way. *American Behavioral Scientist*, 43(9), 1395–1411.

Sanne, C. (2002). Willing consumers – or locked-in? Policies for a sustainable consumption. *Ecological Economics*, 42(1), 273–287.

Schelly, C. (2017). *Dwelling in Resistance: Living with Alternative Technologies in America*. New Brunswick, NJ: Rutgers University Press.

Schor, J. (2008). *The Overworked American: The Unexpected Decline of Leisure*. New York: Basic Books.

Shove, E. (2003). *Comfort, Cleanliness and Convenience*, London, Berg.

Shove, E. (2010). Beyond the ABC: Climate change policy and theories of social change. *Environment and Planning A*, *42*(6), 1273–1285.

Shove, E., and Warde, A. (2002). Inconspicuous consumption: The sociology of consumption, lifestyles and the environment. In R. E. Dunlap, F. H. Buttel, P. Dickens, and A. Gijswijt (eds.) *Sociological Theory and The Environment: Classical Foundations, Contemporary Insights* (230–251). Lanham, MD: Rowman and Littlefield Publishers, Inc.

Soneryd, L., and Uggla, Y. (2015). Green governmentality and responsibilization: New forms of governance and responses to "consumer responsibility." *Environmental Politics*, *24*(6), 913–931.

Soper, K. (2004). Rethinking the "good life": The consumer as citizen. *Capitalism Nature Socialism*, *15*(3), 111–116.

Southerton, D., Warde, A., and M. Hand. (2004). The limited autonomy of the consumer: Implications for sustainable consumption. *Sustainable Consumption: The Implications of Changing Infrastructures of Provision*, 32–48. Northampton, MA: Edward Elgar Publishing.

Steffan, W., Richardson, K., Röckström, J. et al. Planetary boundaries: Guiding development on a changing planet. *Science*, 347 (6223) DOI:10.1126/science.1259855.

Stern, P.C., Dietz, T., Abel, T.D., Guagnano, G.A. and L. Kalof. (1999). A value-belief-norm theory of support for social movements: The case of environmentalism. *Human Ecology Review*, *6*(2), 81–97.

Stolle, D., and Micheletti, M. (2013). *Political Consumerism: Global Responsibility in Action*. New York: Cambridge University Press.

Swidler, A. (1986). Culture in action: Symbols and strategies. *American Sociological Review*, *51*, 273–286.

Szasz, A. (2007). *Shopping Our Way to Safety: How We Changed from Protecting the Environment to Protecting Ourselves*. Minneapolis, MN: University of Minnesota Press.

Vergragt, P., Akenji, L., and Dewick, P. (2014). Sustainable production, consumption, and livelihoods: global and regional research perspectives. *Journal of Cleaner Production*, *63*, 1–12.

Warde, A. (2017). Sustainable Consumption: Practices, Habits and Politics. In *Consumption* (181–204). London: Palgrave Macmillan UK.

Warde, A. (2005). Consumption and theories of practice. *Journal of Consumer Culture*, *5*(2), 131–153.

Wiidegren, Ö. (1998). The new environmental paradigm and personal norms. *Environment and Behavior*, *30*(1), 75–100.

14 Sustainability Cultures: Exploring the Relationships Between Cultural Attributes and Sustainability Outcomes

Janet Stephenson

1 Introduction

'We need a culture change'. When achieving sustainability seems impossibly hard, people often reach to the word 'culture' as a shorthand way of referring to the complex and hazy characteristics of societies, groups, and organisations that shape unsustainable behaviour. Using 'culture' as a rationalisation can be a cop-out – a convenient excuse for why individuals and organisations don't change. But, as I will argue in this chapter, a systematically applied cultural analysis offers a useful perspective on unsustainable behaviours. If the world needs culture change, a deeper analysis of culture can help us conceptualise what we need to do to achieve that change.

The concept of 'sustainability cultures' outlined in this chapter offers a structured way of thinking about and investigating culture with a particular focus on sustainability-related outcomes. Originally developed to investigate household energy behaviours (Stephenson et al. 2010), the sustainability cultures approach has since been applied to a variety of sustainability issues including energy consumption, energy production, mobility, water, and greenhouse gas emissions. The cultural lens has been useful for examining questions such as how conventions about resource use become cemented in everyday activities, how unsustainable consumption and production patterns can become normalised, and in identifying opportunities for cultural change that support more sustainable outcomes. Seen in sustainability terms, culture is not a bystander, but a core driver of outcomes that are critical to long-term human survival.

What do we mean by culture? As one of the more slippery words in the English language, it has many applications. It is used to distinguish characteristics of ethnicities (e.g. 'Māori culture') or nations (e.g. 'American culture'); to describe groups of people who display certain characteristics that differentiate them from society in general (e.g. youth subcultures); to describe particular characteristics of businesses (e.g. organisational cultures); and to refer to qualities that are shared pan-nationally (e.g. 'western culture'). Culture is also used to refer to works of art and other creative endeavours or to describe a perceived state of social elevation, although neither of these latter meanings is relevant to our purpose here. All of the former applications, though, typically share at least one common interpretation: that a culture involves some shared subjective characteristics such as beliefs, values, norms and symbolism. A culture also often includes other shared attributes such as practices, material objects, language, and forms of knowledge. For the purposes of this chapter I use a definition that draws on culture's anthropological as well as sociological roots: that culture includes 'not only the beliefs and values of social groups, but also their language, forms of knowledge, and common sense, as well as the material products, interactional practices and ways of life established by these' (Hays 1994:65).

This chapter draws from nearly a decade of research on the links between cultural attributes and sustainability outcomes. The cultural characteristics that I am interested in do not necessarily coincide with nationality, ethnicity, or other more traditional uses of 'culture'. Instead, I am interested in the links between a sustainability outcome (e.g. energy use) and people's cultural attributes. Using an energy example, less sustainable outcomes would include high levels of fossil fuel consumption, while more sustainable outcomes would include greater energy efficiency and more use of renewable fuels. The 'culture' that underpins these outcomes is the interplay between people's norms, practices, material possessions, and other cultural attributes. By investigating these, we can identify how the interplay of cultural attributes can often result in habitual patterns of behaviours that are similar across individuals or groups, and also how and why these patterns can change over time.

Whose culture? I am talking about people in their everyday lives as members of households and also as members of organisations including businesses, non-governmental organisations and governments. Sustainability cultures can be investigated at all of these scales, so I use the term 'actors' to refer to people or organisations at whatever scale is relevant. Actors have a degree of agency in adopting their own cultural attributes, but they are also strongly influenced by broader cultural and structural factors beyond their control. By exploring the interplay between actors' cultures and these broader factors it is possible to gain insights into why change can be supremely difficult, as well as why a cultural change leading to improved sustainability outcomes can sometimes be rapid.

In the following sections I discuss the origins of the sustainability cultures concept as a framework to support interdisciplinary research, and link it back to social theories. Although the concept of sustainability cultures is starting to be applied in many fields, most scholarly work to date has focused on energy ('energy cultures') and transport ('mobility cultures') so most examples will be on these topics. Drawing from over twenty-five studies using this cultural framing undertaken by my team and by others internationally, I outline relevant findings and describe its use as a methodology. I conclude with a brief discussion of the potential of this approach for future studies on sustainability questions.

2 Origins and Theoretical Underpinning

We first used culture as a lens in an interdisciplinary research programme on household energy efficiency, as detailed in Stephenson et al. (2010, 2015a). The research team members, from the disciplines of physics, economics, law, consumer psychology, and sociology, needed a common framework that made sense to all participants. The team was interested in how households and businesses were often locked into seemingly habitual patterns of beliefs, behaviours and/or technologies that might be inefficient and expensive but nonetheless proved hard to alter. The concept of culture encapsulated the range of factors the team were interested in: as a broad concept that helped reveal the internal dynamics of actors' lives. The team set about developing an approach that supported the interdisciplinary study of culture in relation to energy-related outcomes (at that stage called 'energy culture').

The evolving sustainability cultures approach was informed by multiple theories which I have discussed elsewhere in some detail (Stephenson et al. 2010, Stephenson et al. 2015a, Stephenson 2018), but will briefly touch on here. Within the realm of sociological and cultural theory, Bourdieu's concept of habitus (1977) suggests that persistent patterns of thought, perceptions and action can be self-replicating, and are a response to the broader conditions within which people live. Giddens (1984) differentiates between the ability of people to act as relatively free agents and the constraining influences of social, financial, and political structures, while acknowledging that both are continuously in interplay with each other. Latour (2005) writes of the importance of material artefacts in shaping behaviour in actor–network theory. Systems approaches highlight the active interplay between parts of a system, and require that boundaries are drawn to define the system under study (von Bertalanffy 1968; Midgley 2003). Socio-technical systems theory (Geels 2002) identifies the multiple levels and actors involved in systemic change. Together these concepts informed the development of our cultural framing.

With its theoretical roots in Giddens' structuration theory and Bourdieu's theory of habitus, the sustainability cultures approach has a similar genealogy to social practice theory, but diverges at some key points. Practice theory (Reckwitz 2002, Schatzki) builds on the observation that much human behaviour occurs habitually with little in the way of conscious rule-following. Practice, as applied to consumption research, is defined as: 'a routinized type of behaviour which consists of several elements, interconnected to one other: forms of bodily activities, forms of mental activities, "things" and their use, a background knowledge in the form of understanding, know-how, states of emotion, and motivational knowledge' (Reckwitz 2002: 249). Reckwitz (2002) usefully clarifies that 'practice' as used in practice theory refers to behavioural routines (Praktik) rather than all human actions (Praxis).

Practice theorists propose that a social practice is an entity that cannot be reduced to any single element, and that it exists separately from the individuals who engage in it. Practices such as washing and cooking 'historically precede individuals' (Røpke 2009: 2493) while individuals are "carriers" of practice (Reckwitz 2002: 250), and consumption occurs "within and for the sake of practices" (Warde 2005: 145). Although individuals can express agency by withdrawing from or defecting from a practice, practices recruit practitioners to engage and thereby reproduce (Shove & Pantzar 2007) which suggests that practices have a form of agency. Practice theorists have applied this approach to studying a range of sustainability topics such as energy use and greenhouse gas emissions (Shove 2003, Shove & Spurling 2013, Kurz et al.

2015). Outcomes such as patterns of energy use are explained by people's adherence to, and reproduction of, social practices that achieve desired qualities of social life.

The sustainability cultures approach diverges from practice theory at a number of critical junctures. As discussed in detail in Stephenson et al. (2018), the key points of departure for sustainability cultures are: (a) a focus on actors' cultural attributes rather than on social practices; (b) a focus on Praxis (all human actions) rather than a narrower focus on Praktik; (c) the 'unit of enquiry' being the sustainability outcomes of cultural formations, rather than of practices in themselves; (d) a predominant interest in cultural change rather than a predominant focus on cultural replication; and (e) a focus on the interplay between actors' agency and social structures through the medium of culture rather than through the medium of practice.

Culture, as interpreted in sustainability cultures research, is not limited to the more subjective characteristics (e.g. meanings, beliefs, norms, symbolism) that have been the hallmark of the cultural turn in sociology, but includes material and behavioural characteristics in their own right. When studying the sustainability culture of an actor or group of actors, the cultural attributes of interest include their norms, practices (in the sense of Praxis) and their material and immaterial artefacts, along with other cultural characteristics as quoted above from Hays (1994).

3 The 'Cultures' Framework

Culture can be investigated in many ways, but to date most sustainability cultures research has used a heuristic, or simple model, representing key cultural attributes and the interplay between them (Figure 14.1). The framework reduces the complexity of sustainability cultures' underpinning concepts to five key ideas, offering a number of high-level variables to provide for structured, yet nuanced,

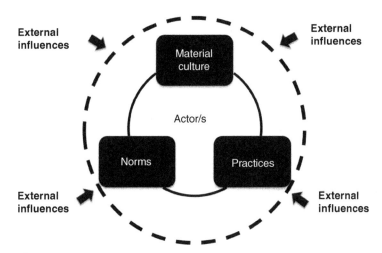

Figure 14.1 *Sustainability cultures framework*

inquiries. The framework remains open to including other cultural attributes that are not specifically named in the figure, such as language, knowledge, and symbolism.

At the centre of the framework (Figure 14.1) is the actor – an individual, household, business, organisation, or any collective thereof. The framework highlights three key aspects of the actor's culture: their norms, practices, and material culture. Norms include actors' expectations and aspirations relating to their practices and material culture, as shaped by their beliefs about what is appropriate or socially desirable. Practices (Praxis) include all actions, activities and processes undertaken by the actor that have a causal relationship with the outcomes of interest. Material culture comprises the technologies, structures, and other physical and non-physical assets that are within scope of the actor's agency (e.g. owned, leased).

The arrows in Figure 14.1 draw attention to how norms, practices, and material culture interact. The arrows indicate, for example, that norms are likely to influence actors in their acquisition of new material items, that actors' material possessions influence their practices, and that new practices may induce new norms (for examples see section 4). 'Sustainability culture' comprises the combination of norms, practices, and material culture (and other relevant cultural attributes) and the way they interact to result in sustainability-related outcomes. A researcher's choice of which cultural attributes to study will depend on the sustainability outcomes that are the topic of inquiry – for example, a study of water-related outcomes is likely to involve different cultural attributes to a study of mobility-related outcomes.

The outer parts of Figure 14.1 represent the structural influences on an actor's culture, over which they have little or no control. These influences may include local and global systems of production and consumption; tangible and intangible infrastructures; laws and policies; institutions and their roles; and the shared rules and principles that underpin these. Studies of sustainability culture involve seeking to understanding how these factors can constrain actors' choices or alternatively influence change in actors' cultures. The broken line represents the usual limits of the actor's agency, but is permeable outwards as well as inwards because in some circumstances structures can be moulded by collective cultural shifts (examples are discussed in section 4).

The actor and their scope of influence informs what is 'inside' or 'outside' the semi-permeable boundary of a sustainability culture. For a household, for example, their material culture will include the mobility technologies that they own (e.g. cars, bicycles), but not the roads or public transport provided by the municipality. These facilities would form part of the external influence on the actor's mobility culture. On the other hand, when investigating a municipality's mobility culture, the transport infrastructure would be a relevant cultural attribute.

The sustainability cultures approach thus offers an actor-centred methodology for considering the role of culture in the social realm, and the associated sustainability outcomes. Starting with an actor or actors as the base unit of analysis, sustainability researchers can build up an understanding of shared or similar cultural patterns across the population under study. Many researchers have used the 'cultures' framework with its pared-down concepts, or an adjusted version of it, to structure their research design and/or analysis. Investigating a broader range of cultural attributes, such as knowledge (Scott et al. 2016) and beliefs (Dew et al. 2017) adds complexity but can repay in richer detail.

In the framework's early stages of development it was mainly applied to describing and distinguishing the energy cultures of households and businesses and how these relate to energy consumption outcomes. The framework's early applications included identification of clusters of similar energy cultures in relation to efficiency (Lawson & Williams 2012), motivations and influences on household energy consumption (Sweeney et al. 2013), business adoption of efficient technologies (Bell et al. 2014), interventions for more efficient household energy behaviours (Scott et al. 2016), and a study of energy poverty (McKague et al. 2016). In the energy field the framework has also been used to study the adoption of solar lighting (Walton et al. 2014) and uptake of solar generation (Ford et al. 2017). Researchers have also applied the sustainability cultures approach to other topics as diverse as individuals' actions to reduce greenhouse gas emissions (Young & Middlemiss 2012), pro-environmental behaviour (Hoicka, 2012), transport transitions (Stephenson et al. 2015b), urban freight efficiency (Hopkins & McCarthy 2016), driver efficiency (Scott & Lawson 2017), urban freight delivery (Hopkins & McCarthy 2016), youth mobility (Hopkins & Stephenson 2016), energy consumption by age cohorts (Bardazzi & Pazienza 2017) and domestic water demand (Manouseli et al. 2018).

4 Insights from Sustainability Cultures Research to Date

Academic research using the sustainability cultures approach and/or framework has now been applied to a sufficiently wide range of topics that it is possible to start to draw some high-level insights from across multiple studies. In some situations, actors' cultures are found to be relatively unchanging, and opportunities to alter to more sustainable (e.g. low-carbon, energy-efficient, healthier) ways of operating are not being taken up. For example, Bell et al.'s (2014) research on firms involved in timber drying revealed that most were locked into using inefficient kilns with significant particulate emissions (rather than more efficient drying methods such as heat pumps) even where this did not appear to make financial sense. This 'culture' was reinforced by the wider timber sector through its own embedded systems of belief and practices. In another example, a study of the US Navy's longstanding refusal to install LED lighting in its ships identified the Navy's immovable energy culture as a key barrier to change. The Navy's 'professional norms, long-institutionalised practices and idiosyncratic material culture' – together with entrenched external influences (e.g. congressional approval practices) – reinforced the status quo (Dew et al. 2017: 63).

Much of the research into sustainability cultures has, however, focused on how even seemingly subtle changes to cultural attributes can have significant sustainability implications. It is rare for a change to occur with only a single cultural attribute, as any alteration (e.g. adopting a new technology) is likely to involve consequential changes to other attributes. For example, Walton et al.'s 2014 study in Vanuatu (a Pacific nation) identified consequential changes from the widespread adoption by households of small solar lamps, which had been spurred by an aid project. When the pico-lamps replaced the ubiquitous kerosene lamps, new practices emerged. Children could use the lamps because they were safe, and since people could afford multiple lamps there was a rise in new evening practices such as women's handicrafts and children doing homework. Not having to earn money to buy kerosene meant less need for families to engage in the cash economy. A normative shift was also evident, whereby people made value statements deriding kerosene and supporting free energy from the sun; 'made by God' as proclaimed in a poster. Families' positive experience with the pico-lamps also drove new aspirations to acquire other solar technologies. The cascading effect of the adoption of pico-lamps was a significant change in families' energy cultures.

At a larger scale, Gnoth (2016) found that when people move house (a change to their material culture) they often undertake consequential changes such as discarding and renewing appliances, adopting new routines, and having different expectations about the energy performance of the home. The changes result in a different energy culture with different energy-use outcomes. Walton et al's (2019) research on small-medium businesses found that their journey towards more sustainable business outcomes started at one of three points: a change in material culture (e.g. replacement of a broken technology); a change in practices (e.g. improving process efficiency); or a change in norms (e.g. an intentional change in business values and expectations towards sustainability). The initial change had often led to consequential changes (e.g. a material change leading to practice changes) and different sustainability outcomes. The most comprehensive changes were with businesses where a normative shift drove intentional changes in all aspects of the organisational culture.

Other research has identified the links between external influences and cultural change. For example, an intervention trial to encourage households to become more energy-efficient compared the effectiveness of individual home energy audits with community energy events. 'Before' and 'after' surveys and interviews captured change in households' energy culture. Both forms of intervention resulted in change, although due to the low income of many of the households their capacity to change was limited. Nevertheless, changes were evident in material culture (e.g. insulation and energy-efficient lightbulbs installed), practices (e.g. more regular closing of curtains, drying clothes outside rather than indoors) and norms (e.g. expectation of more comfort, and feelings of empowerment from community sessions) (Scott et al. 2016).

Some cultural shifts clearly arise from the interaction of multiple factors. Hopkins and Stephenson's (2016) research into the changing mobility cultures of Generation Y (those born in the two decades prior to the 2000s) contrasted the mobility cultures of those who were still embedded in automobility to others who shunned car ownership but instead engaged in multiple forms of mobility (walking, cycling, public transport, car sharing, etc.). Both groups valued freedom, but for the first group freedom was associated with car ownership, whereas the second group aligned freedom with the lack of vehicle ownership and the flexibility which that offered. Those who adopted multi-mobility appeared to be in part influenced by their own cultural attributes such as normative positions that supported sharing, collaboration and dematerialisation, and their concerns about health and greenhouse gas emissions. Their cultures were also in part driven by external factors such as the increasing cost and complexity of driver licencing and insurance, and increasing availability of public and active transport infrastructure and car-sharing schemes. The interactions of these cultural and structural characteristics appear to be driving a shift in mobility culture amongst young people.

Other research has found that external influences can also drive unsustainable cultures despite the aspirations of actors to be more sustainable. A study of urban freight companies found that significant inefficiencies in deliveries were resulting from online shopping and its increase in home deliveries, customer expectations of parcel tracking and signing, just-in-time deliveries, and the additional administration involved. Profit margins in the industry were cut to the extent that investing in low-carbon efficient transport was unachievable, despite firms' interest in becoming more sustainable (Hopkins & McCarthy 2016).

The many applications of the 'cultures' framing shows that it is applicable at any scale. Using the elements of the cultures framework (Figure 14.1) it is possible to describe the cultural attributes of an individual, a household, a community, a business, a sector, city or even at a global level. For example, McKague et al. (2016) undertook a qualitative study of the lives of people in fuel poverty, describing individuals' practices to keep warm, the nature of their homes and appliances, and their expectations of warmth and comfort. At the opposite end of the scale, Stephenson et al. (2015b) interviewed international transport experts and used the sustainability cultures framework to structure the analysis. Widespread norms, practices and materiality that were reinforcing a global culture of automobility included widely shared expectations of individualised mobility; fossil fuel subsidies; vested interests in the status quo; risk averseness; and outdated urban design. Experts identified some cities and countries which were working to develop a different mobility culture, such as London Transport with its 'culture of diversity' (Stephenson et al. 2015b). In third example, Bell et al.'s (2014) research on timber drying firms identified energy cultures at two levels – the cultures of the individual firms, and a broader culture at the sector level whereby industry-wide norms and practices supported the more inefficient technologies (Bell et al. 2014).

Research has also identified clusters of actors with similar sustainability cultures. For example, Bardazzi & Pazienza's (2017) research in Italy compared energy use across representative age cohorts and found differences in cultural characteristics between the war generation and the baby boomer generation. The younger group was more attuned to thermal comfort than energy-saving attitudes, and more active in adopting digital and electronic equipment, with implications for increasing energy use. Lawson and Williams (2012) identified four clusters of energy cultures across the New Zealand population with respect to energy efficiency outcomes, which they called Energy Economic, Energy Extravagant, Energy Efficient, and Energy Easy, which appeared to have some link with family life cycle.

While changes within actors' energy and mobility cultures are important, a bigger question is the potential for wider transformative change. The research into the mobility cultures of Generation Y suggests that the many individuals choosing to not own or drive cars may signal a broad cultural shift away from automobility (Hopkins 2017). Concordant cultural changes across multiple actors may also be able to create structural changes. For example, adoption of solar photovoltaics (PV) in New Zealand has not been supported by government policy, and sector messaging has strongly discouraged PV uptake, but despite there has been a steady growth in household adoption of PV. Ford et al. (2017) interpreted this as a change in households' energy cultures driven in part by a desire for independence. They found that the mass effect of these household-level changes in energy culture has been a change in expectations amongst incumbent and new firms, with the development of new technologies and services to suit the changing material, normative and practice characteristics of PV adoptees. As a result of the widespread shift in household energy cultures, their collective agency has reached beyond the semi-permeable boundary of influence and is starting to re-shape cultures and structures within the electricity sector.

The energy and mobility cultures research in New Zealand has supported the development of policy advice to central government agencies, identifying opportunities for interventions to support more sustainable outcomes. A policy report by Barton et al. (2013) for example, which focused on outcomes from energy efficiency research, recommended different sets policy interventions for improving efficiency crafted to suit four clusters of household energy cultures. Another report covering both energy and transport policy used the approach of characterising relevant cultures and their outcomes, and then identifying options to reduce external barriers to change, or interventions to support change (Stephenson et al. 2016). For example, the research that revealed the 'trigger points for change' for small-medium businesses (Walton et al. 2019) formed the basis of policy advice for assisting these businesses to become more energy efficient. The report suggested that businesses should use the 'trigger points' to help them think about when energy efficiency changes could be made, such as targeting key points in the life of businesses (e.g. equipment breakdowns, moving premises) as the time to alter an aspect of their energy culture. It also proposed a focus on efficiency journeys (i.e. how change in one attribute can stimulate change in other aspects of energy culture) (Stephenson et al. 2016).

A wide variety of methodological approaches have been used to undertake sustainability cultures research. Most studies so far have used qualitative approaches, either designing interviews structured around the elements of the framework (e.g. Hopkins & McCarthy 2016) or applying the framework retrospectively to qualitative and quantitative material (e.g. McKague et al. 2016). Other studies have taken a quantitative approach, using large data sets to analyse variables and/or develop clusters of actors with similar cultural characteristics (e.g. Lawson & Williams 2012, Bardazzi & Pazienza 2017). Still other studies have used a mix of qualitative and quantitative methods (e.g. Scott et al. 2016). Ford et al. (2016) have identified and sought to standardise measures for evaluating behaviour-based energy interventions, based on the three core elements of the cultures framework. Other researchers working on water demand have proposed the framework as a way to bring together multiple data sets which between them provide data on all aspects of a 'cultures' framing (Manouseli et al. 2018).

The framework can also be used as an overarching model to design and integrate research across a multidisciplinary research programme. The energy cultures research programme in New Zealand used multiple qualitative and quantitative research methods derived from the different disciplines involved, each of which specialised in knowledge production that could inform different aspects of the cultures under investigation. As described in Stephenson et al. 2015a, 'a national household survey gathered data about material culture and practices; in-depth interviews gathered data about norms in relation to practices; choice modelling helped understand the link between norms and material culture through variations in householders' willingness to make trade-offs among the attributes to space and water heating systems; and reviews of law and policy explored the external context of household energy cultures' (p. 120). As mentioned above, the framework was also used to integrate the findings and to develop policy recommendations (Stephenson et al. 2016).

5 Conclusion

To achieve a sustainable future, effective theories are needed that can inform the widespread transformations required at all scales of consumption and production. The concept of culture can assist in this journey. The sustainability cultures approach builds from established social theories and offers an actor-centred approach to investigating the interplay between an actor's cultural attributes, broader societal structures, and sustainability outcomes. The approach also offers an integrative method that captures a number of topics that are typically the domain of different disciplines, and 'forces the analyst to recognise that these factors interact in a range of ways' (Manouseli et al. 2018: 441).

Applications of the sustainability cultures approach, as described above, show that it can usefully identify cultural 'units' (Archer 1996: xii) that can be investigated to explore their association with various sustainability outcomes. The research undertaken so far has identified how actors' cultural attributes interplay with external influences to drive outcomes such as energy consumption and mobility choices. Sometimes this results in an unhelpful stability, whereby there is a self-reinforcing interplay between cultural attributes and/or reinforcement by external influences. In other circumstances, cultures are undergoing change, and this can be triggered from within or by factors external to a culture. An actor exercising their agency to change one attribute (e.g. a different material possession) can have cascading effects on other cultural attributes, so that a relatively simple change can have unexpected ramifications. And while some external influences may reinforce actors' unsustainable cultures, other external influences can stimulate cultural change and thereby drive different sustainability outcomes. In a few instances, collective cultural change amongst actors appeared to be influential in wider structural changes.

There is much scope to extend the sustainability cultures approach. Within the energy context, household and business energy cultures can be explored and contrasted to help understand heterogeneity and identify options to develop carefully targeted policy interventions. Studies are needed to examine the extent to which actors' cultural attributes are necessarily consistent with each other, and, if not, whether inconsistency (e.g. aspirations are at odds with practices) indicates opportunities for change interventions. Investigations of multiple co-existing sustainability cultures (e.g. an actor's water culture vs energy culture, or multiple mobility cultures within a single household) could reveal what happens when there is a dynamic interplay between cultures. Future opportunities for methodological extension exist to undertake qualitative studies that are not limited to the elements of the cultures framework, but explore a wider range of cultural characteristics including knowledge, language and symbolism. Studies of under-explored and new topics (e.g. water, carbon, agriculture, food) using the sustainability cultures framing could help address applied questions as well as assist in further theoretical development.

We will not achieve a sustainable future without significant changes to our beliefs, values, norms, and practices as well as our material possessions. The lens of sustainability cultures offers a structured way to explore the causal links between actors' cultural attributes and sustainability outcomes, and to identify opportunities to achieve positive change.

References

Archer, M. S. (1996). *Culture and Agency: The place of culture in social theory.* Cambridge University Press, Cambridge, UK.

Bardazzi, R., & Pazienza, M. G. (2017). Switch off the light, please! Energy use, aging population and consumption habits. *Energy Economics*, 65, 161–71.

Barton, B., Blackwell, S., Carrington, G. et al. (2013). Energy cultures: Implications for policymakers. Centre for Sustainability. Retrieved from https://ourarchive.otago.ac.nz/handle/10523/3747

Bell, M., G. Carrington, R. Lawson, J. Stephenson (2014). Socio-technical barriers to the use of low-emission timber drying technology in New Zealand. *Energy Policy*, 67, 747–55.

Bourdieu, P. (1977). *Outline of a Theory of Practice.* Cambridge University Press, Cambridge, UK.

Dew, N., Aten, K., & Ferrer, G. (2017). How many admirals does it take to change a light bulb? Organizational innovation, energy efficiency, and the United States Navy's battle over LED lighting. *Energy Research & Social Science*, 27, 57–67.

Ford, R., Karlin, B., Frantz, C. (2016). Evaluating Energy Cultures: Identifying and validating measures for behaviour-based energy interventions. International Energy Policies and Programmes Evaluation Conference, Amsterdam.

Ford, R., Walton, S., Stephenson, J. et al. (2017). Emerging energy transitions: PV uptake beyond subsidies. *Technological Forecasting and Social Change*, 117, 138–50.

Geels, F.W. (2002). Technical transitions as evolutionary reconfiguration processes: a multi-level perspective and a case-study. *Research Policy* 31, 1257–74.

Giddens, A. (1984). *The Constitution of Society*. University of California Press, Berkeley.
Gnoth., D. (2016). Residential mobility and changing energy related behaviour (Thesis, Doctor of Philosophy). University of Otago.
Hays, S. (1994). Structure and agency and the sticky problem of culture. *Sociological Theory*, 12 (1), 57–72.
Hoicka, C. (2012). Understanding Pro-Environmental Behaviour as Process: Assessing the Importance of Program Structure and (Doctoral dissertation, University of Waterloo).
Hopkins, D. (2017). Destabilising automobility? The emergent mobilities of generation Y. *Ambio*, 46(3), 371–83.
Hopkins, D., & McCarthy, A. (2016). Change trends in urban freight delivery: A qualitative inquiry. *Geoforum*, 74, 158–70.
Hopkins, D., & Stephenson, J. (2014). Generation Y mobilities through the lens of energy cultures: a preliminary exploration of mobility cultures. *Journal of Transport Geography*, 38, 88–91.
Hopkins, D., & Stephenson, J. (2016). The replication and reduction of automobility: Findings from Aotearoa New Zealand. *Journal of Transport Geography*, 56, 92–101.
King, G., Stephenson, J., & Ford, R. (2014) *PV in Blueskin: Drivers, barriers and enablers of uptake of household photovoltaic systems in the Blueskin communities. Centre for Sustainability*, University of Otago, New Zealand.
Kurz, T., Gardner, B., Verplankem, B., Abraham, C (2015). Habitual behaviour or patterns of practice? Explaining and changing repetitive climate-relevant actions. *WIREs Climate Change*, 6,113–28.
Latour, B. (2005). *Reassembling the Social: An introduction to Actor–Network Theory*. Oxford University Press, Oxford.
Lawson, R., Williams, J. (2012). Understanding energy cultures. Annual conference of the Australia and New Zealand Academy of Marketing (ANZMAC), December 2012, University of New South Wales, Adelaide
Manouseli, D., Anderson, B., & Nagarajan, M. (2018). Domestic water demand during droughts in temperate climates: Synthesising evidence for an integrated framework. *Water Resources Management*, 32 (2), 433–47.
McKague, F., Lawson, R., Scott, M. and Wooliscroft, B. (2016). Understanding the energy consumption choices and coping mechanisms of fuel poor households in New Zealand. *New Zealand Sociology*, 31(1), 106–26.
Midgely, G., (2003). *Systems Thinking*. Sage publications, London.
Moezzi, M., Janda, K. B., & Rotmann, S. (2017). Using stories, narratives, and storytelling in energy and climate change research. *Energy Research & Social Science*, 31, 1–10.
Reckwitz, A. (2002). Toward a theory of social practices: a development in culturalist theorizing. *European Journal of Social Theory*, 5(2), 243–63.
Røpke, I. (2009). Theories of practice—new inspiration for ecological economic studies on consumption. *Ecological Economics*, 68 (10), 2490–2497.
Schatzki, T. (2002). *The Site of the Social: A philosophical account of the constitution of social life and change*. Pennsylvania State University Press, University Park, PA.
Scott, M. G., McCarthy, A., Ford, R., Stephenson, J., & Gorrie, S. (2016). Evaluating the impact of energy interventions: home audits vs. community events. *Energy Efficiency*, 9(6), 1221–40.
Scott, M.G. and Lawson, R. (2017). The road code: Encouraging more efficient driving practices in New Zealand, *Journal of Energy Efficiency*, 1–10

Shove, E. (2003). *Comfort, Cleanliness and Convenience: The social organisation of normality*. Berg: Oxford.

Shove, E., & Pantzar, M. (2007). Recruitment and reproduction: the careers and carriers of digital photography and floorball. *Human Affairs*, 17 (2), 154–167.

Shove, E. & Spurling, N. (2013). *Sustainable Practices: Social theory and climate change*. Routledge, Abingdon.

Stephenson, J. (2018). Sustainability cultures: An actor-centred interpretation of cultural theory. *Energy Research and Social Science*, 44, 242–249.

Stephenson, J., Barton, B., Carrington, G. et al. (2010). Energy cultures: A framework for understanding energy behaviours. *Energy Policy*, 38: 6120–9.

Stephenson, J., Barton, B., Carrington, G. et al. (2015a). The energy cultures framework: Exploring the role of norms, practices and material culture in shaping energy behaviour in New Zealand. *Energy Research & Social Science*, 7, 117–23.

Stephenson, J., Hopkins, D., Doering, A. (2015b). Conceptualizing transport transitions: Energy Cultures as an organizing framework. *Wiley Interdisciplinary Reviews: Energy & Environment*, 4, 354–64.

Stephenson, J., Barton, B., Carrington, G. et al. (2016). Energy Cultures Policy Briefs. Centre for Sustainability, University of Otago. Retrieved from https://ourarchive.otago.ac.nz/handle/10523/7104

Sweeney J. C., Kresling, J., Webb, D., Soutar, G. N., Mazzarol, T. (2013). Energy saving behaviours: Development of a practice-based model. *Energy Policy*, 61, 371–381.

von Bertalanffy, L. (1968). *General System Theory: Foundations, development, applications*. George Braziller, Inc., New York.

Walton, S., Doering, A., Gabriel, C., Ford, R. (2014). Energy Transitions: Lighting in Vanuatu. Report prepared for The Australian Aid – Governance for Growth Programme. Retrieved from https://ourarchive.otago.ac.nz/handle/10523/4859

Walton, S., Zhang, A., & O'Kane, C. (2019). Energy eco-innovations for sustainable development: Exploring organizational strategic capabilities through an energy cultures framework. *Business Strategy and the Environment*, 29(3), 812–826.

Warde, A. (2005). Consumption and theories of practice. *Journal of Consumer Culture*, 5(2), 131–53.

Young, W., & Middlemiss, L. (2012). A rethink of how policy and social science approach changing individuals' actions on greenhouse gas emissions. *Energy Policy*, 41, 742–7.

15 Socio-Ecological Sustainability and New Forms of Governance: Community Forestry and Citizen Involvement with Trees, Woods, and Forests

Bianca Ambrose-Oji

Introduction

The complexity of forest protection and management issues at global, regional, national, and local scale levels mitigates against simple governance solutions. Forests produce valuable goods and services that cross the private (e.g. timber) and public (e.g. biodiversity, hydrological regulation) domains, and are at increasing risk from global and local pressures including climate change, new pests and diseases, urbanisation, and over-exploitation. Ever since sustainable forestry appeared to reach the top of the international agenda in 1992 at the United Nations Conference on the Environment and Development (the Rio Conference) the story of how sustainability has been sought through these challenges has been well-rehearsed. Whether looking at either the industrialised or the developing nations of the world, many commentators agree that the last thirty years or so have seen parallel trends of continuing penetration and integration of the market and the commoditisation of 'free' or public goods and services, alongside the de-centralisation of natural resource decision making and management.

The shortcomings of instrumental top-down forest planning and conservation processes, a continuing decline in government or public sector departmental budgets for the care and management of the environment and natural resources, and a failure of state agencies to deliver all the services and functions demanded by society, means that a key component of the dominant narratives is that sustainable forest management and sustainable forest governance have a strong decentralised 'people' component. Not only are the 'people' identified as the primary beneficiaries of the benefits of forestry, but they are conceived of as active and engaged citizens or communities with effective agency. The narratives suggest that citizens, local communities, and community-based organisations can make important contributions to the creation, management, and protection of natural resources such as forests, and they may also play an active role in the decision–making processes including the development of forest policy at local and national levels.

Over the same period sociological perspectives on environmental governance have also been influenced by Giddens' (2002) views on globalisation, and what he described as space/time compression, i.e. the intensification of interconnectivities between social and natural systems with significantly increased human induced impact at global and local scales. This move into what many researchers recognise as the Anthropocene, has seen further development of socio-ecological systems science. These approaches incorporate ideas about vulnerability, risk and resilience as key aspects of sustainability that are linked with socio-ecological landscapes, ecosystems and the natural resources humanity depends on (Dearing, 2012). Socio-ecological system perspectives also suggest that a shift in conventional governance practice is required if society is to meet the challenges of changing global systems (Folke et al., 2007). A shift to adaptive governance is conceptualised as one where 'the divide between those governing and those being governed' is eliminated (Berkes, 2017: 1240), and where adaptive practice comes from social learning and innovation that emanates from citizens, communities, and citizen organisations as much as from traditional institutions of government (Folke et al., 2005).

Contested Views of Environmental Governance in Forestry

In his review, Arts (2014) describes how this increasing role for 'the people' in forestry, i.e. the 'shift from government to governance' (Rosenau and Czempiel, 1992), can be interpreted in two different ways. The first is an implication that new forms of governance shift authority and competencies away from the state and into international institutions, NGOs, civil society organisations, and businesses. The second is a broader conceptualisation. This recognises a myriad of different relationships between civil society, public bodies, and businesses which might act autonomously or mutually, and combine conventional with new forms of governance. In other words, 'governance may at the same time refer to governing *by*, governing *with* and governing *without* the state' (Arts, 2014: 19). The latter more polycentric view of governance is reflected in the socio-ecological systems literature, where it is recognised as an emergent property of a complex adaptive system involving many kinds of actors, operating at different scale levels, and one that does not have, indeed cannot have, a central controlling authority over the processes and structures of the system (Ostrom, 2010; Bixler, 2014). These different characterisations contest the degree to which citizens and communities have agency and hold or manage decision making power over natural resources such as forests. For some authors, citizen involvement in forest governance is a form of co-option into a neoliberal agenda which rarely leads to genuine power devolution down to local users, and may even limit citizen or community ability to innovate and adapt to new social, economic, and ecological circumstances (Blaikie, 2006; Parkins et al., 2016; Ribot, 2004; Wagenaar et al., 2015). For others, the mixed assemblage of actors involved in new forms of forest governance 'cannot be resolved into neat binaries that separate power from resistance, or progressive forces from reactionary ones. It is difficult to determine who has been co-opted and who betrayed.' (Li, 2007 quoted in Belsky, 2015: 30). Yet others feel the opposite remains true, and, in particular contexts, local communities are empowered to innovate around adaptation choices that fit their own resource management goals and set their own development trajectories (Chapin et al., 2016). Some researchers find evidence of both the support and development of community power, and the deterioration of community rights and access to resources, at the same time as a host of unintended and unexpected impacts which are not easy to interpret (Blomley et al., 2017).

There is plenty of evidence to suggest that multi-scale, multi-actor polycentric systems of governance are the norm in forestry contexts: When looking at the situations in developing countries, new forms of forest governance that involve forest communities and community forestry arrangements are dominated by market-based forms, the most prominent being Reducing Emissions from Deforestation and Forest Degradation Plus (REDD+) mechanisms, Payment for Ecosystem Services (PES) and certification schemes. These governance systems lie at the intersection between multi-level strategies and regulation, neoliberal economic logic and local level action. With REDD+ and PES, individual small and medium scale private land owners and managers as well as community-based forestry initiatives are encouraged to deliver carbon mitigation or the provision of ecosystem services through alternative financial mechanisms, or promises of secure land tenure, which meets agendas set by other institutions and interest groups (Cadman et al., 2017; Primmer et al., 2015). In the case of certification schemes, social regulation of the market might be led by citizen organisations, pressure groups and NGOs, but they combine with private and public sectors to regulate business, so that community-based forestry complies with externally enforced modalities.

Researchers interested in a sociology of transnational social movements have shown that one innovative response to these globally connected, multi-scale forms of governance has been the rise of transnational self-help networks. These networks help to connect local and national scale community forestry organisations with the political and decision making sphere at regional and global scale, and help to mediate and implement the conditionalities of mechanisms such as REDD+ that uphold community rights and access to forests (Dupuits, 2015).

In industrialised country contexts new forms of forest governance have also developed. Citizens are encouraged to engage with public agencies and larger NGOs, through platforms for consultation and dialogue that contribute to forest management decision making and policy issues on public and open access forest land (see for example Eckerberg and Buizer, 2017; Teder and Kaimre, 2017; Weber, 2017). There has been a growth in active citizenship through environmental volunteering in woodlands and forests where individuals and groups have varying degrees of influence over forests owned by a variety of organisations including local government, public agencies, NGOs, and community-based organisations (Buijs et al., 2019; Buijs et al., 2016; Jerome et al., 2017; van der Jagt et al., 2016). The evolution of governance for forest commons that may also include citizens beyond the traditional right holding 'commoners' in resource use and decision making has been evidenced (Gatto and Bogataj, 2015; Kluvánková and Gezík, 2016; Premrl et al.,

2015). There is also an increasing amount of effort placed in the engagement of 'the people' in the decision making and management of woods and forests in urban and peri-urban locations (Konijnendijk van den Bosch, 2012; Lawrence and Dandy, 2012).

New Forms of Governance with Citizens and Community Forestry Groups in the UK

Over the last twenty years a process of devolution in the UK has decentralised autonomy in specific policy areas to the national governments (i.e. England, Northern Ireland, Scotland, and Wales). Forestry is one such 'devolved matter', so each country has distinct institutional arrangements for implementing their national forest policies and overseeing management of their public forest estates. This means that opportunities for citizen innovation and inclusion in forest governance differ, so that the development of community forestry has followed different trajectories in each country (Lawrence et al., 2009; Lawrence and Molteno, 2012; Lawrence, Van der Jagt, and Ambrose-Oji, 2014; O'Brien et al., 2008). In Scotland a policy and legislative context favours civil society involvement in local and national level forest governance, and supports the transfer of forestry assets to community ownership and management. In Wales there is a focus on inter-generational sustainability and socio-economic resilience, so forest policy promotes community involvement in environmental governance. Forest policy in England has seen a move towards localism and the inclusion of the third sector and community groups supporting the functions and institutions of the state, as well as an increased emphasis on natural capital based enterprise development in private woodlands (Defra, 2013; Independent Panel, 2012). In Northern Ireland because more than half of the forests are in state and public ownership, maintaining local community access for recreation is a policy priority, with a nascent community forestry emerging on woodlands owned by third sector organisations such as the Woodland Trust.

Characterisations of community forestry in the UK distinguish differences between initiatives and community groups by features which have a bearing on forest governance and social innovation. These include: variations in outcome attributable to institutional arrangements, collective values, and social practice (Ambrose-Oji et al., 2015; Lawrence and Ambrose-Oji, 2013; Lawrence and Ambrose-Oji, 2014); the degree of decision-making power afforded actors (Tidey and Pollard, 2010); or the form and function of engagement with public agencies (Swade et al., 2013).

This evidence suggests that there are distinct overlaps between community-based collective action, and the agency of active citizens. It also shows that social innovation in governance blurs the boundaries between the explicit domains of individuals, community-based and civil society organisations, third sector organisations, and public institutions. Table 15.1 maps the complexity of these overlaps. The matrix plots the type of actor holding power over the resources (i.e. community, civil society group/NGO, public sector) against the different kinds of social forest practice, with a description of the structural institutional forms and discourses that steer the collective action.

Table 15.1 Innovative forest governance in Britain: community and citizen involvement in different kinds of forest practice with different facilitating/leading actors[1]

Practice type	Actor constellation and locus of power over woodland resources		
	Community-based organisations facilitate and lead	Third sector organisations (e.g. NGOs, charities, and civil society organisations), facilitate and/or lead	Government and public agencies facilitate and/or lead
Environmental volunteering	Citizens take part in regular work parties or *ad hoc* actions with, and for, community woodland groups that hold ownership or tenure of woodlands e.g. *Gorham and Admiral Community Woods, Bicknor, England*	Citizens take part in regular work parties or *ad hoc* actions with/on behalf of organisations with ownership or tenure of woodlands such as the Woodland Trust, or the National Trust. This may be organised through a third party institution such as, e.g. *The Conservation Volunteers*. Can also include, e.g. citizen science initiatives such as National Bluebell Survey for Woodland Trust.	Citizens take part in regular work parties or *ad hoc* actions with/on behalf of state forestry services or other government public agencies, e.g. *i-Tree or Observatree citizen scientist volunteers* for the Forestry Commission in London, or **volunteering on the public forest estate in Wales or Scotland**
Management of woodlands	A defined, constituted or legally incorporated community woodland group run by local people, with ownership or formal tenurial rights, managing one or more specific woodland sites according to their own objectives e.g. **Blean Bran Community Woodland Group**, in south Wales or Gordon Community Woodland, Gordon, Scottish Borders, Scotland	Formal or informal arrangement to undertake management on woodland in tenure of civil society organisation e.g. **Arkaig Community Forest and the Woodland Trust co-governance** of forest land Lochaber, Scotland	Formal or informal arrangement to undertake management on woodland in public ownership to maintain ecological and social benefits e.g. *Ashford Community Woodland*, for Ashford Council in England, or **Friends of Tower Hamlets Cemetery Park**, for Tower Hamlets Council in London, or on state forest land, e.g. *Friends of Newton Hill Woodland in Scotland*
Woodland-based social enterprise	Enterprise-oriented endeavour, may be a community woodland group or another formally constituted or legally incorporated	Enterprise-oriented endeavour undertaken on woodland in civil society organisation ownership, to produce social and	Enterprise-oriented endeavour undertaken on woodland in public ownership, to produce social and

1 Examples discussed in this chapter indicated in Bold

Table 15.1 (cont.)

Practice type	Community-based organisations facilitate and lead	Third sector organisations (e.g. NGOs, charities, and civil society organisations), facilitate and/or lead	Government and public agencies facilitate and/or lead
	group using and/or managing woodland with ownership or formal tenurial rights to generate social and environmental benefits e.g. **ARC CIC at Foundry Wood** Leamington Spa, England providing an innovative service offer to the local community	environmental benefits, e.g. **Wood Matters on National Trust land**, Cumbria, England engaging the community in an income generating woodshare scheme	environmental benefits, e.g. *Neroche Woodlanders, near Taunton, England*
Public engagement	Community woodland groups consult the wider local community about forest management, may be as part of asset transfer processes or as more general attempt to involve community in governance, e.g. *Hill Holt Wood, England*	Active citizenship though involvement in third sector organisation campaigning and protest which may be traditional, or using social media platforms, e.g. the Woodland Trust's Enough is Enough campaign, or 38 Degrees fight against sell-off of public forest estate in England	Citizen and community group involvement in formal public consultation processes, e.g. *Harnessing the Energy of the Community, Thetford Forest, England or Central Scotland Forest Design Planning processes on public forest estate woodlands*
Hybridised practice through partnerships and collaborative working	Formal and informal associations of community woodland groups, working in partnership with each other or with other private and public agencies to deliver benefits for community woodlands e.g. **Llais Y Goedwig**, *community woodland association in Wales delivering Government policy priority actions through formal partnership*	Community forestry groups or active citizens involved in the governance of woodland with, through or for third sector organisations, often involving other stakeholders, e.g. *Mersey Forest and other Community Forests in England* working with citizen groups and other third sector organisations and public agencies to deliver new Northern Forest bringing private and public sector funding together to achieve Government policy ambitions for regional tree cover	Formal partnership or other arrangements for citizen groups or community woodland groups to support publicly owned and managed sites with public agencies and others, e.g. **Friends of Westonbirt**, *running Westonbirt National Arboretum with the Forestry Commission near Tetbury, England*

Innovation in Governance with and amongst Communities

Where communities take the lead, the following examples illustrate that collective problem solving is focused on issues of importance to those communities. These issues are often connected with a failure in existing governance arrangements. Social innovation may lead to the formation of new organisations and dynamic institutional forms better placed to adapt social forestry practice in response to socio-ecological change.

Blaen Brân Community Woodland Group: Innovation to Manage Woodland for Social and Environmental Benefits

The community around Torfaen and Cwmbran in South Wales were frustrated about Blaen Brân their local woodland having no formal public access to meet the increasing community demand for recreation, as well as it becoming a magnet for antisocial behaviour because the owner was unable maintain the property. After a public meeting was held with the Local Authority to discuss local people's concerns, Coed Gwaun-y-fferiad Community Trust was formed by the community to help the woodland owner better manage the woods. Volunteer working groups were enlisted to meet the most pressing practical management issues. Two years later, in 2005, the group formalised its legal status and became Blaen Brân Community Woodland (BBCW) a company limited by guarantee. This change enabled the group to access financial resources available to bona fide community groups and lead to them purchasing the woodland. Through their practice of social forestry and active woodland management BBCW members developed their forestry skills and knowledge. When the discovery of a serious tree disease (*Phytophthora ramorum*) meant part of the woodland had to be felled under a statutory felling notice, BBCW were able to innovate, replanting with culturally important tree species that would also confer some resilience to the expected impacts of climate change. The capital received from the felling operations was also used to innovate around the types of services that the woodland provides. BBCW now uses the site for forest-based education, and health and wellbeing activities, some of which generate additional income for the group. The governance and practice instituted by BBCW has been so well regarded that they have been asked to consider scaling up and taking on the management of other neglected woodland sites in Local Authority ownership.

Achieving Results in Communities at Foundry Wood: Developing an Urban Green Space for Local Communities

Working in Leamington Spa, England, Achieving Results in Communities (ARC) realised that a small and neglected brownfield site in the centre of town on which a secondary woodland had grown up represented an important urban greenspace that could provide local people with a place to connect with nature. Formally constituting themselves as a Community Interest Company (CIC) in 2011 meant that ARC could find a way of acquiring Foundry Wood as a community asset. ARC were able to use a planning system conditionality (Section 106 agreement) brokered by the Local Authority to gain the site and a small capital sum from the owners, a development company. ARC used the resources they acquired to improve access into the site and improve essential infrastructure such as boundary fencing. They also built a small shelter that could be used as a central meeting point. Their governance of the site involved the local community from the outset. The community acted as consultees expressing their opinions about what they felt the woodland should be providing, as well as coming as volunteers to help clear the site, plant additional trees and establish woodland management. ARC realised the best way to ensure sustainability of governance and woodland management, as well as responding to changing demands for services and benefits the woodland might provide, was to involve the community even further. So ARC facilitated the development of the Friends of Foundry Wood (FoFW) run by local volunteer supporters, as a charitable incorporated organisation (CIO) formalised in 2014. This provided a form of co-governance for the wood between ARC and FoFW. Since 2015 a programme of regular events has grown and innovated in the provision of on-site ecotherapy sessions, woodland-based education for schools and after-school groups, arts, and cultural activities. In addition to providing a wide range of social and environmental benefits for the community, many of these activities generate income for local practitioners, as well as a small financial surplus that is reinvested in woodland management. In 2017 the Royal Forestry Society recognised the significant flow of benefits that had been achieved from such a small site, and the innovation in urban community forestry governance that ARC-FoFW had realised by presenting them with an Excellence in Forestry Award.

Llais Y Goedwig Community Woodland Association in Wales: Putting Community Forestry into Policy and Politics

Llais Y Goedwig (LYG) was established as a community woodland association for Wales by a small group of active citizens who had become involved in forming and managing community woodland groups themselves and felt there was no umbrella organisation providing a voice for them. Constituted as a company limited by guarantee in 2006, it's membership has grown to over 50 community woodland groups in Wales, with more than 250 active members. The voluntary directors realised that they would need to innovate if their network was to leverage significant impact in the Welsh forestry sector and policy arena. They spotted opportunities for increasing their influence through three mechanisms. Supporting the development of new community woodland groups increased the size of the sector and influence of the membership. Horizontal scaling out connected their network with other community groups, community focused organisations, and partners. Vertical scaling up built relationships with policy makers through existing institutions of governance. By 2011 one member of LYG was able to say:

> We have connections now to policy. With the policy divisions, it's a very different relationship because we have more e-mail communication, more of an ongoing expectation and a greater developing relationship now ... We have a different type of

contact, more formalised and more of an expectation that we can formally help to deliver (Ambrose- Oji et al., 2010: 26)

By 2012, the success of their vertical integration into existing governance structures led to a formal delivery partnership with the Forest Commission Wales, and then the succeeding government body, Natural Resources Wales' (NRW) Action Plan for Woodlands. Latterly, and up until 2017, the Welsh Programme for Government Action Plan has involved LYG as a named partner in the delivery of several priority actions in the annual NRW Business Plans. In 2014 LYG used this relationship to advocate for innovation in use of the public forest estate, working with NRW to identify opportunities to open up public forest land that could be used by community groups and support community woodland growth. The growth of the network and the increasing advocacy potential, was helped by LYG opening up an office co-locating with other social forestry and forest business groups in a developing 'Forestry Hub'. Late in 2015 LYG was provided with three year's worth of substantial Welsh Government core funding to continue their horizontal growth, building new network partnerships with other third sector organisations, strengthening the position of community forestry within the regional economy as well as increasing community representation in environmental governance. NRW is currently in the process of formulating Area Statements. These represent the evidence-based governance instruments that the Welsh Government will use to implement natural resources policy outlined in the Environment Act. The Welsh Government expects to align their forest resource planning process with these Statements. With LYG closely involved in the delivery of policy and given a remit to encourage community engagement, this should mean the organisation will be able to encourage communities to identify opportunities where local economic development goals can be achieved through community management of forests.

Innovation in Governance with the Third Sector (NGOs, Charities, and Other Civil Society Organisations)

Some of the largest private owners of woodland in the UK are charitable organisations, two amongst them are particularly important. The National Trust is the largest private landowner in the UK, and the largest natural and cultural heritage conservation charity in Europe. It has over 5 million members and an annual income around £500 million. The Woodland Trust is the UK's largest conservation organisation with half a million members, owning and managing 26,000 hectares of woodland and an annual income of around £50 million. These NGOs are values driven and have clear missions to conserve and manage the natural heritage of the country. In recent years they have recognised the value of involving active citizens as volunteers. For example, the National Trust has 60,000 active volunteers contributing the equivalent in hours to 1,590 full-time staff, and it has a 'Going Local' strategy to increase civil society engagement in its work. Partnerships involving community woodland groups are a more recent innovation.

Loch Arkaig Community Forest: Working with the Woodland Trust to Purchase and Restore an Area of Native Caledonian Pine Woods

The Achnacarry, Bunarkaig and Clunes (ABC) group SCIO (Scottish Charitable Incorporated Organisation) made an application in 2014 to acquire two woodland blocks of ancient Caledonian pine forest around Loch Arkaig. These were being disposed of by Forestry Commission Scotland and made available for community purchase through the National Forest Land Scheme. ABC recognised that on their own they were unlikely to raise the full purchase price of the woodlands valued at over £50,000. ABC's innovative answer to this problem was to approach the Woodland Trust as a potential partner because both organisations had a mission to conserve and restore Scotland's native woodlands. Joining forces raised enough money amongst the local community and the wider Woodland Trust membership to purchase the woodland in 2016. ABC evolved in form becoming Arkaig Community Forest SCIO as the institutional vehicle most suited to representing community perspectives in a co-governance arrangement with the Woodland Trust. Having consulted with the community, the woodland will be managed to create benefits for the local economy as well as improve forest biodiversity.

Woodmatters: A Social Enterprise Working with the National Trust and Local Communities for Social and Environmental Benefits

The National Trust has managed the historic Sizergh Estate, near Kendal in Cumbria since 1952. In recent decades, one of the major challenges on the estate has been to find ways to ensure the sustainable management of the coppice woodland and other forest areas for conservation benefits. This kind of management is usually labour intensive and as such presents high operating costs. Although the National Trust had successfully applied for a Woodland Improvement Grant from the Forestry Commission, it was conditional on establishment of a community wood fuel project in line with regional forest policy objectives.

Two local social entrepreneurs who had already established Woodmatters – an enterprise primarily concerned with establishing community connections with nature – believed this challenge could be met through an innovative governance scheme. They proposed a Woodshare Community Membership Scheme. This was inspired by models of Community Supported Agriculture (CSA), where a direct and active partnership is built between community, landowner, and land manager in the production of goods members can buy themselves or sell on the open market. The National Trust felt comfortable entrusting woodland operations and volunteer management to Woodmatters who had demonstrated the skills and capability to realise the work over the estate.

Members of the local community who wish to join the Woodshare scheme pay an annual subscription which entitles them to take part in agreed coppicing and other forestry operations. By taking part in the woodland work, community members earn rights to the products/materials produced which vary according to the Woodshares they have bought. This reduces the cost of firewood quite substantially for those taking part, as well as providing them with an opportunity for social interaction, woodland-based learning and skills development, and physical exercise. Since the early years of establishment, the scheme has contributed to the health and wellbeing of the community, become financially sustainable, and most importantly has restored neglected coppice stands and woodland to improved biodiversity condition As a consequence of this success, Woodmatters has been invited to offer community-based activities in other National Trust woods, where it has continued to innovate in nature-based education and therapeutic experiences e.g. immersive forest mindfulness.

Innovation in Governance with the State or Public Sector

As well as forming innovative governance arrangements with private organisations, community woodland groups and active citizens also work with public agencies. Some of these governance arrangements may not be particularly innovative from the community or citizen perspective, but it is important to acknowledge how innovative these arrangements might be to the public organisations involved. One important consequence of this is the potential to normalise community–state relationships and establish a positive discourse about citizen and community involvement in forestry.

Volunteering on the Public Forest Estate in Wales and Scotland

Active citizens have been given the opportunity to volunteer to help maintain Welsh woodlands for NRW for some years. This kind of engagement might be viewed as rather instrumental or functional involvement with little innovation or governance impact. However, as far as public forestry staff are concerned, the involvement of volunteers is often regarded as an innovation, particularly in the sense that it builds relationships between local communities or interest groups and state forestry agencies which can have some significant and positive impacts. Friction between citizens, communities, and state forest agencies can often be attributed to either party having little knowledge about the other, particularly their woodland governance aims and objectives and the constraints to achieving them. From the point of view of government agencies volunteering schemes can help overcome this:

> One of the key things we wanted to do with the [volunteering] scheme is to draw local people in get people to meet and then know each other, and us, and to have fun ... there is a sense of community built up, with people who really know what is happening on the site and why, and you can manage the conflicts and difficulties better as a result. If you can get buy in like this it makes our jobs much easier in the end (Ambrose-Oji, 2011: 24–5)

The social policy context in Scotland provides many opportunities for community groups to take part in the governance and management of woodlands on the public forest estate. For some community groups their innovations are frustrated, as finding an adaptive space within the rules of the game remains elusive, as one person put it:

> I think genuinely from high up, there's a commitment to community engagement [and] many of the individuals on the ground also feel that ... But, where it's falling down somewhere in between, either the resources haven't been put into the staff to make it happen or the processes haven't been set up for some of the procedures (Lawrence, Ambrose-Oji, and O'Brien 2014: 42).

In both Wales and Scotland, better communication and relationships can also lead to government forest agencies acknowledging the positive impacts of engaging with communities. A discourse of success has helped achieve normalisation of community involvement so that it is perceived as a common and practicable situation rather than a suspicious exception, as a representative of Forest Enterprise Scotland said:

> You become a bit more comfortable when you see a community group who is being successful ... you move away from 'we must be in control' and 'we need things in triplicate'. (Lawrence et al., 2014a: 44)

The Friends of Tower Hamlets Cemetery Park Innovating for Environmental and Community Benefits

After the abolition of the Greater London Council in 1986, the London Borough of Tower Hamlets (LBTH) became owners of Tower Hamlets Cemetery. Having been closed to burials in 1966 a period of neglect meant woodland and other semi-natural vegetation established across the site leading to the cemetery's designation as a Local Nature Reserve and a Site of Metropolitan Importance for Nature Conservation. In 1990 the Friends of Tower Hamlets Cemetery Park (FoTHCP) were formed out of concern for the site, with the objective of preserving, promoting, improving, and caring for Tower Hamlets' only woodland park. They registered as a charity in 1993, employing their first staff member in 2002, and their second in 2012.

Since 2004 FoTHCp have realised their objectives by taking on the governance of the park on behalf of LBTH via a Service Level Agreement. This hybrid arrangement does not confer ownership or strict tenurial rights but facilitates governance by the community group not only to manage the site for wildlife but to service some of the demands and needs felt in the local community, e.g. for environmental education. The most recent innovation has been for FoTHCp to begin running their endeavour as a social enterprise, raising around two thirds of their income independently of LBTH from events on the site and active corporate and public fundraising. The FoTHCp fulltime staff and the efforts of over three thousand local and corporate volunteers per year continue the renovation and maintenance of the park, and strategic planning and expansion of the park into adjacent areas of neglected land.

Friends of Westonbirt Arboretum (FoWA): Developing Hybrid Co-Governance Arrangements to Secure the Future of Woodland in Public Ownership

Westonbirt Arboretum was established during the middle of the nineteenth century and is a nationally important collection of more than 15,000 labelled specimens of trees from temperate areas of the globe. The tree collection is situated in a parkland and wooded landscape covering around 240 hectares near Tetbury in Gloucestershire, England. Westonbirt was given over to the Forestry Commission as part of the public forest estate in 1956, when it was opened to the public for the first time as a resource for conservation, recreation, and education. In 1985 a small group of active citizens who were looking to ensure the sustainability of the site decided to focus their efforts on supporting the scientific and educational activities associated with the trees.

The legal form of the group has developed over the years from a Trust and an Association to a public benefit entity in the form of a charity and a company limited by guarantee. This change has accommodated the groups response to new societal demands for the cultural and social benefits Westonbirt could provide. It has also allowed Friends of Westonbirt Arboretum (FoWA) to meet the opportunities offered by a policy push for increased civil society involvement in the management of sites in public ownership. A formal partnership agreement with the Forestry Commission means that today FoWA are not only concerned with realising scientific and learning objectives, but have innovated to expand the strategic and ongoing management activities they take part in.

Expanding and finding new roles for volunteers has been critical to FoWA operations, who delivered over 25,000 hours of support in 2017. In a co-governance arrangement the Forestry Commission works with a fulltime FoWA volunteer coordinator to oversee and direct the working of about 250 volunteers recruited mainly from the 30,000 members of FoWA. The volunteers help the FoWA executive contribute to managing the woodland and tree collection, designing and managing new infrastructure projects (such as the Great Hall, a tree canopy walkway, and the Wolfson Tree Management Centre), and providing visitor services through a uniformed volunteer staff. Because FoWA is able to access a range of funding streams open to charities and companies which are not accessible to the Forestry Commission as a public body, FoWA has been particularly innovative developing income generation strategies, raising between £1.2 and £1.9 million per annum between 2012 and 2017.

Conclusions

All the examples presented in this chapter illustrate just how far community forestry involves different constellations of local actors who come together to practice community-focused, community-supported, or community-directed forestry in very different ways, across multiple contexts. Community forestry through community or citizen involvement in forest governance is therefore shown to be pluralistic and multi-scaled. Innovation in governance comes from a drive to capitalise on new opportunities involving woodlands, or to solving local socio-environmental problems through woodland management. The challenges that communities and citizens have addressed through innovative governance arrangements and approaches to management are also diverse and include: providing greater community and public access to local greenspaces; providing socio-ecological system services communities want and need; using woodland as a way to provide local employment and other aspects of social cohesion; and developing sustainable forms of place-keeping, including for organisations and institutions facing resource limitations and less able to do this for themselves. Such variety underscores how difficult it is to define community forestry, and the broad scope of social and governance innovations that arise.

The boundaries between active citizenship and collective action, and between community focus on local sites and community networking for vertical integration into traditional governance structures are all very fluid. The British case studies also demonstrate that social innovation in community forestry is not necessarily tied to the actual ownership of woodlands. Innovation can come through integration into governance systems over land that other agencies or organisations have ownership of. The examples also show how the agency of government emerges as important. The cases demonstrate that even when considering self-mobilised community innovation, Local Authorities and state forest services continue to play an important role supporting innovation. They may provide access to resources whether land, woodland, or finance through government-administered funding schemes; advice and facilitation along pathways navigating the rules of the game; they help build a discourse of success that can weave community forestry into policy and strategy.

Social innovation in governance may mean creating something new, but as our cases show most innovations are tied to context and will build on governance forms and process which are already there (Lawrence and Dandy 2012). As Arts et al. (2014) agree governance processes are historically contingent. However, Arts et al. also emphasise that associated practices cannot be steered, and innovative and emerging ideas and adaptations to new social and ecological circumstances will bring both intended and unintended changes. This was true in the cases presented and included such things as the unexpected growth of the Friends of Westonbirt into active managers of a state forest service site.

The discourses of success built by local government and public agencies were important to the normalisation of community and citizen involvement in forest governance. These rested on knowledge not only of self-mobilised community forestry but also of the more functional and passive forms of innovative practice with state forest agencies. The development of storylines by government and public agents provides discursive influence over the shaping and determination of what successful community forestry practice 'looks like', justifying its contribution to the sustainability of forest resources, and how policy can contribute to it being achieved (Bäckstrand and Lövbrand, 2006; Hajer and Versteeg, 2005). Discourse can also build legitimacy within a community too. Discourse was important to building unified storylines to support community groups in their collective problem solving, to generating messages that legitimised group management of the woodland resource in the eyes of local communities and stakeholders. As Depuits (2015) identified, the emergence of unified storylines was key to building connectivity between community forestry initiatives at different scales, and lifting the impact of governance innovations from a focus on purely local concerns and local practices and into processes of policy change and regional or national level politics (Wydra and Pülzl, 2013).

Overall, the conclusion is that although place-based sustainability strategies rely on the inclusion of active citizens or community involvement, different forms of community forestry are likely to achieve adaptive capacity through social innovation that engages with a constellation of actors and multiple centres of decision making (Carlisle and Gruby, 2017). There is not, and will not be, a single definition of community forestry, nor of the governance arrangements or innovations associated with it. Combinations of practice that manage differing and evolving combinations of resources, forest functions and uses, and organisational structures and actors have the best potential to innovate for sustainability. Engaging with other community groups and networks, with large organisations and the state, across different scales and sectors, allows for local practices and ideas to be incorporated within existing institutions and to transform and develop in response to changing socio-ecological conditions (George and Reed, 2017; Wilson and Cagalanan, 2016). That is where the hope of community forestry lies.

References

Ambrose- Oji, B., Wallace J., Lawrence A., and Stewart, A. (2010) Forestry Commission working with the Third Sector. Farnham, Surrey: Forest Research report to Forestry Commission England.

Ambrose-Oji, B. (2011) Volunteering and Forestry Commission Wales: scope, opportunities and barriers. Farnham, Surrey: Forest Research report to Forestry Commission Wales.

Ambrose-Oji, B., Lawrence, A., and Stewart, A. (2015) Community based forest enterprises in Britain: two organising typologies. *Forest Policy and Economics* 58: 65–74.

Arts, B. (2014) Assessing forest governance from a 'Triple G' perspective: government, governance, governmentality. *Forest Policy and Economics* 49: 17–22.

Bäckstrand, K., and Lövbrand, E. (2006) Planting trees to mitigate climate change: contested discourses of ecological modernization, green governmentality and civic environmentalism. *Global Environmental Politics* 6: 50–75.

Belsky, J. M. (2015) Community forestry engagement with market forces: a comparative perspective from Bhutan and Montana. *Forest Policy and Economics* 58: 29–36.

Berkes, F. (2017) Environmental governance for the Anthropocene? Social-Ecological systems, resilience, and collaborative learning. *Sustainability* 9: 1232.

Bixler, R. P. (2014) From community forest management to polycentric governance: assessing evidence from the bottom up. *Society and Natural Resources* 27: 155–169.

Blaikie, P. (2006) Is small really beautiful? Community-based natural resource management in Malawi and Botswana. *World Development* 34: 1942–1957.

Blomley, T., Edwards, K., Kingazi S., et al. (2017) When community forestry meets REDD+: has REDD+ helped address implementation barriers to participatory forest management in Tanzania? *Journal of Eastern African Studies* 11: 549–570.

Buijs, A., Hansen, R., Van der Jagt A., et al. (2019) Mosaic governance for urban green infrastructure: upscaling active citizenship from a local government perspective. *Urban Forestry & Urban Greening* 40: 53–62.

Buijs, A. E., Mattijssen, T. J. M, Van der Jagt A. P. N., et al. (2016) Active citizenship for urban green infrastructure: fostering the diversity and dynamics of citizen contributions through mosaic governance. *Current Opinion in Environmental Sustainability* 22: 1–6.

Cadman, T., Maraseni, T., Ok Ma H., and Lopez-Casero, F. (2017) Five years of REDD+governance: the use of market mechanisms as a response to anthropogenic climate change. *Forest Policy and Economics* 79: 8–16.

Carlisle, K. and Gruby, R. L. (2017) Polycentric systems of governance: a theoretical model for the commons. *Policy Studies Journal* 47: 927–52.

Chapin, F. S., Knapp, C. N., Brinkman, T. J., et al. (2016) Community-empowered adaptation for self-reliance. *Current Opinion in Environmental Sustainability* 19: 67–75.

Dearing, J. A. (2012) Navigating the perfect storm: research strategies for social-ecological systems in a rapidly evolving world. *Environmental Management* 49: 767–75.

Defra. (2013) Government Forestry and Woodlands Policy Statement. London: Department for Environment Food and Rural Affairs.

Dupuits, E. (2015) Transnational self-help networks and community forestry: a theoretical framework. *Forest Policy and Economics* 58: 5–11.

Eckerberg, K., and Buizer, M. (2017) Promises and dilemmas in forest fire management decision-making: exploring conditions for community engagement in Australia and Sweden. *Forest Policy and Economics* 80: 133–40.

Folke, C., Hahn, T., Olsson P., and Norberg, J. (2005) Adaptive governance of social-ecological systems. *Annual Review of Environment and Resources* 30: 441–73.

Folke, C., Pritchard, L., Berkes, F, Colding, J., and Svedin, U. (2007) The problem of fit between ecosystems and institutions: ten years later. *Ecology and Society* 12(1): 30.

Gatto, P., and Bogataj, N. (2015) Disturbances, robustness and adaptation in forest commons: comparative insights from two cases in the Southeastern Alps. *Forest Policy and Economics* 58: 56–64.

George, C., and Reed, M. G. (2017) Operationalising just sustainability: towards a model for place-based governance. *Local Environment* 22: 1105–23.

Giddens, A. (2002) *Runaway World: How globalisation is shaping our lives*, Oxford: Routledge.

Hajer, M., and Versteeg, W. (2005) A decade of discourse analysis of environmental politics: Achievements, challenges, perspectives. *Journal of Environmental Policy & Planning* 7: 175–84.

Independent Panel on Forestry. (2012) Independent Panel Report on Forestry: Final Report. London: report to Department for Environment Food and Rural Affairs.

Jerome, G., Mell, I., and Shaw, D. (2017) Re-defining the characteristics of environmental volunteering: creating a typology of community-scale green infrastructure. *Environmental Research* 158: 399–408.

Kluvánková, T. and Gezík, V. (2016) Survival of commons? Institutions for robust forest social-ecological systems. *Forest Economics* 24: 175–85.

Konijnendijk van den Bosch, C. (2012) Innovations in urban forest governance in Europe. In Johnston, M. and Percival, G. (eds) *Trees, People and the Built Environment. Proceedings of the Urban Trees Research Conference, 13–14 April 2011.* Edinburgh: Forestry Commission, pp. 141–47.

Lawrence, A. and Dandy, N. (2012) Governance and the urban forest. In: Johnston, M. and Percival, G. (eds) *Trees, People and the Built Environment. Proceedings of the Urban Trees Research Conference, 13–14 April 2011.* Edinburgh: Forestry Commission, 148–58.

Lawrence, A. and Ambrose-Oji, B. (2013) A framework for sharing experiences of community woodland groups. *Research Note*. Edinburgh: Forestry Commission.

Lawrence, A. and Ambrose-Oji, B. (2014) Beauty, friends, power, money: navigating the impacts of community woodlands. *Geographical Journal* 181: 268–79.

Lawrence, A. and Molteno, S. (2012) Community forest governance: a rapid evidence review. Farnham, Surrey: Forest Research, 141.

Lawrence, A., Ambrose- Oji, B. and O'Brien, E. (2014) Current approaches to supporting and working with communities on the National Forest Estate: feedback from community organisations and FES delivery staff Roslin, Midlothian: Forest Research.

Lawrence, A., Anglezarke, B., Frost, B., et al. (2009) What does community forestry mean in a devolved Great Britain? *The International Forestry Review* 11: 281–297.

Lawrence, A., Van der Jagt, A., Ambrose- Oji, B., and Stewart, A. (2014) Local authorities in Scotland: a catalyst for community engagement in urban forests? *Trees, People and the Built Environment II. Proceedings of the Urban Trees Research Conference, 2-3 April 2014*. University of Birmingham. Institute of Chartered Foresters, Birmingham.

O'Brien, L., Townsend, M. and Ebden, M. (2008) 'I'd like to think that when I'm gone I will have left this a better place': environmental volunteering – motivations, barriers and benefits. Report to the Scottish Forestry Trust and Forestry Commission.

Ostrom, E. (2010) Polycentric systems for coping with collective action and global environmental change. *Global Environmental Change* 20: 550–57.

Parkins, J. R., Dunn, M., Reed, M. G., and Sinclair, A. John. (2016) Forest governance as neoliberal strategy: a comparative case study of the Model Forest Program in Canada. *Journal of Rural Studies* 45: 270–78.

Premrl, T., Udovč, A., Bogataj, N., and Krč, J. (2015) From restitution to revival: a case of commons re-establishment and restitution in Slovenia. *Forest Policy and Economics* 59: 19–26.

Primmer, E., Jokinen, P., Blicharska, M., et al. (2015) Governance of ecosystem services: a framework for empirical analysis. *Ecosystem Services* 16: 158–66.

Ribot J. (2004) *Waiting for Democracy: The Politics of Choice in Natural Resource Decentralisation*, Washington, DC: World Resource Institute.

Rosenau, J., and Czempiel, E. (1992) *Governance Without Government: Order and Change in World Politics*. Cambridge: Cambridge University Press.

Swade, K., Walker, A., Walton M., and Barker, K. (2013) Community management of Local Authority woodland in England: A report to Forest Research. Shared Assets report to Forest Research.

Teder, M., and Kaimre, P. (2017) The participation of stakeholders in the policy processes and their satisfaction with results: a case of Estonian forestry policy. *Forest Policy and Economics* 89: 54–62.

Tidey, P. and Pollard, A. (2010) Characterising community woodlands in England and exploring support needs. Small Woods Association report to Forest Research.

van der Jagt, A., Elands B., Ambrose- Oji, B., Gerőházi, É., and Steen Møller, M. (2016) Participatory governance of urban green spaces: trends and practices in the European Union. *Nordic Journal of Architectural Research* 3: 11–34.

Wagenaar, H., Healey, P., Laino, G., et al. (2015) The transformative potential of civic enterprise. *Planning Theory and Practice* 16: 557–585.

Weber, N. (2017) Participation or involvement? Development of forest strategies on national and sub-national level in Germany. *Forest Policy and Economics* 89: 98–106.

Wilson, S. J. and Cagalanan, D. (2016) Governing restoration: strategies, adaptations and innovations for tomorrow's forest landscapes. *World Development Perspectives* 4: 11–15.

Wydra, Doris, and Pülzl, Helga (2013). Sustainability governance in democracies. *International Journal of Social Ecology and Sustainable Development* 4(1): 86–107. doi:10.4018/jsesd.2013010105

16 Carbon Markets and International Environmental Governance

John Chung-En Liu and Mark H. Cooper

Introduction

Carbon markets are arguably the highest-profile and most common tool within the diverse array of approaches that governments use to regulate greenhouse gas emissions and mitigate climate change. Since the introduction of the first national-level carbon market in 1999, emissions trading schemes have been adopted in more than forty-two countries and twenty-five sub-national jurisdictions (Voß 2010).[1] As of 2017, the annual value of global carbon pricing initiatives was $52 billion, up from $30 billion in 2013. These existing initiatives, however, cover only 15 percent of annual global greenhouse gas emissions. The ratification of the Paris Agreement has seen renewed ambitions for greenhouse gas mitigation policy; eighty-one countries – responsible for 55 percent of annual global emissions – have indicated an intention to use carbon markets or carbon taxes to meet their national contribution to greenhouse gas mitigation (World Bank 2017). As Newell and Paterson (2009, 77) put it, "Climate politics are increasingly conducted by, through and for markets."

The creation of carbon markets is premised on creating financial costs for emitting greenhouse gases – putting a price on carbon. The basic description offered by economists and public policy researchers for how carbon markets function holds that the state sets a limit on the volume of greenhouse gases that can be emitted in a particular year. This is the "cap" in well-known "cap and trade" systems. The state then allocates permits (either free or by auction) to responsible parties, such as power stations, factories, and automotive fuel suppliers. These parties buy and sell permits (i.e. the "trade") depending on the volume of their emissions. The price of emissions permits – and therefore the financial cost of producing emissions – results from the supply and demand of permits in the carbon market. Individuals and firms are expected to respond to this price signal by changing their behaviors or adopting technologies that produce less emissions. Proponents of carbon markets frequently

---

1 Within this chapter we use the terms "carbon market" and "emissions trading scheme" interchangeably. Carbon markets are a subset of emissions trading schemes and refer to programs that regulate carbon dioxide (CO_2) and other greenhouse gases. Emissions trading schemes also exist for sulphur dioxide (SO_2) and nitrous oxides (NO_x). In climate change discussions "carbon" is commonly used as a synecdoche for carbon dioxide or greenhouse gases.

claim that, compared to other forms of regulation, they deliver emissions reductions at the least cost (Lane 2012). While carbon markets rely on the market mechanism to coordinate the buying and selling of emissions permits, their design, implementation, and operation, are thoroughly products of the state. In addition to conventional cap and trade systems, there are offset schemes that generate carbon credits from hypothetical baseline scenarios (i.e. "baseline and credit" schemes). These schemes are either sanctioned by various levels of governments (e.g. the Clean Development Mechanism) or sometimes provided entirely by private companies.

The academic literature on carbon markets has traditionally been dominated by economists. This is likely due, in part, to the relative closeness between economists and policymakers, and economists' engagement with experimental policy design. Political scientists, through their attention to public policy and the state, and geographers, through their attention to human–environment interactions, have also developed a substantial body of research on carbon markets. Compared to these disciplines, sociologists have been slow to engage this topic. However, as we describe in this chapter, the influence of sociologists and sociological perspectives in debates about carbon markets has grown substantially in recent years. More importantly, the potential contributions of sociological research to debates and development of carbon markets are numerous and sizeable, as is the potential for research on carbon markets to contribute to problems central to environmental sociology, economic sociology, political sociology, and sociology of science.

In this chapter, we review how sociological research has helped advance understanding of carbon markets as a new type of socio-economic institution and new form of environmental governance. We also reflect on how studying carbon markets can contribute to concepts and theories central to environmental sociology. It is not possible, given the remit of this chapter, to review the breadth and diversity of research on carbon markets by researchers in other disciplines such as political science, geography, and economics. Similarly, it is impossible to fully separate sociology's engagement with carbon markets from the questions, debates, and conclusions originating in other disciplines.

Our aim here will be to discuss the key concepts and insights sociologists have developed in the study of carbon markets, situate these within the broader transdisciplinary landscape of related research, and suggest further opportunities to use sociology's methodological, analytical, and theoretical toolkit in the study of carbon markets, climate change mitigation, and international environmental governance. We structure the chapter around a few fundamental questions. First, we address the questions "Where do carbon markets come from?" and "How are carbon markets made?" We review research on the role of policy networks, various stakeholders, the states, the discourses, and various metrics, standards, and expertise in producing this new type of socio-economic institution. We then explore the issue "What do carbon markets do?" and discuss how researchers have characterized and critiqued the effectiveness, socio-economic outcomes, and environmental justice dimensions of market-based climate policies. Finally, we conclude by considering the question "How does understanding carbon markets contribute to environmental sociology?"

Where Do Carbon Markets Come from and How Are They Made?

Many sociologists contend that carbon markets, along with other market-based environmental policies, gained popularity because they fit the dominant framework of global governance that dominated the 1990s and 2000s. Spaargaren and Mol (2013, 117) note that "when climate policies took off in the post-Rio 1990s ... the policies suggested fitted the neoliberal and global framework of the capitalist world order. The flexible, global and network-like character of carbon markets, and their promise of bringing 'climate' and 'capitalism' together in ways that benefit both, explains why globally operating carbon markets became the dominant approach in global climate governance in the new millennium." The rise of carbon markets, however, has often been contingent upon particular political and institutional circumstances. For example, the European Union was initially averse to carbon trading but later dropped its opposition; some have claimed that this occurred to appease the United States (Bailey and Wilson 2009, 2335). Within the EU, the creation of a carbon market rather than a carbon tax was aided by an "idiosyncratic" feature of the EU's institutional procedure: tax measures require unanimity from all countries, but an emissions trading scheme was defined as an environmental measure which could be established through "qualified majority voting" (MacKenzie 2009a, 155). While carbon markets are clearly one aspect of the "carbon economy" (Brown and Corbera 2003) or "climate capitalism," (Newell and Paterson 2010) there is a range of theoretical approaches for understanding the concrete and ideological origins of carbon markets.

Ecological modernization theory seeks to describe how environmental problems are engaged as politically, economically, and technologically solvable through reform and management that occurs within existing institutions, power structures, and economic systems. Spaargaren and Mol (2013, 190–191) have argued that "carbon markets can be regarded as exemplary new, ecomodernist institutions" and are "potentially radical instruments for a further eco-modernisation of production and consumption." The key elements in the spread of carbon markets are those similar to other modernist reforms, such as the formation of discourse coalitions, action by pioneer countries, policy learning, and adaptive governance (Bailey et al. 2011). From the perspective of ecological modernization, the use of market-based instruments to reform high-carbon societies is justified not only by the expectation of achieving emissions reductions, but also because "there is no short-term alternative available" (Spaargaren and Mol 2013, 177, 186).

In stark contrast to ecological modernization, scholars of critical or Marxian political economy argue that states opted for carbon markets over other alternatives not merely because market-based policy is compatible with the dominant logics of capitalism, but because carbon markets promise new opportunities for capital accumulation that can benefit powerful political-economic interests such as fossil fuel producers, manufacturers, and the financial sector (Bumpus and Liverman 2008; Paterson and P-Laberge 2018). Many scholars within the critical political economy approach argue that the creation of carbon markets begins with commodification of carbon into emissions permits, which introduces a new circuit through which the capitalization of nature can occur (Smith 2006; Paterson 2010; Böhm et al. 2012; Bryant 2018). There exist both optimistic and pessimistic accounts within this approach. Optimistic accounts (Meckling 2011; Newell and Paterson 2010) recognize that the popularity of carbon markets depends in part on their potential to open up new domains of accumulation, but argue that they nonetheless represent a potential route toward a new form of capitalism based on low-carbon development. Pessimistic accounts (Böhm and Dabhi 2009; Bond 2011; Lohmann 2005; 2006; 2012; Spash 2010), however, argue that carbon markets function only as a delaying tactic against real regulation or as a means to transfer wealth, and that the mitigation offered by carbon markets is insignificant or illusory. Many scholars within critical political economy would agree with ecological modernization scholars' claims that carbon markets fit within the dominant capitalist, or neoliberal, global governance framework. However, critical political economy scholars argue that the reason carbon markets have become popular is because they enable powerful political-economic interests to capture rents and exert influence over how carbon is regulated (Paterson and P-Laberge 2018).

Sociologists have sought to explain why particular policy trajectories that favor carbon markets have emerged at global, national, and subnational scales. A number of studies have analyzed how corporate elites mobilize resources and form networks in order to shape environmental policy (Downey and Strife 2010; Bonds 2011; Bonds 2016a). The processes of corporate mobilization and influence are evident in climate politics on issues such as the "organization of denial" (Jacques et al. 2008), "merchants of doubt" (Oreskes and Conway 2010) and the "climate opportunism" of extending territorial control in the Arctic (Bonds 2016b), and researchers have documented the critical role that organizations such as think tanks (Dunlap and Jacques 2013; Bonds 2016b) and private foundations (Brulle 2014) play in the development – or obstruction – of climate change mitigation policy. Regarding the emergence and spread of carbon markets, Meckling (2011) described how a transnational business coalition – consisting primarily of energy firms and energy-intensive manufacturers – gathered crucial support from governments and business-friendly environmental groups to facilitate the rise of carbon trading. Using a social network approach, Sapinski (2015; 2019) traced the contours of business support of the ecomodernist, pro-carbon market agenda through corporate-funded climate and environmental policy groups and found that these corporate policy groups maintain substantial ties to key international organizations, often mediating relationships between these organizations. These networks have contributed to the dominance of "climate capitalism" within climate politics, but possess a "thin architecture" which suggests that further movement toward ecological modernization of greenhouse gas emissions remains uncertain (Sapinski 2016). Other studies have identified differences within corporate engagement with carbon markets. For example, Pulver

(2007) noted that the world's three largest oil companies – ExxonMobil, BP, and Shell – adopted divergent responses to carbon regulations despite broadly similar operational structures. Such studies demonstrate the importance of attention to both mid-range and macro-level processes in the development of environmental governance.

Other research has examined how broader networks of individuals and organizations have promoted carbon markets as the preferred approach to climate policy (Braun 2009; Betsill and Hoffman 2011; Jordan and Huitema 2014). For example, Paterson et al. (2014) employed a social network approach to show that the diffusion of carbon markets as a policy instrument cannot be explained by the narrative of U.S. coercion, or by learning and emulation, but is rather a case of a "polycentric" diffusion based on networks of corporate, NGO, and academic actors. Indeed, Lohmann, a pointed critic of carbon markets, argued that "pollution trading itself is no corporate conspiracy but rather a joint invention of civil society, business and the state. Non-governmental organisations (NGOs) have been nearly as prominent in its development as private corporations" (2006, 58). In contrast to research that frames the rise of carbon markets as one of policy choice or policy diffusion, Voß and Simons (2014; Simons and Voß 2018) analyzed carbon markets as an example of how "instrument constituencies" arise which both sustain, and are sustained by, a particular policy instrument. From this perspective, policymakers do not first define a problem and then identify the most appropriate policy solution to meet their demand; instead, actors within an instrument constituency frame the problem in a way that suits their preferred instrument and are then ready to supply a solution – along with the economic models, legal advice, and consultants to manage policy development. In a similar vein, Knox-Hayes (2010) noted that climate legislation depends strongly on prior policy efforts, and carbon markets need to be "embedded" in social relations, as well as in common beliefs and practices, to be fully operationalized (Knox-Hayes 2012). On the whole, this literature demonstrates that the rise of carbon markets did not occur in an intellectual vacuum, guided only by economic theory. It is the "social lives" of policy alternatives that determine which options take hold and which fall out of favor.

Beyond corporate interests and policy coalitions, scholars have also given attention to the central role of state in the making of carbon markets. Lederer (2012) highlighted the role of the state in creating markets and noted that much of the literature on carbon markets underestimates the importance of their institutional underpinnings. Based on analyses of four different forms of carbon market – the EU ETS, the CDM, the voluntary market, and REDD+ – Lederer clarifies that carbon markets, while injecting a market mechanism, are still a form of governmental regulation. In the short history of carbon markets, states and international agreements have been essential for lending legitimacy to their establishment; attempts to create "private" regulations and markets free of state involvement have been limited in scope, fallen short of even modest expectations, or failed entirely. Ironically, while carbon markets are often criticized as being too friendly to private capital, climate change deniers often dwell on the role of state – to the extent that it forms "regulatory cartel" – to voice their opposition to these markets (Bohr 2016). The statist regulatory origin of carbon markets is even more salient in newer markets created in the global South. As Lederer (2014) noted, carbon markets in the South are often part of a state-led industrial policy program and, compared to Northern carbon markets, they are much less reliant on business and civil society for their creation. This claim is further supported by case studies in South Korea (Yun et al. 2014) and China (Lo 2015), in which government entities are clearly the principal actors shaping these new markets. The trajectory of these new carbon markets resembles developmental state processes more than neoliberalization.

Whether fueled by the corporates or the state, the expertise to account for carbon is absolutely critical, yet often overlooked, in the making of carbon markets. As Engels (2009) identified, many companies have limited capacity to monitor their emissions and engage in permit trading, much less incorporate carbon mitigation into their firm's broader strategies. The specific forms of expertise necessary for market participants to become "rational" cannot be taken for granted. Lovell and MacKenzie (2009) examined the role of professional accounting organizations in governing climate change and characterized how accountants moved from a first stage of "reluctant engagement" around the turn of the century to a stage of "strategic engagement" after 2005. The result of this process has been a narrow framing of the issue which sees climate change as "corporate problem" to be solved by careful application of existing accounting approaches and techniques rather than a reconsideration of firms' relationship to carbon emissions. Other research within this literature has shown that there are multiple modes of carbon accounting expertise needed to "activate" the market (Ascui and Lovell 2011; 2012). Expertise is also central to Foucauldian studies of carbon markets, as they focus on the processes used to make carbon "legible" and make state agencies, firms, and individuals "carbon capable." Research in this vein highlights "green governmentality" discourses that turn stocks and flows of carbon into objects of governance (Backstrand and Lovbrand 2006; Lovbrand and Stripple 2011) and directs attention to how greenhouse gases, and the processes that generate them, are "measured, quantified, demarcated and statistically aggregated" into administrative domains.

Researchers using the performativity approach from economic sociology have made a significant contribution to research on carbon markets. This approach sees the economy as a performative process, meaning that economics not only describes how the economy works, but actively shapes economic outcomes (MacKenzie 2009a). In this view, a market is a social-technical *agencement* – a collective of human beings, technical devices, algorithms, knowledge, and so on – which together engender the capacity to act and to give meaning to action. Callon, a leading theorist of this approach, described carbon markets as an active experiment. He especially emphasized the market's evolving character: "[at] the heart of markets we find debates, issues, feelings, matters of concern, dissatisfaction, regrets, and plans to alter existing rules, which cannot be internalized once and for all because they are linked to irreducible uncertainties, to what I have called framings which are never either definitive or unquestionable" (Callon 2009: 541). Note that ecological modernization theorists also express a similar view: "current carbon markets must be regarded as institutions-in-the-making, in need of fine tuning, correction and repair."

(Spaargaren and Mol 2013, 191) The performativity approach often focuses on the process of framing, or drawing boundaries between things that enter into consideration, and things that are ignored in the calculations within market exchange. Empirical research inspired by this framework includes MacKenzie's (2009b) study on how different greenhouse gases are "made the same" in emissions inventories and accounting records. MacKenzie urged sociologists to focus on the nuts and bolts of carbon market design, as the technical details often determine a policy's outcomes. Researchers have also employed perspectives informed by actor–network theory's attention to the role of non-humans in social processes. In particular, this approach directs attention to the material properties of greenhouse gases, the processes that produce them, and the relevance of carbon's materiality in the function of carbon markets (Cooper 2015). Some researchers using this approach have argued that materiality is often a relatively insignificant issue (Lansing 2012), while others have contended that carbon's divergent materialities have distinct effects on market outcomes (Bumpus 2011; Liu 2017).

Finally, there is a distinctively sociological literature examines carbon markets through the lens of moral meanings. This agenda can be traced back to Viviana Zelizer's cultural approach to economic sociology, as well as sociologists such as Fourcade and Healy (2007), who examine connections between morality and the economy. Descheneau (2012) argued that carbon credits can been seen as a form of currency that works across carbon's "social meanings" and the differentiated forms of carbon – akin to how Zelizer treated money (Zelizer 1997). This line of research focuses on how social actors interpret and generate meanings in the carbon market. In this view, carbon markets are made possible because participants view "carbons" as desirable because of their perceived moral virtues (Goodman and Boyd 2011; Descheneau and Paterson 2011; Paterson and Stripple 2012). Jackson and his colleagues (2017) provided an example based on Australian carbon farming projects, where carbon mitigation is conceptualized with "reciprocal relationships and logic of care." Conversely, there have been many vocal critics of carbon markets who based their argument on the moral meanings of carbon. Harvard philosopher Michael Sandel (1997) argued that "turning pollution into a commodity to be bought and sold removes the moral stigma that is properly associated with it," and undermines a sense of shared responsibility. Therefore, for Sandel, carbon markets are morally reprehensible. Other scholars have critiqued carbon markets, especially offsets, as being "indulgences" for privileged populations to assuage their guilt and continue overconsumption (Smith et al. 2007).

What Do Carbon Markets Do?

The first criteria employed when analyzing "what carbon markets do" is often the extent to which they are effective in bringing about greenhouse gas emissions mitigation. A number of studies have cast doubt on the effectiveness of carbon markets to date, noting that carbon prices in regulatory schemes such as the EU ETS have remained too low to incentivize significant change, and that offset schemes have included too many projects with ambiguous mitigation outcomes and perverse incentive structures (Wara 2007; Gilbertson and Reyes 2009; Stephan and Paterson 2012; Bryant 2016). These issues represent a failure to establish the conditions that economic theory claims are necessary to achieve effective mitigation through carbon markets. Research examining firms' experience within carbon markets has also found a range of behaviors that conflict with the ideal of market efficiency. For example, Engels (2009) examined the early experiences of firms in the EU ETS, and found substantial differences in how companies in different industries and countries dealt with new carbon market obligations. Many companies were also slow to adjust to the new regulation, with some failing to establish systems for knowing their own carbon market fundamentals. Similarly, Hultman et al. (2012) examined investment in select CDM projects and found that while anticipated revenue was central to most investment decisions, projects that did not require carbon revenue to be viable were favored. Other studies have raised fundamental questions about assumptions of economic theory and the design of carbon markets, such as whether actors will change their behavior in keeping with price signals, or whether other values may prevail (Lo 2016; Cooper and Rosin 2014). These studies highlight the importance of understanding the motivations, opportunities, and barriers for firms' engagement with carbon markets. At a more fundamental level, research on the effectiveness of carbon markets in different contexts can be used to examine key points of difference in ongoing debates about carbon markets: Can the underperformance of carbon markets be overcome through policy reform? Is the underperformance of carbon markets due to deep-rooted problems within climate governance? Or is the underperformance of carbon markets due to faults endemic within the function of market-based governance?

Ecological modernization theory holds that carbon markets can contribute to the transition to a zero-carbon economy by encouraging efficiency and technological change. The underperformance of carbon markets is therefore seen primarily as a problem of reforming institutions and governance, and developing deeper individual and civil society commitments to greenhouse gas mitigation. Spaargaren and Mol (2013, 190) admit that "carbon markets as they developed in the first decade of this century are (still) economically unstable, do not live up to environmental promises and bring with them social risks and inequalities." Likewise, the approach welcomes some forms of critique because they provide "starting points for a reflection on how to improve carbon markets" (Spaargaren and Mol 2013, 190). Other critiques of carbon markets, however, are seen as fundamentally incompatible with ecological modernization. Mol (2012, 23) claims that "it is currently less necessary to advocate for the use of financial and market institutions to mitigate climate change, as ecological modernizationists did for two decades. That 'battle' is won: economic institutions are widely used in climate governance, even at the global level. It is now vital to ensure that climate change mitigation rationalities remain the dominant logic in these institutions." This exemplifies a problem that arises in how ecological modernization scholars have approached carbon markets as a regulatory instrument: because ecological modernization theory assumes the merit of market-based approaches to climate policy, it necessarily remains blind to the possibility that the underperformance of carbon markets might arise from their incompatibility with the real-world challenges of global climate governance or with faults endemic to the theory or operation of carbon markets. Ecological modernization therefore provides meager assistance for analyzing one of the essential questions for research on carbon markets: the extent to which the "carbon economy" represents a genuine path toward decarbonizing our economies and societies.

Researchers working within the broad range of political economy approaches often treat the suitability and potential of carbon markets to contribute to greenhouse gas mitigation as a key issue in their analyses. The effectiveness, or potential effectiveness of carbon markets, however, is also commonly situated within discussion of processes, tensions, or contradictions within capitalism and capitalist economies and their influence on climate change and decarbonization. For some critics of carbon markets, "what carbon markets do" is less about climate change – upon which they expect carbon markets will have little effect – and primarily about the extension of capitalism into new domains of nature and society via the commodification of the atmospheric commons and the right to pollute (Lohmann 2010; Bohm et al. 2012) Numerous studies have examined the effectiveness and effects of carbon offset projects and programs. Particularly notable is the examination of CDM offsets by Bumpus and Liverman (2008), which they characterized as an example of "accumulation by dispossession" whereby collective property is transferred to private ownership. Criticisms of carbon offsets frequently noted that they constitute a form of carbon colonialism by enabling polluters in developed countries to benefit financially by paying for emissions reductions in developing countries rather than undertaking emissions reductions themselves. While offsets were one of the more distinctive features of early carbon markets, critical scholars are effectively unanimous that offsets have made little contribution toward reorienting capitalism to a low-carbon pathway, and have likely delayed or undermined efforts for more stringent mitigation. The origin of carbon offsets was premised on the Kyoto Protocol's establishment of binding commitments for emissions reductions by developed countries, but not developing countries. The post-2020 climate framework embodied within the Paris Agreement, however, leaves substantial uncertainty about the future for carbon offsets.

Although the critical political economy literature on carbon markets has been dominated by studies on carbon offsets, a growing number of studies have examined the effectiveness and effects of compliance-based carbon markets such as the EU ETS. Carton (2014) described the EU ETS as having a "lamentable history," but acknowledged that a degree of "learning" was occurring in EU climate policy – albeit at a rate inadequate for achieving necessary progress toward decarbonization. Similarly, researchers have found little support for the idea that the current form of carbon markets offers significant potential for the circulation and accumulation of capital. As Bigger (2018, 525) noted, carbon markets and other tradable permit systems "do not result in institutional forms that facilitate the circulation of capital through nature." It is clear that "climate capitalism" and the alignment of the circulation and accumulation of capital with low-carbon development has not yet arrived. Whether the blame for this situation sits with carbon markets, or with the capitalist political economies within which carbon markets are embedded, remains an issue of debate.

Alongside the critique of transnational carbon offsets based on their reproduction of inequalities between developed and developing countries, the environmental justice critique of carbon markets has argued that carbon markets (and other market-based programs) are in fundamental tension with an environmental justice paradigm. From a distributive justice perspective, carbon markets are formally indifferent to place, but the market mechanism may incentivize the concentration of polluting activities in areas of lower land value which often correlate with minority communities and lower socio-economic status communities. From a participatory justice perspective, carbon markets are likely to increase industry flexibility and reduce the state's role in direct regulation which may run counter to the environmental justice movement's call for increased participatory engagement (Kaswan 2013). Fundamentally, the issue of environmental justice and carbon markets is broadly similar to other political issues around carbon markets: "who gains and who bears the costs of responding to climate change in this way" (Boyd et al. 2011, 609).

Carbon dioxide mixes and distributes quickly in the atmosphere and is not locally damaging to the environment or human health. Carbon dioxide emissions, however, are usually accompanied by reductions in co-pollutants such as particulates, sulphur oxides, ozone precursors as well as toxic pollutants (London et al. 2013). Market-based regulation of carbon dioxide emissions could therefore reinforce or create hotspots for these co-pollutants and their effects on local communities. The question of causality of environmental harm via direct or indirect action is a major concern of environmental justice; appeals to 'market forces' as being responsible for environmental injustice have long been used as a defense against responsibility being placed on any particular party. Such a defense is inviable for carbon markets though: carbon markets are always products of the state and the state therefore invitably bears responsibility for designing programs in a manner consistent with environmental justice principles. Studies have examined how environmental justice organizations were involved in both the formal and discursive aspects of the legislative process (Sze et al. 2009) and the role of place and race and the anticipated distributional effects associated with the development of California's carbon market (London et al. 2013). There remains significant potential for further study of how environmental justice principles might be incorporated into the design and reform of carbon markets and the extent to which environmental justice and carbon markets are fundamentally incompatible in an important topic for further inquiry.

How Does Understanding Carbon Markets Contribute to Environmental Sociology?

The study of carbon markets and the relationship between nature, the state, and the economy offers researchers a number of opportunities for making theoretical advancements in environmental sociology. Carbon markets, and other market-based environmental policy instruments such as species conservation banking, are "state-directed but market-oriented" regulatory institutions (Rea 2017). Environmental sociology is still grappling with how best to understand and explain these new forms of governance. Future studies of carbon markets are likely to further the fundamental debate between ecological modernization and critical or eco-Marxian approaches. The burgeoning number of emerging carbon markets, and the diversity of countries and sub-national territories where markets are developing, means that there is no shortage of excellent cases and processes in need of empirical investigation. Several fundamental questions around carbon markets have received little attention to date from environmental sociologists or researchers in related fields. These include: to what extent are carbon markets making a genuine and substantial contribution to greenhouse gas mitigation, and what are the processes through which such successes occur? Who are the financial beneficiaries of carbon markets, and what is the scale of profit, rent, or capital accumulation that occurs via particular markets? As MacKenzie (2009b) suggested, studies of carbon markets should account for the "nuts and bolts" of these markets as well as the relationship between institutions and policy instruments. Such research requires attending to the variegated nature of the carbon markets and being sensitive to the potential for cross-case comparisons as what may be seemingly technical, apolitical, design features (e.g. free allowances versus auction, cap and trade versus offsets) often lead to divergent outcomes.

Carbon markets also demand sociologists to bring environmental sociology closer with the insights from Polanyian economic sociology. In recent years, environmental sociologists have sought to advance theory through introducing the environment and environmental issues into classical Polanyian approaches. For example, Kaup (2015) argues that, in addition to the embeddedness in political institutions and social fabrics, markets are also embedded in ecology. According to this view, market societies, in the Polanyian sense, both shape and are shaped by nature. Malin (2015), through her studies of resource extraction communities, developed the term "triple movement" to document the "normalization of neoliberal ideologies and policies at the grassroots level, where ... actors support commodification of fictitious commodities, like markets in land and mineral wealth, and even defend those markets as part of their communities' social fabrics." This analysis demonstrates why social actors in the "triple movement" support corporate self-regulation, as corporations are regarded as the main social protection providers alongside the retreat of the state.

Carbon markets, with their entirely "artificial" nature, remind us of Polanyi's argument that market formation often goes hand in hand with state-building processes. Carbon markets are not static institutions, but are constantly reshaped by new regulations and public pressures. Environmental sociologists could benefit from not only focusing on how these markets work in the current forms, but also on how they come into being and how they might evolve in the future. We are likely to see scholars continue to debate whether carbon market represents a "double movement" – an increased social protection from climate change (Spaargaren and Mol 2013; Carton 2014), a further deepening of market ideology (Lohmann 2010; Stuart et al. 2017), or whether the verdict depends on the local contexts (Osborne and Shapiro-Garza 2018). These endeavors will help sociologists better understand how theoretical tools such as the "double movement," "embeddedness," and "fictitious commodity" may apply to the environmental realm more generally.

The study of carbon markets may also help advance the scholarship on environmental justice. As discussed in the previous section, the existing literature has documented various ways that carbon projects may lead to negative consequences in communities in developing countries, often referred a form of "carbon colonialism" (Bachram 2004; Lyons and Westoby 2014; Camrody and Taylor 2016). While progressive social movements often critique carbon markets as environmentally and socially unjust, there is significant room for empirical investigation to be undertaken based on new contexts and comparisons, and directed at carbon markets and the marketization of carbon beyond offsets. While existing critiques of carbon offsets have illuminated the dark side of the new carbon economy, they have often missed the larger and more conventional cap and trade systems. So far, very few, if any sociological research deals with potentially unequal impacts from these conventional, regulatory cap and trade systems. We know very little about who the winners and losers in the existing cap and systems are, or why particular market outcomes have transpired the way they have. Environmental sociologists, with their diverse research methods and theoretical frameworks, are well positioned to make critical contributions to inform the policy debates beyond conventional economistic approaches.

One of the important ways that environmental sociologists might develop future studies of carbon markets is to follow Lederer's (2017) suggestion to "go beyond dichotomous pro/con argumentation about whether carbon markets should be used at all, by focusing instead on actual practices." Reframing studies of carbon markets around "practices" would mean treating carbon markets as more than a theoretical abstraction and as an opportunity to accumulate solid empirical work about a multitude of contexts and market-making and market-sustaining practices. At the macro level, there are opportunities for sociologists to further pursue the question of "where markets come from" by doing fine-grained policy network analysis. There is also potential to connect the literature on carbon markets with various social movement theories to explain their emergence, and their failures and limitations. At the meso level, environmental sociologists can benefit from organizational/institutional research to examine private sector engagement with regulatory carbon markets or market design within the private sector. As the World Bank noted in a recent report, "momentum is also building for carbon pricing in the private sector, where an increasing number of companies are using internal carbon pricing to actively manage climate-related risks" (World Bank Group, State and Trends of Carbon Pricing 2017). There is much more to be learned about how companies learn to calculate and govern carbon. Sociologists are well positioned to use ethnography or social network analysis to develop unique insights that would be difficult to produce in other disciplines. Finally, the existing literature has tended to focus on the commodification processes of carbon – the "production" dimension of the carbon market, but at the micro level, it is also worthwhile to explore the "consumption" dimensions and how social actors make sense (or resist making sense) of new ways of valuing and monetizing carbon. Sociology's disciplinary orientation for understanding social meaning is likely to open up a number of fruitful research programs on carbon markets that have not yet emerged.

Conclusion

The empirical research and theoretical approaches discussed in this chapter have, by necessity, been highly interdisciplinary. Political scientists have been active in explaining the origins and political processes involved in the development of carbon markets; geographers have developed critical analyses that question whether carbon markets really serve their stated purposes. Compared to these fields, sociology has been slow to take up the intellectual challenge of critically examining carbon markets. Recent content analysis of environmental sociology research found that topics on climate change policies have not had a significant presence within the subfield (Bohr and Dunlap 2017) and in top sociology journals (Scott and Johnson 2017). This is, to some degree, a symptom of the general neglect of environmental issues, climate change issues, and policy issues within the broader discipline of sociology. There is no reason that this should remain the norm. For example, the subfield of environmental sociology has led the development of analyses that dissect the climate change denial movement. Perhaps the field could again use more of the "environmental pragmatism" described by Dunlap (2010) to make important contributions to ongoing debates about carbon markets, climate change mitigation, and global environmental governance.

Existing research on carbon markets, like many topics within environmental sociology, is marked by an uneven geographical focus. The majority of the literature on carbon markets has focused on experiences in the European Union, especially on the development of the EU ETS. To date, little research, if any, has examined the development of carbon markets in contexts such California, the Northeastern United States (RGGI), or Alberta. It is telling that only a few of the authors cited in this chapter work in sociology departments in the United States. Similarly, there has been a lack of engagement by sociologists with other emerging carbon markets – China, South Korea, Mexico, etc. – outside of developed countries. These contexts offer exciting intellectual questions and have the potential to challenge our understanding of carbon markets that has been based on Western experiences and paradigms. Research on carbon markets and carbon pricing in these new geographical sites is likely to not only of notable theoretical value, but will – given the growth of greenhouse gas emissions in the developing world – be of significant political importance as well.

References

Ascui, Francisco, and Heather Lovell. "As frames collide: Making sense of carbon accounting." *Accounting, Auditing & Accountability Journal* 24.8 (2011): 978–999.
Ascui, Francisco, and Heather Lovell. "Carbon accounting and the construction of competence." *Journal of Cleaner Production* 36 (2012): 48–59.

Bachram, Heidi. "Climate fraud and carbon colonialism: The new trade in greenhouse gases." *Capitalism Nature Socialism* 15.4 (2004): 5–20.

Bäckstrand, Karin, and Eva Lövbrand. "Planting trees to mitigate climate change: Contested discourses of ecological modernization, green governmentality and civic environmentalism." *Global Environmental Politics* 6.1 (2006): 50–75.

Bigger, Patrick. "Hybridity, possibility: Degrees of marketization in tradable permit systems." *Environment and Planning A* 50.3 (2018): 512–530.

Bailey, Ian, Andy Gouldson, and Peter Newell. "Ecological modernisation and the governance of carbon: A critical analysis." *Antipode* 43.3 (2011): 682–703.

Bailey, Ian, and Geoff A. Wilson. "Theorising transitional pathways in response to climate change: Technocentrism, ecocentrism, and the carbon economy." *Environment and Planning A* 41 (2009): 2324–2341.

Betsill, Michele and Matthew J. Hoffman. "The contours of 'cap and trade': The evolution of emissions trading systems for greenhouse gases." *Review of Policy Research* 28.1 (2011): 83–106.

Böhm, Steffen, and Siddhartha Dabhi (eds.) *Upsetting the Offset: The Political Economy of Carbon Markets.* (2009) London: MayFly Books.

Böhm, Steffen, Maria Ceci Misoczky, and Sandra Moog. "Greening capitalism? A Marxist critique of carbon markets." *Organization Studies* 33.11 (2012): 1617–1638.

Bohr, Jeremiah, and Riley E. Dunlap. "Key topics in environmental sociology, 1990–2014: Results from a computational text analysis." *Environmental Sociology* (2017): 1–15.

Bohr, Jeremiah. "The 'climatism' cartel: Why climate change deniers oppose market-based mitigation policy." *Environmental Politics* 25.5 (2016): 812–830.

Bond P. "Carbon capital's trial, the Kyoto protocol's demise, and openings for climate justice." *Capitalism Nature Socialism* 22 (2011): 3–17

Bonds, Eric. "The knowledge-shaping process: Elite mobilization and environmental policy." *Critical Sociology* 37.4 (2011): 429–446.

Bonds, Eric. "Beyond denialism: Think tank approaches to climate change." *Sociology Compass* 10.4 (2016a): 306–317.

Bonds, Eric. "Losing the Arctic: The US corporate community, the national-security state, and climate change." *Environmental Sociology* 2.1 (2016b): 5–17.

Boyd, Emily, Maxwell Boykoff, and Peter Newell. "The 'new' carbon economy: What's new?" *Antipode* 43.3 (2011): 601–611.

Braun, Marcel. "The evolution of emissions trading in the European Union: The role of policy networks, knowledge and policy entrepreneurs." *Accounting, Organizations and Society* 34.3–4 (2009): 469–487.

Brown, Katrina, and Esteve Corbera. "Exploring equity and sustainable development in the new carbon economy." *Climate Policy* 3S1 (2003): S41–S56.

Brulle, Robert J. "Institutionalizing delay: Foundation funding and the creation of US climate change counter-movement organizations." *Climatic Change* 122.4 (2014): 681–694.

Bryant, Gareth. "The politics of carbon market design: Rethinking the techno-politics and post-politics of climate change." *Antipode* 48.4 (2016): 877–898.

Bryant, Gareth. "Nature as accumulation strategy? Finance, nature, and value in carbon markets." *Annals of the Association of American Geographers* 103.3 (2018): 605–619.

Bumpus, Adam G., and Diana M. Liverman. "Accumulation by decarbonization and the governance of carbon offsets." *Economic Geography* 84.2 (2008): 127–155.

Bumpus, Adam G. "The matter of carbon: Understanding the materiality of tCO2e in carbon offsets." *Antipode* 43.3 (2011): 612–638.

Callon, Michel. "Civilizing markets: Carbon trading between in vitro and in vivo experiments." *Accounting, Organizations and Society* 34.3–4 (2009): 535–548.

Carmody, Padraig, and David Taylor. "Globalization, land grabbing, and the present-day colonial state in Uganda: Ecolonization and its impacts." *The Journal of Environment & Development* 25.1 (2016): 100–126.

Carton, Wim. "Environmental protection as market pathology?: Carbon trading and the dialectics of the 'double movement'." *Environment and Planning D: Society and Space* 32.6 (2014): 1002–1018.

Cooper, Mark H., and Christopher Rosin. "Absolving the sins of emission: The politics of regulating agricultural greenhouse gas emissions in New Zealand." *Journal of Rural Studies* 36 (2014): 391–400.

Cooper, Mark H. "Measure for measure? Commensuration, commodification, and metrology in emissions markets and beyond." *Environment and Planning A* 47.9 (2015): 1787–1804.

Descheneau, Philippe. "The currencies of carbon: Carbon money and its social meaning." *Environmental Politics* 21.4 (2012): 604–620.

Descheneau, Philippe, and Matthew Paterson. "Between desire and routine: Assembling environment and finance in carbon markets." *Antipode* 43.3 (2011): 662–681.

Downey Liam, and Susan Strife. "Inequality, democracy, and the environment." *Organization and Environment* 23.1 (2010): 155–188

Dunlap, Riley E. "The maturation and diversification of environmental sociology: From constructivism and realism to agnosticism and pragmatism." In *The International Handbook of Environmental Sociology* ed. Michael R. Redclift and Graham Woodgate (2010) Edward Elgar Publishing, Cheltenham UK and Northampton, MA, pp. 15–32.

Dunlap, Riley E., and Peter J. Jacques. "Climate change denial books and conservative think tanks: Exploring the connection." *American Behavioral Scientist* 57.6 (2013): 699–731.

Engels, Anita. "The European Emissions Trading Scheme: An exploratory study of how companies learn to account for carbon." *Accounting, Organizations and Society* 34.3–4 (2009): 488–498.

Fourcade, Marion, and Kieran Healy. "Moral views of market society." *Annual Review of Sociology* 33 (2007): 285–311.

Gilbertson, Tamra, and Oscar Reyes. "Carbon trading, how it works and why it fails." *Critical Currents: Dag Hammarskjöld Foundation Occasional Paper Series*. (2009) Uppsala: Dag Hammarskjöld Foundation.

Goodman, Michael K., and Emily Boyd. "A social life for carbon? Commodification, markets and care." *The Geographical Journal* 177.2 (2011): 102–109.

Hultman, Nathan E., Simone Pulver, Leticia Guimarães, Ranjit Deshmukh, and Jennifer Kane. "Carbon market risks and rewards: Firm perceptions of CDM investment decisions in Brazil and India." *Energy Policy* 40 (2012): 90–102.

Jackson, Sue, Lisa Palmer, Fergus McDonald, and Adam Bumpus. "Cultures of carbon and the logic of care: The possibilities for carbon enrichment and its cultural signature." *Annals of the American Association of Geographers* 107.4 (2017): 867–882.

Jacques, Peter J., Riley E. Dunlap, and Mark Freeman. "The organisation of denial: Conservative think tanks and environmental scepticism." *Environmental Politics* 27.3 (2008): 349–385.

Jordan, Andrew, and Dave Huitema. "Policy innovation in a changing climate: Sources, patterns and effects." *Global Environmental Change* 29 (2014): 387–394.

Kaup, Brent Z. "Markets, nature, and society: Embedding economic & environmental sociology." *Sociological Theory* 33.3 (2015): 280–296.

Kaswan, Alice. "Environmental justice and environmental law." *Fordam Environmental Law Review* 24.2 (2013): 149–179.

Knox-Hayes, Janelle. "Negotiating climate legislation: Policy path dependence and coalition stabilization." *Regulation & Governance* 6.4 (2012): 545–567.

Knox-Hayes, Janelle. "Creating the carbon market institution: Analysis of the organizations and relationships that build the market." *Competition & Change* 14.3–4 (2010): 176–202.

Lane, Richard. "The promiscuous history of market efficiency: The development of early emissions trading systems." *Climate Policy* 21.4 (2012): 583–603.

Lansing, David M. "Performing carbon's materiality: The production of carbon offsets and the framing of exchange." *Environment and Planning A* 44.1 (2012): 204–220.

Lederer, Markus. "Market making via regulation: The role of the state in carbon markets." *Regulation & Governance* 6.4 (2012): 524–544.

Lederer, Markus. "The politics of carbon markets in the global south." *The Politics of Carbon Markets* (2014): 133–149.

Lederer, Markus. "Carbon trading: Who gets what, when, and how?" *Global Environmental Politics* 17.3 (2017): 134–140.

Liu, John Chung-En. "Pacifying uncooperative carbon: Examining the materiality of the carbon market." *Economy and Society* 46.3–4 (2017): 522–544.

Lo, Alex Y. "National development and carbon trading: The symbolism of Chinese climate capitalism." *Eurasian Geography and Economics* 56.2 (2015): 111–126.

Lo, Alex Y. "Challenges to the development of carbon markets in China." *Climate Policy* 16.1 (2016): 109–124.

Lohmann, Larry. "Marketing and making carbon dumps: Commodification, calculation and counter-factuals in climate change mitigation." *Science as Culture* 14.3 (2005): 203–235.

Lohmann, Larry. "Carbon trading: A critical conversation on climate change, privatisation and power." *Development Dialogue* 48 (2006).

Lohmann, Larry. "Uncertainty markets and carbon markets: Variations on Polanyian themes." *New Political Economy* 15.2 (2010): 225–254.

Lohmann Larry. "A rejoinder to Matthew Paterson and Peter Newell." *Development and Change* 43 (2012): 1177–1184.

London, Jonathan, Alex Karner, Julie Sze, Dana Rowan, Gararardo Gamiazzio, and Deb Niemeier. "Racing climate change: Collaboration and conflict in California's global climate change policy arena." *Global Environmental Change* 23 (2013): 791–799.

Lövbrand, Eva, and Johannes Stripple. "Making climate change governable: Accounting for carbon as sinks, credits and personal budgets." *Critical Policy Studies* 5.2 (2011): 187–200.

Lovell, Heather, and Donald MacKenzie. "Accounting for carbon: The role of accounting professional organisations in governing climate change." *Antipode* 43.3 (2011): 704–730.

Lovell, Heather, and Diana Liverman. "Understanding carbon offset technologies." *New Political Economy* 15.2 (2010): 255–273.

Lyons, Kristen, and Peter Westoby. "Carbon colonialism and the new land grab: Plantation forestry in Uganda and its livelihood impacts." *Journal of Rural Studies* 36 (2014): 13–21.

MacKenzie, Donald. "Making things the same: Gases, emission rights and the politics of carbon markets." *Accounting, Organizations and Society* 34.3–4 (2009b): 440–455.

MacKenzie, Donald. *Material Markets: How Economic Agents are Constructed.* (2009a) Oxford: Oxford University Press.

Malin, Stephanie A. *The Price of Nuclear Power: Uranium Communities and Environmental Justice.* (2015) New Brunswick, NJ: Rutgers University Press.

Meckling, Jonas. *Carbon Coalitions: Business, Climate Politics, and the Rise of Emissions Trading.* (2011) Cambridge, MA: MIT Press.

Mol, Arthur P. J. "Carbon flows, financial markets and climate change mitigation." *Environmental Development* 1 (2012): 10–24

Newell, Peter and Matthew Paterson. "The politics of the carbon economy." In *The Politics of Climate Change: A Survey* ed. Maxwell T. Boykoff. (2009) Routledge: London. pp.77–95.

Newell, Peter, and Matthew Paterson. *Climate Capitalism: Global Warming and the Transformation of the Global Economy.* (2010) Cambridge: Cambridge University Press,

Oreskes, Naomi and Erik M. Conway. *Merchants of Doubt: How a Handful of Scientists Obscured the Truth on Issues from Tobacco Smoke to Global Warming.* (2010) New York: Bloomsbury.

Osborne, Tracey, and Elizabeth Shapiro-Garza. "Embedding carbon markets: Complicating commodification of ecosystem services in Mexico's forests." *Annals of the American Association of Geographers* 108.1 (2018): 88–105.

Paterson, Matthew. "Legitimation and accumulation in climate change governance." *New Political Economy* 15.3 (2010): 345–368.

Paterson, Matthew and Xavier P-Laberge. "Political economies of climate change." *Wiley Interdisciplinary Reviews: Climate Change* 9.2 (2018): e506.

Paterson, Matthew, and Johannes Stripple. "Virtuous carbon." *Environmental Politics* 21.4 (2012): 563–582.

Paterson, Matthew, Matthew Hoffmann, Michele Betsill, and Steven Bernstein. "The micro foundations of policy diffusion toward complex global governance: An analysis of the transnational carbon emission trading network." *Comparative Political Studies* 47.3 (2014): 420–449.

Pulver, Simone. "Making sense of corporate environmentalism: An environmental contestation approach to analyzing the causes and consequences of the climate change policy split in the oil industry." *Organization & Environment* 20.1 (2007): 44–83.

Rea, Christopher M. "Theorizing command-and-commodify regulation: The case of species conservation banking in the United States." *Theory and Society* 46.1 (2017): 21–56.

Sandel, Michael J. "It's immoral to buy the right to pollute." *New York Times* December 17, (1997): A29.

Sapinski, Jean Philippe. "Climate capitalism and the global corporate elite network." *Environmental Sociology* 1.4 (2015): 268–279.

Sapinski, Jean Philippe. "Constructing climate capitalism: Corporate power and the global climate policy-planning network." *Global Networks* 16.1 (2016): 89–111.

Sapinski, J. P. "Corporate climate policy-planning in the global polity: A network analysis." *Critical Sociology* 45.4-5 (2019): 565–582.

Scott, Lauren N., and Erik W. Johnson. "From fringe to core? The integration of environmental sociology." *Environmental Sociology* 3.1 (2017): 17–29.

Simons, Arno, and Jan-Peter Voß. "The concept of instrument constituencies: Accounting for dynamics and practices of knowing governance." *Policy and Society* 37.1 (2018): 1–22.

Smith, Neil. "Nature as accumulation strategy." *Socialist Register* 43 (2006): 19–41.

Smith, Kevin, Oscar Reyes, and Timothy Byakola. *The Carbon Neutral Myth: Offset Indulgences for Your Climate Sins*. (2007) Amsterdam: Carbon Trade Watch.

Spaargaren, Gert and Arthur P. J. Mol. "Carbon flows, carbon markets, and low-carbon lifestyles: Reflecting on the role of markets in climate governance." *Environmental Politics* 22.1 (2013): 174–193.

Spash, Clive L. "The brave new world of carbon trading." *New Political Economy* 15.2 (2010): 169–195.

Stephan, Benjamin and Matthew Paterson. "The politics of carbon markets: An introduction." *Environmental Politics* 21.4 (2012): 545–562.

Stuart, Diana, Ryan Gunderson, and Brian Petersen. "Climate Change and the Polanyian counter-movement: Carbon markets or degrowth?" *New Political Economy* 24.1 (2017): 1–14.

Sze, Julie, Gerardo Gambirazzio, Alex Karner, Dana Rowan, Jonathan London, Deb Niemeier. "Best in show? Climate and environmental justice policy in California?" *Environmental Justice* 2.4 (2009): 179–184.

Voß, Jan-Peter, and Arno Simons. "Instrument constituencies and the supply side of policy innovation: The social life of emissions trading." *Environmental Politics* 23.5 (2014): 735–754.

Voß, Jan Peter. "Innovation of governance: The case of emissions trading." In *Governance of Innovation: Firms, Clusters and Institutions in a Changing Setting*. Maarten J. Arentsen, Wouter van Rossum, and Albert E. Stenge (2010) Cheltenham: Edward Elgar, pp. 125–148.

Wara, M. "Is the global carbon market working?" *Nature* 445. 8 February (2007): 595–596.

World Bank. *State and Trends of Carbon Pricing 2017* (2017) Washington, DC: World Bank Group.

Yun, Sun-Jin, Dowan Ku, and Jin-Yi Han. "Climate policy networks in South Korea: Alliances and conflicts." *Climate policy* 14.2 (2014): 283–301.

Zelizer, Viviana A. *The Social Meaning of Money*. (1997). Princeton, NJ: Princeton University Press.

17 The Multi-Level Governance Challenge of Climate Change in Brazil

Leila da Costa Ferreira

Introduction

Climate change risks can be interpreted as contemporary environmental risks, since they result from the development process of industrial society and are a direct consequence of the globalization process. They have a strict connection to the future and uncertainties. Knowing precisely what the world will be like in the next few decades is impossible. Nevertheless, it is possible to calculate probabilities and scenarios. These risks are hard to predict, to avoid, to calculate, to reverse and to see, being both local and global, and threaten all humanity. Their causes and consequences are all over the world, surpassing geographical limits (Beck, 1992; 1995; 2000; 2009; 2010; Giddens, 1990; 2000; 2009; Basso and Viola, 2014; Dunlap and Brulle, 2015; Ferreira, 2017).

Governments are important stakeholders in the process of responding to climate change risks, since they are relevant agents for the outlining of appropriate rules, institutions and modes of governance in order to deal with these risks at different levels and scales (Giddens, 2000; 2009; Bulkeley and Kern, 2006; Bulkeley and Newell, 2010; Ferreira et al, 2012; Ferreira and Barbi, 2016).

Without undermining the role of climate policy at a national level, it is important to highlight the role of subnational governments (state and municipal), as they represent important forums and enablers of global climate governance. Thus, since this issue permeates different interconnected levels of government, we should consider climate change as a multilevel challenge (Ferreira and Barbi, 2016).

In addition to the roles played by the government in its different levels, it is necessary to emphasize the need to involve civil society, and all the different types of organizations, enabling the development of so-called multi-stakeholder governance, in which the different social spheres contribute to soften the problems linked to climate change (Newell, Patterberg and Schroeder, 2012).

Most of the literature on climate policy is occupied by research on global and regional levels of governance, focusing on norms, rules and decision processes regarding the international climate regime (Betsill and Bulkeley, 2007; Okereke, Bulkeley, and Schroeder, 2009).

Despite these necessary contributions, taking account of the regional and local dimension of the theme is also relevant, since most human activities that contribute to global climate changes take place at these levels and, at the same time, these levels are the most affected by the impacts of these changes (Storbjörk, 2007; Dodman, 2009; Satterthwaite, 2010; Hoornweg et al, 2011; Ferreira et al, 2012; Barbi and Ferreira, 2013; Ferreira, 2018).

The nation-states no longer monopolize the formulation of policies, which also happens because of the need for collective decision-making on complex problems and because subnational governments are interconnected by policy networks. Thus, the multilevel governance perspective becomes an alternative approach to the analysis of the role of subnational governments in the development of climate policy (Bulkeley and Betsill, 2003; Gupta, 2007).

The reduction of greenhouse gases (GHG) emissions – i.e. mitigation and/or adaptation to impacts – can minimize climate change risks. But actions are necessary as well, and should be complementary (IPCC, 2007; Renn and Klinke, 2012; Barbi, Ferreira and Guo, 2016).

Biofuels, renewable energy, energy efficiency, low carbon agriculture, carbon market, changes in consumption patterns and waste reduction help to minimize and stabilize GHG emissions. Hence, mitigation actions include all human activities intended to reduce GHG emissions or increase GHG sinks (IPCC, 2007; Klein et al, 2005; IPCC, 2001). The main obstacles to the implementation of these policies include the emission reduction costs and political will (Winkler et al, 2007). Other influences on mitigation policies include expert–non-expert relations, regulator–industry relations, risk perception, power and influence of interest groups and historical culture (Burch and Robinson, 2007; Dunlap and Brulle, 2015).

Literature on this issue (Bulkeley and Kern, 2006; Betsill and Bulkeley, 2007; IPCC, 2007; Winkler et al, 2007; Burch and Robinson, 2007; Bizikova et al, 2010; UN-HABITAT, 2011) identifies five key sectors that concentrate the responses to climate change mitigation at the subnational level: urban development (territory planning strategies), built environment, urban infrastructure (power systems, water, sanitation and solid waste), transport and carbon sequestration (conservation, reforestation).

Considering climate change risks, the key sectors for adaptation responses at the subnational level are: urban development (land use management); built environment; urban infrastructure and services; environment; health and disaster management. This shows the diversity of possible and necessary actions at subnational level and also the variety of government sectors involved in these two aspects of climate policy.

Addressing such a multifaceted challenge as climate change, the expectation is of multifaceted solutions, including diverse fields of human activity and several stakeholders and sectors of society: for instance, multilateral agencies, governments, private sector, research institutes and organized civil society groups aiming to elucidate the factors that generate the climate change risks and to come upon the conditions for their confrontation (Bulkeley and Newell, 2010; Ferreira and Barbi, 2016).

Brazil is an important player when it comes to climate change governance because of its significant GHG emissions (Ferreira, 2017). Brazilian GHG emissions accounted for 5 percent of global emissions in 2008, when the country was the fifth largest world emitter (OC, 2014; Ferreira, 2017). At the same time, national, state and municipal GHG emissions are different within the country and come from different sectors of activity.

Therefore, mitigation policies should consider these differences across scales. Moreover, climate change impacts are felt differently depending on the scale and locality and involve different sectors of activity depending on the level of governance.

Methodological Aspects

Methodological aspects include three main points of analysis: (i) the trajectory of GHG emissions in Brazil, based on secondary data; (ii) political and institutional structures mobilized around the climate issue, focusing on mitigation (based on a review of the literature); (iii) policy responses to climate change, based on the analysis of policies, plans, programs and major projects related to climate change mitigation through official government documents and reports.

The global changes in climate, environment, economy, societies, governments, institutions and cultures converge in different locations. The effects at local level, in turn, contribute to the global changes and are affected by them. As a result, the connection of the local and global scales, across a broad range of disciplines and issues – integrated assessments of population, economy, technology, and environmental changes – allows a potentially deeper understanding of global environmental change, including climate change in all its complexity (Betsill and Bulkeley 2007; Okereke et al, 2009). In the age of risk, Nation State frontiers are remodeled by the Anthropocene logic, meaning that the frontiers would be dynamic as the future scenarios point to changes in territories shared by different countries.

Nevertheless, the local governments could, for example, develop policies to deal with about 30–50 percent of the national GHG. Local governments are usually responsible for transport, local development planning and energy management. Since a large number of people live and work within the limits of the cities and are under the jurisdiction of local governments, these governments are in a privileged position to influence a diverse range of activities responsible for the emission of large amounts of GHG.

Thus, the quality of local governments has an important effect on the production of responses to the risks of climate change. Currently, many local governments of emerging countries exhibit low institutional and response capability to deal with the various problems that affect the quality of life of these populations, especially the poorest ones, who feature lower coping capacity as well as weak links in their networks of social protection. However, climate change does not hit only poor populations; its impacts affect different sectors of society, the public realm and also the market.

Climate Issues in Brazil

The evolution of Brazilian GHG emissions is here divided into three phases, compared to the dynamics of global emissions. First, between 1990 and 1997,

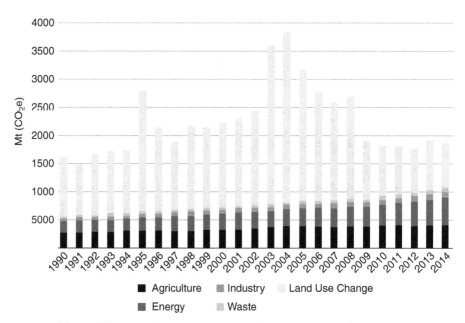

Figure 17.1 *Brazilian GHG emissions, by activity sector (1990–2014)*
Source: Observatório Clima

Brazilian emissions grew at a faster pace than global emissions. Second, between 1998 and 2004, emissions grew at a similar pace to global emissions. Finally, after 2005, Brazilian emissions showed a sharp decline as global emissions continued to grow. Nevertheless, in 2008, Brazilian GHG emissions accounted for 5 percent of global emissions, when the country was the fifth-largest world emitter (UN-HABITAT, 2011; OC, 2014).

For most of the last twenty years land use change and the forestry sector were largely responsible for Brazilian GHG emissions (see Figure 17.1). In 2005, this sector accounted for 58 percent of national emissions. However, in 2005–2007 and then 2008–2012, emissions from this sector decreased significantly, reaching 15 percent of total emissions in 2012. This decrease in emissions was mainly due to the reduction of deforestation in the Amazon since 2004. Nevertheless, emissions from other sectors of activity increased in this period. Energy and agriculture were the ones that had the greatest increase in emissions from 2005 to 2012.

The energy sector stands out due to the rapid growth and high levels of emissions from 1970 to 2013 in comparison to other sectors. During this period, GHG levels in that sector quadrupled, representing 29 percent of Brazilian emissions in 2013 (OC, 2014).

Access to data on the contribution of each sector to national emissions was possible only in 2004, when the first Brazilian inventory of GHG emissions was launched. In this publication emissions referred to the period 1990–1994, that is, with a delay of 10 years regarding the data. The second inventory, published in 2010, showed emissions data for 1990, 1994, 2000, 2005, still with a delay of five years.

The lack of data and the delay in the publications regarding GHG emissions compromise the policy planning, measures and actions to mitigate climate change. Inventories and official estimates are fundamental not only to fulfill the obligations of the country to the UNFCCC, but also to assess the situation of GHG emissions from different sources and the progression of these emissions over time, in order to support public policies aimed at reducing and controlling emissions.

At the same time, civil society has organized and gathered efforts to understand the trajectory of emissions in the country through studies and reports. These initiatives are part of the Climate Observatory, especially through the Estimation System of Greenhouse Gases Emissions.

The analysis of Brazilian GHG emissions trajectory showed a significant reduction for a period. However, there was an 8 percent increase in emissions, despite the stagnation of the economy between 2012 and 2013 (Ferreira and Martinelli, 2016; OC, 2014). It also showed that there has been a change in the profile of GHG emissions in the country, which means that mitigation measures should accompany this change. Besides continuing to focus on the land use change and forestry sector, and hence on deforestation, the Brazilian government must consider energy and agriculture sectors, which have increased their contribution to GHG emissions.

Political Responses to Climate Change in Brazil

Brazilian political responses to climate change were divided into four main stages, as a result of data analysis. The main results of each stage are detailed below.

This first phase of mobilization for the climate issue, from 1992 to 2002, establishes political, institutional and scientific structures engaged with the issue. Because of this, few actions are highlighted in this initial stage when the issue enters the national political agenda. That is, a moment of articulation between international and national negotiation level, among institutions and stakeholders and of political and scientific needs identification.

The first internal response to the Climate Convention took place in 1999 with the creation of a governmental structure, the Inter Ministerial Committee for Sustainable Development. The first political-scientific mobilizations on climate change were realized with the implementation of the Center for Weather Forecasting and Climate Studies, under the National Institute for Space Research, improving weather and climate modeling systems in Brazil (Ferreira, 2017).

After that, a more specific framework related to the climate issue was created, with the Inter-Ministerial Commission on Global Climate Change, under the Ministry of Science Technology and Innovation, in 1999. The remit of this commission resided in the articulation of government actions under the Climate Convention, with the participation of several ministries involved in the issue, and in issuing feedback on eligible projects for the Clean Development Mechanism (CDM).

The articulation of non-governmental actors around the climate issue came in 2000 with the Brazilian Forum of Climate Change (BFCC), which had as its main objective the assistance of the federal government in the incorporation of climate change issues in public policy by involving stakeholders from several ministries and government agencies, representatives of private sector, research institutions and civil society. One of the positive effects of the forum was the creation of fourteen state forums on climate change, enabling the involvement of other levels of government in the debate. Another relevant mobilization in this regard was the Climate Observatory, established in 2002. It integrates the most important civil society organizations in the discussion of climate change issues within the Brazilian context (Barbi, Ferreira and Guo, 2016).

This period, in which political-institutional and scientific structures were established, was strongly influenced by the debate at the international level. These structures were relevant to the construction of climate policy at the national level, as discussed below.

The second phase of mobilization for the climate issue extends from 2003 to 2008 and stands out for greater understanding of the issue in the country. It is marked by the development of a political and scientific agenda around the theme and planning of mitigation actions in the country. The construction of this agenda played a key role in preparing the foundation for the national policy on climate change.

During this period, in 2004, the "First Brazilian Inventory of Anthropogenic GHG Emissions" was released. At the same time, the "National Action Plan to Tackle Climate Change," resulting from BFCC contributions culminated in the establishment of the Inter-Ministerial Committee on Climate Change and its executive group. This group played a significant role in the elaboration of the National Plan on Climate Change, which provided the basis for the national policy on climate change (Ferreira, 2018).

The generation and dissemination of knowledge focused on the challenges of climate change in Brazil has been enhanced by the Brazilian Network for Research on Global Climate Change (CLIMATE Network), founded in 2007. Another important scientific mobilization was the Centre of Earth System Science in 2008. The center expanded the climate change agenda by integrating global environmental changes and development issues for the country. It also made it possible to generate specific scenarios of global environmental change over the next 50 to 100 years and their effects on the national territory (Nobre, 2011; Barbi, Ferreira and Guo, 2016). Until then, there were not specific scenarios to Brazil, but rather to the South American region.

The establishment of the scientific agenda for climate change also included the National Institute of Science and Technology for Climate Change, which enabled the formation of an interdisciplinary research network. The articulation and integration of these structures occurs through the participation of researchers who comprise the majority of these networks.

The establishment of these political and scientific structures contributed to the foundations of the National Plan on Climate Change in 2008. The relevance of the Plan lies in guidelines for climate change mitigation in sectors such as: energy, transport, construction, industry, agriculture, forestry and solid waste. All actions in progress in these sectors, with GHG emission reduction potential, integrated the plan. However, the established goals did not constitute specific targets to reduce emissions.

Climate change mitigation was reinforced with CDM projects during this period. During the first commitment period of the Kyoto Protocol, from 2008 to 2012, Brazil received 300 projects. The reduction of GHG emissions associated with these projects is estimated at 351 million tCO2eq, corresponding to 4.8 percent of the world total in 2012 (PBMC, 2013).

The development and adoption of climate policies and the strengthening of the scientific agenda around the theme marks the third phase of the climate issue internalization process in Brazil from 2009 to 2012.

Despite having important scientific structures set out in the second phase, as previously presented, there was a gap in the science–policy interface integration. The Brazilian Panel on Climate Change (BPCC) was launched in 2009 with the aim of filling this gap, based on the protocols of the IPCC, in order to provide scientific assessments on impacts, vulnerability, adaptation actions and climate change mitigation. The "First National Assessment Report" of the Panel was published in 2012, during the Rio+20 Conference. The importance of this report lies in the elements that it can provide for the implementation of climate policies in the country, related to both mitigation and adaptation.

Mitigation actions have been reinforced with the regulation of CDM actions through the "Program of Activities under the Clean Development Mechanism." Finally, in 2009, the National Policy on Climate Change (NPCC) was approved. Its approval was strongly related to the international context, prior to the Conference of the Parties under UNFCCC (COP 15) in Copenhagen. This was a turning point in the discussions on climate change around the world, when the theme was in the international political agenda and featured political interest of national governments of many countries. The Brazilian policy had been under internal discussion and had its basis in the National Plan, as a result of articulation and mobilization presented in previous sessions. This policy highlighted the establishment of voluntary specific targets and deadlines for GHG emissions reduction.

Although the foundations of the policy were already on the national agenda, the establishment of mitigation targets was a result of private sector and civil society pressure as the core of the government was supposed to meet quantified emissions targets up until July 2009. Therefore, there were no systematic and consistent studies to determine the emissions targets (UN-HABITAT, 2011; Ferreira and Barbi, 2016).

An important instrument of NPCC was the National Fund on Climate Change, which uses part of the proceeds from oil exploitation revenue to reinvest in mitigation and adaptation to climate change. The rules for its operation were established in 2011.The fund is managed by a steering committee with the participation of representatives from several federal agencies, civil society, states and municipalities.

The NPCC stipulated the preparation of Sectoral Plans, elaborated from 2012 to 2013. The national policy agenda is focused on the implementation of these sectorial plans.

The presented data show a significant increase in the production and systematization of studies and reports, especially since 2000, which can assist the preparation and implementation of policies related to the theme and reduce uncertainty related to climate change. Scientific knowledge is fundamental to the development of GHG emission inventories, which guide the implementation of plans and reduction targets.

Regarding policy responses related to climate change the results show that climate change mitigation was present in the plans leading up to the national policy. The specific GHG emission reduction targets were a high point of Brazilian policy. Although the country decreased emissions for a while, the fact that they have been increasing since 2012 represents a new series of challenges that the present political and institutional structure has to deal with.

The fourth phase of political and institutional actions to climate change in Brazil begins with some concern regarding GHG emissions, which have been increasing since 2012. If no additional mitigation measures involving other sectors of activity take place, Brazilian GHG emissions may rise after 2020 (Okereke, Bulkeley and Schroeder, 2009). It is possible that the energy sector emissions may become the largest in Brazil by 2050.

Concurrently, within the national development plan, the country has invested in the oil and gas industry, confronting global discussions on low carbon development. There are few studies on the scenarios of future GHG emissions with the exploration of the Brazilian pre-salt oil. There are however decreased chances of oil demand, especially by the transport sector through regulations on fuel efficiency, developing alternative cleaner energy sources and especially reducing the private use of motor vehicles with improved public transportation. In addition, the potential for renewable energy such as wind, solar and biomass is large in the country (Olivier, 2014). Thus, the challenge for climate change mitigation in Brazil is to reconcile the advances in the energy sector and GHG emissions reduction targets set out in the NPCC and to maintain attention on the reduction of deforestation rates.

Subnational Level in Brazil: Political Responses to Climate Changes in Brazilian States and the Local Level

Subnational emissions come from different sectors of activities all over the country. Therefore, national climate change mitigation policies should be articulated with subnational mitigation guidelines concerning regional emissions' profiles. This is also essential to ensure that Brazilian emissions have a reduction and that the national policy on climate change succeeds.

The first state to approve their climate change policy was Amazonas, in 2007, followed by Tocantins, in 2008, and Goiás, Santa Catarina and São Paulo, in 2009, even before the approval of the National Policy on Climate Change in December, 2009. Since this period, the approval of climate policies has been intensified: fourteen out of the twenty-seven Brazilian states have approved their climate policy (see Figure 17.2). Among the states with approved climate policies, only Paraíba, Rio de Janeiro and São Paulo have established targets to reduce GHG emissions. However, some other policies also have the intention of stabilizing or reduce GHG emissions.

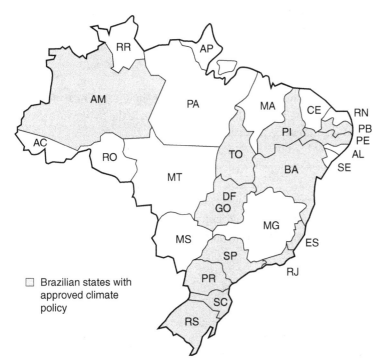

Figure 17.2 *Brazilian states with approved climate change policies.*
Source: Barbi and Ferreira, 2013.

In such cases, the policies provide for the development of mitigation plans that will set GHG emissions reductions goals. Most policies, nine of them, have the intention of developing an adaptation plan to the impacts of climate change. The only policies that cover these two aspects of climate policy are Distrito Federal, Paraíba, Rio de Janeiro and São Paulo (Barbi and Ferreira, 2013).

In the case of state policies, the State Forums of Climate Change had a fundamental role in their approval: sixteen states created their Forum between 2005 and 2009, and twelve out of fourteen states with climate legislation had a climate forum before the approval of their policy. Only Amazonas created its Forum after the approval of the state policy on climate change. The State Forums are planned in the Brazilian Forum of Climate Change, the national forum established in 2000. They focus on regional complementation and acting in accordance with state specifications. The importance of this institutional arrangement is the possibility of dialogue between the government and society in the search for incorporating climate change issues in the different stages of public policy.

The President of the country chairs the Brazilian Forum. In the case of states, the governors preside over them. They also include the participation of civil society organizations, universities and research institutes and the private sector.

Regarding the climate policy implementation, eleven out of the fourteen states have assumed a multi-sectorial perspective on climate governance by creating multi-thematic spaces of sectorial coordination, which involve several departments and stakeholders from different segments of society and consider climate change a cross-action issue. Amazonas, Espírito Santo and Paraná have chosen to create specific institutional structures to address the climate issue and coordinate the implementation of the policy.

Brazilian subnational policies are isolated initiatives in the national context. According to Barbi and Ferreira (2013) not all policies include the two main aspects of climate policy, that is, mitigation and adaptation.

As previously mentioned, Brazil is a significant country in the global climate change scenario. The biggest responsible sector for Brazilian emissions over twenty years was land use change and forestry. At the same time, emissions from other sectors of activity increased during this time. The greatest increase in emissions from 2005 to 2012 took place in energy and agriculture. Energy emissions represented 29 percent of Brazilian emissions in 2013 (Barbi and Ferreira, 2016; Ferreira, 2017; Ferreira, 2018).

Brazilian Local Climate Policies

Brazilian local climate policies are isolated initiatives in the national context. By 2016, six municipalities out of 5,570 had a specific approved legislation related to the climate issue. Most of them were approved from 2009 onwards, at a time when the climate issue was a priority in the international political agenda, by the time of the Conference of the Parties (COP) in Copenhagen. Other cities with climate policies in Brazil are: Palmas, in Tocantins, Belo Horizonte, in Minas Gerais, Feira de Santana, in Bahia, Recife, in Pernambuco and Rio de Janeiro, in Rio de Janeiro. The policies in Belo Horizonte, Recife, Rio de Janeiro and São Paulo regulate strategies with specific targets to both mitigate and to adapt to climate change.

Regarding adaptation, the policies of Belo Horizonte, Recife, Rio de Janeiro and São Paulo plan the establishment of an adaptation policy. The promotion of adaptation strategies in Belo Horizonte ought to involve civil defense, land use and health sectors. Recife's adaptation plan is yet to be defined. In the case of Rio de Janeiro, the municipal civil defense is in charge of activities related to adaptation. In São Paulo, adaptation strategies involved mainly the requalification of housing in risk areas and the recovery of permanent preservation areas, in order to prevent or minimize the risks of extreme weather events. Feira de Santana's policy is the vaguest, determining only that it aims to "define and implement measures to promote adaptation." Palmas' policy also makes no mention of actions or adaptation plans.

Belo Horizonte, Recife, Rio de Janeiro and São Paulo counted on institutional arrangements that allowed the articulation of different agents from different segments of society in the policy-making process. In Belo Horizonte, debates regarding the climate policy started in 2006 when the Municipal Committee on Climate Change and Eco-Efficiency was created under the city government, connected to the Department of Environment. The Committee was created to give advice, consultancy and articulate existing environmental policies in different municipal agencies to reduce GHG emissions.

In Recife, the articulation around the policy started in 2013, with the formation of two municipal forums: Comclima and Geclima. Comclima is comprised of stakeholders from the local, state and federal government, academia and organized civil society. Geclima counts on representatives from the local government from different sectors of activity. The Department of Environment coordinates both groups, which subsidize the municipality in issues involving climate change.

In Rio de Janeiro, negotiations on the policy were reinforced by the municipal Forum on Climate Change in 2009 (Barbi and Ferreira, 2013). The institutional structure of the Forum follows that of the Brazilian Forum on Climate Change, with the participation of several agents. The main purpose of these forums is the establishment of a climate policy. Currently, the Forum of Rio de Janeiro is one of the key agents of management and consolidation of the climate policy.

In São Paulo, the discussion about the climate policy began in 2005 with the formation of the Municipal Committee on Climate Change and Sustainable Eco-Economy, an initiative of the local government, with the aim of promoting and encouraging actions related to the mitigation of GHG emissions (Barbi and Ferreira, 2013). The elaboration process of the policy took four years and involved the participation of agents who were actively involved, such as the Department of Environment, the Research Center for Sustainability of Foundation Getulio Vargas, ICLEI, Local Governments for Sustainability and Fabio Feldmann Consultants (Barbi, 2015).

Feira de Santana did not have this kind of institutional arrangement before the law was passed. However, the policy provides for the establishment of the Forum on Climate Change "for the manifestation of social movements, scientific sector, the business sector and all others interested in the subject, in order to promote transparency of the process and social participation in the development and implementation."

In the case of Palmas, the approval of the law took place without the participation of other agents, at a time when the city was looking for a tool that would allow the sale of carbon credits and the hire of consultants to carry out projects in the area of climate change, which was possible through the law (Barbi and Ferreira, 2013).

Regarding the implementation of the policies in Belo Horizonte, Recife, Rio de Janeiro and São Paulo, climate governance can be considered multi-sectoral. In this case, these municipalities counted on an institutional arrangement prior to the approval of the law, with a multi-sectoral profile for conducting the policy. In Belo Horizonte, the Municipal Committee on Climate Change and Eco-Efficiency is responsible for implementing the policy, with the participation of other sectors of the municipal and state governments, representatives of the City Council, universities, NGOs and organizations representing industry and commerce.

In Recife, the implementation of the policy counted on the elaboration of a low-carbon plan, which had the participation of the civil society, the private sector and several sectors of municipal and state governments.

In the case of Rio de Janeiro, although the policy is coordinated by the Department of Environment, its implementation is through its Climate Change and Sustainable Development Unit, in a crosscutting manner and with the participation of several areas of the municipal administration and partnerships with academic institutions. For example, the Vulnerability Map of Rio de Janeiro Metropolitan Area, which identifies the impacts on the physical environment and their respective vulnerabilities in socio-economic and natural systems, was elaborated in partnership with the National Institute for Space Research and the University of Campinas (Barbi and Ferreira, 2013).

As a result of the policy in São Paulo, the following work groups were created under the Municipal Committee of Climate Change and Eco-Economy: Transport, Energy, Construction, Land Use, Waste and Health. They were responsible for the preparation of the Guidelines for Mitigation and Adaptation to Climate Change, in order to detail the strategies prescribed by the policy.

In Feira de Santana, the execution of the policy is connected to the preparation of the Municipal Plan on Climate Change by the Department of Environment and Natural Resources, under the coordination of the Municipal Council of Environmental Defense. Its preparation relies on public consultations through the Climate Change Forum. In Palmas, the implementation of the policy is responsibility of the Department of Environment and Public Services.

Finally, Belo Horizonte, Palmas, Recife, Rio de Janeiro and São Paulo are members of the cooperation network ICLEI – Local Governments for Sustainability – and were members of ICLEI's CCP campaign (Cities for Climate Protection). Palmas participated in the campaign between 2002 and 2004, when its policy was approved. The other cities are still part of the network. Under the CCP campaign, Belo Horizonte and São Paulo joined the project "Sustainable Construction Policies (PoliCS)" with the goal of determining the commitment of these cities in the elaboration and implementation of sustainable building policies, focusing on energy efficiency and the promotion of low-carbon technologies. They also participated in the project "Model Communities in Local Renewable Energies (Rede Elo)" in order to reinforce the generation and use of energy from renewable sources and energy efficiency, focusing on the roles and responsibilities of local government as a driving force for technological innovation and investment in sustainable development. São Paulo also joined the projects "Green and Healthy Environment (Pavs)" and "Promoting sustainable public procurement in Brazil (CPS-Brazil)" aimed to change consumption patterns by the government. Recife participated in the project "Urban Leds," aimed at urban development based on low GHG emissions (Barbi and Ferreira, 2013).

Rio de Janeiro hosted ICLEI in Brazil for six years and São Paulo hosted the network after that, from 2007 until 2012. These cities are also members of the C-40 network, and thus must set targets and goals regarding climate change mitigation and adaptation.

Conclusion

Climate change constitutes an unprecedented challenge to contemporary societies in the transition to the Anthropocene. Responding to this problem, or not doing so, will greatly influence future life on Earth, as climate change mitigation in the coming decades will be crucial to determine the amount of long-term warming and the risks posed by these changes to populations and ecosystems.

Within social and political dimensions of the climate issue, this chapter has highlighted governments as relevant stakeholders in proposing appropriate forms of climate change governance understanding that they are not the only ones facing this challenge. Emphasizing the importance of this actor does not mean defending a top-down governance system. In this sense, this study analyzed political responses to the climate issue in Brazil.

In Brazil, the results showed the participation of non-governmental actors in data elaboration and systematization regarding GHG emissions and in developing mitigation policy strategies to climate change. The significant production and systematization of scientific studies on the subject in the country were highlighted as instruments that can support policy responses to climate change.

According to scientists, mitigation efforts needed to limit warming to less than 2°C above pre-industrial levels are economically viable (Dunlap and Brulle, 2015). The question that remains, however, for both cases presented here is whether the proposed actions and the existing political and institutional structures are sufficient to respond effectively to the magnitude and complexity of the challenge, especially in time to prevent irreversible climate change.

In addition, climate policy focused on mitigation must have public involvement, encouraging changes in lifestyle and consumption. The real climate policy is not just about the weather, but about the transformation of basic concepts and institutions established during the development of contemporary society (Beck, 2010). It is about the transformation of human activities responsible for causing the problem.

In this sense, the transposition of climate issues into the political agenda can be considered as a first response to the challenge (Giddens, 2009). The next move must involve the introduction of this issue in institutions and daily concerns of citizens. And there is certainly much to be done in this direction as well.

Following the growing international movement of local responses to climate change (Bulkeley and Newell, 2010), in Brazil, this type of action was more expressive first at the city level. Then, it reached the state level and finally, the federal level. In this movement, the climate policy of the city of São Paulo had a leading role, since it was the first one to establish GHG reduction targets, influencing the approval of the state policy, which also exerted influence on the federal policy process of approval.

References

Barbi, F., & Ferreira, L. 2013. Climate change in Brazilian cities: Policy strategies and responses to global warming. *International Journal of Environmental Science and Development* 4(1): 49–51.

Barbi, F.; Ferreira, L. & Guo, S. 2016. Climate change challenges and China's response: mitigation and governance. *Journal of Chinese Governance* 1(2): 324–339.

Basso, L., & Viola, E. 2014. O Progresso da política energética chinesa e os desafios na transição para o desenvolvimento de baixo carbono, 2006–2013. *Revista Brasileira de Política Internacional* 57: 174–192.

Beck, U. 1992. *Risk Society: Towards a New Modernity*. Beverly Hills, CA: Sage.

Beck, U. 1995. *Ecological Politics in an Age of Risk*. Cambridge: Polity.

Beck, U. 2000. Risk Society Revisited: Theory, Politics and Research Programmes. In *The Risk Society and Beyond: Critical Issues for Social Theory*, ed. B. Adam, U. Beck, & J. V. Loon. London: Sage Publications, pp. 211–239.

Beck, U. 2009. *World at Risk*. Cambridge: Polity.

Beck, U. 2010. Climate for change, or how to create a green modernity? *Theory, Culture & Society* 27(2–3): 254–266.

Betsill, M. M., & Bulkeley, H. 2007. Looking back and thinking ahead: A decade of cities and climate change research. *Local Governments* 12(5): 447–456.

Bizikova, L., Burch, S., Cohen, S. & Robinson, J. 2010. Linking sustainable development with climate change adaptation and mitigation. In *Climate Change, Ethics and Human Security* eds. K. L. O'Brien, A. L. St.Clair & B. Kristoffersen. Cambridge: Cambridge University Press.

Bulkeley, H., & Betsill, M. 2003. *Cities and Climate Change – Urban Sustainability and Global Environmental Governance*. New York: Routledge.

Bulkeley, H., & Kern. K. 2006. Local government and the governing of climate change in Germany and the UK. *Urban Studies* 43(12): 2237–2259.

Bulkeley, H., & Newell, P. 2010. *Governing Climate Change*. New York: Routledge.

Burch, S., & Robinson, J. 2007. A framework for explaining the links between capacity and action in response to global climate change. *Climate Policy* 7(4): 304–316.

Dunlap, R., & Brulle, R. 2015. *Climate Change and Society. Sociological Perspectives.* New York: Oxford University Press.

Dodman, D. 2009. Blaming cities for climate change? An analysis of urban greenhouse gas emissions inventories. *Environment and Urbanization* 21(1): 185–198.

Ferreira, L. (ORG). 2017. *O Desafio das mudanças climáticas. Os casos Brasil e China*. Jundiaí, Brazil: Paco Editorial.

Ferreira, L. 2018. *The Sociology of Environmental Issues. Theoretical and Empirical Investigations*. Curitiba, Brazil: CRV Editor.

Ferreira, Leila, and Martinelli, M. 2016. Anthropocene: Governing Climate Change in China and Brazil. *Sociology and Anthropology* 4(12): 1084–1092. www.hrpub.org DOI:10.13189/sa.2016.041207.

Ferreira, Leila, R. D. Martins, F. Barbi, et al. 2012. Risk and Climate Change in Brazilian Coastal Cities. In *Risk and Social Theory in Environmental Management* eds. T. G. Measham & S. Lockie. Collingwood: CSIRO Publishing, pp. 133–146.

Ferreira, Leila, and F. Barbi. 2016. The challenge of global environmental change in the Anthropocene: An analysis of Brazil and China. *Chinese Political Science Review* 1(4): 685–697.

Giddens, A. 1990. *The Consequences of Modernity*. Stanford, CA : Stanford University Press.

Giddens, A. 2000. *Runaway World. How Globalization Is Reshaping Our Lives*. New York: Routledge.

Giddens, A. 2009. *The Politics of Climate Change*. Cambridge: Polity Press.

Gupta, J. 2007. The multi-level governance challenge of climate change. *Journal of Integrative Environmental Sciences* 4(3): 131–137.

Hoornweg, D., L. Sugar, L., and C. L. T. Gomez. 2011. Cities and greenhouse gas emissions: moving forward. *Environment and Urbanization* 23(1): 207–227.

IPCC – Intergovernmental Panel on Climate Change. 2007. *Summary for Policymakers*. In *Climate Change 2007: Mitigation. Contribution of Working Group III to the Fourth Assessment Report of the Intergovernmental Panel on Climate Change*, eds. B. Metz, O. R. Davidson, P. R. Bosch, R. Dave & L. A. Meyer, Cambridge,: Cambridge University Press, pp. 1–23.

IPCC – Intergovernmental Panel on Climate Change. 2001. Summary for Policymakers. In *Climate Change 2001: Impacts, Adaptation, and Vulnerability. Contribution of Working Group II to the Third Assessment Report of the Intergovernmental Panel on Climate Change* eds. J. J. McCarthy, O. F. Canziani, N. A. Leary, D. J. Dokken, & K. S. White. Cambridge: Cambridge University Press, pp. 1–17.

Klein, R. J. T., Schipper, E. L. F., & S. Dessai. 2005. Integrating mitigation and adaptation into climate and development policy: three research questions. *Environmental Science & Policy*, 8(1): 579–588.

Newell, P., Patterberg, P., & Schroeder, H. 2012. Multiactor Governance and the Environment. *Annual Review of Environment and Resources*, 37: 365–387.

Nobre, Carlos A. 2011. *Vulnerabilidades das Megacidades Brasileiras às Mudanças Climáticas: Região Metropolitana de São Paulo: Relatório Final*. São José dos Campos, SP Brasil: Instituto Nacional de Pesquisas Espaciais.

OC – Observatório do Clima. 2014. *Análise da evolução das emissões de GEE no Brasil (1990–2012)*. Documento Síntese. São Paulo: Instituto de Energia e Meio Ambiente.

Okereke, C., Bulkeley, H., & Schroeder, H. 2009. Conceptualizing Climate Governance Beyond the International Regime. *Global Environmental Politics* 9(1): 58–78.

PBMC. Painel Brasileiro de Mudanças Climáticas. 2013. *Contribuição do Grupo de Trabalho 2 ao Primeiro Relatório de Avaliação Nacional do Painel Brasileiro de Mudanças Climáticas. Sumário Executivo do GT2*. Rio de Janeiro, Brasil: PBMC.

Renn, O., & Klinke, A. 2012. Complexity, uncertainty and ambiguity in inclusive risk governance. In *Risk and Social Theory in Environmental Management*, eds. T. G. Measham & S. Lockie. Collingwood: CSIRO Publishing, pp. 59–76.

Satterthwaite, D. 2010. The contribution of cities to global warming and their potential contributions to solutions. *Environment and Urbanization Asia* 1(1): 1–12.

Storbjörk, S. 2007. Governing climate adaptation in the local arena: Challenges of risk management and planning in Sweden. *Local Environment* 12(5): 457–469.

UN-HABITAT (United Nations Human Settlements Programme). 2011. *Cities and Climate Change: Global Report on Human Settlements*. Earthscan.

Winkler, H., K. Baumert, K., O. Blanchard, S. Burch, and J. Robinson. 2007. What factors influence mitigative capacity? *Energy Policy* 35(1): 692–703.

PART V

Resources

18 Enclosing Water: Privatization, Commodification, and Access

Daniel Jaffee

In the past decade, popular culture and news media have spread the message that the world now faces a crisis of water scarcity, and that fresh water constitutes the "new oil" over which future wars will be fought (Barlow and Clarke 2002). Missing from this formulation, however, is the insight that the "global water crisis" is socially produced, and exacerbated by anthropogenic climate change, ecological degradation, and overextraction by agribusiness and industry (Zwarteveen and Boelens 2014). Nonetheless, the lack of access to potable water for many does indeed constitute a grave emergency. As of 2017, 2.3 billion people had no access to improved sanitation and 844 million people worldwide lacked access to safe drinking water (WHO and UNICEF 2017), although some argue the latter figure is actually closer to 1.8 billion (Goff and Crow 2014). Waterborne diseases including diarrhea and cholera account for approximately 5 percent of all deaths worldwide (UNDP 2006). The increasing inequalities between "the water 'haves' and 'have nots'" (Zwarteveen and Boelens 2014: 143) have been described as "one of the greatest crimes of the twenty-first century" (Mehta 2016). Lack of access to potable water is a key marker of growing social and environmental inequality on a global scale.

These trends overlap with a major shift in water management regimes over the last three decades, from a "state hydraulic paradigm" in which provision of clean drinking water was understood as a public good and a right of citizenship, to a neoliberal or "market environmentalist" paradigm (Bakker 2005) in which – to cite the influential Dublin Principles of 1992 – water "has economic value in all its competing uses and should be recognized as an economic good" (United Nations 1992). Since the 1990s, a constellation of actors, led by the World Bank and large private water service firms, have pressed the argument that states in the global South had failed to provide access to water to growing and urbanizing populations, and that only the private market could effectively expand water service (Bakker 2013b; Goldman 2007). Only by pricing water at its "true value," the argument continues, can this scarce resource be conserved. To accomplish these goals, private firms should take over the task of providing drinking water from states, making water users into customers, and to do so they would need "full-cost recovery" – guaranteed profit margins backed up by substantial increases in water bills (Castro 2007).

To state the obvious, when a good or resource is commodified or privatized, access is based on the ability to pay. Allowing the market to set a price for water may indeed reduce use, by pushing it out of the reach of those who lack "effective demand"

(Bond 2008). Yet water is essential for life, and unsubstitutable. This generates a fundamental contradiction that explains the intensity, size, and breadth of opposition to the neoliberalization of water that has emerged in recent decades (Subramaniam and Williford 2012). Struggles over water commodification are often framed in terms of justice – i.e. the injustice of economic or physical barriers to accessing clean drinking water – and this collective resistance is often termed the global water justice movement (Mehta et al. 2014; Zwarteveen and Boelens 2014).

In 2010, the United Nations Assembly and UN Human Rights Council declared water to be a human right, a major victory for water justice advocates (Lederer 2010). However, this formal declaration – while a valuable tool for citizens aiming to hold states accountable – has yet to change substantially the dynamics of access to water. Actually operationalizing the human right to water remains an enormous challenge, one further complicated by market control. Yet private water firms and even some UN officials have argued that the "rights" framing is not incompatible with private provision (Bakker 2010; Sultana and Loftus 2012). Ultimately, debates over the desirability of market versus public (or community) provision of drinking water come down to a basic question: Should affordable access to clean water for all be an inviolable part of the social contract or not?

This chapter examines the global political economy of access to *drinking water*, with particular attention to the implications for environmental and social justice. (While water for irrigation is also subject to many of the same dynamics of commodification and their negative externalities, it is beyond the scope of this contribution.) The following section reviews some key theoretical approaches to the privatization and commodification of drinking water. The next two sections examine the institutional and ideological drivers, dynamics, and effects of the enclosure of municipal (tap) water supplies, as well as the substantial countermovements it has generated, drawing on case studies from both the global South and the North. After briefly reviewing the present status of municipal water privatization, the chapter then turns to another major modality of water commodification: bottled water. It briefly explores the dramatic growth of this relatively new commodity, its environmental and social externalities, and the grassroots movements opposing water extraction by the global bottled water industry in specific localities. The concluding section discusses the linkages between these various modes of water commodification, and the implications for the future of access to affordable, safe drinking water for all.

Privatization, Commodification, and Accumulation

The commodification or neoliberalization of nature has generated major social contention, centering both on social justice issues of access to life- and livelihood-sustaining resources and on questions of sustainability and ecological impact. Water exemplifies these tensions. Karl Polanyi (1944) identified what he termed the "fictitious commodities" of land (nature), labor, and money, highlighting their centrality to the destructive tendencies of a "self-regulating" market economy. Neoliberal globalization has since expanded the fictitious commodities even further into areas such as (patented) seeds, genes, and life forms (Kloppenburg 2004; Laxer and Soron 2006). Water, as an element of Polanyi's land, offers a compelling contemporary illustration of the dangers of the commodity fiction.

Some scholars have characterized the transformation of water into a marketable commodity as a form of primitive accumulation, drawing on Marx's analysis of the process by which capitalism has separated producers from the social means of subsistence and production (Glassman 2006; Marx 1867; Roberts 2008). Harvey (2003) has usefully extended Marx's framework to emphasize the ongoing nature of these dynamics in the present day, developing the influential concept of "accumulation by dispossession." Harvey argues that accumulation by dispossession is capital's response to an overaccumulation crisis, in which it must expand into new terrains in order to return to or retain profitability (2003: 149). The essence of this process is commodification, the dynamic by which formerly public goods and other non-market functions and services are incorporated into the market. Privatization, Harvey argues, forms the "cutting edge" of accumulation by dispossession, as illustrated by the role of international financial institutions in imposing privatization of public services on debtor nations in the global South. These modern enclosures combine commodification with exclusion, whether that exclusion is economic or physical. Harvey also describes the broad range of social movements against accumulation by dispossession globally, noting a key commonality in their emphasis on "reclaiming the commons" (2003: 166).

Several scholars have applied Harvey's framework explicitly to the commodification of water, examining specific case studies of privatization and the countermovements that they have generated (e.g., Ahlers 2010; Perreault 2013). Spronk and Webber, examining water privatization in Bolivia, observe that accumulation by dispossession involves not only the "privatization of formerly state or public resources but their acquisition by transnational capital in the US and other core economies" (Spronk and Webber 2007: 32). Swyngedouw (2005) argues (echoing Polanyi) that active state intervention is essential to facilitate such accumulation by dispossession. Nonetheless, scholars have documented that the privatization of municipal (tap) water systems poses many obstacles to profitability, a point I return to below.

A dynamic and valuable literature on the political ecology of water has also emerged in recent years, overlapping with many concerns of environmental sociologists. Alex Loftus writes that political ecology "seeks to politicize understandings of the distribution of water," stemming from an understanding that "water and social power are ... mutually constitutive" (Loftus 2009: 953, 959). One of the central concerns of this body of work is to illustrate "the fundamentally socially produced character of ... inequitable hydro-social configurations" (Swyngedouw 2009: 57). Political ecologists examine a range of issues including water infrastructures, urban water, and irrigation, as well as water injustice (Ranganathan and Balazs 2015), the gendered nature of access to and conflicts over water (Ahlers 2005; O'Reilly 2011; Sultana 2011), and municipal water privatization and its countermovements (Bakker 2005; Bakker 2007; Roberts 2008).

Privatizing Public Water

In many urban areas of the global South, only upper- and upper-middle-income residents are served by the public tap water networks common in the North, and even where such piped systems are available, supplies may be contaminated or insufficiently treated (Bakker 2010). Middle- and lower-income urban and peri-urban residents typically rely on a mix of less formal sources, including refilling stations and water vendors selling trucked or locally bottled water, which costs far more than their wealthier neighbors pay for tap water (Godoy 2010; Malkin 2012). Others turn to communal standpipes, household wells, or even untreated surface water (Liddle, Mager and Nel 2014; Narain 2014). In part due to this growing crisis of quantity and quality, drinking water is increasingly seen as a profitable commodity to be sold at market rates. This has facilitated the emergence of a nearly $500 billion global water services industry, one-fourth of which is presently controlled by two French-based multinational corporations, Veolia and Suez (Arup 2015).

While the wholesale sell-off of public water utilities in the United Kingdom under Margaret Thatcher in 1989 is often cited as the functional start of the global push to privatize water (Bakker 2005), this dynamic actually dates back to Chicago School-influenced policies implemented by the Pinochet dictatorship in Chile beginning in 1980 (Harris 2013). The privatization of tap water systems in the global South was largely imposed through lending conditionality by international financial institutions, primarily the World Bank and IMF, under so-called structural adjustment programs. Starting in the late 1980s and 1990s, these agencies mandated that debtor governments offer up public utilities for concession or sale as a condition of debt renegotiation, arguing that states had failed to provide "water for all" and that the private sector was uniquely qualified to fill this role (Conca 2008; Goldman 2007). It is true that in the neoliberal era, many Southern states have indeed been hard-pressed to extend piped water networks to keep pace with rapid urbanization, yet that is often due to the very same strictures of debt servicing (Bakker 2013b). Nonetheless, water is also essential for life, and privatization has in some cases made it "so unaffordable that citizens are forced to drink water from contaminated sources" (Kurland and Zell 2011: 329). Therefore, the struggle over public versus market provision of water carries profound implications for human rights, health, and social justice. When access to clean water is based solely on the ability to pay, some people will lose access, which can be deadly.

In reality, the term "privatization" subsumes a wide variety of forms of market involvement, ranging from (relatively uncommon) outright asset sales like those in the UK, to long-term leases or concession contracts – "outsourcing" arrangements in which water systems remain publicly owned but private firms are responsible for operations, maintenance, and sometimes infrastructure investments – to "corporatized" public water utilities that now operate on strict market principles (Bakker 2014; Castro 2007; McDonald 2016). The large majority of the "privatizations" that have caused major social protest in the global South were in fact multi-decade concessions – often termed "public–private partnerships" (PPPs) by proponents – which typically guarantee the corporate operator a minimum profit margin and the freedom to set water rates (Bakker 2014). All of these constitute forms of enclosure, because access to water is mediated by the market. Yet scholars disagree on the extent to which privatization, commodification, and marketization overlap. Bakker, for example, argues that due in part to water's biophysical and sociocultural qualities, "private ownership and the introduction of markets do not necessarily entail commodification," leading her to characterize water as an "uncooperative commodity" that poses barriers to its full conversion into a market good (Bakker 2005: 559). However, scholarly debates over the boundaries between commodification and privatization are usually of less concern to local communities and grassroots water justice activists, many of whom employ "privatization" as a broad term encompassing all forms of market control over water, which they view as antithetical to human rights and the public interest (Campero Arena 2016). In this chapter, I use the term "privatization" to refer to multiple arrangements that cede control over key functions of public (or community) water management to private firms, ranging from PPPs (concessions and leases) to outright system sales.

The World Bank, private water firms, and other key players have since the early 1990s manufactured a global "consensus" around the private provision of water, creating "transnational policy networks" that incorporated Southern bureaucrats, global water firms, NGOs, and putative "civil society" entities such as the World Water Council (Goldman 2007). Goldman (2005) argues that water privatization is emblematic of the World Bank's newest policy regime, which he terms "green neoliberalism" – an approach fusing global discourses of sustainable development with the extension of market fundamentalism. Bakker (2014: 474–475) describes this paradigm as market environmentalism, a "doctrine premised on the synergies between environmental conservation and protection, economic growth, market economies, and neoliberal governance." By whatever name, this regime led to a dramatic increase in concessions and facilitated the growth of a global "cartel" of private water service firms, two of which – French-based Suez and Veolia – at their peak controlled over 70 percent of the private water sector worldwide (Barlow 2007).

A substantial body of literature examines the results of privatization in the global South – particularly the dominant model of long-term concession contracts with large transnational firms. While this literature partly reflects the ideological divide over privatization, and some research has found increases in water connections under private concessions in certain cities, the large majority of studies concur that market involvement has failed to substantially expand water service, let alone meet the stated goal of providing "water for all" (Castro 2008; Swyngedouw 2005). The World Bank itself has acknowledged that concession contracts did not generate significant numbers of new water connections (World Bank 2005). The track record of municipal water privatization across the South includes large rate hikes, deviations from promised levels of service, increased instances of unsafe water, the spread of diseases including cholera, pollution and sewage spills, and increased water shutoffs for nonpayment that have forced users to turn to contaminated water sources (Food and Water Watch 2011).

Some of the touted financial efficiencies gained from these arrangements resulted from substantial layoffs of public utility employees under private management, such as in the 1993 concession of the Buenos Aires water system to a consortium led by Suez (Casarin, Delfino, and Delfino 2007). When privatization is imposed by creditors under a tight time frame, the resulting "fire sales" (often with very few bidders) can generate far less revenue for debtor states than projected, highlighting the dynamics of accumulation by dispossession at work (Rankin and Smith 2015; Vilas 2004).

Privatization also has major gendered effects. Ahlers writes that "unequal gendered access to resources is perpetuated and legitimized by the introduction of market mechanisms in the water sector" (Ahlers 2005: 57). Increased water tariffs in pursuit of "full-cost recovery" can reconfigure gendered household economies and divisions of labor, particularly if families lose access to potable water supply due to disconnection (O'Reilly 2011).

Contesting and Reversing Privatization

Public opposition to the neoliberalization of water has accompanied privatization around the globe. Substantial protests have erupted not only in the well-studied Bolivian "water wars" in Cochabamba and El Alto (e.g., Spronk 2015), but also in Ecuador, Uruguay, Argentina, Nicaragua, El Salvador, Tanzania, South Africa, Nigeria, and many other countries (Castro 2008; Romano 2012, Terhorst, Olivera and Dwinell 2013). Many of the citizens and grassroots activists in these uprisings have received support from a broad network of civil society groups and NGOs that collectively constitute the global water justice movement (Barlow 2014; Lobina, Terhorst, and Popov 2011; Nelson 2017). This constellation of actors can be conceptualized as a Polanyian countermovement advocating decommodification of the water commons.

Water justice activists and other opponents of privatization argue that public utilities like water systems – typically understood as natural monopolies – are inherently unsuited to market control, since the profit motive is incompatible with the need for long-term investments in water quality and system maintenance (Barlow 2007; Snitow, Kaufman and Fox 2007). Because the rate hikes needed to ensure profit margins cause the exclusion of those unable to pay, they insist that subjecting a life-sustaining resource to market discipline is inimical to public well-being, and that it must be managed on a non-profit basis (Barlow, 2007).

This critique extends to the environmental implications of market control as well. Zwarteveen and Boelens write that "water justice includes but transcends questions of distribution ... and is intimately linked to the integrity of ecosystems" (Zwarteveen and Boelens 2014: 143). Mehta and her coauthors similarly observe that "struggles to access water are also struggles about environmental justice and sustainability" (Mehta et al. 2014: 158).

Beginning in the early 2000s, many water concession contracts were terminated by municipal or national governments, due to firms' noncompliance with contract terms, failure to invest in infrastructure or system extension, public opposition, or changing state political ideologies (especially in Latin America). Other contracts were ended by the private firms themselves due to public protest and lack of profitability (Food and Water Watch 2011; Hall, Lobina, and Corral 2011; Vidal 2006). The German firm RWE, formerly the third largest private water corporation, has sold off much of its $10 billion worth of water holdings worldwide, including the major subsidiary Thames Water (Esterl 2006). Overall, municipal water privatization has proved to be insufficiently profitable in predominantly poor urban areas in the South (Swyngedouw 2005). Loftus writes that "it is incredibly difficult to make the profits expected by private investors from the large, long-term needs of infrastructural development for poor people" (Loftus 2009: 957).

Public opposition and the poor overall track record of privatizations (including poor water quality, soaring rates, and major labor conflict) has led hundreds of cities to "remunicipalize" their water systems in the past ten to fifteen years, ending private contracts and returning to public management. As of 2017 there were 267 documented cases of water remunicipalization worldwide (Kishimoto and Petitjean 2017). Also stemming from these water justice struggles, several "pink tide" Latin American states have incorporated explicit bans on water privatization into their constitutions, including Uruguay, Bolivia, and Ecuador (Hall, Lobina and De la Motte 2009; Harris 2013). On the other hand, the right-wing Brazilian government has recently embarked on a privatization push. Some remunicipalized water utilities in the South continue to struggle with low investment, indebtedness, and a limited ability to expand water networks (Driessen 2008; Spronk 2015). One model advocated by water justice groups to address such challenges is "public–public partnerships," in which water utilities partner with utilities, NGOs, or labor unions in other cities and nations to share expertise, pool resources, and increase efficiencies (Food and Water Watch 2012b).

South Africa represents a dramatic case of the harmful social consequences of the neoliberalization and marketization of water, even in the absence of outright privatization. Despite inclusion of the right to water in its post-Apartheid constitution and the adoption of a "free basic water" policy guaranteeing access to a minimal amount (25 liters per person daily) without charge, the ANC government has implemented pre-paid water metering on a broad scale, leading to shutoffs of up to 10 million people for nonpayment. These policies have forced many to turn to unsafe water sources, leading to outbreaks of cholera and generating substantial protest movements against metering and shutoffs (Goldman 2007; Ruiters 2007; Smith and Hanson 2003). In a landmark 2009 decision in a lawsuit brought by poor residents of Phiri, Soweto against the recently deprivatized Johannesburg Water, the Constitutional Court upheld the metering policy and ruled that the meager basic water allocation was sufficient (Dugard 2010). Many observers argue this decision constitutes a clear denial of the human right to water (e.g., Bond 2012).

Austerity, Affordability, and Contamination in the North

Market involvement in tap water in the global North exhibits some of the same dynamics as in the South. The first drinking water systems were largely developed by private firms in the eighteenth and nineteenth centuries, but in North America and much of Europe, city governments had largely taken over operating water systems by the early twentieth century, due to major problems with disease, poor water quality, pollution, inadequate pressure and supply, and other factors. However, since the 1990s, pressure to return to market provision has resurfaced, generating vocal countermovements (Food and Water Watch 2012a; Varghese 2007). Several large US cities entered into private concession contracts with Suez, Veolia, and other firms to manage drinking water and sewerage systems, including Atlanta, Indianapolis, New Orleans, Indianapolis, and Stockton, CA. Due partly to their poor track record (large rate hikes, more frequent water quality emergencies, and diminished service levels), virtually all of the high-profile concessions in major cities have since been reversed, adding to the global remunicipalization trend (Esterl 2006; Robinson 2013; Snitow, Kaufman and Fox 2007).

Nevertheless, management of public water systems in the United States is increasingly hampered by deteriorating public water and sewer infrastructure resulting from decades of underinvestment and fiscal austerity, leading to an infrastructure maintenance backlog estimated at $1 trillion (American Water Works Association 2012). The dramatically declining US federal investment in state and local water infrastructure since the early 1980s is emblematic of the retreat of states from service provision under neoliberal restructuring, which has especially come to the fore since the Great Recession (Peck 2015; Szasz 2007). The rising burden of infrastructure maintenance, combined with local fiscal crises causing budget cuts to water utilities, has made "public–private partnerships" appealing to many cash-strapped cities, such as Baltimore, where Suez has worked to persuade public officials to approve a concession of the city's water system (Sloan 2016). Private equity firms such as the Carlyle Group have increasingly attempted to buy out both private and public water utilities in the United States in a search for high short-term returns (Food and Water Watch 2012a). A few US states, including New Jersey, have passed legislation incentivizing private sector involvement in upgrades of public water systems (Augenstein 2015). Some observers have analyzed these phenomena as instances of "disaster capitalism" (Klein 2007), emphasizing how political elites and investors can seize upon crises, whether natural disasters or fiscal shocks, to advance the privatization or commodification of public resources.

Despite these trends, the proportion of the US population served by public drinking water systems has actually risen slightly, from 83 percent in 2007 to 87 percent in 2015 (Food and Water Watch 2016). Studies have documented that customers of privately owned utilities pay on average 59 percent more than those of public utilities (Food and Water Watch 2016), due to both higher borrowing costs and the imperative to generate returns for investors (Food and Water Watch 2017b; Wait and Petrie 2017).

However, although tap water provided by the public sector is less expensive, the United States is now experiencing a burgeoning crisis of water affordability. Municipalities forced to take over the burden of water and sewer system maintenance from the federal government have passed the costs on to users through rising bills. A 2017 study found that nearly 12 percent of US households have unaffordable water and wastewater bills (defined as more than 4.5 percent of household income) – a figure projected to rise to a startling 35 percent by 2022 if current trends continue (Mack and Wrase 2017).

While the affordability crisis is nationwide, its epicenter is arguably Detroit, where mass shutoffs of up to 100,000 poor households for nonpayment of water bills since 2014 have drawn international media coverage, condemnation from the UN Special Rapporteur on the Right to Water, and dramatic protests by residents (Kornberg 2016; Rushe 2014). Some critics allege this practice is aimed at reducing debt to prepare the water utility for privatization (Rector 2016). Baltimore and Pittsburgh also have experienced soaring water bills and large-scale water shutoffs (Food and Water Watch 2017b). These policies raise obvious and troubling parallels to the struggles over unaffordability, shutoffs, and violations of the human right to water in South Africa and elsewhere in the South.

Concerns about unsafe drinking water in the United States have also risen sharply in recent years, drawing national attention to the unequal racial and class distribution of the risks associated with decaying water infrastructure (Miller and Wesley 2016). The case of Flint, Michigan represents a particularly dramatic and troubling illustration of these dynamics. Flint was one of several Michigan cities that the Republican governor placed under the control of appointed emergency fiscal managers (EFMs), a practice critics charge eviscerates local sovereignty and constitutes a form of disaster capitalism, by bypassing elected local officials and promoting the sale of public assets (Rector 2016). Flint's EFM in 2014 mandated that the city switch its water supply from the Detroit water system to the Flint River, which was both highly polluted and corrosive. The EFM, in a move to save $5 million, made the fateful decision not to add anti-corrosion chemicals to the water, resulting in widespread contamination from corroding pipes and causing Flint residents to drink highly toxic drinking water for over a year, with lead levels in tap water as high as 10,000 times over the EPA legal limit (Blow 2016; Butler, Scammell, and Benson 2016). This has caused severe lead poisoning, especially of Flint's children, which can lead to irreversible organ and developmental damage (Bellinger 2016). Costs to repair the system are now estimated in the billions of dollars.

Scholars have also linked water privatization more broadly to environmental injustice. Greiner (2016) finds that water utility privatization in the United States is correlated with higher populations of racial minorities, arguing that "communities that have traditionally been politically marginalized are at greater risk of being subjected to neoliberal processes," and that water system privatization should be considered "an instance of environmental inequality" (16–17). The strong racial and class element in the Flint case – the city is 63 percent African-American and disproportionately low-income – has led many critics to charge that this represents an extreme case of environmental injustice (Pulido 2016; Ranganathan 2016). Recent studies also show that far from being limited to a few cases such as Flint, lead contamination has become a widespread problem in US water systems, affecting at least 18 million people (Milman 2016).

In Europe, many municipal tap water systems had also long been in public hands, with exceptions including France, home to Veolia and Suez. Beginning in the 1980s, the UK became the site of the most dramatic experiment in water privatization, and cities in Spain and other nations also began to privatize water systems. Since the financial crisis, however, the strategy of water privatization as part of enforced austerity conditionality for debtor states – long an element of structural adjustment policies in the global South – has now been extended throughout the Eurozone periphery, with privatization mandates imposed by the "troika" of the IMF, European Union, and European Central Bank on Portugal (Bieler and Jordan 2017), Ireland (Mercille and Murphy 2015), Spain (Gonzalez-Gomez, Garcia-Rubio, and Gonzalez-Martinez 2014), and Greece (Pempetzoglou and Patergiannaki 2017). Implementation of "full-cost recovery" policies has led to substantial water shutoffs here too, and privatization has generated major countermovements (Zurita et al. 2015), particularly in Greece and in Italy, where a 2011 national referendum banned water privatization (Carrozza and Fantini 2016). In the 2019 UK General Election, the Labour Party promised to renationalize water and other privatized services if elected, a move polling shows is supported by a sizeable majority of the British public (The Guardian 2017).

Countercurrents: The Present Status of Water Privatization

The overall panorama as of this writing is mixed, with substantial geographic variation. As of 2015, private firms supplied tap water and/or sewerage services to 1.05 billion people, or 16 percent of the global population – up from 50 million people in 1990 (Arup 2015). At the same time, there has been a substantial retreat by the large multinational water firms in Latin America (with exceptions including Mexico and Colombia), in some large Southern cities including Jakarta, Indonesia and Accra, Ghana, and in western Europe. Pierce argues that this "strategic retreat" in the early- and mid-2000s has been followed by a "shallow expansion" of privatization since the Great Recession, with the multinational firms playing a reduced role, and much greater involvement by national capital, as well as private equity investors (Pierce 2015). This expansion of private water provision is bimodal: one the one hand, there is substantial growth in middle-income nations, notably China, India, and (as of 2017) Brazil, while on the other hand, it is being imposed from above through austerity in the Eurozone periphery (Pierce 2015). The enforced privatization of water in southern Europe offers a stark parallel to the earlier wave of market expansion through structural adjustment in Southern debtor states, pointing to strong continuities in the dynamics (and the agents) of accumulation by dispossession. In fact, structural adjustment in the global South has been renamed but not rescinded: nations participating in the World Bank's Highly Indebted Poor Countries (HIPC) initiative still must implement utility privatization as a precondition for securing debt relief. Indeed, Bakker (2013a: 257) terms the current state of affairs a "refinement, rather than a retrenchment, of the neoliberal project" in water.

To sum up, there are substantial countercurrents at work here. A genuine remunicipalization and deprivatization trend in some regions coexists with a countertendency toward new privatizations, most pronounced in semiperipheral nations and now in the core under neoliberal austerity, whether imposed by creditors or transmitted through fiscal crises of the state. The number of people served by privatized municipal drinking water has increased, but more gradually than many boosters or critics had expected. However, another major modality of water commodification is advancing far more rapidly: bottled water.

Bottled Water

In a few short decades, bottled water has been transformed from a niche product into a ubiquitous consumer commodity. In the United States, bottled water consumption grew from two gallons per capita per year in 1978 to over 39 gallons in 2016 (Gleick 2010; Rodwan 2017), putting it in fourth place behind Mexico, Thailand, and Italy, and surpassing soda as the nation's most-consumed beverage (Beverage Marketing Corporation 2017). China's annual bottled water consumption has doubled to over 22 billion gallons since 2011, representing one-quarter of the global total (Rodwan 2017). The international bottled water market, dominated by four transnational agrifood corporations (Nestlé, Coca-Cola, Pepsico, and Danone), was estimated at $198 billion in 2017 and is expected to reach $307 billion by 2024 (Transparency Market Research 2018). This fairly young commodity is situated at the intersection of several major sociopolitical trends, including the shift toward individualized protection from environmental risk (Szasz 2007), the effects of neoliberal austerity on public infrastructure, and falling trust in tap water.

The dynamics driving bottled water's growth differ by world region. In parts of the South, many states have been unable to extend tap water systems sufficiently as urbanization outpaces the existing infrastructure. In this context, corporations, consumers, and governments have increasingly turned to bottled water – both in single-serving containers and large multi-gallon jugs – as the solution to actual or perceived drinking water scarcity (Hawkins 2017). In many cities, there is a two-tier market: the transnational bottled water firms are buying local companies and targeting middle- and upper-income consumers with branded water, while the remaining local vendors supply working-class residents with water of often uncertain origin and quality (Girard 2009). Yet even the cheapest options are prohibitive for the poorest residents (Greene 2014), illustrating the troubling implications of this commodity for the human right to water.

In the North, the reasons for bottled water's rise are different. Bottled water firms promote their product by appealing to consumer concerns with purity, social status, and health. Industry advertising campaigns have also focused on disparaging tap water, both capitalizing on and contributing to public fears about water quality (Gleick 2010; Parag and Roberts 2009). News coverage of instances of toxic tap water, such as in Flint or in Walkerton, Ontario, further increases the demand for bottled water, despite the reality that bottled water in the United States and Canada is far less regulated than tap water (Gleick 2010). Ironically, those bottles increasingly contain purified tap water, rather than groundwater or spring water: the proportion of US bottled water drawn from already-treated municipal sources rose from 49 percent in 2008 to 63 percent in 2017 (Food and Water Watch 2017a). Among US adults, bottled water now accounts for a remarkable 44 percent of total drinking water consumption (Drewnowski, Rehm, and Constant 2013), even though it typically costs thousands of times more per unit volume than tap water, and is far less regulated (NRDC 1999).

Bottled water also generates substantial negative environmental externalities, including energy footprints up to 2,000 times higher than tap water (Gleick and Cooley 2009), major water waste in manufacturing, and the enormous terrestrial and marine waste problems generated worldwide by the disposal of approximately 500 billion bottles annually, an issue some observers have termed an ecological threat on par with climate change (Laville and Taylor 2017).

A range of movements has emerged in opposition to bottled water, which can be divided into two broad categories: those opposing water extraction by the industry, and those challenging bottled water consumption. On the consumption end, movements to "take back the tap," primarily in the North, have successfully lobbied a number of city governments, universities, and other institutions to ban purchases of bottled water, reinvest in public water infrastructure (including water fountains), and launch campaigns to promote the high quality of their tap water (Gentile 2008; Lagos 2014).

Movements against the bottling industry's groundwater extraction have developed in both South and North. While the volumes extracted are often insignificant relative to total groundwater usage, they can have damaging hydrological and ecological effects in specific localities, including depletion, drawdown, or contamination of aquifers (Gleick 2010; Barlow 2014). In North America, the majority of this facet of activism has been focused on Nestlé, the industry leader, which has found it increasingly challenging to establish new spring water and groundwater extraction sites. A decade-long attempt to establish a bottling plant in Oregon's Columbia Gorge (Jaffee and Newman 2013a) was defeated in 2017 by a coalition of grassroots groups and environmental NGOs, and in Ontario, Canada, the company has faced sustained public opposition to new and existing wells as well as increased extraction fees and regulatory scrutiny from the provincial government (Jaffee and Case 2018). Conflicts over water extraction by the bottled water and beverage industry have also emerged in the global South, notably in India, Pakistan, and Brazil (Rosemann 2005; Drew 2008; Sitisarn 2012).

Bakker has influentially framed water as an "uncooperative commodity," whose biophysical qualities – it is heavy and difficult to transport – render it a poor fit with the demands of capital accumulation (2005). In contrast, bottled water's plastic package and extreme portability, along with the political-economic shifts that have enabled its rapid growth, have allowed it to avoid many of these constraints (Jaffee and Newman 2013a). While municipal tap water systems constitute natural monopolies with substantial sunk infrastructure costs, are inherently tied to place, and have low price elasticity, bottled water is both more profitable and far more mobile, making it a "more perfect commodity" for capital accumulation (Jaffee and Newman 2013b). However, while bottled water represents an unambiguous case of commodification and a clear instance of accumulation by dispossession, it typically does not constitute outright privatization, emphasizing that the latter is far from the only way the market can assert control over water.

In closing this section, it is worth stepping back to acknowledge the social-justice and ecological costs of bottled water's dramatic growth. Considering its vastly higher carbon and water footprints and the enormous waste problems caused by disposal of half a trillion single-use plastic bottles annually, despite the ready availability of a far cheaper and less damaging alternative (tap water), it would be hard to identify another commodity that encapsulates the concept of unsustainability so dramatically. Given the prominent role it has played in filling water needs in the wake of austerity-induced crises of toxic water in Flint and elsewhere, bottled water can also be viewed as an important marker of water injustice. While it may have a legitimate short-term role to play in genuine disasters, Gleick et al. (2002: 12) argue that sales of bottled water "must not be considered acceptable substitutes for adequate municipal water supply. Bottled water rarely provides adequate volumes of water for domestic use, and the costs of such water are typically exorbitant."

Conclusion

In February 2018, municipal authorities in Cape Town, South Africa made the stunning announcement that by mid-summer, the city would reach "Day Zero" – the moment its municipal water supply would effectively run dry and water service would be cut off to over one million homes (Frederick 2018). After three years of extreme drought, the city's reservoir was virtually depleted. Government officials implemented severe water rationing – with residents allowed to fill containers at collective taps with 25 liters per day per person – but also acknowledged they were preparing for "anarchy" if the taps ran dry. In this quasi-apocalyptic context, the wealthy dug personal wells and built water storage tanks, while most residents stocked up on bottled water to fill the void, with sales (and imports) booming (Watts 2018). While rains in mid-2018 postponed "Day Zero" for the moment, other large cities, including Delhi, are now approaching similar crises. (Oaten and Patidar, 2020).

The fates of these two modes of water provision – tap and bottled – are closely linked. The growing embrace of bottled water as an acceptable, or even a preferred, mode of delivery has troubling implications not only for the century-old project of universal public provision of safe drinking water in much of the North, but also for ensuring access to drinking water worldwide. To the extent that the increasing prevalence of bottled water enables further disinvestment in public drinking water infrastructure, this commodity arguably poses a greater threat to drinking water access than privatization of municipal tap water itself (Jaffee and Newman 2013a). According to Barlow (2007: 100–101), bottled water "allows people to view water as a commodity and sets the stage – one bottle at a time – for the complete corporate takeover of water."

Water's unique status as unsubstitutable and essential for life helps to explain the intensity and vitality of movements for water justice worldwide. These countermovements have proven partially successful at reversing, slowing, or preventing privatization, and in posing obstacles to the further commodification of water. As such, they can be understood as movements for decommodification (Vail 2010), which is closely tied to the affirmation of water as a human right.

Finally, we must return to the question posed at the outset: Do we agree that affordable access to clean drinking water for all is an essential part of the social contract, or do we not? If the answer is yes, it appears highly unlikely that this right will be achieved or ensured through market provision. Although capital has asserted substantial if highly uneven control over water worldwide, hope for restoring and expanding equitable access to clean, safe drinking water for all ultimately rests in the hands of states, citizens, communities, and social movements.

References

Ahlers, Rhodante. 2005. "Gender Dimensions of Neoliberal Water Policy in Mexico and Bolivia: Empowering or Disempowering?" pp. 53–71 in *Opposing Currents: The Politics of Water and Gender in Latin America*, edited by V. Bennett, S. Davila-Poblete, and M. Nieves Rico. Pittsburgh, PA: University of Pittsburgh Press.

Ahlers, Rhodante. 2010. "Fixing and Nixing: The Politics of Water Privatization." *Review of Radical Political Economics* 42(2):213–230.

American Water Works Association. 2012. *Buried No Longer: Confronting America's Water Infrastructure Challenge*. Denver, CO: American Water Works Association.

Arup. 2015. *In-Depth Water Yearbook 2014–2015*. London: Arup.

Augenstein, Seth. 2015. "Christie Signs Law Greenlighting Fast Track Sale of N.J. Public Water Systems." *NJ.com* (February 5). Retrieved February 15, 2018 (www.nj.com/politics/index.ssf/2015/02/christie_signs_law_greenlighting_sales_of_public_water_systems.html).

Bakker, Karen. 2005. "Neoliberalizing Nature? Market Environmentalism in Water Supply in England and Wales." *Annals of the Association of American Geographers* 95(3):542–565.

Bakker, Karen. 2007. "The 'Commons' Versus the 'Commodity': Alter-Globalization, Anti-Privatization and the Human Right to Water in the Global South." *Antipode* 39(3):430–455.

Bakker, Karen. 2010. *Privatizing Water: Governance Failure and the World's Urban Water Crisis*. Ithaca, NY: Cornell University Press.

Bakker, Karen. 2013a. "Neoliberal Versus Postneoliberal Water: Geographies of Privatization and Resistance." *Annals of the Association of American Geographers* 103(2):253–260.

Bakker, Karen. 2013b. "Constructing 'Public' Water: The World Bank, Urban Water Supply, and the Biopolitics of Development." *Environment and Planning D-Society & Space* 31(2):280–300.

Bakker, Karen. 2014. "The Business of Water: Market Environmentalism in the Water Sector." *Annual Review of Environment & Resources* 39(1):469–494.

Barlow, Maude. 2007. *Blue Covenant: The Global Water Crisis and the Coming Battle for the Right to Water*. New York: The New Press.

Barlow, Maude. 2014. *Blue Future: Protecting Water for People and the Planet Forever.* New York: The New Press.

Barlow, Maude and Tony Clarke. 2002. *Blue Gold: The Fight to Stop the Corporate Theft of the World's Water.* New York: The New Press.

Bellinger, David C. 2016. "Lead Contamination in Flint – an Abject Failure to Protect Public Health." *New England Journal of Medicine* 374 (March 24):1101–1103.

Beverage Marketing Corporation. 2017, "Bottled Water Becomes Number-One Beverage in the US, Data from Beverage Marketing Corporation Show." Retrieved March 10, 2018 (www.beveragemarketing.com/news-detail.asp?id=438).

Bieler, Andreas, and Jamie Jordan. 2017. "Commodification and 'the Commons': The Politics of Privatising Public Water in Greece and Portugal During the Eurozone Crisis." *European Journal of International Relations* 24(4):934–957. Doi:1354066117728383.

Blow, Charles M. 2016. "The Poisoning of Flint's Water." *New York Times* (January 21), www.nytimes.com/2016/01/21/opinion/the-poisoning-of-flints-water.html.

Bond, Patrick. 2008. "Macrodynamics of Globalisation, Uneven Urban Development and the Commodification of Water." *Law, Social Justice and Global Development (An Electronic Law Journal)* 11:1–14.

Bond, Patrick. 2012. "The Right to the City and the Eco-Social Commoning of Water: Discursive and Political Lessons from South Africa." pp. 190–205 in *The Right to Water*, edited by F. Sultana and A. Loftus. London: Earthscan.

Butler, Lindsey J., Madeleine K. Scammell and Eugene B. Benson. 2016. "The Flint, Michigan, Water Crisis: A Case Study in Regulatory Failure and Environmental Injustice." *Environmental Justice* 9(4):93–97.

Campero Arena, Claudia. 2016. Water Campaigner, Blue Planet Project (Personal Communication). Mexico City, January 30.

Carrozza, Chiara and Emanuele Fantini. 2016. "The Italian Water Movement and the Politics of the Commons." *Water Alternatives-an Interdisciplinary Journal on Water Politics and Development* 9(1):99–119.

Casarin, Ariel A., Jose A. Delfino and Maria Eugenia Delfino. 2007. "Failures in Water Reform: Lessons from the Buenos Aires's Concession." *Utilities Policy* 15 (4):234–247.

Castro, José Esteban. 2007. "Poverty and Citizenship: Sociological Perspectives on Water Services and Public-Private Participation." *Geoforum* 38:756–771.

Castro, José Esteban. 2008. "Water Struggles, Citizenship and Governance in Latin America." *Development* 51:72–76.

Conca, Ken. 2008. "The United States and International Water Policy." *Journal of Environment & Development* 17(3):215–237.

Drew, Georgina. 2008. "From the Groundwater Up: Asserting Water Rights in India." *Development* 51(1): 37–41.

Drewnowski, Adam, Colin D. Rehm, and Florence Constant. 2013. "Water and Beverage Consumption Among Adults in the United States: Cross-Sectional Study Using Data from NHANES 2005–2010." *BMC Public Health* 13(1):1–19 doi:10.1186/1471-2458-13-1068.

Driessen, Travis. 2008. "Collective Management Strategies and Elite Resistance in Cochabamba, Bolivia." *Development* 51(1):89–95.

Dugard, Jackie. 2010. "Can Human Rights Transcend the Commercialization of Water in South Africa? Soweto's Legal Fight for an Equitable Water Policy." *Review of Radical Political Economics* 42(2):175–194.

Esterl, Mike. 2006. "Dry Hole: Great Expectations for Private Water Fail to Pan Out." *Wall Street Journal* (June 26), www.wsj.com/articles/SB115128641717890452.

Food and Water Watch. 2011. *Water = Life: How Privatization Undermines the Human Right to Water*. Washington, DC: Food and Water Watch.

Food and Water Watch. 2012a. *Private Equity Public Inequity: The Public Cost of Private Equity Takeovers of U.S. Water Infrastructure*. Washington, DC: Food and Water Watch.

Food and Water Watch. 2012b. *Public–Public Partnerships: An Alternative Model to Leverage the Capacity of Municipal Water Utilities*. Washington, DC: Food and Water Watch.

Food and Water Watch. 2016. *The State of Public Water in the United States*. Washington, DC: Food and Water Watch.

Food and Water Watch. 2017a. *Take Back the Tap: The Big Business Hustle of Bottled Water*. Washington, DC: Food and Water Watch.

Food and Water Watch. 2017b. *Water Injustice: Economic and Racial Disparities in Access to Safe and Clean Water in the United States*. Washington, DC: Food and Water Watch.

Frederick, Franklin. 2018. "Water: The Warning Coming from South Africa." *Dawn News* (February 15). Retrieved February 28, 2018 (www.thedawn-news.org/2018/02/27/water-the-warning-coming-from-south-africa/).

Gentile, Annie. 2008. "Mayors Push Benefits of Cities' Tap Water." *American City & County* 123(9):18–20.

Girard, Richard. 2009. "Bottled Water Industry Targets a New Market: The Global South." *AlterNet*, June 15. Accessed September 1, 2011. www.alternet.org/module/printversion/140671.

Glassman, Jim. 2006. "Primitive Accumulation, Accumulation by Dispossession, Accumulation by 'Extra-Economic' Means." *Progress in Human Geography* 30(5):608–625.

Gleick, Peter H., Gary Wolff, Elizabeth L. Chalecki, and Rachel Reyes. 2002. *The New Economy of Water: The Risks and Benefits of Globalization and Privatization of Fresh Water*. Oakland, CA: Pacific Institute.

Gleick, Peter H. 2010. *Bottled and Sold: The Story Behind Our Obsession with Bottled Water*. Washington, DC: Island Press.

Gleick, Peter H., and Heather S. Cooley. 2009. "Energy Implications of Bottled Water." *Environmental Research Letters* 4(1):1–6. doi:10.1088/1748-9326/4/1/014009.

Godoy, Emilio. 2010. "Mexico: Soaring Bottled Water Use Highlights Mistrust of Tap Water." *Inter Press Service*, www.globalissues.org/news/2010/09/23/7056.

Goff, Matthew and Ben Crow. 2014. "What Is Water Equity? The Unfortunate Consequences of a Global Focus on 'Drinking Water'." *Water International* 39(2):159–171.

Goldman, Michael. 2005. *Imperial Nature: The World Bank and Struggles for Social Justice in the Age of Globalization*. New Haven CT: Yale University Press.

Goldman, Michael. 2007. "How 'Water for All!' Policy Became Hegemonic: The Power of the World Bank and Its Transnational Policy Networks." *Geoforum* 38:786–800.

Gonzalez-Gomez, Francisco, Miguel A. Garcia-Rubio, and Jesús Gonzalez-Martinez. 2014. "Beyond the Public-Private Controversy in Urban Water Management in Spain." *Utilities Policy* 31:1–9.

Greene, Joshua Cullen. 2014. "The Bottled Water Industry in Mexico." Master of Global Policy Studies Master's thesis, University of Texas at Austin.

Greiner, Patrick. 2016. "Social Drivers of Water Utility Privatization in the United States: An Examination of the Presence of Variegated Neoliberal Strategies in the Water Utility Sector." *Rural Sociology* 81(3):387–406.

Hall, David, Emanuele Lobina and Robin De la Motte. 2009. *Making Water Privatisation Illegal: New Laws in Netherlands and Uruguay*. London: Public Service International Research Unit.

Hall, David, Emanele Lobina and Violeta Corral. 2011. *Trends in Water Privatization*. London: Public Services International Research Unit.

Harris, Leila M. 2013. "Variable Histories and Geographies of Marketization and Privatization." pp. 118–32 in *Contemporary Water Governance in the Global South: Scarcity, Marketization and Participation*, edited by L. M. Harris, J. A. Goldin, and C. Sneddon. London: Routledge.

Harvey, David. 2003. *The New Imperialism*. New York: Oxford University Press.

Hawkins, Gay. 2017. "The Impacts of Bottled Water: An Analysis of Bottled Water Markets and their Interactions with Tap Water Provision." *WIREs Water* 4 (3). doi:10.1002/wat2.1203.

Jaffee, Daniel, and Soren Newman. 2013a. "A Bottle Half Empty: Bottled Water, Commodification, and Contestation." *Organization & Environment* 26(3):318–335.

Jaffee, Daniel, and Soren Newman. 2013b. "A More Perfect Commodity: Bottled Water, Global Accumulation, and Local Contestation." *Rural Sociology* 78(1):1–28.

Jaffee, Daniel, and Robert A. Case. 2018. "Draining Us Dry: Scarcity Discourses in Contention over Bottled Water Extraction." *Local Environment* 23(4):485–501.

Kishimoto, Satoko, and Olivier Petitjean. 2017. *Reclaiming Public Services: How Cities and Citizens Are Turning Back Privatisation*. Amsterdam: Transnational Institute.

Klein, Naomi. 2007. *The Shock Doctrine: The Rise of Disaster Capitalism*. Toronto: Knopf Canada.

Kloppenburg, Jack R. 2004. *First the Seed: The Political Economy of Plant Biotechnology, 1492–2000 (2nd Ed.)*. Madison, WI: University of Wisconsin Press.

Kornberg, Dana. 2016. "The Structural Origins of Territorial Stigma: Water and Racial Politics in Metropolitan Detroit, 1950s-2010s." *International Journal of Urban and Regional Research* 40(2):263–283.

Kurland, Nancy B. and Deone Zell. 2011. "Water and Business: A Taxonomy and Review of the Research." *Organization & Environment* 23(3):316–353.

Lagos, Marisa. 2014. "S.F. Supervisors Back Ban on Sale of Plastic Water Bottles." *San Francisco Chronicle* (March 5), www.sfgate.com/bayarea/article/S-F-supervisors-back-ban-on-sale-of-plastic-5289089.php.

Laville, Sandra, and Matthew Taylor. 2017. A Million Bottles a Minute: World's Plastic Binge "as Dangerous as Climate Change." *The Guardian* (June 28).

Laxer, Gordon, and Dennis Soron. 2006. *Not for Sale: Decommodifying Public Life*. Peterborough, Canada: Broadview.

Lederer, Edith. 2010. "Access to Clean Water Is 'Human Right,' Says U.N." *The Independent* (July 30), www.independent.co.uk/news/world/politics/access-to-clean-water-is-human-right-says-un-2039083.html.

Liddle, Elisabeth S., Sarah M. Mager, and Etienne L. Nel. 2014. "The Importance of Community-Based Informal Water Supply Systems in the Developing World and the Need for Formal Sector Support." *The Geographical Journal*. doi:10.1111/geoj.12117.

Lobina, Emanuele, Philipp Terhorst and Vladimir Popov. 2011. "Policy Networks and Social Resistance to Water Privatization in Latin America." *Procedia Social and Behavioral Sciences* 10:19–25.

Loftus, Alex. 2009. "Rethinking Political Ecologies of Water." *Third World Quarterly* 30 (5):953–968.

Mack, Elizabeth A. and Sarah Wrase. 2017. "A Burgeoning Crisis? A Nationwide Assessment of the Geography of Water Affordability in the United States." *PLoS One* 12(1). Doi:10.1371/journal.pone.0169488.

Malkin, Elisabeth. 2012. "Bottled-Water Habit Keeps Tight Grip on Mexicans." *New York Times* (July 17), www.nytimes.com/2012/07/17/world/americas/mexicans-struggle-to-kick-bottled-water-habit.html.

Marx, Karl. 1867. *Capital: A Critique of Political Economy*. Chicago, IL: Charles H. Kerr & Co.

McDonald, David A. 2016. "To Corporatize or Not to Corporatize (and If So, How?)." *Utilities Policy* 40:107–114.

Mehta, Lyla, Jeremy Allouche, Alan Nicol, and Anna Walnycki. 2014. "Global Environmental Justice and the Right to Water: The Case of Peri-Urban Cochabamba and Delhi." *Geoforum* 54:158–166.

Mehta, Lyla. 2016. "Why Invisible Power and Structural Violence Persist in the Water Domain." *IDS Bulletin* 47(5):31–42.

Mercille, Julien and Enda Murphy. 2015. "Conceptualising European Privatisation Processes after the Great Recession." *Antipode* 48(3):685–704.

Miller, DeMond Shondell and Nyjeer Wesley. 2016. "Toxic Disasters, Biopolitics, and Corrosive Communities: Guiding Principles in the Quest for Healing in Flint, Michigan." *Environmental Justice* 9(3):69–75.

Milman, Oliver. 2016. "Millions Exposed to Dangerous Lead Levels in US Drinking Water, Report Finds." *The Guardian* (June 28), www.theguardian.com/environment/2016/jun/28/lead-drinking-water-level-nrdc-report-flint-crisis.

Narain, Vishal. 2014. "Whose Land? Whose Water? Water Rights, Equity and Justice in a Peri-Urban Context." *Local Environment* 19(9):974–989.

Nelson, Paul. 2017. "Citizens, Consumers, Workers, and Activists: Civil Society During and after Water Privatization Struggles." *Journal of Civil Society* 13(2):202–221.

NRDC. 1999. *Bottled Water: Pure Drink or Pure Hype?* Washington, DC: Natural Resources Defense Council.

Oaten, James, and Som Patidar. 2020. Delhi is Facing a Water Crisis. Ahead of Day Zero, the City's Residents Have Turned to the Mafia and Murder. *ABC News* (February 8). Accessed March 11, 2020, www.abc.net.au/news/2020-02-08/delhi-water-crisis-leads-to-mafia-murder-and-mutiny/11931208

O'Reilly, Kathleen. 2011. "'They Are Not of This House': The Gendered Costs of Drinking Water's Commodification." *Economic and Political Weekly* 46(18):49–55.

Parag, Y., and J. T. Roberts. 2009. "A Battle against the Bottles: Building, Claiming, and Regaining Tap-Water Trustworthiness." *Society & Natural Resources* 22(7):625–36.

Peck, Jamie. 2015. *Austerity Urbanism: The Neoliberal Crisis of American Cities*. New York: Rosa Luxemburg Stiftung.

Pempetzoglou, Maria and Zoi Patergiannaki. 2017. "Debt-Driven Water Privatization: The Case of Greece." *European Journal of Multidisciplinary Studies* 5(1):102–111.

Perreault, Tom. 2013. "Dispossession by Accumulation? Mining, Water and the Nature of Enclosure on the Bolivian Altiplano." *Antipode* 45(5):1050–1069.

Pierce, Gregory. 2015. "Beyond the Strategic Retreat? Explaining Urban Water Privatization's Shallow Expansion in Low- and Middle-Income Countries." *Journal of Planning Literature* 30(2):119–131.

Polanyi, Karl. 1944. *The Great Transformation*. Boston, MA: Beacon Press.

Pulido, Laura. 2016. "Flint, Environmental Racism, and Racial Capitalism." *Capitalism Nature Socialism*. 17(3): 1–16. Doi:http://dx.doi.org/10.1080/10455752.2016.1213013.

Ranganathan, Malini and Carolina Balazs. 2015. "Water Marginalization at the Urban Fringe: Environmental Justice and Urban Political Ecology across the North-South Divide." *Urban Geography* 36:403–423.

Ranganathan, Malini. 2016. "Thinking with Flint: Racial Liberalism and the Roots of an American Water Tragedy." *Capitalism Nature Socialism* 27(3):17–33.

Rankin, Jennifer and Helena Smith. 2015. "The Great Greece Fire Sale." *The Guardian* (July 24), www.theguardian.com/business/2015/jul/24/greek-debt-crisis-great-greece-fire-sale

Rector, Josiah. 2016. "Neoliberalism's Deadly Experiment: In Michigan, Privatization and Free-Market Governance Has Left 100,000 People Without Water." *Jacobin* (October 21), www.jacobinmag.com/2016/10/water-detroit-flint-emergency-management-lead-snyder-privatization/.

Roberts, Adrienne. 2008. "Privatizing Social Reproduction: The Primitive Accumulation of Water in an Era of Neoliberalism." *Antipode* 40(4):535–560.

Robinson, Joanna. 2013. *Contested Water: The Struggle Against Water Privatization in the United States and Canada*. Cambridge, MA: MIT Press.

Rodwan, John G. 2017. "Bottled Water 2016: No. 1 and Growing: U.S. And International Developments and Statistics." *Bottled Water Reporter* 57(4) (July/Aug):12–21.

Romano, Sarah T. 2012. "From Protest to Proposal: The Contentious Politics of the Nicaraguan Anti-Water Privatization Social Movement." *Bulletin of Latin American Research* 31(4):499–514.

Rosemann, Nils. 2005. *Drinking Water Crisis in Pakistan and the Issue of Bottled Water: The Case of Nestlé's "Pure Life."* Berne, Switzerland: ActionAid.

Ruiters, Greg. 2007. "Contradictions in Municipal Services in Contemporary South Africa: Disciplinary Commodification and Self-Disconnections." *Critical Social Policy* 27 (4):487–508.

Rushe, Dominic. 2014. "Blow to Detroit's Poorest as Judge Rules Water Shutoffs Can Continue." *The Guardian* (September 29), www.theguardian.com/world/2014/sep/29/detroit-water-shutoffs-legal-judge-bankruptcy-revenue.

Sitisarn, Savarin. 2012. "Political Ecology of the Soft Drink and Bottled Water Business in India; a Case Study of Plachimada." Master's thesis, Lund University.

Sloan, Carrie. 2016. "How Wall Street Caused a Water Crisis in America's Cities." *The Nation* (March 11), www.thenation.com/article/how-wall-street-caused-a-water-crisis-in-americas-cities/.

Smith, Laila and Susan Hanson. 2003. "Access to Water for the Urban Poor in Cape Town: Where Equity Meets Cost Recovery." *Urban Studies* 40(8):1517.

Snitow, Alan, Deborah Kaufman and Michael Fox. 2007. *Thirst: Fighting the Corporate Theft of Our Water*. San Francisco, CA: Wiley & Sons.

Spronk, Susan and Jeffery R. Webber. 2007. "Struggles against Accumulation by Dispossession in Bolivia: The Political Economy of Natural Resource Contention." *Latin American Perspectives* 34(2):31–47.

Spronk, Susan. 2015. "Roots of Resistance to Urban Water Privatisation in Bolivia: The 'New Working Class', the Crisis of Neoliberalism, and Public Services." pp. 29–51 in *Crisis and Contradiction: Marxist Perspectives on Latin America in the Global*

Political Economy, edited by S. J. Spronk and J. R. Webber. Leiden, Netherlands: Brill.

Subramaniam, Mangala and Beth Williford. 2012. "Contesting Water Rights: Collective Ownership and Struggles against Privatization." *Sociology Compass* 6(5):413–424.

Sultana, Farhana. 2011. "Suffering for Water, Suffering from Water: Emotional Geographies of Resource Access, Control and Conflict." *Geoforum* 42:163–172.

Sultana, Farhana, and Alex Loftus, eds. 2012. *The Right to Water: Politics, Governance, and Social Struggles*. London: Earthscan.

Swyngedouw, Erik. 2005. "Dispossessing H2O: The Contested Terrain of Water Privatization." *Capitalism Nature Socialism* 16(1):81–98.

Swyngedouw, Erik. 2009. "The Political Economy and Political Ecology of the Hydro-Social Cycle." *Journal of Contemporary Water Research & Education* 142(1):56–60.

Szasz, Andrew. 2007. *Shopping Our Way to Safety: How We Changed from Protecting the Environment to Protecting Ourselves*. Minneapolis, MN: University of Minnesota Press.

Terhorst, Philipp, Marcela Olivera and Alexander Dwinell. 2013. "Social Movements, Left Governments, and the Limits of Water Sector Reform in Latin America's Left Turn." *Latin American Perspectives* 40:55–69.

The Guardian. 2017. "Jeremy Corbyn's Nationalisation Plans Are Music to Ears of Public." *The Guardian* (October 1), www.theguardian.com/business/2017/oct/01/jeremy-corbyn-nationalisation-plans-voters-tired-free-markets

Transparency Market Research. 2018. "Global Bottled Water Market is Projected to Reach US$307.2 Billion by 2024." Press Release (April). Retrieved July 2, 2018 (www.transparencymarketresearch.com/pressrelease/bottled-water-market.htm).

UNDP. 2006. "Human Development Report 2006: Beyond Scarcity: Power, Poverty and the Global Water Crisis." New York: United Nations Development Program.

United Nations. 1992. "Dublin Statement on Water and Sustainable Development." New York: United Nations.

Vail, John. 2010. "Decommodification and Egalitarian Political Economy." *Politics & Society* 38(3):310–346.

Varghese, Shiney. 2007. Privatizing U.S. Water. Minneapolis, MN: Institute for Agriculture and Trade Policy.

Vidal, John. 2006. "Big Water Companies Quit Poor Countries." *The Guardian* (March 22), www.theguardian.com/world/2006/mar/22/globalisation.water.

Vilas, Carlos. 2004. "Water Privatization in Buenos Aires." *NACLA Report on the Americas* 38(1):34–42.

Wait, Isaac W. and William A. Petrie. 2017. "Comparison of Water Pricing for Publicly and Privately Owned Water Utilities in the United States." *Water International* 42 (8):967–980.

Watts, Jonathan. 2018. "Cape Town Faces Day Zero: What Happens When the City Turns Off the Taps?" *The Guardian* (February 3), www.theguardian.com/cities/2018/feb/03/day-zero-cape-town-turns-off-taps.

WHO and UNICEF. 2017. Progress on Drinking Water, Sanitation and Hygiene, 2017. Geneva: World Health Organization and United Nations Children's Fund.

World Bank. 2005. Infrastructure Development: The Roles of the Public and Private Sectors; World Bank Group's Approach to Supporting Investments in Infrastructure. Washington, DC: World Bank.

Zurita, Maria D. M., Dana C. Thomsen, Timothy F. Smith et al. 2015. "Reframing Water: Contesting H2O Within the European Union." *Geoforum* 65: 170–178.

Zwarteveen, Margreet Z. and Rutgerd Boelens. 2014. "Defining, Researching and Struggling for Water Justice: Some Conceptual Building Blocks for Research and Action." *Water International* 39(2):143–158.

19 Speech Is Silver, Silence Is Gold in the Fracking Zone

Debra J. Davidson

1 Introduction

Speech is silver and silence is golden, or so the old proverb says. A good story can offer rewards, but sometimes those rewards are much larger if we keep our mouth shut. The rewards, however, may not necessarily go to the one who is silenced. While the original scribe who put this proverb to pen likely never knew of the fossil fuels that would change the course of history, this proverb has been given new meaning in extraction zones, where stories told about the vast wealth and wellbeing such resources allow are shiny as silver, but these stories only reward investors with gold when the victims of the resulting environmental and health costs are silenced.

In rural, agricultural communities, silence is indeed gold, but not the cold hard cash variety. The silence of the Prairies is the quiet rustle through the fields, also golden, at harvest time. It is the sound of a woodpecker far off in the towering spruce trees, and the slow, quiet chatter of folks not inclined to complain at the local diner. Here, many locals keep to themselves as a matter of course, and they hold dear the slow quiet of rural living. In communities confronting hydraulic fracturing in the southern Prairies of Alberta, however, a new noise has erupted, followed by a new form of silence. Following on the cacophonous onslaught of heavy equipment, the relentless pounding of compressor stations, and the scream of truck brakes at all hours of the day and night, the voices of concerned residents, unused to being raised in the first place, are being silenced.

This story is not unique to southern Alberta. The flip side of manufacturing consent (Hermann and Chomsky 1988), after all, is silencing dissent, and we can observe escalating efforts on both fronts among the fossil fuel elite. Continuation of wealth generation among fossil fuel elites requires the development of increasingly marginal resources. Tensions in extraction zones and transportation corridors around the globe have escalated as conventional reserves of oil and gas have become depleted, combined with technological developments that have rendered previously "unmarketable" low quality reserves like shale, bitumen, and deep-sea reserves, referred to as non-conventional fuels, attractive to investors. All of which are associated with lower returns, and higher risks and consequences, however. Davidson (2017; Davidson and Andrews 2013) documents a steep increase in the marginal impact of oil and gas development in Alberta, even when only conventional drilling is included, but those impacts increase substantially with growing reliance on non-conventional fuels.

Manufacturing consent for such enterprises requires ever-taller tales of riches; Jasanoff and Kim (2009; 2013) refer to those tales with a particularly strong influence on the course of future pathways as "sociotechnical imaginaries." But it also requires muzzling the growing chorus of people who have come to recognize the folly of such tales – many learning first-hand of such folly. Silencing tactics can include formal measures, such as new legislation that constrains who has "standing" to voice their concerns about new development, to other more subtle and personal means, including cooptation, invalidation, or intimidation. In this chapter, I provide a selection of findings from an ongoing case study initiated in 2016 of the social impacts of and responses to hydraulic fracturing in Alberta, in which each of these strategies appears to have been deployed. The study has included mass media and documentary analysis, including *Hansard,* a publicly available online database containing the transcripts of all sessions of the Legislative Assembly of Alberta. In addition, I conducted qualitative, in-person interviews with a snowball sample of twenty rural residents who have experienced the impacts of hydraulic fracturing near their homes, identified initially through publicly available sources. All interviewees, about equal numbers of men and women, are long-term or lifetime residents, and most are farmers. Interviews took anywhere between 1 and 5 hours, and most took place at the interviewee's residence. These were supplemented with a small set of key informant interviewees in government (3), elected officials (2), environmental organizations (2), and industry (1), again identified through publicly available sources, and subsequently through snowball sampling.

2 Conceptual Background

Jasanoff's sociotechnical imaginaries are "collectively imagined forms of social life and social order reflected in the design and fulfillment of nation-specific scientific and/or technological projects" (Jasanoff and Kim 2009: 120). Such storylines are powerful forces in many forms of collective decision making, including decisions regarding energy development, particularly in their characterization of risks and benefits (Jasanoff and Kim 2013). Kuchler (2017) elaborates in his application of socio-technical imaginaries to the means by which shale reserves in Europe were essentially brought to life. Socio-technical imaginaries, according to Kuchler (2017: 33), are "capable of facilitating and/or influencing techno-scientific trajectories through projections of what is considered desirable and attainable in terms of current and anticipated knowledge." Sociotechnical imaginaries are supported by scientific and technical knowledge, such as reserve estimates, which make a resource visible and calculable, sometimes erroneously. They can also, ironically, be supported by the *absence* or *concealment* of knowledge, in order to justify claims of safety, through "scientific certainty argumentation strategies" (Freudenburg, Gramling, and Davidson 2008). Socio-technical imaginaries are neither true nor false, but they can be more or less supported by strongly held beliefs that may well be more imaginary than real, and consequently contain contradictions.

As a result, such storylines are vulnerable to discrediting by anyone who has the ability to notice the stories' cracks and contradictions, and the means to share those observations with others. According to Habermas (1970), concealing those cracks and contradictions poses a fundamental challenge for advanced industrial societies that proclaim to be democratic. Gramsci (1971) paid particular attention to the means by which such legitimacy battles are avoided, arguing that the ruling class goes to great lengths to generate consent for its dominant position among the dominated, particularly involving processes of cultural reproduction and representation. While Gramsci (1971) viewed such processes as involving continuous struggle, Murray Edelman (1964), argued that this concealment in most cases is not actually that difficult. Politics for the majority is rather more like a "spectator sport," (p. 5), involving rituals, myths and symbols that can be enormously effective at eliciting continued allegiance, while processes that allocate material resources to particular elite groups remain unremarked upon.

Environmental hazards are a particular quandary for democracy (e.g. Adam 1996; Latour 2004), and thus particularly susceptible to such concealment strategies. As environmental sociologists have catalogued in numerous research studies, the "material resources" at stake are doubly disproportional, involving the disproportionate accumulation of wealth associated with development activities, and the disproportionate accumulation of environmental disruption as a consequence of those activities, disruption that in almost all cases is borne by those who have *not* enjoyed the material benefits.

One key vehicle through which symbols, myths and rituals, and ultimately political power materialize is discourse – who are seen as legitimate participants in that discourse; the information available to inform that discourse; and the framing of that information (Foucault 1979). A long research record dating at least back to Maarten Hajer's foundational book (1995) describes state and corporate discursive efforts to conceal the private accumulation of wealth generated by environmentally risky development, and the imposition of environmental costs onto publics, in order to avoid constraining economic development. The work of the late William Freudenburg has been particularly noteworthy in this regard. In 2005, Freudenburg articulated the two faces of disproportionality as a "double diversion": the diversion of rights and resources into the coffers of the privileged, combined with the diversion of attention by members of that elite group, through skilful dissemination of "privileged accounts." Privileged accounts may include, for example, the claim that environmental regulations would impose undue burdens on our economies. Or, as consumers we are all (equally) responsible for anthropogenic climate change. Sociotechnical imaginaries can be viewed as accounts with a particular form of privilege, based as they often are on scientific expertise.

In a separate paper, Freudenburg and Alario (2007) elaborate on the mechanics behind diversionary reframing, which they liken to "Magicianship." Diversionary reframing involves both the vociferous proclamation of privileged accounts, and the diverting of attention away from challenges to those accounts. Such efforts are most effective if contradictions and inequities can fairly readily be kept dormant or made to disappear from view, as is the case with invisible contaminants, for example (Edelstein 2004). The more this concealment becomes difficult, however, the more acute is the need for the magician to find ever more elaborate ways to shift attention. One particularly effective means of doing so is to question the legitimacy of the whistle blowers. In effect, say Freudenburg and Alario (2007: 161), a skillful response to the question "will your industry threaten the health of local children?" would be to raise a different question entirely, like "are these critics against economic development?" The end result – attention is drawn *away* from the shaky support for the claims of polluters and *toward* the (il)legitimacy of dissenters. According to Blüdorn (2007; 2013), these efforts appear to have been enormously effective at stifling ecological critique, to such an extent that we now embrace a sort of simulative democracy, involving narratives of reassurance that hazards are taken seriously, and being addressed (with technological expertise), with the benefit of rituals of democratic participation. Meanwhile, the more substantive state actions undertaken behind the scenes disperse political responsibility, obscure chains of accountability, and delimit the rights of citizens to participate.

3 Silver and Gold on Fracking's Frontlines

Fracking, short for hydraulic fracturing, refers to forcing open cracks in solid substrates with the injection of liquid at high pressure, in order to release the trapped oil or gas. It is not a new technology, but it has taken a few significant turns over the past decade, toward horizontal, multi-stage hydraulic fracturing. Shale deposits are characteristically wide but shallow, so technological advances allowing a horizontal approach to the deposit have greatly enhanced the marketability of these deposits. Drilling is also typically multi-stage, referring to the fact that several fracks take place from the same well site. Beneath the surface of a well site, in other words, is hidden a spiderlike organism with a central, vertical body, and several legs spreading from the centre that can be several kilometres long. Other significant technological shifts include vast increases in the pressure applied, necessary given the tremendous distances from the well bore that the drill extends, and the introduction of new suites of chemicals for everything from emulsification to biocides.

Proponents hail fracking as an economic "game changer" that creates jobs, personal wealth, tax revenues, rural revitalization, and energy independence, among other things (Ladd and York 2017). Even some environmental groups are on side, depicting natural gas as a necessary "bridge fuel" to a renewable energy future that will lower carbon emissions (Cosgrove et al. 2015; Gold, 2014; Yergin, 2011 [cited in Ladd and York 2017]). The impacts of fracking, however, are starting to receive scientific attention, including concern about air pollution and greenhouse gases (e.g. Roy, Adams, and Robinson 2014; Howarth, Santoro, and Ingraffea 2012); water quantity and quality (e.g. Warner et al. 2013; Myers 2012); earthquakes (e.g. Davies et al. 2013); and human health implications (e.g. Kassotis et al. 2014; Colborn et al. 2011). The prevailing finding from a national study by the Canadian Water Network further raised concerns about the lack of knowledge about fracking impacts (Ryan et al. 2015). Canada's Council of Canadian Academies (2014) has also raised concerns about known risks, and the extent of unknowns regarding potentially serious hazards. In a broad assessment of the current state of scientific knowledge, Small and colleagues (2014: 8299) concur, identifying what they define as "acute deficits of knowledge," including in relation to public health. Despite these gaps, Rahm and Riha (2014) note that the scientific assessments of the impacts of fracking in the Marcellus shale region in the United States are now extensive, and confirm a number of risks posed by fracking, to water supplies in particular. These authors also provide evidence of the consistently high numbers of violations of existing standards for protecting water supplies committed by fracking companies in the Marcellus region.

Studies of social impacts are few in number, but noteworthy. Gullion (2015) catalogues social impacts and responses to fracking in an urban region in Texas. Residents expressed deep anxiety over the uncertainty and invisibility of the contaminants to which they were exposed, and outrage at the lack of acknowledgement of their concerns, leading to loss of confidence that their governments will serve the interests of citizens, and disruptions in family and community life. Perry (2012) focuses on a rural Pennsylvania region subject to fracking, in which residents have been dealing with increases in traffic, noise, strangers in town, changes in the smell and appearance of drinking water, chemical spills, and increased violence. Fracking often takes place in previously quiet rural enclaves, and thus residents also complain that their community has become an industrial park, and "the place that was once their sanctuary became a disaster area" (Jerolmack and Berman 2016: 199). Studies also note the intensity of community divisions that can emerge (Jerolmack and Berman 2016; Perry 2012; Willow and Wylie 2014). Jerolmack and Berman (2016) describe how many residents consider fracking's impacts an acceptable price for much-needed economic development, but others disagree. These divisions are further exemplified by the fact that some landowners receive rent for the facilities placed on their land, while others receive no income but nonetheless may be exposed to the impacts of those facilities.

Citizen alarm and opposition to fracking are consequently on the rise in several regions (Szeman 2012; Elliott and Short 2014; O'Riordan 2015). According to Elliott and Short (2014), the anti-fracking movement is the fastest growing social movement in the UK, currently with over 180 local groups, up from around 30 just a year earlier. In many jurisdictions, these voices have been met with a counteroffensive involving several silencing tactics, both by overt means, such as the use of excessive intimidation against protestors (O'Riordan 2015; Elliott and Short 2014), and by more subtle means, through the privileging of "expertise," and the crafting of new laws that delimit who has the right to participate (Hudgins and Poole 2014).

4 Alberta's Fracking Frontline

Albertans are accustomed to the noise, occasional spills, industrial infrastructure, and wealth associated with oil and gas extraction, and the slow introduction of modern fracking onto the landscape in many ways simply blended in. Generating between 20 and 30 percent of provincial GDP in any given year (it's substantially lower as of this writing due to a downturn in the sector), government revenues rely heavily upon royalties from oil and gas, and are directly affected by changes in the prices of those commodities. In an effort to boost production at a time of low oil prices combined with declining reserves in recent years, the Province initiated an "Enhanced Approval Process" to speed up the processing of new license applications,[1] and further lowered what were already among the lowest royalties in the world.

This political-economic structure and its institutional logic have painted Alberta's landscape with a broad brush, but that landscape was far from empty to begin with. Combined with this energy identity is a cultural system that celebrates Alberta's ranching history. The ranching families that still populate the countryside are part of an old economic order that has dwindled in fiscal importance. Agriculture still holds a respectable place in the provincial economy, but supports far fewer jobs. As with other parts of the Western world, the number of farms in Alberta has been steadily and precipitously declining for decades. Yet the cultural antecedents of this economic order linger. This heritage is also reflected in politics. Right-wing populism was born on the range in Alberta's ranching country over a century ago, and continues to characterize Alberta's politics today. Right wing populism is embodied in the Progressive Conservative (PC) Party (recently re-vamped and re-named the United Conservative Party), which remained at the helm of Alberta's government for over four decades. While urban residents voted in a surprising majority New Democratic Party (NDP) in 2015, among ranching communities conservative populist political beliefs remain strong, and the NDP was ousted in 2019 after a single term.

4.1 Silver-Lined Socio-Technical Imaginaries

Albertans are no strangers to official story-telling, particularly stories about oil and gas as leading protagonists in the larger story that is Alberta. With the rapid expansion of oilsands production shortly after the turn of the twenty-first century, development proponents who had previously taken comfort in the province's quiet backwater status were blindsided by the glaringly divergent interpretations of their stories of heritage, enterprise, and most of all, wealth. In response to international outcries following the Internet spread of graphic images of oilsands mines and tailings ponds, the government launched a well-resourced public relations campaign in an attempt to shift the story back onside, this time in celebration of the purported "ethical" and (relatively) clean character of Alberta crude (Davidson and Gismondi 2011). Such efforts only met with limited success on the international stage, but then the success of those international protest efforts could not be said to have been resounding either. As illustrated in Figure 19.1, the recent wildfires appear to have had much greater effect on production than international protest.

---

[1] https://open.alberta.ca/publications/enhanced-approval-process-eap-manual

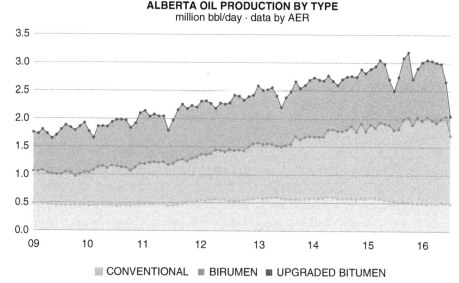

Figure 19.1 *Alberta oil production by type over time.*
Source: Oilsands Magazine, available at www.oilsandsmagazine.com/news/ 2016/8/3/alberta-oil-production-declined-sharply-during-may-wildfires. Accessed December 27, 2017.

Outside the oilsands' limelight, the creation of the socio-technical imaginary that supports hydraulic fracturing in Alberta rests upon three pillars, each of which has a number of cracks beneath the surface. These include: 1) Alberta has abundant supplies of natural gas; 2) natural gas development is necessary; and 3) the new technologies required to access natural gas are safe. Below, each of these imaginaries is presented, and the cracks noted.

4.1.1 Alberta Has Abundant Supplies of Natural Gas

In industry lingo, the decline of conventional reserves is euphemistically referred to as maturation, suggesting forward new horizons of wisdom and quality. The Provincial Energy Ministry, for example, says "as conventional gas plays in Alberta continue to mature, industry is looking towards other potential gas sources for development. Shale has the potential to make a significant contribution to Alberta's future natural gas supply."[2] According to the Canadian Association of Petroleum Producers, Canada has enough natural gas to supply Canada's consumption needs for 300 years (CAPP 2015). Fortunately, those "other" sources are seemingly even more plentiful than the initial conventional reserves. On page 4 of the Alberta Oil and Gas Industry Quarterly Update for Winter 2017,[3] the text describing a colourful map showing the extent of natural gas

---

2 Accessed at www.energy.alberta.ca/OurBusiness/944.asp, December 18, 2017.
3 Accessed at www.albertacanada.com/files/albertacanada/OilGas_QuarterlyUpdate_Fall2017.pdf, December 18, 2017.

basins beneath the province states that Alberta has "about 33 tcf (trillion cubic feet) and estimated potential of up to 500 trillion cubic feet," of natural gas trapped coalbed methane alone. But even this, the report assures readers, may be an underestimate: "in addition, a large-scale resource assessment of shale gas potential is underway and could significantly add to the natural gas prospects for the province."

A not so hidden but nonetheless not remarked upon crack in this pillar is the 15-fold margin of uncertainty between the numbers 33 and 500 (or more). This margin has to do with the difficulties of estimating the volumes of subsurface reserves, and the future technological developments that may be in store for accessing them, rendering reserve size estimation a tricky and politically volatile business indeed, one that leaves open the temptation to exaggerate. Doing so can have very real immediate benefits, in terms of investments and stock prices today, whether or not their accuracy materializes tomorrow. A study by researchers at the University of Texas (Inman 2014) re-examined the projections of share productivity provided by four authoritative bodies, such as the US Energy Information Administration, and found that all four offered gross over-estimates of future production rates. What few petro-geologists can dispute, however, is that the volume of oil or gas that can be extracted from a reserve is in all cases substantially less than the size of the reserve itself. Conventional oil wells typically "recover" around 35 percent of a reserve, but fracking technology has not been able to match even this recovery rate. In the neighbouring Bakken oil field beneath North Dakota, for example, wells in the oil shale field have an average recovery factor of just 1.2 percent (Sandreas 2012).

Curiously, the Alberta Energy Regulator's own historic production data and future forecast paint a far more modest picture than the colourful storylines depicted above, as represented in Figure 19.2 below. The significant shifts in fracking technology

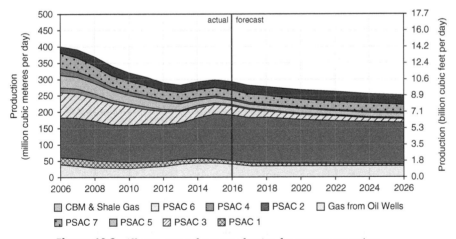

Figure 19.2 *Alberta natural gas production by reserve over time.*
Source: Alberta Energy Regulator, available at https://aer.ca/data-and-publications/statistical-reports/natural-gas-production, accessed December 27, 2017.

introduced this century appear to have had little effect on the precipitous decline in total annual production, and future forecasts only offer a modest tempering in that decline.

4.1.2 Natural Gas Is Needed

This pillar is not a tough sell in Alberta, particularly in January when temperatures can drop well below freezing, and nearly every building is kept warm with a natural gas furnace. Additional stories are also heard, such as the number of jobs supported by natural gas development, and most recently, the necessity of natural gas as the least climate offensive fossil fuel, to "bridge" our pathway toward a low carbon transition. All of my elected official and civil servant interview participants, all of whom expressed concern for climate change and support for renewable energy, spoke of natural gas in glowing terms, as a necessary "bridge" or "transition" fuel, with "huge potential," that is "easily accessible," "drives our economy," and "we're lucky to have it." Even the Ministry of Culture and Tourism hails natural gas as "a significant part of our energy resources heritage" (www.history.alberta.ca/energy heritage/gas/default.aspx).

While there is an element of truth to this pillar – a sudden reduction in natural gas supply would pose a serious hazard in such an extreme climate – there are also a number of cracks. First, research suggests that natural gas production emits far more greenhouse gases than originally thought, in the form of fugitive emissions of methane, a gas with a far higher global warming potency than CO_2 (Nikiforuk 2017). This may pose a significant problem for the Government of Alberta, which promised to reduce methane emissions substantially.

The concept of "need," however, is rather subjective. Again, there is some truth to the assertion that a transition to a renewable energy-based economy, even if taken seriously, will take some time, and a massive overhaul of energy infrastructure. On the other hand, ironically perhaps, a large proportion of this "need" was created by Alberta's Climate Leadership Plan. The plan calls for the phase-out of coal-fired electricity, 35 percent of which is to be replaced by natural gas.[4] Current demand also warrants unpacking. According to the Alberta Energy Regulator,[5] Alberta's residences consumed 13.6 million cubic metres per day on average in 2016, a not insignificant amount for 3.3 million people. On the other hand, the oilsands and petrochemical industrial sectors combined consumed 57.4 million cubic metres that year, suggesting that the biggest "bridge" built by natural gas is toward expansion of fossil fuel production, not renewable energy. This gap is expected to continue to grow, as oilsands production outpaces the population. By 2025, the residential sector is expected to consume 14.1 mcm/d. The slice of pie to be consumed by the petrochemical and oilsands sectors by then will be 88.2 mcm/d.

---

4 www.alberta.ca/climate-leadership-plan.aspx
5 Alberta Energy Regulator Table ST98: ST98: Alberta's Energy Reserves and Supply/Demand Outlook, Figure S5.6.

4.1.3 Fracking Is Safe

The third pillar – lauding the safety of fracking, is expressed in at least two separate storylines. One states that fracking fluids are 99 percent water and sand (so it must be safe). Second, there has never been a proven case of water contamination, and therefore it is safe. Both of these storylines in their turn are cloaked in a glossy cover story about Alberta's top-notch environmental regulations.

According to the Canadian Association of Petroleum Producers (CAPP), "hydraulic fracturing fluids are comprised primarily of water and sand and a small amount of *additives*;" in fact "fracking fluid is made up of ~98.5% water and sand."[6] The term "additives" is not the only one used to describe what is a cocktail of chemicals. One of my stakeholder interviewees referred to "product," while *FracFocus,* set up by the BC Oil and Gas Commission as an industry platform for fracking companies to self-report the chemicals they use in Canada, simply refers to "Other." *FracFocus* goes on to describe the composition of fracking fluid with the graph shown in Figure 19.3, depicting the seemingly inconsequential amount of "Other" in a large, clean drop of water:

Even more effective, however, are the assertions by elected officials that they have our backs. In Legislative Assembly in 2011, for example, Member of Legislative Assembly (MLA) and then-Energy Minister Ted Morton proclaims: "I'm happy to report that for this government and this Premier protecting water is the number one – number one – priority of this government. That's reflected in the

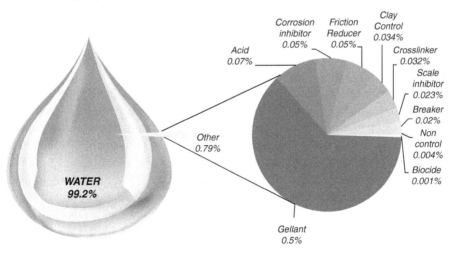

Figure 19.3 *Fracking fluid composition as depicted by* FracFocus. Available at https://fracfocus.org/water-protection/drilling-usage, *accessed December 22, 2017.*

---

6 www.canadasnaturalgas.ca/en/explore-topics/drilling-fluids.

strict regulatory regime that we have in place. The track record proves this. In the past thirty years 167,000 wells have been fracked. There's no proven record – no proven record – of any contamination of groundwater through that fracturing" (Hansard, November 29, 2011). Because of this diligence, if there was anything there we would hear about it, says MLA Knight: "If there is any indication of the types of surfactants or release agents that actually provide lubricant to push frac fluids and frac solids into fractures in production facilities underground, if there was any contamination, cross-contamination, it would not be difficult for the ERCB [predecessor to the Alberta Energy Regulator] to be able to determine what those contaminants and chemicals were" (Hansard, April 6, 2009). On the other hand, the breaches that cannot be concealed were there all the time, proclaimed MLA Knight back in 2007: "I'll just take a minute to address a couple of the situations, particularly the one around the fact that there's migration of methane to potable water. That's absolutely true. It's been true for I don't know how many years but certainly thousands of years. ... They're just there. Methane. Natural escape of methane." (Hansard, May 16, 2007)

This third pillar of the socio-technical imaginary would seem to be the shakiest of all, since all one needs to do is resort to the peer-reviewed and grey scientific literature to find evidence of its folly, referred to above. But there are two particular cracks painted over by these stories that I would like to highlight. The first has to do with the difficulty of finding things when one is not searching. My daughters often state with high-decibel certainty that they can't find their socks, or their backpack, for example, *before* they begin to look for them. It is even more difficult to locate a contaminated water well when one is not looking, and by all accounts, the search for contaminated water wells in Alberta has yet to begin. One essential requirement that is not taking place, as noted by numerous affected residents interviewed for this study, is baseline testing prior to drilling. This lack of testing has served as a valuable means of enabling development, since the complaints of residents, even those who undertake the costs of independent water quality testing themselves, can be disregarded for lack of baseline data.

The perseverance of the story about the "99%" of fracking fluid being sand and water is even more curious, as the statement itself should be alarming, not comforting. Even many of the chemicals used by fracking companies that are self-reported in *FracFocus*, have toxicity levels that are measured in parts per billion, and thus the prospect that such chemicals may be present in concentrations of parts per hundred would hardly seem to be a bragging point. These fluids, the Alberta Energy Regulator assures, are not allowed to be released into natural water bodies,[7] but such regulations do little to prevent leaks and spills, to which Albertans have been exposed regularly. One recent study by a University of Alberta biologist found fracking fluids, even at very low concentrations, to be harmful to fish (Weber 2017).

---

7 www.aer.ca/about-aer/spotlight-on/unconventional-regulatory-framework/what-is-hydraulic-fracturing. Accessed December 22, 2017.

4.2 Voices of Concern

The three pillars supporting Alberta's natural gas sociotechnical imaginary are certainly compelling. This has not prevented the emergence of concern, however. Several concerned residents have filed complaints, and with the help of the Council of Canadians have formed an online information sharing platform, called the Alberta Fracking Resource and Action Coalition. A handful of residents with resources and perseverance have also pursued legal avenues of recourse. Alberta's fracking victims have also been highlighted in a number of media projects, including coverage by award-winning journalist Andrew Nikiforuk, a video series called Alberta Voices, and a CBC documentary. And over the past decade there have been voices of concern among minority members of the legislature. For example, from MLA Swann:

> The Minister of Environment talked about and indicated that the vast majority have not been able to conclude any impacts from resource development. I guess I would have to ask if isotope testing was done in all those cases. Without isotope testing it's impossible, again, to say whether there is or there isn't evidence of industry impact. It's not good enough to say that we found methane and that there's some bacteria and therefore the cause is bacteria. People are also getting very tired of being told that they don't maintain their wells very well, when they've had dramatic increases in gas and they've had dramatic changes in volume since oil companies have moved into town. (Hansard, May 16, 2007).

My own interview participants shared numerous concerns with me, suggesting an enormous emotional toll; on top of the health and economic impacts they have endured (Davidson 2018). Just as prevalent in our conversations, however, were stories of silence, described in the next section.

4.3 Silencing

The departure from Alberta's natural gas sociotechnical imaginary presented by the stories of residents who have experienced the negative effects of fracking is glaring. The tactics subsequently used to silence these voices are several, including legislative restrictions on democratic rights; a confusing, onerous complaint process; cooptation; intimidation; invalidation and shaming.

The most formidable silencing tactic is of course the imposition of authority, granted through legislation. At the federal level, indicating the degree of relevance the oil and gas industry has for the entire country, the Harper government installed new rules that effectively outlawed protest of oil and gas infrastructure projects, and allowed authorities to treat protestors in much the same way as they would suspected terrorists (Le Billon and Carter 2012; Carter, Fraser and Zalik 2017). At home, a number of pieces of legislation have been passed in recent years by the Alberta Legislature that infringe upon the rights of citizens to express their views about new energy development, including in particular the *Responsible Energy Development Act*, passed in 2012, which, among other things, determined that only individuals who are deemed "directly and adversely affected" by a proposed development have a right to appeal such development. This Act also created a one-stop energy industry promotion and regulation shop, touted the Alberta Energy Regulator (AER), which had the power to decide how to interpret "directly and adversely," and define and oversee the appeals process. As affected residents are now fully aware, the only avenue available to voice a concern is the AER, and many, not surprisingly, have little confidence that doing so will lead to any remedies for their concerns.

The creation of the AER did at least partially address another silencing mechanism: a confusing, onerous complaint process that left many residents baffled and frustrated, and less than enthusiastic about getting results. Here is an anecdote from one such resident:

> Well first of all you phone the Energy Utilities Board (predecessor to the AER), and tell them there's this problem. And they say "oh you phone Alberta Environment, they're the ones who deal with this." So I phoned Alberta Environment and I told them what happened, and he said um, "you get in touch with the EUB and they'll set up any testing that you want." So back to the EUB and then they sent me back to Alberta Environment, and then they said "ok you know what? [Name of company] will come and test it."

The AER did not completely eliminate such sources of frustration, however. Other interview participants have shared stories of continued frustrations involving emergency contact numbers that do not work, notifications for multiple proposed new developments simultaneously, information documents that are undecipherable by anyone not in the industry, and short notification periods.

Residents have also shared with me various forms of cooptation, including the infiltration of community meetings by industry representatives who do not always disclose their identity, as well as befriending efforts, including gift offers. One interviewee shared her discomfort with these tactics, as industry staff attempt to enter her private space:

> You see, they wanna get in your house, they want to get in your face. They want to, and I've already had people sit in my kitchen table. These bland men, and we didn't understand the process. And they always come, and they come in your space. And it's your space. You're not certain, is this a friend of mine? Is this guy? They try to, oh you know, like, like kitchen table stuff.

Cooptation may have been effective with some local citizens, but certainly not all. Those who persisted in voicing their concerns were met by intimidation, sometimes with tangible threats like job loss, but in other cases on a more general, personal level. The two interviewees below share stories of each:

> I have relatives near [nearby town]. My nephew is a heavy-duty mechanic (main business is with oil and gas). And they had earth tremors just like we had here. And he went to the oil and gas company, and they basically said, "you'll get no business from us if you speak out about this."
>
> We were sitting there talking, and we were having a nice conversation, like we're having now. And we brought this thing up about the gas in the water. And we talked, calmly for about 5 minutes, maybe 10. And then he slammed his fist on the table and said "there will be no more talk about this." And this was the CEO of the company.

Unquestionably the most painful – and possibly most effective – tactic is shame. Concerned residents have been subjected to a number of shaming devices, sometimes from their own neighbours, who accuse them of being against development. Others suggest their tales are just that, as MLA McQueen implies here:

> We are also in the new year going out and having a water conversation in this province. One of the four topics that we will be talking about, because it's important to Albertans, is hydraulic fracturing. We'll hear from Albertans, and we'll be able to tell them the story and the facts about what we do in this province *to make sure that they have the facts out there with regard to mapping and with regard to baseline testing and not some myths that some people like to tell.* (Hansard, November 27, 2012. Emphasis added).

Here are the experiences two interviewees shared:

> They'd send young people oftentimes, who had been to university and knew a lot. Unlike stupid people who had just lived here, knew nothing. We got that card thrown at us many times.
>
> We've heard from various meetings ... they'll be way up north, or down south, [and they'll say] "oh those [surname of participants]. Oh those people are not credible."

Blaming the victim is a tried and true silencing device used in many contexts, and it has certainly come up among several interview participants in this study. Here, two residents interviewed together share a particularly colourful effort to do so:

FEMALE : And so, they made a mess of this road, like it wrecked my tires. I called up the Councillor ... he's like, same thing.
MALE : [quoting the Councillor] And what are you, and what are you doing living out there and driving a car like that?
FEMALE : No, he said why did you move there?

The one shaming device that cuts most deeply, particularly for a farmer, is the accusation that any water contamination they are experiencing is due to their own poor management and ignorance. As one interviewee shared, she might as well be told she was dirty, because that is how it felt. Here is more from this interviewee:

> I phoned Alberta Environment, and I told them what happened, and that I suspected it was related to fracking. And he said "there's no proof that anything happens with fracking! [loudly] You've got bacteria in your well!" And I said, "well, I think you should come and test that?" [laughs] And he says "I don't need to test that, that's bacteria! Sounds like bacteria."

Based on the following statement by MLA Renner, this was by no means an isolated event:

> I want to talk very briefly about this whole issue of monitoring of wells and following up on investigations with some interesting statistics.... To date we have received 95 complaints. We have investigated each and every one of them. Seventy-six of those files have now been closed, and the reason that they're closed is because there was found to be no connection between coal-bed methane and the

issue that the complainant was dealing with. In most cases the issue came down to an issue of well maintenance.... It is critical that the owner of a well on a regular basis shock that well with chlorine bleach and chlorine and do so in an appropriate manner. We encourage, again, all owners of wells to consult an expert and find out exactly how that should be done on an ongoing basis. (Hansard, May 16, 2007).

5 Conclusions

The golden days of fossil fuel-derived economic growth are coming to a close, and no manner of silver-tongued story-telling can eliminate the mounting costs of our insistent continuation of fossil fuel extraction. However, while many hail the transition toward renewable resources as a time of excitement, hope, and innovation, the transition *away* from fossil fuels has to date also been fraught with conflict and tragedy. Those conflicts and tragedies tend to play out in extraction zones, where the remaining unconventional sources of oil and gas are wrested from the ground in increasingly consequential ways.

As with so many other chapters in Western history, official story-telling coincides with silencing – of women, of all manner of brown and indigenous peoples, of the poor – and the storytelling that enables continued fossil fuel production despite the escalating folly of doing so once again rests on the ability to silence. Those silencing efforts have been bolstered by the fact that so many of those extraction zones are located in remote, sparsely populated and rural spaces, where the voices are few and far from the podiums of political influence.

On the other hand, the pillars supporting sociotechnical imaginaries like that describing natural gas in Alberta are replete with cracks that, with just a little weathering, can be exposed, rendering the entire project vulnerable to collapse. And those remote voices are increasingly allying themselves with each other, and with louder voices, such as those in the climate and fossil fuel divestment movements. Their unseating of fossil fuel imaginaries is only a matter of time; but how much time, and how many losses we incur before we get there, is the real question.

References

Adam, B. (1996). "Beyond the present": Nature, technology and the democratic ideal. *Time and Society*, **5**(3), 319–338.

Blüdorn, I. (2013). The governance of unsustainability: Ecology and democracy after the post-democratic turn. *Environmental Politics*, **22**(1), 16–36.

Blüdorn, I. (2007). Sustaining the unsustainable: Symbolic politics and the politics of simulation. *Environmental Politics*, **16**(2), 251–275.

Canadian Association of Petroleum Producers. (2015). The Facts on Canada's Natural Gas. Available at: www.capp.ca/publications-and-statistics/publications/272337

Carter, A.V., Fraser, G. S., & Zalik, A. (2017). Environmental policy convergence in Canada's fossil fuel provinces? Regulatory streamlining, impediments, and drift. *Canadian Public Policy*, **43**(1), 61–76.

Colborn, T., Kwiatkowski, C., Schultz, K., & Bachran, M. (2011). Natural gas operations from a public health perspective. *Human and Ecological Risk assessment*, **17**(5), 1039–1056.

Cosgrove, B. M., LaFave, D. R., Dissanayake, S. T. M., & Donihue, M. R. (2015). The economic impact of shale gas development: A natural experiment along the New York/Pennsylvania border. *Agricultural and Resource Economics Review*, **44**(2), 20–39.

Council of Canadian Academies (CCA), 2014. *Environmental Impacts of Shale Gas Extraction in Canada*. Ottawa (ON): The Expert Panel on Harnessing Science and Technology to Understand the Environmental Impacts of Shale Gas Extraction, Council of Canadian Academies.

Davidson, D. J. (2017). The effort factor: An adjustment to our understanding of social-ecological metabolism in the era of Peak Oil. *Social Problems*, **66**(1), 69–85.

Davidson, D. J. (2018). Evaluating the effects of living with contamination from the lens of trauma: A case study of fracking development in Alberta, Canada. *Environmental Sociology*, **4**(2), 196–209.

Davidson, D. J., & Andrews, J. (2013). Not all about consumption. *Science*, 339(15 March), 1286–1287.

Davidson, D. J., & Gismondi, M. (2011). *Challenging Legitimacy at the Precipice of Energy Calamity*. New York: Springer.

Davies, R., Foulger, G., Bindley, A., & Styles, P. (2013). Induced seismicity and hydraulic fracturing for the recovery of hydrocarbons. *Marine and Petroleum Geology*, **45**, 171–185.

Edelman, M. (1985 [1964]). *The Symbolic Uses of Politics*. Urbana and Chicago, IL: University of Illinois Press.

Edelstein, M.R. (2004). *Contaminated Communities: Coping with Residential Toxic Exposure*. Boulder, CO: Westview Press.

Elliot, J., & Short, D. (2014). Fracking Is Driving UK Civil and Political Rights Violations, *The Ecologist* (30 October), https://theecologist.org/2014/oct/30/fracking-driving-uk-civil-and-political-rights-violations.

Foucault, M. (1979) *The History of Sexuality*. Vol. 1. Harmondsworth: Penguin.

Freudenburg, W. R. (2005). Privileged access, privileged accounts: Toward a socially structured theory of resources and discourses. *Social Forces*, **84**(1), 89–114.

Freudenburg, W. R., & Alario, M. (2007). Weapons of Mass Distraction: Magicianship, misdirection, and the dark side of legitimation. *Sociological Forum*, **22**(2), 146–173.

Freudenburg, W.R., Gramling, R., & Davidson, D.J. (2008). Scientific certainty argumentation methods (SCAMs): Science and the politics of doubt. *Sociological Inquiry*, **78**(1), 2–38.

Gold, R. (2014). *The Boom: How Fracking Ignited the American Energy Revolution and Changed the World*. New York: Simon & Schuster.

Government of Alberta. 2000. Environmental Protection and Enhancement Act. RSA 2000, c. E–12.

Gramsci, A. 1971. *Selections from the Prison Notebooks of Antonio Gramsci*. Hare, Q., & Nowell-Sm, G. (Ed. and trans). London: Lawrence and Wishart.

Gullion, J. S. (2015). *Fracking the Neighborhood: Reluctant Activists and Natural Gas Drilling*. Cambridge, MA.: MIT Press.

Haarstad, H., & Wanvik, T. I. (2017). Carbonscapes and beyond: Conceptualizing the instability of oil landscapes. *Progress in Human Geography*, **41**(4), 432–450.

Habermas, J. (1970). *Toward a Rational Society*. North Palm Beach, FL: Beacon.

Hajer, M. (1995). *The Politics of Environmental Discourse: Ecological Modernization and the Policy Process*. Leicester, UK: Clarendon Books.

Hansard (n.d.). www.assembly.ab.ca/net/index.aspx?p=han&section=doc&fid=1

Herman, E. S. & Chomsky, N. (1988). *Manufacturing Consent: The Political Economy of the Mass Media*. New York, NY: Pantheon Books.

Howarth, R. W., Santoro, R. & Ingraffea, A. (2012). Venting and leaking of methane from shale gas development: Response to Cathles et al. *Climatic Change* **113**, 537–549.

Hudgins, A., & Poole, A. (2014). Framing fracking: Private property, common resources, and regimes of governance. *Journal of Political Ecology*, **21**, 222–348.

Inman, M. (2014). The fracking fallacy. *Nature*, **516**, Dec. 4.

Jasanoff, S., & Kim, S-H. (2013) Sociotechnical imaginaries and national energy policies. *Science as Culture*, **22**(2), 189–196.

Jasanoff, S., & Kim, S-H. (2009) Containing the atom: Sociotechnical imaginaries and nuclear regulation in the U.S. and South Korea. *Minerva*, **47**(2), 119–146.

Jerolmack, C. & Berman, N. (2016). Fracking communities. *Public Culture*, **28**(2), 193–214.

Kassotis, C. D., Tillitt, D.E., Davis, J. W., Hormann, A. M. & Nagel, S. C. (2014). Estrogen and androgen receptor activities of hydraulic fracturing chemicals and surface and ground water in a drilling-dense region. *Endocrinology*, **155**, 897–907.

Kuchler, M. (2017). Post-conventional energy futures: Rendering Europe's shale gas resources governable. *Energy Research and Social Science*, http://dx.doi.org/10.1016/j.erss.2017.05.028.

Ladd, A. E., & York, R. (2017). Hydraulic fracking, shale energy development, and climate inaction: A new landscape of risk in the Trump era. *Human Ecology Review*, **23**(1), 65–79.

Latour, B. 2004. *The Politics of Nature: How to Bring the Sciences into Democracy*. Cambridge, MA: Harvard University Press.

Le Billon, P., & Carter, A. (2012). Securing Alberta's tar sands: Resistance and criminalization on a new energy frontier. pp. 170–192 in Schnurr M., & Swatuk, M. (eds.), *Natural Resources and Social Conflict: Towards Critical Environmental Security*. London: Palgrave Macmillan.

Mitchell, T. (2011). *Carbon Democracy: Political Power in the Age of Oil*. London: Verso.

Myers, T. (2012). Potential contaminant pathways from hydraulically fractured shale to aquifers. *Ground Water*, **50**, 872–882.

Nikiforuk, A. (2017). Canada's methane leakage massively under-reported, studies find. *The Tyee*, April 27. Available at: https://thetyee.ca/News/2017/04/27/Canada-Methane-Leakage-Under-Reported/, accessed December 27, 2017.

O'Riordan, T. (2015). Fracking, sustainability, and democracy. *Environment*, **57**(1), 2–3.

Perry, S.L. (2012). Development, land use, and collective trauma: The Marcellus Shale gas boom in rural Pennsylvania. *Culture, Agriculture, Food and Environment*, **34**(1), 81–92.

Rahm, B.G. & Riha, S.J. (2014). Evolving shale gas management: Water resource risks, impacts, and lessons learned. *Environmental Science Processes & Impacts*, **16**, 1400–1412.

Roy, A.A., Adams, P.J. & Robinson, A.L. (2014). Air pollutant emissions from the development, production, and processing of Marcellus Shale natural gas. *Journal of the Air & Waste Management Association*, **64**, 19–37.

Ryan, M.C. Alessi, D., Mahani, A.B., et al. (2015). *Subsurface Impacts of Hydraulic Fracturing: Contamination, Seismic Sensitivity, and Groundwater Use and Demand Management*. Report prepared for the Canadian Water Network, October.

Sandreas, R. (2012). "Evaluating Production Potential of Mature U.S. Oil, Gas Shale Plays." *Oil and Gas Journal Online*, Retrieved December 3, 2012 (www.ogj.com/articles/print/vol-110/issue-12/exploration-development/evaluating-production-potential-of-mature-us-oil.html).

Small, M.J., Stern, P.C., Bomberg, E., Christopherson, S.M., et al. (2014). Risks and risk governance in unconventional shale gas development. *Environmental Science and Technology*, **48**, 8289–8297.

Szeman, I. (2012) Crude aesthetics: The politics of oil documentaries. *Journal of American Studies*, **46**, 423–439.

Warner, N.R., Christie, C.A., Jackson, R.B. & Vengosh, A. (2013). Impacts of shale gas wastewater disposal on water quality in Western Pennsylvania. *Environmental Science and Technology*, **47**, 11849–11857.

Weber, B. (2017). Alberta research shows fracking fluids cause 'significant' harm to fish. CBC News, Jan. 24. Available at: www.cbc.ca/news/canada/edmonton/alberta-research-shows-fracking-fluids-cause-significant-harm-to-fish-1.3950539, accessed December 22, 2017.

Willow, A.J. & Wylie, S. (2014). Politics, ecology, and the new anthropology of energy: exploring the emerging frontiers of hydraulic fracking. *Journal of Political Ecology*, **21**(12), 222–236.

Yergin, D. (2011). *The Quest: Energy, security, and the remaking of the modern world*. New York: Penguin Press.

20 Environmental Sociology and the Genomic Revolution

Valerie Berseth and Ralph Matthews

> With the tools and the knowledge, I could turn a developing snail's egg into an elephant. It is not so much a matter of chemicals because snails and elephants do not differ that much; it is a matter of timing the action of genes.
> – Barbara McClintock, quoted in *The Search for the Gene* (Wallace, 1992)

Sociology and biology have historically had a complex and sometimes inimical relationship. As an emerging discipline, early sociologists distinguished their work from the natural sciences on the basis that social actions, institutions, and identities should be explained through references to social, and not biological, factors. As Meloni et al. (2016) note, this was not a total rejection of the natural sciences. For example, Durkheim adopted the metaphors of biology in his explanation of social cohesion but maintained that social facts can and should be studied independently of natural forces. In *The Conservation of Races*, Du Bois (1897: 7) puts forth a definition of race that references but moves away from its biological origins. He writes, 'What, then, is a race? It is a vast family of human beings, generally of common blood and language, always of common history, traditions and impulses, who are both voluntarily and involuntarily striving together for the accomplishment of certain more or less vividly conceived ideals of life'. For the most part, both disciplines have progressed fruitfully through this separation, though problems have occurred when social scientists have attempted to incorporate biology in the explanation of social differences. Social Darwinism, eugenics, biological theories of criminality, and sociobiology have provided warnings of the dangers of biological determinism and reinforced disciplinary divisions.

However, this divide has also meant that sociologists and biologists have rarely engaged in cross-disciplinary conversations and sociologists have yet to fully engage with the developments that have taken place in biology over the past several years (Meloni et al., 2016). Contemporary biology is now recognized as being so different from previous periods that it is said to have undergone a revolution (Richardson and Stevens, 2015). The term post-genomics is used to distinguish between the periods before and after sequencing the human genome in 2003. The 'post' in post-genomics does not signal a move away from the genome, but a shift in how the concept of the genome is understood. The conclusion reached through the Human Genome Project and in subsequent years is that genomes are not discrete structures of genes that directly manifest as physical characteristics, such as eye colour or disease. Instead, genomes are networks of diverse biological actors that shape gene expression and inheritance in response to environmental signals. What fundamentally distinguishes the post-genomic era from the period that preceded it is the epistemological shift from thinking about genomes as stable, permanent biological fixtures to understanding them as fluid and dynamic (Griffiths and Stotz, 2013; Meloni and Testa, 2014).

This new way of thinking about genomes has also been accompanied by emerging applications and debates of several questions that are directly relevant for social science: How and when should gene editing occur? What are the social, cultural, and ecological consequences of doing so? How should legal and regulatory systems adapt to address the potential impacts of new genomic technologies on both humans and the environment? And, how should contemporary decisions account for intergenerational interests and well-being?

A growing number of sociologists have called for a sociological approach to the new era of genomics. There have been two special issues devoted to sociology and genomics (Guo, 2008; Parry and Dupré, 2010), a chapter in the 2009 *Blackwell Companion to Social Theory* (Corrigan, 2009), and an article in the *Annual Review of Sociology* on the sociology of epigenetics. The recent history of genomics has also been explored in Science and Technology Studies (Wynne, 2005; Meloni, 2015; Rheinberger and Muller-Wille, 2017). A central concern is how to integrate current knowledge about human genetics with our understanding of social action, as well as what effects genomic applications will have on cultures, institutions, bodies, identities, and well-being. We agree that these are fundamental issues for sociologists and biologists alike. Nonetheless, we have found that in these discussions the environment has often played a secondary role in favour of human genomics. Most references to 'environment' refer to any non-biological factors that play a role in human gene expression, including social surroundings.[1]

This is significant for three reasons. First, at a fundamental level, humans and social systems are connected to other species and natural processes that make up the environment. In a widely-cited article, Ceballos et al. (2017: E6089) conclude that human activity has initiated the sixth mass extinction through 'massive anthropogenic erosion of biodiversity and of the ecosystem services essential to civilization'. Biodiversity and genetic variation are essential to the survival of species and the ecosystems they inhabit. For sociologists studying environmental change and degradation, this is a pressing and significant issue at both global and local scales. Second, new genomic tools have also been applied to a wide range of complex social-environmental problems with the hopes of developing sustainable solutions. Climate change, extractive industries, global food security, and wildlife conservation are some of the issues that are being pursued actively within genomic science. It is critical to understand the impacts of changes in these areas for ecosystems and for human societies, cultures, and economies. Finally, as we will demonstrate, advanced genomic technologies such as gene editing have also been accompanied by new and evolving risks to humans and the environment. For sociologists interested in the dynamics of the 'risk society' (Beck, 1992), how these risks are evaluated by people

---

1 Of the works listed above, we note that the Special Issue on 'Nature After the Genome' in *The Sociological Review Monographs* (Parry and Dupré, 2010) is an exception. Several of the chapters deal directly with the relationship between human and non-human natures, how boundaries are drawn between what is natural or unnatural, and how understandings of nature are socially situated.

with different knowledges, cultures, interests, and environmental values will affect the acceptance and regulation of new genomic technologies. There are also ethical issues relating to data ownership. For example, patenting genetic sequences and even whole genomes for some staple crops by private entities has increased relative to public IP ownership in recent years, prompting concerns that private interests are benefiting from and influencing the direction of genomic science (O'Malley et al., 2005; Jefferson et al., 2015).

Environmental sociology is a field that is ideally suited methodologically and theoretically to address the ecological gap in sociological studies of genomics. Since its inception, environmental sociologists have been calling for sociology writ-large to 'bring nature back in' to the sociological imagination (Catton and Dunlap, 1978; Carolan, 2005). This focus, as well as the field's past marginalization from mainstream sociology, has pushed environmental sociologists to engage with a diverse range of disciplines, from geography and anthropology to climatology and biology (Pellow and Brehm, 2013). Indeed, many scholars have been engaged with the controversies surrounding the production and regulation of genetically modified organisms (GMOs) (Yearley, 2010). However, the field of genomics has changed so rapidly in recent years that it warrants a closer examination of the sociological relevance of these developments. Today, the meaning of 'genetic modification' is changing. High resolution knowledge and technologies such as gene editing have increased our ability to modify organisms faster and more precisely than before. This has prompted us to consider how the social construction of nature relates to the reconstruction of genetic material. From a genomics perspective, nature is fundamentally material – from microscopic chromosomes, to genetically diverse populations. At the same time, genomic applications introduce possibilities for humans to intervene in ways that reflect socially constructed and sometimes conflicting values and cultural understandings about what forms of nature are acceptable or desired. This poses a challenge to the materialist-constructionist divide in environmental sociology. There is also a need to interrogate the effect that social systems have on the genetic makeup of organisms and species, particularly in the case of natural resources such as forestry or marine life. The social, political, and economic dimensions of the post-genomic era are as vital to understand as the biological dimensions themselves. Where are genomic technologies used, and by whom? How does access to genomic technologies map onto existing governance regimes and power dynamics at the regional, national, and global levels? What knowledges are produced, validated, or denied in this process? And, how are public understandings of risk and nature evolving alongside these interventions?

In this chapter, we discuss these issues as they relate to the post-genomic era and consider their relevance for existing debates within the field of environmental sociology. In doing so, we aim to address some of the challenges identified above that must be faced in order to bring together the social and life sciences. We begin by tracing the recent evolution of genetic and genomic research as such knowledge is necessary to provide context for our evaluation of the post-genomic era. We then consider the reconstruction of nature through genomic interventions and the implications of this for the realism–constructionism debate in environmental sociology. In the third section, we investigate the potential for genomic technologies to diagnose and address environmental degradation. Finally, we investigate the ethical and power dynamics surrounding the new genomics and consider the implications for environmental justice.

Critical Developments in Genomic Science

In this section, we turn to three critical developments in contemporary genomic science: the shift from studying genes to genomes, epigenetics, and gene editing technologies. While the discussion here is somewhat lengthy and technical, our aim is to provide a solid foundation in genome science for social scientists, to which we can then bring core theories and approaches from environmental sociology to bear on relevant aspects of the post-genomic era.

From Genes to Genomes: The Rise of 'Network' Thinking

Few discoveries have so fundamentally transformed our understanding of and interactions with nature as that of the gene.[2] The concept shed light on two fundamental biological questions: 'What is the basis for variation in physical characteristics among organisms?' and 'How are these variable traits passed on to subsequent generations?' In 1953, James Watson and Francis Crick presented the helical structure of DNA, demonstrating that DNA functions as the molecular vehicle for heredity. This gave rise to a focus in the 1960s and 1970s on molecular biology, where researchers sought to decode how variations in nucleic acid were associated with physical characteristics. As sequencing and computational tools progressed geneticists could identify and reference particular 'genes' based on the order of nucleotides (adenine, guanine, cytosine, or thymine).

The primacy of genes as an explanatory factor in biology and sociobiology, often termed 'genetic reductionism' or 'gene centrism,' was challenged by the picture of human genetic material provided by the Human Genome Project. There turned out to be far more than genes that controlled proteins and enzymes. As Evelyn Fox Keller (2015: 9) writes, 'The genome is not the organism. But neither is it any longer a mere collection of genes'. One of the first surprises was that scientists found roughly one fifth of total genes, or 'coding sequences', they expected (Rheinberger and Müller-Wille, and Bostanci, 2017). Of these, 98 per cent seemingly had no purpose and were labelled 'junk DNA' (Stevens and Richardson, 2015). Genes viewed from the perspective of the genome appeared to be neither identifiable sequences of DNA, nor could they sufficiently explain how nucleotide sequences were coded and transcribed in the process of gene expression. To explain the mechanics of genes, genomes, and gene expression, additional actors and processes are required. In short,

---

2 The meaning of the 'gene' has evolved significantly over the last century. Though the concept is still undergoing revisions, contemporary biologists broadly refer to genes as sequences of DNA or RNA that provide codes or blueprints for cells, instructing them to produce proteins or enzymes.

biology had to be reoriented around the logic of networks to account for the complex and dynamic nature of genomes.

This move to 'network thinking' was helped by the creation of the Encyclopedia of DNA Elements (ENCODE) in 2003, which sought 'to build a comprehensive parts list of functional elements in the human genome,' as well as other species (ENCODE, 2018:1). By 2012, ENCODE was able to 'sound the death knell' for the idea of junk DNA, demonstrating that 80 per cent of the genome functions in a variety of roles, such as organizing genetic materials and assisting as intermediaries in the process of protein coding (Pennisi, 2012: 1). Much of the DNA previously thought useless is involved in a sophisticated, overlapping, multi-layered process of transcription and regulation. The picture of the genome that emerges is not a line of stretched out nucleotides sectioned off neatly into 'genes' that produce phenotypical traits, but a three-dimensional network– what Gerstein and others (2012: 91) have termed a 'human transcriptional regulatory network'.

Epigenetics

Moving from genes to genomes has renewed biologists' interests in 'epigenetics,' a field of biology that studies how outside factors alter gene regulation and expression without changing DNA. These factors can be related to the environment, including exposure to pollutants, carcinogens, endocrine disruptors, and metals, or to lifestyle, such as diet and nutrition, behavior, stress, physical activity, and work habits (Alegría-Torres et al., 2011). Additionally, exposure to these variants is also disproportionately distributed along gendered, economic, and racial lines, as well as cultural patterns of consumption (Landecker and Panofsky, 2013). The 'classic epigenetic mark' is DNA methylation, where hydrocarbon groups called 'methyls' attach to DNA segments and multiply, interrupting the normal coding process (Schübeler, 2015: 321). More commonly described as 'turning genes on and off,' the outcomes range from the temporary suppression of protein, to the permanent inactivation of an X chromosome, or the silencing of repetitive DNA strands. There is a possibility for this process to be passed to subsequent generations through transgenerational epigenetic inheritance even in individuals not exposed to that environment (Nelson and Nadeau, 2010).

Current research on epigenetics has considerable potential for environmental scholars investigating the impacts that human activities have on humans and non-humans, as well as ecological processes. It also opens up issues of both the efficacy of regulatory schemes that deal with environment degradation and their ethical consequences. The picture that emerges here is that changes in social and/or environmental conditions can have enduring biological consequences for humans and non-humans, but the permanence of these changes and their underlying causes remain uncertain and difficult to interpret. This uncertainty can be helpful for innovation in science, but also creates problems for how the risks and implications of epigenetic science is communicated in the public sphere (Pickersgill, 2016). For social scientists, this will require not only new theoretical understanding of epigenetics and human–nature interactions, but also new methodological approaches that integrate the methods of big data analysis with conventional methods so as to analyse large amounts of biological and social data at multiple scales (Liu and Guo, 2016).

Technological Advancements and the CRISPR Revolution

The shift in focus from genes to genomes required the simultaneous development of 'whole genome' technologies. This encompasses a whole suite of varying new tools and techniques that enable biologists to process, analyse, and interpret the large amounts of information that are produced during sequencing. A key development has been 'next-generation high-throughput sequencing', which allows for billions of DNA strands to be sequenced simultaneously (Skowronski and Lipkin, 2011).

One of the most explosive and controversial technological developments in recent years is the ability to carry out genome engineering. This method, called 'CRISPR-Cas9', uses a DNA cutting protein (Cas9) and a guide molecule of RNA to sever desired points in DNA (Jinek et al., 2012). As the host cell attempts to repair this break, the new segments of DNA are often mutated, effectively 'knocking out' the corresponding segment of DNA. The repair process can also be modified, such that a desired genetic sequence can be inserted, deleted, mutated, or corrected. While conventional genetic modification of plants and animals involves transferring genes from similar species, the simplicity of the CRISPR-Cas9 gives anyone with sufficient knowledge unprecedented abilities to make far more precise and extensive changes, not just to DNA but to the processes that regulate gene expression and heritability (Pontin, 2015). CRISPR-Cas9 can be used to insert precise codes into plants and animals to produce ideal traits for specific purposes like food production. For instance, genome editing (or gene editing) is being used on some staple crops like maize and cassava to mitigate climate change effects, increase disease resistance and nutrition, and usher in a new Green Revolution (Blumenstein, 2016; Arora and Narula, 2017; Mahfouz, 2017).

CRISPR Cas-9 can also be used in reproductive cells to modify the DNA responsible for transmitting information through mitosis, spreading the changes to subsequent generations of cells and organisms. This process of 'germline editing' has potentially far-reaching social and environmental consequences, such as permanently altering entire species, and has complicated the debates around gene editing and genetic modification. Some commonly cited risks of this process include irreversible changes, unintended effects, lack of informed consent from future generations, spiritual or cultural impacts, and inequality among those who can and cannot access the technology. For example, researchers working with mosquitos have also shown that making particular changes to the genome can spread these changes rapidly through whole populations, overriding natural heredity and evolution (Hammond et al., 2016). Most genes have a 50 per cent chance of being passed on to offspring depending on whether they are dominant or recessive, but scientists can create 'gene drives' that programs genes to cut segments of DNA and insert copies of themselves, forcing the next generation of organisms to inherit only the desired trait(s). This has sparked an international debate among scientists, governments, legal and medical communities, and the public about the risks of genome editing.

In short, CRISPR-Cas9 promises to revolutionize how we interact with the building blocks of nature and intensify existing controversies over the ethical applications and governance of genomic technologies. However, there is also a tension between genomic technologies such as CRISPR-Cas9 and the hype surrounding them. Fortun (2005) suggests that promises are an enduring characteristic of genomics as a field that is oriented towards providing solutions to future problematic genetic realities. The 'promissory technologies' of genomics 'cannot be reduced to empty hype, or to formal contract, but occup(y) the uncertain, difficult space in between' (Fortun, 2005: 158). Thus, there are areas where genomics will likely have ground-breaking impacts, and areas where it will fail to deliver or be met with resistance.

(Re)Constructing Nature

Initially, environmental sociology was formed in response to the human exemptionalism that characterized mainstream sociological research (Catton and Dunlap, 1978; Dunlap and Catton, 1979). As the field developed, two distinct theoretical orientations to the human–nature relationship emerged: the 'realist/materialist' approach and the 'social constructivist' approach. Early work in environmental sociology wrestled with the material reality of ecological resources and systems, and the constraints they placed on human social systems (Spaargaren and Mol, 1992; Beus and Dunlap, 1994; Foster, 1999). In the 1990s, a new subfield of 'social constructivism' emerged, drawing on insights from symbolic interactionism and science studies, that engaged with the social and cultural processes through which nature is imagined and interpreted (Buttel and Taylor, 1992; Macnaghten and Urry, 1995). The realist–constructivist debate became contentious, with materialists defending the field against assertions that nature only exists as a symbolic construction, and constructivists arguing against objectivist science. Though the realist/materialist and social constructionist camps have made peace and few scholars today adopt extreme positions on either side, the issue of how to incorporate nature into sociological analysis continues to be debated. Van Koppen (2017) identifies three issues that have yet to be fully resolved: (1) the role of natural objects and processes in social dynamics; (2) natural science explanations of social phenomena; and (3) the suitability of core sociological concepts and theories for dealing with natural processes.

The growth of genomic technologies aimed at addressing environmental problems presents a challenge to dichotomous materialist-constructionist approaches. While the field of genomics is undoubtedly grounded in the material, present applications of genomic technologies offer the ability to quickly and precisely reshape that material according to needs and desires that are socially constructed. As Hannigan (1995: 2) writes, 'environmental problems do not materialise by themselves; rather they must be 'constructed' by individuals or organizations who define pollution or some other objective condition as worrisome and seek to do something about it'. A central task for environmental sociologists investigating genomics is not just understanding the way that change is occurring to organisms and their environments, but also how and why certain 'material' conditions become identified as targets for genomic intervention. The new genomics disrupts conventional binaries of nature/culture, material/language. As a result, reality and cannot be divided into either the social or material construction of nature (Woodgate and Redclift, 1998; Pellizzoni, 2016). Genome editing provides a clear demonstration of this point.

While several scientists have called for a moratorium on human germline editing, the practice is already being used on plants, animals, and insects (Caplan et al., 2015; Wade, 2015). In an article published in *Nature,* Reardon (2016) describes this body of work as the 'CRISPR zoo,' a menagerie of animals that are in the process of having their genomes 'reimagined'. Disease-resistant bees, mammoth-like elephants brought back from extinction, female cows with more muscle for food production – these are some examples of ongoing work that will inform future applications. Some of these examples represent an acceleration of existing trait-based breeding methods that have occurred for centuries, where others, such as the use of gene editing to select for mammoth versions of elephant genes involve a greater level of human intervention into genetic processes.

One area that germline editing has had a lot of attention is in pest control. For example, in New Zealand the arrival of gene drives has coincided with the growth of a grassroots environmental movement, Predator-Free New Zealand (Yong, 2017; MWLR, 2018). The country has historically had a rich diversity of unique bird species that have evolved to become flightless in the absence of natural predators. The arrival of humans brought invasive, non-native predators including rats, stoats, weasels, and possums that are devastating bird populations, including the country's beloved kiwi bird. The predator-free movement, championed by the late New Zealand physicist Paul Callaghan, has rapidly progressed in the last five years. In 2016, Prime Minister John Key announced Project Free 2050, a government program that would devote $20 million to ridding the nation of eight predator species by the year 2050. The plan relies on aerial drops of poison, trapping, and hunting, which are costly, can affect other fauna, and challenging in urban areas (Roy, 2016). However, CRISPR-Cas9 presents a faster, cheaper solution: if gene drives inserted in predators' genomes force their offspring to inherit genes for sterility, entire species could be eradicated within just a few generations.

The social and ecological risks of this technology are as powerful as its promises. Once released into the environment, a gene drive can quickly spread beyond political and natural borders, for example rewriting the genes of rats or stoats all over the world. Although these species are invasive and harmful in New Zealand, they may be vital to the cultures, ecosystems, and economies of other places. Thus, decisions that might normally take place at the local or national levels have become potential global problems. Additionally, the history of invasive species in New Zealand and elsewhere is testament to the difficulty of preventing transmission. Kevin Esvelt, one of the biologists who initially suggested gene drives could eradicate invasive species, has since reversed course, saying, 'It was profoundly wrong of me to even suggest it, because I badly misled many conservationists who are desperately in need of hope' (Yong, 2017: 1). Together with Neil Gemmell, Esvelt has publicly called for New

Zealand to abandon its plans to use gene drives, stating that 'now is the time to be bold in our caution' and develop alternatives (Esvelt and Gemmell, 2017: 4). However, the political process for reviewing the technology is already underway. New Zealand's top scientific body, the Royal Society Te Apārangi has convened a multidisciplinary panel to consider the social, cultural, legal, and economic implications of gene drives for pest control, and public engagement is currently underway (RSTA, 2017). Cases like New Zealand's Project Free 2050 demonstrate that science alone does not control decisions made about which risks are acceptable and what priorities will inform environmental management. An important and fertile area for future research will involve studies that theorize the relation between understandings of scientific risk, trust in science, and regulatory change in the context of new genomic knowledge.

Genomics and Ecological Degradation

A central concern for environmental sociologists is the connection between human activity and ecological change. Genomics is playing an increasing role in both in diagnosing and presenting technological solutions. However, issues of environmental risk and destruction cannot be approached as solely technical or scientific problems because they are rooted in human activities. As such, understanding the causes and solutions must examine 'how our social order developed its current treatment of the natural world' (Brulle, 2000: 16). Several schools of thought have been developed to explain the social processes that contribute to environmental degradation. These have ranged from micro-level studies of individual environmental attitudes and values, most notably Catton and Dunlap's (1978) New Ecological Paradigm, to macro-level theories of risk and modernity (Beck, 1992; Beck et al., 1994) and the political-economic causes of environmental declines, such as the 'treadmill of production' (Schnaiberg et al., 2002; Gould et al., 2008) or 'metabolic rifts' (Marx, 1976; Foster, 1999). Scholars have also identified institutions and cultural factors as key mechanisms through which the human–nature relationship is mediated (Yearley, 2005; Buttel, 2010). These frameworks can provide important insights into the social and ecological changes that are taking place in the post-genomic era. To explore this further, we turn to the ocean and consider some of the connections between capital accumulation, genetics, and salmon conservation as examples of metabolic rifts.

The world's ocean sustains a rich biodiversity of plants, animals, and phytoplankton that in turn provide food, oxygen, and economic benefits for human populations globally, and in coastal areas in particular. However, human exploitation of these resources and degradation of the marine environment through climate change, pollution, and habitat destruction has changed the abundance and distribution of marine life (Barange et al., 2010; Halpern et al., 2015). This problem is exacerbated by high demands and market values for large, carnivorous fish, such as tuna, halibut, and salmon. The ability for prized fish populations to reproduce in sufficient numbers to replace harvested amounts is strained by the current rate of exploitation, at

80 million tonnes of fish harvested annually (Hazin et al., 2016). Extending Marx's theory of metabolism to the marine environment, Clausen and Clark (2005) argue that the capital-intensive exploitation of fish stocks has created a 'metabolic rift' that has disrupted the natural cycles that sustain marine ecosystems and the human–ocean relationship. As Foster (1999: 381) has noted, the concept of 'metabolism' is both social and ecological, referring to the 'the complex, interdependent process linking human society to nature'. Rifts are worsened over time as accumulation intensifies and spreads to other geographic regions and technological solutions are needed to sustain the economic system of accumulation.

Two methods have been pursued to artificially produce salmon that can sustain social, economic, and ecological demands. The first is aquaculture, which applies the principles and methods of traditional agriculture to produce fish, plants, or shellfish. For salmon, this involves breeding smolts and raising them in nets that are suspended in the ocean for harvest. While traditionally this has been done on a small scale, technological advancements have facilitated the 'blue revolution' of aquaculture to large-scale, multinational operations that produced over 167 million tons of fish in 2016 (Young and Matthews, 2010; FAO, 2016). The second method is through salmon hatchery programs, also termed 'hatch and release', which breed juvenile salmon and release them into the environment where they mingle with their non-hatchery counterparts. These programs tend to be government-run but are viewed as critical supports of the commercial and recreational fishing industries, as well as a means of rehabilitating wild populations. In the United States alone, the National Fish Hatchery System generates over $900 million annually (Charbonneau and Claudill, 2010).

Of these two approaches, aquaculture has received far more attention and opposition in the public sphere, in part because its methods of production, such as the use of growth hormones and pen rearing, alter the natural life cycle of salmon (Clausen and Clark, 2005). Conversely, hatchery programs are operated in a way that mimic natural processes, as the salmon they release spend the vast majority of their lives outside of hatchery environments. However, hatchery salmon have been shown to fair poorer in the marine environment than fish that spawn in the 'wild', suggesting that there is something about hatcheries that may be affecting their chances of survival. A growing body of research is demonstrating that the hatchery environment can cause 'epigenetic reprogramming', or changes in gene expression that can reduce genetic fitness of offspring, and decrease the likelihood of survival (Frankham 2008; Christie et al., 2012; Le Luyer et al., 2017: 12967). This reduced fitness can also be inherited by future generations if hatcheries continue to use hatchery-born salmon to breed the next generation of salmon as is sometimes the practice (Araki et al., 2007, 2008). The issue is further complicated because hatchery salmon can spawn with naturally spawning salmon, challenging our understanding of what can be considered wild or natural. In that sense, the issue of 'What is wild?' becomes a central one for production and conservation of fish stocks. This is a considerable issue that also demonstrates the importance of genetic knowledge for 'governing the commons' (Dietz et al., 2003: 1907).

Emerging genomic science is helping shed light on causes of and potential solutions to environmental degradation. As the case of salmon hatcheries demonstrates, human efforts to manage nature can cause genetic rifts in ecosystems. Genomics can illuminate disruptive effects about which we were previously unaware and that may actually reduce survival of the very organisms we were intending to conserve. Put more generally, genomics can also prescribe courses of action for public and private institutions seeking to recover endangered species. At the same time, the science of conservation biology cannot be regarded as a neutral or objective 'guiding hand'. Rather, as Bierman and Mansfield (2014) remind us, conservation is biopolitical – it provides a logic for distinguishing between different types of 'life,' what is normal and abnormal, and what elements of nature to 'make live, or let die'.

Power, Ethics, and Environmental Justice

Thus far, we have explored the relevance of genomics for human efforts to understand and manage other organisms and the natural environment. In this section, we consider what the post-genomic era may mean for human-nature relationships and the overlapping issues of power, ethics, and justice. Environmental justice emerged as a social movement in the 1980s as activists in the United States fought against the disproportionate environmental burdens experienced by people of colour, Indigenous people, and low-income communities. The movement prompted academic investigations into the distributions of environmental threats, such as pollution, hazardous waste, landfills, and power plants, along racial, class, and gendered divisions (Mohai et al., 2009). While the sensory and physical effects of exposure to toxic pollutants are relatively easy to document, it is more difficult to identify the pathways through which exposure causes harm at a genetic level. However, doing so is necessary to understanding the intergenerational harms associated with exposure. Over fifty years ago, Rachel Carson (1962: 208) described this imperative in *Silent Spring*, writing, 'a possession infinitely more valuable than individual life is our genetic heritage, our link with past and future ... Yet genetic deterioration through man-made agents is the menace of our time, "the last and greatest danger to our civilization."'

Genomics can contribute to epidemiological studies of exposure to environmental harms in two respects: gene-environment interactions, and epigene–environment interactions (Bollati and Baccarelli, 2010). With gene–environment interactions, the primary concern is how exposure to certain substances causes genetic damage that can be transmitted through genetic inheritance to the next generation, as well as whether some in the population are more susceptible to harms from exposure. As an example, Lourenço et al. (2013) found that mice near an abandoned uranium mine experienced serious damage to their DNA, causing genomic instability. Uranium mining has been a persistent environmental justice issue, particularly for Indigenous peoples in Canada and the United States (Lovelace, 2009; Hoover et al., 2012). As we noted above, epigenetic interactions can also be inherited and successive generations may still experience disease-related effects of exposure from their ancestors even if the source of contamination has been removed (Wolffe and Guschin, 2000). Recent research has shown that DDT, the pesticide at the heart of *Silent Spring*, can cause epigenetic transgenerational transmission of obesity, kidney disease, and ovary disease (Skinner et al., 2013; Manikkam et al., 2014). Kabasenche and Skinner (2014) argue that this poses an ethical dilemma in places such as Africa, where DDT is currently used as a cost-effective weapon against malaria. They suggest that decision-making processes and risk analyses need to be re-framed to account for threats to future generations at the most fundamental genetic level of their being.

The issue of how decisions are made and by whom has been a central theme in the environmental justice movement. One of the Principles of Environmental Justice developed at the 1991 First National People of Color Environmental Leadership Summit is 'the right to participate as equal partners at every level of decision-making' (UCCCRJ, 1991: 1). Careful attention to procedural justice issues is especially relevant to current advances in genomics that may impact socio-politically on marginalized groups. Evaluations of risk are historically, culturally, and socially situated, meaning that there may be fundamental differences in the way that people perceive genomic knowledge and technologies. For some, this may be based on past experiences with unethical research, such as the use of blood samples or genetic materials from Indigenous peoples for research or profit without their consent (Smith, 2013; TallBear, 2013a).

There may also be apprehension towards genomic 'solutions' to environmental problems because of cultural or epistemological differences in how human–ecological relationships are understood. An important illustrative case is New Zealand's planned use of gene drives to eradicate invasive predators. For years, Māori have engaged with scientific and political debates regarding the use of genetic technologies in environmental management to protect biodiversity and maintain their guardianship rights over the land (Rixecker and Tipene-Matua, 2003). For many Māori, human intervention in genomic processes contradicts core spiritual and material values. To inform decisions in the current gene-drive debate, a panel of Māori and non-Māori experts from law, biology, health, and mathematics has been formed. In a recent discussion paper, the panel outlined several Māori values that must be considered including: 'whakapapa (of the organism, as well as the relationship/kinship between humans and other species), tika (what is right or correct), manaakitanga (cultural and social responsibility/accountability, for example to other nations who value [the invasive species]), mana (justice and equity), [and] tapu' (RSTA, 2017: 5). Though Māori have fought to participate in environmental decision-making processes, their role as 'equal partners' has been challenged by the complexities of genomic science (Roberts et al., 2004). This point is echoed by TallBear (2013a). In her book *Native American DNA*, TallBear provides a review of basic genomic science because otherwise 'we in Indian Country will find it difficult to accurately assess the risks and benefits ... [of] genetic research. Indian Country will remain largely at the mercy of non-Native technical advisers ... that stand to benefit from the geneticization of Native American identity' (TallBear, 2013a: 40).

Another barrier to environmental justice is the position of power held by genomic science. As Kent (2013) and TallBear (2013b) argue, when there are discrepancies in knowledge claims, states tend to privilege claims made using genomic science over those made using Indigenous knowledge. This is not just an issue for genomic science but concerns Western science more broadly (Nadasdy 1999; Simpson, 2004). However, it does support the assertion that the application of science in environmental management is a cultural and political process. Debates regarding the environmental and social risks of genomic knowledge and new technologies are sites of contestation that are embedded in a larger struggle for epistemological authority and environmental justice.

Conclusion

Environmental sociology has a vital role to play in the post-genomics era. While much of the public and scholarly attention has focused on human applications, there have been rapid and significant environmental reforms across many sectors. Armed with greater knowledge about genetic processes and greater technological and computational resources, modern genomics is enabling researchers and practitioners to examine and manipulate nature in ways that are fundamentally different from those of the past. At the same time, scientific and political institutions are being confronted with the moral and ethical ramifications of new technologies like CRISPR-Cas9. These developments have wide-ranging social and ecological implications, from climate adaptation to the conservation of culturally significant species, to understanding the epigenetic effects of exposure to toxic pollutants in humans and animals. Much work remains to understand how the genomic revolution is impacting and intensifying longstanding debates in the field of environmental sociology about risk, technological progress, the social construction and regulation of nature, and the production of knowledge.

References

Alegría-Torres, J. A., Baccarelli, A., and Bollati, V. (2011). Epigenetics and Lifestyle. *Epigenomics*, **3**(3), 267–77.
Araki, H., Berejikian, B. A., Ford, M.J., and Blouin, M. S. (2008). Fitness of Hatchery-Reared Salmonids in the Wild. *Evolutionary Applications*, **1**(2), 342–55.
Araki, H., Cooper, B., and Blouin, M. S. (2007). Genetic Effects of Captive Breeding Cause A Rapid, Cumulative Fitness Decline in the Wild. *Science*, **318**(5847), 100–3.
Arora, L. and Narula, A. (2017). Gene Editing and Crop Improvement Using CRISPR-Cas9 System. *Frontiers in Plant Science*, **8**(1932), 1–22.
Barange, M., Field, J. G., Harris, R. P., et al. (2010). *Marine Ecosystems and Global Change*. Oxford: Oxford University Press.
Beck, U. (1992). *Risk Society: Towards a New Modernity*. London: SAGE Publications.
Beck, U., Giddens, A., and Lash, S. (1994). *Reflexive Modernization: Politics, Tradition and Aesthetics in the Modern Social Order*. Stanford, CA: Stanford University Press.
Beus, C.E. and Dunlap, R. E. (1994). Agricultural Paradigms and the Practice of Agriculture. *Rural Sociology*, **59**(4), 620–35.

Biermann, C. and Mansfield, B. (2014). Biodiversity, Purity, and Death: Conservation Biology as Biopolitics. *Environment and Planning D: Society and Space*, **32**(2), 257–73.

Blumenstein, R. (2016). Bill Gates: GMOs Will End Starvation in Africa. *Wall Street Journal*. www.wsj.com/video/bill-gates-gmos-will-end-starvation-in-africa/3085A8D1-BB58-4CAA-9394-E567033434A4.html

Bollati, V. and Baccarelli, A. (2010). Environmental Epigenetics. *Heredity*, **105**(1), 105–12.

Brulle, R.J. (2000). *Agency, Democracy, and Nature: The U.S. Environmental Movement from a Critical Theory Perspective*. London: MIT Press.

Buttel, F. H. (2010). Social Institutions and Environmental Change. In Redclift, M. and Woodgate, G., eds, *The International Handbook of Environmental Sociology*. Cheltenham: Edward Elgar Publishing, pp. 33–47.

Buttel, F. H., and Taylor, P.J. (1992). Environmental Sociology and Global Environmental Change: A Critical Assessment. *Society & Natural Resources*, **5**(3), 211–30.

Caplan, A. L., Parent, B., Shen, M., and Plunkett, C. (2015). No Time to Waste: The Ethical Challenges Created by CRISPR. *EMBO Reports*, **16**(11), 1421–6.

Carolan, M. S. (2005). Society, Biology and Ecology: Bringing Nature Back into Sociology's Disciplinary Narrative Through Critical Realism. *Organization & Environment*, **18**(4), 393–421.

Carson, R. (1962). *Silent Spring*. Mariner, New York.

Catton, W. R. and Dunlap, R. E. (1978). Environmental Sociology: A New Paradigm. *The American Sociologist*, **13**(February), 41–9.

Ceballos, G., Ehrlich, P.R., and Dirzo, R. (2017). Biological Annihilation Via the Ongoing Sixth Mass Extinction Signaled by Vertebrate Population Losses and Declines. *Proceedings of The National Academy of Sciences*, **114**(30), E6089–E6096.

Charbonneau, J.J. and Caudill, J. (2010). *Conserving America's Fisheries: An Assessment of Economic Contributions from Fisheries and Aquatic Resource Conservation*. Arlington, VA: US Fish and Wildlife Service.

Christie, M. R., Marine, M. L., French, R. A., Waples, R. S., and Blouin, M. S. (2012). Effective Size of a Wild Salmonid Population Is Greatly Reduced by Hatchery Supplementation. *Heredity*, **109**(4), 254–60.

Clausen, R., and Clark, B. (2005). The Metabolic Rift and Marine Ecology: An Analysis of the Ocean Crisis Within Capitalist Production. *Organization and Environment*, **18**(4), 422–44.

Corrigan, O. (2009). Genetics and Social Theory. In Turner, B. S., ed., *The New Blackwell Companion to Social Theory*. West Sussex: Blackwell Publishing Ltd, pp. 343–59.

Dietz, T., Ostrom, E., and Stern, P. C. (2003). The Struggle to Govern the Commons. *Science*, **302**(5652), 1907–12.

Du Bois, W. (1897). *The Conservation of Races*. Washington, DC: American Negro Academy.

Dunlap, R. E. and Catton, W. R. (1979). Environmental Sociology. *Annual Review of Sociology*, **5**(1), 243–73.

ENCODE. (2018). *ENCODE: Encyclopedia of DNA Elements*. Stanford University. www.encodeproject.org/

Esvelt, K.M. and Gemmell, N.J. (2017). Conservation Demands Safe Gene Drive. *PLOS Biology*, **15**(11), e2003850.

Food and Agriculture Organization of the United Nations (FAO). (2016). *The State of World's Fisheries and Aquaculture 2016*. Rome.

Fortun, M. (2005). For an Ethics of Promising, Or: A Few Kind Words About James Watson. *New Genetics and Society*, **24**(2), 157–73.

Foster, J.B. (1999). Marx's Theory of Metabolic Rift: Classical Foundations for Environmental Sociology. *American Journal of Sociology*, **105**(2), 366–405.

Frankham, R. (2008). Genetic Adaptation to Captivity in Species Conservation Programs. *Molecular Ecology*, **17**(1), 325–33.

Gerstein, M. B., Kundaje, A., Hariharan, M. et al. (2012). Architecture of the Human Regulatory Network Derived from ENCODE Data. *Nature*, 489 (7414), 91–100.

Gould, K. A., Pellow, D. N., and Schnaiberg, A. (2008). *The Treadmill of Production*. New York: Routledge.

Guo, G. (2008). Society and Genetics [Special Issue]. *Sociological Methods & Research*, **37**(2), 159–63.

Griffiths, P., and Stotz, K. (2013). *Genetics and Philosophy: An Introduction*. Cambridge: Cambridge University Press.

Halpern, B.S., Frazier, M., Potapenko, J., et al. (2015). Spatial and Temporal Changes in Cumulative Human Impacts on The World's Ocean. *Nature Communications*, **6**(1), 7615.

Hammond, A., Galizi, R., Kyrou, K., et al. (2016). A CRISPR-Cas9 Gene Drive System Targeting Female Reproduction in The Malaria Mosquito Vector Anopheles Gambiae. *Nature Biotechnology*, **34**(1), 78–83.

Hannigan, J. A. (1995). *Environmental Sociology: A Social Constructionist Perspective*. New York: Routledge

Hazin, F., Marschoff, E., Ferreira, B.P., Rice, J., and Rosenberg, A. (2016). Capture Fisheries. In Inniss L., Simcock, A., Ajawin, A. et al., eds *First Global Integrated Marine Assessment*. United Nations Division for Ocean Affairs and The Law of the Sea, pp. 1–24.

Hoover, E., Cook, K., Plain, R., et al. (2012). Indigenous Peoples of North America: Environmental Exposures and Reproductive Justice. *Environmental Health Perspectives*, **120**(12), 1645–9.

Jefferson, O.A., Köllhofer, D., Ehrich, T.H., and Jefferson, R.A. (2015). The Ownership Question of Plant Gene and Genome Intellectual Properties. *Nature Biotechnology*, 33(11), 1138–43.

Jinek, M., Chylinski, K., Fonfara, I., et al. (2012). A Programmable Dual-RNA-Guided DNA Endonuclease in Adaptive Bacterial Immunity. *Science*, **337**(6096), 816–21.

Kabasenche, W.P. and Skinner, M.K. (2014). DDT, Epigenetic Harm, and Transgenerational Environmental Justice. *Environmental Health*, **13**(1), 1–5.

Keller, E.F. (2015). The Postgenomic Genome. In Richardson, S. S. and Stevens, H., eds, *Postgenomics: Perspectives on Biology after the Genome*. Durham, NC: Duke University Press, pp. 9–31.

Kent, M. (2013). The Importance of Being Uros: Indigenous Identity Politics in the Genomic Age. *Social Studies of Science*, **43**(4), 534–56.

Landecker, H. and Panofsky, A. (2013). From Social Structure to Gene Regulation, and Back: A Critical Introduction to Environmental Epigenetics for Sociology. *Annual Review of Sociology*, **39**(1), 333–57.

Le Luyer, J., Laporte, M., Beacham, T.D., et al. (2017). Parallel Epigenetic Modifications Induced by Hatchery Rearing in a Pacific Salmon. *Proceedings of the National Academy of Sciences*, **114**(49), 12964–9.

Liu, H. and Guo, G. (2016). Opportunities and Challenges of Big Data for the Social Sciences: The Case of Genomic Data. *Social Science Research*, **59**(2016), 13–22.

Lourenço, J., Pereira, R., Gonçalves, F., and Mendo, S. (2013). Metal Bioaccumulation, Genotoxicity and Gene Expression in the European Wood Mouse (Apodemus sylvaticus) Inhabiting an Abandoned Uranium Mining Area. *Science of The Total Environment*, **443**, 673–80.

Lovelace, R. (2009). Notes from Prison: Protecting Algonquin Lands from Uranium Mining. In *Speaking for Ourselves: Environmental Justice in Canada*. Vancouver: UBC Press, pp. ix–xix.

Macnaghten, P. and Urry, J. (1998). *Contested Natures*. London: Sage Publications.

Mahfouz, M.M. (2017) Genome Editing: The Efficient Tool CRISPR-Cpf1. *Nature Plants*, **3** (17028), 1–2.

Manaaki Whenua Landcare Research (MWLR). (2018). *Predator-Free New Zealand*, www.landcareresearch.co.nz/science/plants-animals-fungi/animals/pfnz.

Manikkam, M., Haque, M. M., Guerrero-Bosagna, C., Nilsson, E. E., and Skinner, M. K. (2014). Pesticide Methoxychlor Promotes the Epigenetic Transgenerational Inheritance of Adult-Onset Disease through the Female Germline. *PLoS ONE*, **9**(7), e102091.

Marx, K. (1976). *Capital: A Critique of Political Economy* (Vol. 1). Vintage, New York.

Meloni, M. (2015). Epigenetics for The Social Sciences: Justice, Embodiment, and Inheritance in the Postgenomic Age. *New Genetics and Society*, **34**(2), 125–51.

Meloni, M. and Testa, G. (2014). Scrutinizing the Epigenetics Revolution. *BioSocieties*, **9**(4), 431–56.

Meloni, M., Williams, S., and Martin, P. (2016). The Biosocial: Sociological Themes and Issues. The *Sociological Review Monographs*, **64**(1), 7–25.

Mohai, P., Pellow, D., and Roberts, J.T. (2009). Environmental Justice. *Annual Review of Environment and Resources*, **34**(1), 405–30.

Nadasdy, P. (1999). The Politics of Tek: Power and the 'Integration' of Knowledge. *Arctic Anthropology*, **36**(1/2), 1–18.

Nelson, V.R. and Nadeau, J.H. (2010). Transgenerational Genetic Effects. *Epigenomics*, **2**(6), 797–806.

O'Malley, M.A., Bostanci, A., and Calvert, J. (2005). Whole-Genome Patenting. *Nature Reviews Genetics*, **6**(6), 502–6.

Parry, S. and Dupré, J. (2010). Nature after the Genome [Special Issue]. *Sociological Review*, **58**(1).

Pellizzoni, L. (2016). Catching up with Things? Environmental Sociology and the Material Turn in Social Theory. *Environmental Sociology*, **2**(4), 312–21.

Pellow, D. N. and Brehm, H. N. (2013). An Environmental Sociology for the Twenty-First Century. *Annual Review of Sociology*, **39**(1), 229–50.

Pennisi, E. (2012). ENCODE Project Writes Eulogy for Junk DNA. *Science*, **337**(6099), 1159–61.

Pickersgill, M. (2016). Epistemic Modesty, Ostentatiousness and the Uncertainties of Epigenetics: On the Knowledge Machinery of (Social) Science. *The Sociological Review Monographs*, **64**(1), 186–202.

Pontin, J. (2015). Editing Human DNA. *MIT Technology Review*. www.technologyreview.com/s/536696/editing-human-dna/

Reardon, S. (2016). Welcome to the CRISPR Zoo. *Nature*, **531**(7593), 160–3.

Rheinberger, H.-J., Müller-Wille, S., and Bostanci, A. (2017). *The Gene: From Genetics to Postgenomics*. Chicago, IL: University of Chicago Press.

Richardson, S. S. and Stevens, H. (2015). *Postgenomics: Perspectives on Biology after the Genome*. Chicago, IL:University of Chicago Press.

Rixecker, S. S. and Tipene-Matua, B. (2003). Maori Kaupapa and the Inseparability of Social and Environmental Justice: An Analysis of Bioprospecting and Peoples' Resistance to Bio(cultural) Assimilation. In Agyeman, J., Bullard, R. D., and Evans, B., eds, *Just Sustainabilities: Development in an Unequal World*, London: Earthscan Publications Ltd, pp. 252–68.

Roberts, M., Haami, B., Benton, R. A., et al. (2004). Whakapapa as a Maori Mental Construct: Some Implications for the Debate over Genetic Modification of Organisms. *The Contemporary Pacific*, **16**(1), 1–28.

Roy, E. A. (2017). No More Rats: New Zealand to Exterminate All Introduced Predators. *The Guardian*. www.theguardian.com/world/2016/jul/25/no-more-rats-new-zealand-to-exterminate-all-introduced-predators

Royal Society Te Apārangi (RSTA). (2017). *The Use of Gene Editing in Pest Control*. Wellington. www.landcareresearch.co.nz/science/plants-animals-fungi/animals/pfnz

Schnaiberg, A., Pellow, D. N., and Weinberg, A. (2002). The Treadmill of Production and the Environmental State. In Mol, P. and Buttel, F., eds, *The Environmental State Under Pressure*. Amsterdam: Elsevier Science, pp. 15–32.

Schübeler, D. (2015). Function and Information Content of DNA Methylation. *Nature*, **517**(7534), 321–6.

Simpson, L.R. (2004). Anticolonial Strategies for the Recovery and Maintenance of Indigenous Knowledge. *The American Indian Quarterly*, **28**(3), 373–84.

Skinner, M.K., Manikkam, M., Tracey, R., et al. (2013). Ancestral Dichlorodiphenyltrichloroethane (DDT) Exposure Promotes Epigenetic Transgenerational Inheritance of Obesity. *BMC Medicine*, **11**(1), 228.

Skowronski, E. and Ian Lipkin, W. (2011). Molecular Microbial Surveillance and Discovery in Bioforensics. In Budowle, B. Schutzer, S. E. Breeze, R. G. Kleim, P. S. and Morse, S. A., eds, *Microbial Forensics*. Burlington, VT: Academic Press, pp. 173–85.

Smith, L. T. (2013). *Decolonizing Methodologies: Research and Indigenous Peoples*. 2nd edition. London: Zed Books Ltd.

Spaargaren, G. and Mol, A.P. (1992). Sociology, Environment, And Modernity: Ecological Modernization as a Theory of Social Change. *Society & Natural Resources*, **5**(4), 323–44.

Stevens, H., and Richardson, S. S. (2015). Beyond the Genome. In Richardson, S. S. and Stevens, H., eds., *Postgenomics: Perspectives on Biology after the Genome*. Durham, NC: Duke University Press, pp. 1–8.

TallBear, K. (2013a). *Native American DNA: Tribal Belonging and the False Promise of Genetic Science*. Minneapolis, MN: University of Minnesota Press.

TallBear, K. (2013b). Genomic Articulations of Indigeneity. *Social Studies of Science*, **43**(4), 509–33.

United Church of Christ Commission for Racial Justice (UCCCRJ). (1991). Principles of Environmental Justice. In *The Proceedings of the First National People of Color Environmental Leadership Summit*, Washington, DC: United Church of Christ.

van Koppen, C. K. (2017). Incorporating Nature in Environmental Sociology: A Critique of Bhaskar and Latour, and a Proposal. *Environmental Sociology*, **3**(3), 173–85.

Wade, N. (2015). Scientists Seek Moratorium on Edits to Human Genome That Could Be Inherited. *The New York Times*. www.nytimes.com/2015/12/04/science/crispr-cas9-human-genome-editing-moratorium.html

Wallace, B. (1992). *The Search for the Gene*. Ithaca, NY: Cornell University Press.

Woodgate, G. and Redclift, M. (1998). From a 'Sociology of Nature' to Environmental Sociology: Beyond Social Construction. *Environmental Values*, **7**(1), 3–24.

Wolffe, A. P. and Guschin, D. (2000). Review: Chromatin Structural Features and Targets that Regulate Transcription. *Journal of Structural Biology*, **129**(2–3), 102–22.

Wynne, B. (2005). Reflexing Complexity: Post-Genomic Knowledge and Reductionist Returns in Public Science. *Theory, Culture and Society*, **22**(5), 67–94.

Yearley, S. (2005). *Cultures of Environmentalism: Empirical Studies in Environmental Sociology*. Palgrave Macmillan, New York.

Yearley, S. (2010). Science and The Environment in the Twenty-First Century. In Redclift M. R., and Woodgate, G., eds. *The International Handbook of Environmental Sociology*. 2nd edition. Cheltenham: Edward Elgar Publishing, pp. 212–23.

Yong, E. (2017). New Zealand's War on Rats Could Change the World. *The Atlantic*. www.theatlantic.com/science/archive/2017/11/new-zealand-predator-free-2050-rats-gene-drive-ruh-roh/546011/

Young, N. and Matthews, R. (2010). *The Aquaculture Controversy in Canada: Activism, Policy, and Contested Science*. Vancouver: UBC Press.

21 The Future Is Co-Managed: Promises and Problems of Collaborative Governance of Natural Resources

Nathan Young

Introduction

Natural resources play a central role in the global economy, and in the livelihoods of billions of people. The extraction and processing of natural resources are big business, as a cursory glance at the balance sheets of the world's largest energy, mining, and forestry companies attests. At the local scale, economic activities based on natural resources support communities and households across the wealthy and developing worlds. Despite rapid urbanization, half the globe's population lives in rural areas. According to the FAO (2015; 2016), 1.6 billion people worldwide rely on forests for their daily sustenance and/or economic livelihood, while 250 million similarly rely on freshwater and marine fisheries.

Despite this outsized role in global wealth and human well-being, natural resources receive little attention from environmental sociologists (Buttel 2002). While natural resource extraction plays a significant part in neo-Marxist theories such as the treadmill of production and ecological rift, these perspectives take a macro political-economic view of natural resource issues, and rarely descend into the nitty-gritty of resource governance, rights, and regulation. The incomplete attention granted by environmental sociology to natural resources is particularly regrettable because, as I will argue in this chapter, natural resource issues have become sites of significant experimentation and struggle over issues of environmental justice, equity and rights, and democratic governance – themes that are central to the sociological project. The time has come to start paying closer attention.

Among the most significant of these experiments are those involving "co-management," or more accurately "adaptive co-management" (ACM), that are being undertaken by governments in both wealthy and developing countries with increasing regularity (see Plummer et al. 2012). These schemas are intended to connect state and local institutions, capacities, and knowledges in an effort to enhance the efficiency, accuracy, and legitimacy of environmental governance and decision-making (Schultz et al. 2011). They also imply a devolution or sharing of authority over territories, ecosystems, and/or resources, thus connecting the coercive and technical powers of the state with the normative and relational powers of groups and communities. These are potentially radical ideas that, I will argue, deserve the attention and contributions of environmental sociologists.

Like most state-sponsored experiments, the majority of ACM schemas fall well short of their transformative ambitions or potential (Plummer et al. 2012). However, the increasing popularity of ACM in policy circles should not be seen as a flash in the pan either. Experiments in ACM have broad political appeal. The concept is flexible and polymorphous, and it is (at least nominally) consistent with arguments from both social justice and neoliberal perspectives: that distant states ought to defer to community preferences, that experts in the centre may not know better than local people, and that empowering or responsibilizing local groups can create efficiencies and improve key management outcomes such as compliance, monitoring, and enforcement (see Jentoft et al. 1998; Pinkerton et al. 2008). These cross-cutting ideological affinities suggest that we are likely at the beginning of a long period of experimentation with ACM that may see it extended beyond its current focus on natural resources.

The chapter is organized as follows. First, I outline in broad strokes the intellectual history of ACM and discuss why this idea has found purchase within diverse academic and policy networks. I argue that ACM has achieved this popularity in part because of its thematic congruence with some of the most important intellectual movements in contemporary politics and the social sciences. Second, I discuss the issues and challenges involved in translating ACM ideals into workable practice across diverse settings. The political and social world is inherently complex and messy, and much can be learned about ACM by investigating the necessary compromises that occur with implementation. Third, I explore critiques that have emerged of the ACM literature, particularly ongoing conceptual and empirical challenges that are hindering our ability to assess and improve ACM processes. Fourth, I delve into more profound critiques that have been voiced about ACM as an ideal and policy goal. The most potent critiques of ACM revolve around unresolved problems of power, representation, and legitimacy – issues that must be addressed for ACM to evolve and realize its transformative potential. Finally, I consider why ACM matters for environmental sociology, and vice versa. If the future is indeed co-managed, it is critical for environmental sociologists to learn from the existing literature on ACM and to bring our own concepts and ideas into these discussions. My aim in this final section, and in the chapter as a whole, is to present a vision for how this might occur.

Adaptive Co-Management as Idea and Ideal

Over the past several years, ACM has become a bit of a star concept in academic and policy circles. Like most social science concepts, however, it has a genealogy that continues to affect its current uses and emphases. According to Berkes (2009), ACM has a dual heritage that is rooted in both the natural and social sciences. The concept of adaptive management emerged in the applied ecology literature in the late 1970s, as part of the "new ecology" thinking that sees ecosystems as complex, dynamic, and ever-changing, as opposed to static and predictable. Managing dynamic ecosystems requires constant, open-ended learning to inform flexible decision-making (rather than relying on a prescriptive command-and-control approach). The concept of co-management is grounded in the commons literature that was emerging at around the same time. In response to Hardin's (1968) fatalistic analysis of the tragedy of the commons, this literature recast local people and their customs and institutions as assets rather than obstacles to responsible environmental governance (Dietz et al. 2003). Berkes (2009) argues that since the 1970s, "adaptive management and co-management have been evolving toward a common ground because adaptive management without collaboration lacks legitimacy, and co-management without learning-by-doing does not develop the ability to address emerging problems." In today's terms, ACM therefore refers to schemas that involve both cross-scale collaboration and learning-based responsiveness to dynamism and change (Armitage et al. 2007).

I argue that one of the reasons why ACM is generating such interest today is that it encompasses and unites a range of contemporary academic and policy ideas. First, ACM speaks to longstanding ideas and debates about deliberative democracy. Schultz et al. (2011) argue that ACM is grounded in a "participation paradigm" that sees direct public involvement in governance processes as a means of enhancing the both the accuracy and social acceptance of resulting decisions. Habermasian ideas about communicative rationality underscore much of the ACM literature (often implicitly), in that public deliberation is seen as a means of critically evaluating different options and ideas, locating one's positions in a network of interests, and working rationally and progressively towards consensus (Dietz 2013). Similarly, Butler et al. (2015) argue that the participative processes engaged in ACM are important mechanisms for conflict resolution, because enhanced contact among traditional authorities (regulators) and diverse stakeholder groups creates a novel basis for mutual understanding.

Second, ACM speaks to current academic and political interest in learning, plural knowledges, and adaptation to change. In principle, most ACM frameworks make room for multiple ways of knowing natural systems – typically scientific/technical, local/experiential, and customary/traditional knowledge – thus implicitly or explicitly committing to considering each in decision making. In considering different knowledges and perspectives, it is hoped that participants in ACM will engage in a form of deep or transformative learning – what the literature terms "double-loop" or "triple loop" learning that leads participants to question the fundamental assumptions of their perspectives and of governance processes themselves (Armitage et al. 2008). Such learning is seen as enhancing adaptive capacity, or the ability of participants in the ACM process to respond to change and unforeseen challenges by drawing on a range of social, material, and knowledge resources (Armitage et al. 2011).

Third, ACM is connected to the broad intellectual shift away from questions of government and towards those of governance (Wyborn 2015a, 2015b). Theories of governance see power, influence, and authority as pluri-locational and pluri-directional, found not only in state apparati but also in networks, customs, knowledge, and practices (Savoie 2010). Experiments in ACM are experiments in governance, and provide a window into the exercise and interactions of these different types and sources of power. According to Wyborn (2015a, 2015b), while much ACM practice is oriented towards the evaluation of evidence, these should be considered moments of "co-production," where knowledge and social interventions are generated together (Jasanoff 2004). The design of ACM processes is therefore critical for sharing and distributing this co-productive power across institutions and among groups. Not surprisingly, this can be a major point of disagreement and conflict.

Fourth, and this point transects each of those raised above, ACM speaks to the political and intellectual movement to recognize and validate locality in its many forms. Embedded in ACM is the notion that users of a natural resource have special status compared to other members of the citizenry. In some cases, this status involves the recognition of rights, particularly when indigenous communities are involved, or when private property rights are encompassed by the territory or resource in question. In other cases, this status is granted (implicitly or explicitly) on a mix of principles such as adjacency, customary use, and/or stewardship. In nearly all cases, however, the line between who can and cannot participate in ACM is drawn territorially in a way that is meant to privilege local people.

Adaptive Co-Management in Practice: Issues and Challenges

ACM programs are intended to foster "learning and linking" vertically (across multiple scales) and horizontally (across groups), thereby improving environmental governance. They are intended to mobilize multiple types of knowledge and provide mechanisms for rapid response and collaborative decision making – thus, as was mentioned earlier, enhancing the efficiency, accuracy, and legitimacy of governance outcomes. With such a complicated set of goals and processes, what could go wrong?

In practice, designers and participants in ACM programs face a range of issues and challenges. Bringing diverse participants and interests together to exchange and debate the inputs and outputs of decision-making poses enormous communicative and organizational challenges. Some of these challenges are mundane, while others are profound and vexing. While the challenges are intertwined, I group them in the following categories: practical, epistemological, structural/procedural, and relational.

The practical challenges facing ACM programs and participants include cost, time, and distance. The literature on ACM stresses the need for governments to provide adequate and stable funding for these complex processes (Olsson et al. 2004). For example, community groups require considerable funding to consult members, collect and synthesize knowledge, and hold meetings. Travel costs can be substantial, particularly when the physical distances among participants are vast. Translation and other support services may be required, particularly when indigenous or multilingual communities are involved. Just as importantly, engagement in

ACM processes puts a time burden on participants and their families, and individual compensation for lost work and leisure time is sometimes expected. A further complication is that these costs and burdens are borne throughout a process whose outcomes are deferred (Armitage et al. 2009). ACM depends on the long-term development and maintenance of trust, capacity, connections, and knowledge. Governments have been known to lose patience or interest in ACM programs due to the disjuncture between short-term efforts and long-term results, leaving them adrift or in abeyance for long periods of time (Acheson 2013).

ACM programs also confront numerous epistemological challenges. There is a well-known epistemological gulf between the types of knowledge preferred by governments (scientific, technico-expert, numerical) and the types of knowledge held by user and community groups, which are often presented in narrative or illustrative form (Young et al. 2016). Government representatives often have difficulty getting past the stereotype of local, experiential, and traditional knowledge as being "anecdotal" and are thus consciously or unconsciously unwilling to base policy decisions on non-scientific knowledge. Scholars of local and traditional knowledge argue, however, that the true disjuncture is between different types of empiricism (Agrawal 1995; Berkes 2008). Scientific and local/traditional knowledge systems produce knowledge similarly, via observation, experiment, pattern recognition, and generalization, but epistemological socialization makes it difficult for some government representatives to recognize and accept alternative forms of empiricism. Ashwood et al. (2014) argue that this gulf can be bridged by all parties accepting that their knowledge is contextual and, although different in the details, is nonetheless connected to the knowledge held by others. Using an agricultural case of disagreements over water pollution, Ashwood et al. (2014: 428) submit that "the key in our research was not where actors' knowledge came from (i.e., local or expert), but whether participants were able to understand their knowledge as situated in the ground of their own experience, and able to link that knowledge to the ground of others' experience, building a kind of landscape of knowledge that connects one ground to another."

However, the epistemological gulf is often about more than just knowledge claims. For instance, participants from the cultural majority (government and local) typically have difficulty understanding and accepting spiritual and cosmological claims from cultural minorities, especially indigenous groups. The cosmology of many indigenous groups grants human-like agency to certain fauna and flora, landscapes, and natural features such as mountains and glaciers (Cruikshank 2005). This can contribute to fundamental disagreements over the validity of knowledge. For example, there has been longstanding conflict in Canada's Eastern Arctic over the quota system for Inuit communities to hunt polar bear. Government experts generally advocate a reduced quota, pointing to animal telemetry, satellite imagery, and field data that suggest vulnerability at the individual and population level due to climate change (Dowsley and Wenzel 2008). Inuit representatives are not convinced of this, however, and cite more frequent encounters with polar bears as an indicator of population health (Henri et al. 2010). For the Inuit, moreover, polar bears are endowed with significant agency, and are thought to seek out hunters in order to reward them or their families with their flesh. In this instance, to avoid killing the bear would be a profound insult, both to the animal and to the cosmological order. It is difficult to reconcile a quota system with this intertwining of knowledge and ritual, and the complexity of this conflict demands the elusive transformational "double- and triple-loop" learning discussed earlier.

Structural and procedural challenges include key questions of who is included (or at least invited to participate) in ACM processes, what role or capacity different participants will play, the degree of formality or informality in different processes and interactions, and how learning and collaboration are expected to take place. The design and organization of ACM programs are important legacy decisions that are unfortunately often made centrally to the exclusion of user groups and communities (Parkins and Mitchell 2005). Decisions about who gets a seat at the table raise important issues of representation, particularly for groups that have unclear or contested leadership. The assigning of roles and statuses is also an important political step that is not always achieved consensually or transparently. The decision whether to distinguish between rights-holders and stakeholders is critical, but not always carefully considered. Is a distinction to be made between holders of traditional (indigenous) and private property rights? How are the views and interests of different groups to be weighed and weighted in a collaborative decision-making process? In the broadest sense, what are to be the rules, and who will decide upon them? In some cases, these procedural issues lead to the breakdown, or at least the breaking up, of ACM experiments. In Canada's Fraser River system, for example, the government regulator has had to split its ACM programs into multiple branches – some that involve bilateral exchanges directly with indigenous groups, and some that involve multilateral exchanges with all other rights and stakeholders (sometimes attended by leaders of indigenous groups). The result is a bifurcated and opaque system that many participants find disappointing (Cohen 2012: 93).

Finally, relational challenges refer to both relations among people (social) and relations between the ACM process and its biophysical objects (social-ecological). Considering the first, Berkes (2009: 1692) reminds us that "[adaptive] co-management is not just about resources, it is about managing relationships." ACM involves intense interpersonal and intergroup interactions on a range of sensitive issues. Each person and group brings their own interests into play, and with them the potential for "higher order games," as Irwin et al. (2012) describe the strategic use of delay, disagreement, and withholding of consensus within deliberative settings. Trust and legitimacy are central to ACM, but both are slowly built and quickly lost if relationships sour. Lundmark et al. (2014) argue that legitimacy is a multifaceted idea, involving acceptance of both process and outcomes. In addition, people outside the ACM exercise but affected by it (both the "represented" and the unrepresented) will have views on these legitimacy dimensions, and their acceptance matters tremendously for the effectiveness of this type of governance. Complicating things further, the dimensions of legitimacy are often in tension. For example, the exclusion of "troublesome" groups from the process may enhance internal legitimacy, because remaining participants may find it easier to reach agreement. However, exclusions are likely to harm external legitimacy, and provide a narrative to justify rejection and non-compliance with any decisions reached.

Relational challenges also apply in the social-ecological dynamic. ACM programs are typically designed according to political boundaries such as communities, regions, and traditional territories. It goes without saying that ecosystems and wildlife do not often follow these delineations. Natural boundaries, such as watersheds and migration routes, are obvious targets for ACM but are politically difficult to enact (Nguyen et al. 2016). Highly mobile wildlife pose a particular challenge for ACM processes, given that multiple geographically distant political groups hold legitimate rights and knowledge claims that must be coordinated, sometimes across international borders. Such challenges demand significant political flexibility that is difficult to achieve across jurisdictions.

Critiques of the Literature

The scholarly literature on ACM is varied and multi-disciplinary, with the main contributions coming from ecologists, geographers, and researchers in environmental studies or environmental management programs. This diversity is both a strength and a liability. In a series of systematic reviews of the literature, Plummer (2009) and Plummer et al. (2012; 2013) identify several conceptual and methodological weaknesses in ACM research that need to be addressed for the field to mature academically, and to enhance the utility of the ACM concept for policy-makers.

First, ACM research to date has been predominantly descriptive and case study based. While case research is critical for theory-building, there has been a lack of coordinated conceptual and theoretical development based on cumulative case knowledge (Plummer et al. 2012). For example, ACM researchers have identified several criteria that contribute to the success or failure of ACM programs, such as the involvement and enthusiasm of key leaders, well-defined boundaries to the region or resource, well-functioning and legitimate local institutions, and a supportive policy environment (Armitage et al. 2009). They have also identified key processes, such as "learning and linking" across groups and scales, as well as contributing factors, such as community capacities to synthesize and communicate complex knowledges and positions, and the presence or absence of "bridging organizations" that can reach across social, cultural, and political barriers (Berkes 2009; Matthews and Sydneysmith 2010). These findings are critical for theory-building, but in practice are typically considered only as a form of checklist for ACM program design and evaluation.

Second, and related to this, the majority of ACM research to date is (often implicitly) anchored in systems thinking borrowed from ecology, specifically the notions of resiliency and panarchy that have been developed by advocates of the "new ecology" such as C. S. Holling, Lance Gundersson, and Carl Folke (Plummer 2009; see also Young 2016). While concepts such as resiliency are useful, they apply imperfectly to the deeply sociological processes that are evoked in cross-group deliberation and collaboration. For example, Plummer (2009) argues that ecology-inspired systems thinking underestimates the role of institutions (local and extra-local) in shaping how people interact with one another. In the sociological sense, institutions refer to "the habituated and customary dimensions of social life" (Matthews and Sydneysmith 2010: 224), meaning that they are not synonymous with organizations. Nevertheless, many ACM analyses conflate the two. More importantly, Plummer (2009) found that institutions (read: organizations) are often blamed for failures of ACM (i.e., they are presented as obstacles to collaboration) rather than investigating the complex role institutions actually play. A more sophisticated analysis of institutions presents them as simultaneously enabling and constraining, as mediators between individuals and collectives, and as normative arenas for action – all of which are critical components of ACM processes (Matthews and Sydneysmith 2010).

Third, the existing ACM literature often fails to distinguish between actual and potential observations and outcomes (Plummer et al. 2012). Many ACM scholars are deeply invested in the concept and its potential and want to see these programs and experiments succeed. This can lead to a confusion between *is* and *ought*, or analysis and prescription. Even in the absence of cheerleading, the long time horizon of ACM processes poses an analytical challenge. If processes are in course or have not yet reached their conclusion, a measure of speculation is required about potential or future outcomes. In the best-case scenario, this introduces significant error into evaluations of ACM processes that have not yet concluded and may yet change in multiple ways. In the worst-case scenario, the empirical bases for conclusions are cast into serious doubt because it is unclear whether the analysis is based on actual observations, anticipated outcomes, or a blend of both.

Fourth, ACM scholarship is often uncritical in its use of key concepts and in its approach to social relations and processes. Concepts such as learning, knowledge, expertise, governance, collaboration, and deliberation are often deployed without reflection or reference to the broad critical literatures that have developed around each. For example, deliberative forms of democracy and decision-making have long been studied from a variety of perspectives in political science, sociology, and communications, but these are rarely evoked in ACM analyses. Similarly, traditions such as social studies of science, science and technology studies, and public engagement with science have much to contribute to understanding the social and political dynamics of debates over knowledge and expertise, but are also rarely engaged in ACM scholarship (but see Wyborn 2015a; 2015b). With respect to social relations and processes, ACM research typically take these at face value. Power dynamics, when considered at all, are often handled descriptively as context rather than as central to explanation and analysis. The maxim of political ecology, that "politics is inevitably ecological and that ecology is inherently political" (Robbins 2012: 3) is largely unrecognized or considered only superficially in much of the literature.

These critiques reflect the fact that ACM research is still under development. For the moment, at least, however, ACM research suffers from a measure of "imprecision, inconsistency, and confusion" that has yet to be resolved (Plummer et al. 2012: 11). As I will suggest in later sections, this should be seen as an opportunity for environmental sociologists (among others) to make fundamental conceptual and theoretical contributions to this emerging field of study.

Critiques of ACM in Practice

As I argued earlier, the ascension of ACM in academic and policy circles owes much to its resonance with both social justice and neoliberal ideologies. But the acclaim is far from universal. The most trenchant critiques of ACM as a policy goal and real-world practice revolve around questions of power. This theme reflects a long-running debate in the international development literature about the unintended consequences of participatory approaches to economic development, which were meant to replace top-down mechanisms that excluded affected people from decision-making. In 2001, Cooke and Kothari published a provocative edited volume called *Participation: The New Tyranny?* that included essays about how this approach grafts the assumptions of influential Western academics and state-backed development agencies onto less powerful local processes and institutions that are ill equipped to receive them, in some cases causing significant harm to local relations. Other chapters in this volume addressed the limits of a formulaic approach to participation, potential for abuse and manipulation, and clientism that are directly connected to power imbalances (see also Hickey and Mohan 2004).

With respect to ACM, Nadasdy (1999; 2005; 2007) has taken up several of these themes in an influential series of articles and book chapters. In these, he argues that despite the rhetoric of empowerment, collaboration, and learning, senior governments retain all of their formal and informal decision-making powers. Using ethnographic research in the Canadian Yukon, Nadasdy documents the deep discomfort felt by government agents (bureaucrats, scientists, policy experts, and politicians) when communities advance alternative knowledges, narratives, and policy priorities, and the various ways in which these are ignored, undermined, and de-legitimized within the ACM process so that authorities achieve their desired result. More than this, Nadasdy argues that participation in ACM processes has the unintended effect of "bureaucratizing" participants and communities themselves. Because senior governments are so dominant in these processes, they structure the discursive terrain on which evidence and decisions are considered and debated. As such, the logics of scientific/expert knowledge and bureaucratic rationality (rather than a community-based rationality) define the terms and range of subsequent discussions. While space is granted to local and traditional knowledge, communities feel the obligation to translate these into "data" to conform to the logics of sample sizes, error bars, and statistical significance. This process of translation "leads almost automatically to the bureaucratization of the people and communities who participate in co-management" because they are unable to present their views in their proper political and epistemological context (Nadasdy 2005: 216). The fundamental injustice at play in ACM processes is therefore that "to be empowered, local people must first agree to the rules of the game" that are set by, and profoundly advantage, traditional authorities, within processes that are cloaked in the rhetoric of equality and collaboration (Nadasdy 2005: 220).

A second power-related critique of ACM in practice concerns the (inevitable) tension between representative and deliberative democracy. Experiments in deliberative democracy are always partial, because not all potentially affected parties can or should participate (Goodin and Dryzek 2006). Not everyone can be directly involved in debates and discussions, as they would become unwieldy and paralyzed by difference. However, decisions about selection and exclusion in *deliberative* processes are often based on classic assumptions about *representative* democracy. Governments typically invite known leaders to participate or to select delegates, thus entrenching what Schultz et al. (2011: 663) term the "elite capture" of ACM processes. Governments eager to establish ACM and have the process run smoothly and predictably rarely question or critically assess the relationship between these elites and the groups and populations they are presumed to represent.

Finally, some observers criticize ACM experiments and policies for contributing to political quietism and the co-option of dissent (Young 2016). ACM schemas are but one tool in the deep policy toolboxes of senior governments. In this context, the recent government enthusiasm for ACM can be read somewhat cynically. In an era in which governments find themselves increasingly subject to litigation over indigenous rights, adherence to conservation law, and environmental assessment procedures, engagement in ACM can be seen a long-game strategy for keeping decisions out of the courts and major resource projects on the table (Rusnak 1997). Looking beyond the justice system, ACM processes are intended to channel or translate discontent and dissatisfaction at the local level into useful tools to improve governance (Jentoft 2000). It is legitimate to ask if anything is lost in this translation. Active local resistance to larger forces (economic, political, and environmental) is an important part of long-term community resilience and adaptation to change (Young 2016). Channelling resistance into state-led collaborations has an obvious upside for resolving conflicts and solving management problems over the short-term. The jury is still out, however, on whether the disciplining and responsibilizing effects of ACM "empowerment" projects will contribute to the long-term suppression of more traditional forms of dissent, resistance, and activism that have long been a vital part of rural resilience and collective action.

Why ACM Matters to Environmental Sociology (and Vice Versa)

Experiments in ACM are taking place in a wide variety of countries and contexts. They are underway in a number of wealthy and developing nations, they involve both terrestrial and aquatic environments, and are variously intended to govern specific species, spaces, or entire systems such as coastal zones and watersheds (Armitage et al. 2007). This alone makes ACM experimentation significant for environmental sociology and its overarching academic mission of understanding human–nature dynamics, relations, and entanglement. But what makes ACM compellingly relevant to environmental sociology is that these are experiments with forms of environmental governance that recognize, however partially and imperfectly, living-humans-in-living-nature. By this, I mean that ACM aims to simultaneously restructure the governance of human populations and biophysical natures through their points of intersection and togetherness. In their idealized form, at least,

ACM experiments walk the talk of social-ecological systems thinking. They take the principles of locality, agency, and intertwined social and natural histories seriously. They recognize the weight of past environmental and material injustices and seek to redress them through future practice. They have the potential to serve as a first step towards rethinking the role of the central state towards communities in their glorious social-ecological complexity.

How and why the goals of ACM are defined, pursued, contested, subverted, and/or denied are therefore of direct theoretical significance for environmental sociology. They involve the interplay of the ideal and the material, society's constructions and nature's constructions, local livelihoods and global flows. Experiments in ACM are also, in my reading at least, one of the most ambitious non-market attempts at environmental reform currently underway. As argued by Newell and Patterson (2010), current debates about environmental policy are dominated by neoliberal market-inspired instruments such as taxes, cap and trade, environmental bonds, financialization, and climate exchanges. Toss in parallel debates about ecosystem services and environmental externalities, and the field of environmental policy-making begins to look narrowly economic. In this context, ACM is notable for its central focus on the processes of governing sociable humans in real environments. The decisions and actions of these humans may be based on non-economic priorities and values. Their knowledge base may eschew the accounting logic of market instrumentalization. Their institutions and customs may be based on a communitarian rather than an individualistic ethos. These possibilities speak to the core concerns of a humanist environmental sociology – one that is preoccupied with issues of justice, rights, and equity. Moreover, these possibilities open up fascinating avenues for imagining alternative environmental futures, at the micro scales of current experimentation up to the macro scales of potential (and hoped for) reforms. In other words, ACM experiments can tell us a lot about the "range of the possible" with respect to participative environmental governance. These experiments are happening in front of our eyes, simply in different places than environmental sociologists are accustomed to looking.

Environmental sociology also has much to contribute to understanding and improving ACM practices. ACM experiments are fundamentally sociological experiments, intended to harness the diverse capacities of state and local actors, thereby reducing the friction between environmental policies and environmental practices. Yet as discussed earlier, ACM scholarship is still under development and, for the moment at least, is largely descriptive and case study based (Plummer et al. 2012). It has yet to fully address issues of structural and symbolic power, the complex role of local institutions, and the impacts of decisions about process and procedure on the performance and outcomes of ACM. The sociological literature on environmental justice, for instance, could provide valuable insights and conceptual tools for filling these gaps. So too would the literatures on social movements, political sociology, science and technology studies, and public engagement with science. For example, ACM experiments are currently being designed and conducted largely in isolation from the well-documented experiments in public deliberation about contentious environmental issues.

A good deal of sociological research has been conducted over the past several decades on consensus conferences, dialogic democracy, hybrid forums, mini-publics, and citizen science in the context of environmental governance (e.g., Goodin and Dryzek 2006; Callon et al. 2009; Jasanoff 2011; Stilgoe et al. 2014), but this work has not been systematically integrated into ACM scholarship or practice.

Environmental sociologists are well positioned to build these bridges, applying the methods, concepts, and research questions prominent in our own discipline to this multidisciplinary field to assist in filling in some of the theoretical and empirical gaps identified by critics. This would also help us to overcome another shortcoming in our own domain. Environmental sociology has been strangely isolated from thinking and research in the other "environmental social sciences" with which we share core concerns and subject matter (Buttel 2002; Bennett et al. 2017). Direct attention to and participation in ACM experiments would help us better connect with our intellectual fellow travellers as we work to enhance conservation, sustainability, and equity in the social-natural worlds we study.

Conclusion

In this chapter, I have argued that natural resource governance has become an important site of creative experimentation that encompasses some of the most important social-ecological themes of our times, including questions of justice, knowledge, rights, and democracy. These experiments are highly variable, and are being pursued in diverse social and ecological contexts, but are coalescing around the concept of adaptive co-management, or ACM. In its idealized form, ACM is a governance strategy based on "learning and linking" vertically (across scales and hierarchies) and horizontally (across groups). It is intended to mobilize and coordinate the knowledge, capacities, and authority of senior governments with those of local users and communities. In so doing, it aims to enhance the "efficiency, accuracy, and legitimacy" of environmental governance and decision-making (Schultz et al. 2011).

These experiments matter for many reasons. While a good number of ACM schemas fall short of idealized goals (Plummer et al. 2012), they represent important shifts in social-ecological conversations. They are an attempt, however imperfect, at recognizing locality, sharing power and authority, and redressing past wrongs. They also represent imaginings of environmental futures that make room for local social-ecological knowledges, values, and priorities. ACM research already shows that local contributions frequently depart from the technico-scientific and market-based reasoning that dominates in so many other spheres of environmental governance. The question of how to put these contributions together into coherent governance and decision-making strategies is urgent and frankly exciting. Moreover, if I am correct that the increasing popularity of ACM in policy circles is due to its ideological flexibility and (nominal) affinities with both social justice and neoliberal worldviews, then we may expect to see these schemas extended both geographically and across sectors. We may indeed be at the beginning of a long period of experimentation with these forms of governance that would take us well beyond the realm of natural resources.

To date, however, academic and policy discussions of ACM have proceeded without much input from environmental sociology researchers, concepts, or methods. This is unfortunate, because ACM processes are themselves deeply sociological, and because environmental sociologists have developed sophisticated theories of power (material and symbolic), institutions, social action and social mobilization, and knowledge interactions and deployment, that are directly relevant to further research on ACM and to real-world improvements to ACM design and practice. While there is excellent social science about ACM theory and experimentation (e.g., Armitage et al. 2007; Armitage and Plummer 2010; Wyborn 2015a), my position is that environmental sociology has much to contribute. Two decades ago, Buttel (2002) argued that environmental sociology tended to ignore natural resource issues in part because the latter were seen as less relevant to developing meta-theories of pro-environmental ideas and behaviours on the one hand, and material challenges of degradation and scarcity on the other. Surely the time has come to rethink the role of natural resources in these projects, and to see the experiments underway in many resource regions for the deep theoretical and real-world significance they hold.

References

Acheson, J. M. (2013). Co-management in the Maine lobster industry: A study in factional politics. *Conservation and Society*, *11*(1), 60–71.

Agrawal, A. (1995). Dismantling the divide between indigenous and scientific knowledge. *Development and Change*, *26*, 413–439.

Armitage, D., et al. (2009). Adaptive co-management for social-ecological complexity. *Frontiers in Ecology and the Environment*, *7*(2), 95–102.

Armitage, D., Berkes, F., & Doubleday, N. (eds.). (2007). *Adaptive Co-Management*. Vancouver: UBC Press.

Armitage, D., Berkes, F., Dale, A., Kocho-Schellenberg, E., & Patton, E. (2011). Co-management and the co-production of knowledge: Learning to adapt in Canada's Arctic. *Global Environmental Change*, *21*, 995–1004.

Armitage, D., Marschke, M., & Plummer, R. (2008). Adapative co-management and the paradox of learning. *Global Environmental Change*, *18*, 86–98.

Ashwood, L., Harden, N., Bell, M. M., & Bland, W. (2014). Linked and situated: Grounded knowledge. *Rural Sociology*, *79*(4), 427–452.

Bennett, N., & et al. (2017). Conservation social science: Understanding and integrating human dimensions to improve conservation. *Biological Conservation*, *205*, 93–108.

Berkes, F. (2008). *Sacred Ecology* (2nd ed.). New York: Routledge.

Berkes, F. (2009). Evolution of co-management: Role of knowledge generation, bridging organizations and social learning. *Journal of Environmental Management*, *90*, 1692–1702.

Butler, J. R. A., et al. (2015). Evaluating adaptive co-management as conservation conflict resolution: Learning from seals and salmon. *Journal of Environmental Management*, *160*, 212–225.

Buttel, F. H. (2002). Environmental sociology and the sociology of natural resources: Institutional histories and intellectual legacies. *Society and Natural Resources, 15*, 205–211.

Callon, M., Lascoumes, P., & Barthe, Y. (2009). *Acting in an Uncertain World: An Essay on Technical Democracy*. Cambridge, MA: MIT Press.

Cohen, B. I. (2012). *The Uncertain Future of Fraser River Sockeye: Volume 1, The Sockeye Fishery*. Commission of Inquiry into the Decline of Sockeye Salmon in the Fraser River. Retrieved from www.cohencommission.ca/en/FinalReport/

Cooke, B., & Kothari, U. (eds.) (2001). *Participation: The New Tyranny?* New York: Zed Books.

Cruikshank, J. (2005). *Do Glaciers Listen? Local Knowledge, Colonial Encounters, and Social Imagination*. Vancouver: UBC Press.

Dietz, T. (2013). Bringing values and deliberation to science communication. *PNAS, 110*(3), 14081–14087.

Dietz, T., Ostrom, E., & Stern, P. C. (2003). The struggle to govern the commons. *Science, 302*(12), 1907–1912.

Dowsley, M., & Wenzel, G. (2008). "The time of the most polar bears": A co-management conflict in Nunavut. *Arctic, 61*(2), 177–189.

FAO. (2015). *Forests and poverty reduction*. Rome: Food and Agriculture Organization of the United Nations. Retrieved from www.fao.org/forestry/livelihoods/en/

FAO. (2016). *The state of the world's fisheries and aquaculture 2016*. Rome: Food and Agriculture Organization of the United Nations. Retrieved from www.fao.org/fishery/sofia/en

Goodin, R. E., & Dryzek, J. S. (2006). Deliberative impacts: The macro-political uptake of mini-publics. *Politics & Society, 34*(2), 219–244.

Hardin, G. (1968). The tragedy of the commons. *Science, 162*(3859), 1243–1248.

Henri, D., Gilchrist, H. G., & Peacock, E. (2010). Understanding and managing wildlife in Hudson Bay under a changing climate. In S. H. Ferguson (ed.), *A Little Less Arctic: Top Predators in the World's Largest Northern Inland Sea, Hudson's Bay* (pp. 267–289). New York: Springer.

Hickey, S., & Mohan, G. (eds.) (2004). *Participation: from Tyranny to Transformation?* New York: Zed Books.

Irwin, A., Jensen, T. E., & Jones, K. E. (2012). The good, the bad and the perfect: Criticizing engagement practice. *Social Studies of Science, 43*(1), 118–135.

Jasanoff, S. (2004). *States of Knowledge: The Co-Production of Science and the Social Order*. New York: Routledge.

Jasanoff, S. (2011). *Designs on Nature: Science and Democracy in Europe and the United States*. Princeton NJ: Princeton University Press.

Jentoft, S. (2000). Legitimacy and disappointment in fisheries management. *Marine Policy, 24*, 141–148.

Jentoft, S., McCay, B. J., & Wilson, D. C. (1998). Social theory and fisheries co-management. *Marine Policy, 4–5*, 423–436.

Lundmark, C., Matti, S., & Sandstrom, A. (2014). Adaptive co-management: How social networks, deliberation and learning affect legitimacy in carnivore management. *European Journal of Wildlife Research, 60*, 637–644.

Matthews, R., & Sydneysmith, R. (2010). Adaptive capacity as a dynamic institutional process: conceptual perspectives and their application. In D. Armitage & R. Plummer (eds.), *Adaptive Capacity and Environmental Governance* (pp. 223–242). Berlin: Springer-Verlag.

Nadasdy, P. (1999). The politics of TEK: power and the "integration" of knowledge. *Arctic Anthropology*, *36*(1–2), 1–18.

Nadasdy, P. (2005). The anti-politics of TEK: The institutionalization of co-management discourse and practice. *Anthropologica*, *47*(2), 215–232.

Nadasdy, P. (2007). Adaptive co-management and the gospel of resilience. In D. Armitage, F. Berkes, & N. Doubleday (eds.), *Adaptive Co-Management* (pp. 208–227). Vancouver: UBC Press.

Newell, P., & Paterson, M. (2010). *Climate Capitalism: Global Warming and the Transformation of the Global Economy*. New York: Cambridge University Press.

Nguyen, V. M., Lynch, A. J., Young, N. et al. (2016). To manage inland fisheries is to manage at the social-ecological watershed scale. *Journal of Environmental Management*, *181*, 312–325.

Olsson, P., Folke, C., & Hahn, T. (2004). Social-ecological transformation for ecosystem management: The development of adaptive co-management of a wetland landscape in southern Sweden. *Ecology and Society*, *9*(4), 1–26.

Parkins, J., & Mitchell, R. (2005). Public participation as public debate: A deliberative turn in natural resources management. *Society and Natural Resources*, *18*, 529–540.

Pinkerton, E., Heaslip, R., Silver, J. J., & Furman, K. (2008). Finding "space" for comanagement of forests within the neoliberal paradigm: rights, strategies, and tools for asserting a local agenda. *Human Ecology*, *36*, 343–355.

Plummer, R. (2009). The adaptive co-management process: an initial synthesis of representative models and influential variables. *Ecology and Society*, *14*(2), 24.

Plummer, R., Armitage, D., & de Loe, R. C. (2013). Adaptive comanagement and its relationship to environmental governance. *Ecology and Society*, *18*(1), 1–21.

Plummer, R., Crona, B., Armitage, D., et al. (2012). Adaptive comanagement: A systematic review and synthesis. *Ecology and Society*, *17*(3), 11–32.

Robbins, P. (2012). *Political Ecology* (2nd ed.). Malden, MA: Wiley-Blackwell.

Rusnak, G. (1997). *Co-management of natural resources in Canada: a review of concepts and case studies* (Minga working papers No. 2) (pp. 1–23). Ottawa, ON: International Development Research Centre.

Savoie, D. (2010). *Power: Where Is It?* Montreal: McGill-Queen's University Press.

Schultz, L., Duit, A., & Folke, C. (2011). Participation, adaptive co-management, and management performance in the world network of biosphere reserves. *World Development*, *39*(4), 662–671.

Stilgoe, J., Lock, S. J., & Wilsdon, J. (2014). Why should we promote public engagement with science? *Public Understanding of Science*, *23*(1), 4–15.

Wyborn, C. A. (2015a). Connecting knowledge with action through coproductive capacities: Adaptive governance and connectivity conservation. *Ecology and Society*, *20*(1), 1–11.

Wyborn, C. A. (2015b). Co-productive governance: A relational framework for adaptive governance. *Global Environmental Change*, *30*, 56–67.

Young, N. (2016). Responding to rural change: adaptation, resilience and community action. In M. Shucksmith & D. L. Brown (eds.), *The International Handbook of Rural Studies* (pp. 638–649). New York: Routledge.

Young, N., Nguyen, V. M., Corriveau, M., Cooke, S. J., & Hinch, S. G. (2016). Knowledge users' perspectives and advice on how to improve knowledge exchange and mobilization in the case of a co-managed fishery. *Environmental Science & Policy*, *66*, 170–178.

PART VI

Food and Agriculture

22 Future and Food: New Technologies, Old Political Debates

Michael Carolan

Big data and precision tools are frequently referred to as the next "big thing" in conventional agricultural circles (Dawson 2016). This is in no small part due to the expanding amount of information collected relating to farm-level crop production (big soil data) combined with extensive weather data (big climate data), which together form the backbone of precision agriculture technology. According to the United States Department of Agriculture's (USDA) Agricultural Resource Management Survey (ARMS), in 1997, only 17 percent of corn acres were cultivated using precision agriculture equipment. In 2010, the most recent year available for corn production practices, 72 percent of corn acres were planted with this technology (USDA 2015).

The global precision farming market reached US$2.3 billion in 2014, with an estimated annual growth rate of 12 percent through 2020 (Michalopoulos 2015). In a recent report, the Joint Research Centre of the European Commission wrote that "precision agriculture can play a substantial role in the European Union in meeting the increasing demand for food, feed, and raw materials while ensuring sustainable use of natural resources and the environment" (Zarco-Tejada et al. 2014: 9). In the case of the Netherlands, precision techniques are now used to manage 65 percent of the country's arable farmland, a figure that was only 15 percent in 2007 (Michalopoulos 2015). Such trends indicate that agri-food firms are taking notice of these applications. Monsanto, for example, has acquired numerous farm data analytic companies since 2012, most notably Climate Corporation for US$930 million. Climate Corporation produces two popular software platforms: Climate Basic and Climate Pro. Monsanto has stated that its Climate Pro sensors on harvesting equipment generate roughly seven gigabytes of data per acre (Bobkoff 2015). With roughly 1 million acres of farmland in the United States alone, we are talking about a lot of data.

This "big thing," however, is not without its critics. One source of concern lies in what is called the data divide (Andrejevic 2014: 1673), which speaks to the asymmetries between the data *haves* and *have-nots*; though, admittedly, some of these "have-nots" are big data *don't wants*, as you will see shortly. This asymmetry is largely a function of farm scale, geography, capital, credit, and position within value chains – e. g., the highly concentrated farm equipment sector gives firms like John Deere tremendous market power.

Not surprisingly, then, farmers express unease about who owns "their" data. To quote from a report based on a survey conducted by the American Farm Bureau

(2015), "Fully 77.5 percent of farmers surveyed said they feared regulators and other government officials might gain access to their private information without their knowledge or permission. Nearly 76 percent of respondents said they were concerned others could use their information for commodity market speculation without their consent." Moreover, while 81 percent believed they retain ownership of their farm data, 82 percent said they had no idea what companies were doing with it.

Yet technological platforms are never innocent (Foucault 2007; Rose 1999). This is not, however, to make the case for the wholesale rejection of technology *per se* in agriculture. This chapter pushes against a narrative that cast critics of the status quo as "culinary Luddites" (see e.g., Laudan 2015), though such criticisms are understandable given critical theory's penchant for criticizing technology while saying far less about the types we ought to be encouraging. Technologies – as an assemblage of human and non-human actors – should be evaluated by what they do and the politics they practice. This chapter makes a modest contribution to the literature by exploring how these socio-technical artifacts have politics by way of exploring the worldviews they are rooted within by exploring their divergent attachments to what it means to be a "good citizen."

To do this, I look at two case studies: two groups that are part of "smart" farming assemblages in quite divergent ways. These data come from (1) twenty employees (technicians, sale reps, and engineers) from various big data companies located from around North America and the UK, and (2) eighteen farmers from around the United States engaged to various degrees with the loosely organized group called Farm Hack. First, I briefly introduce each empirical case. Next, I explore the findings from an instrument used to generate word clouds for each of the two populations based on respondents' understanding of *good citizen*. The chapter concludes by discussing what it means for data and code in an agricultural context to have politics and how we might think about prioritizing some techniques over others.

Setting the Stage and Outlining the Problem

Monsanto has acquired numerous farm data analytic companies since 2012, most notably Climate Corporation for US$930 million. Climate Corporation produces two popular software platforms: Climate Basic and Climate Pro. The company has stated that its Climate Pro sensors on harvesting equipment generate roughly seven gigabytes of data per acre (Bobkoff 2015).

Monsanto is well poised to benefit from this rush to use large data sets and predictive analytics in agriculture because of its earlier foray into biotechnology. Searching genes for favourable (read: profitable) traits in plants in order to create new seed varieties requires sifting through the billions of base pairs in a genome. Even before acquiring Climate Corporation the firm had arguably *already* assembled the world's most extensive agricultural databases, built on thousands of field tests using countless seed varieties grown under every imaginable experimental field condition.

More than just big, these technologies and techniques are also *fast* – a characteristic known as velocity in big data circles. For example, cloud databases, with their lightning-fast download times and overall greater interconnectivity – replacing slow memory sticks – allow farmers today to make yield maps and variable rate fertilizer prescriptions (based on nutrients removed with a crop, soil characteristics, etc.) in almost no time at all. The process now takes one or two hours, compared to what used to take one or two weeks in the memory stick age.

Adoption rates of precision technologies, such as global positioning satellite (GPS) soil mapping, GPS guidance systems, and variable-rate input application technology (VRT) (allowing farmers to customize input applications using GPS data), are on the rise. In the U.S., adoption rates are well above 50 per cent among producers of major crops – e.g. corn, soybeans, cotton, and spring wheat (Schimmelpfennig 2016).

The (US) Digital Millennium Copyright Act (DMCA) was passed in 1998 to prevent digital piracy. The DMAC, to put the matter plainly, made it illegal for a tractor's owner to access the "brains" of their smart equipment. Unfortunately, this meant farmers could do few repairs to equipment they allegedly owned, save for things like changing oil and repairing belts, as most repairs require accessing their equipment's engine control unit (ECU). According to John Deere, in a 2015 letter to the US Copyright Office, farmers do not actually "own" the equipment they buy from the company's dealerships. Rather, they receive "an implied license for the life of the vehicle to operate the vehicle" (as quoted in Wiens 2015), an argument with remarkable parallels to those made by biotechnology firms on the question of who owns patented seeds (Carolan 2010).

In October 2015, the Librarian of Congress ruled in favour of an exception to the DMCA that would allow anyone who owned a tractor (or car, truck, etc.) to tinker with its code. (The Librarian of Congress has a curious position, in that she or he both oversees the operation of the world's largest library, with a staff that numbers in the thousands and a collection in the millions, and oversees the US Copyright Office, the government office that manages the register of all copyrighted materials.) While seemingly a blow against farmer dependency, by freeing farmers from their implement dealerships and their "approved" technicians, the win was hollow. Only the individual who purchased the equipment can access and tinker with its software under the exception. The moment she brings her (non-approved) tech savvy mechanic to look at it, a line in the legal sand will have been crossed and they all become liable for copyright infringement. Moreover, the exemption does nothing to free up diagnostic equipment and manuals, which remain in the control of firms like John Deere. It is hard to fix something if you do not know what exactly is wrong with your equipment, due to a lack of equipment, manuals, and expertise. And so this dependency continues, unabated even with the exemption.

This brief review of precision agriculture and big data techniques ought to offer sufficient evidence justifying it as a subject of further analysis.

Setting the Empirical Stage

This section provides an overview of two projects that make up the empirical backbone of this chapter.

Big Data Industry

The sample interviewed from big data industry consisted of twenty individuals; a group composed of engineers, sales personnel, and tech specialists – those who develop, sell, and fix the software and hardware, respectively. Specifically, nine worked for implement firms, nine worked for firms that produce and sell predictive analytic software, and two worked as technology/big data consultants. Interview guides were designed to elicit a conversation about the technologies respondents produced, sold, and repaired – e.g., views toward them, their perceived benefits and potential risks, their role in future foodscapes. Interviews lasted between 60 and 120 minutes and were tape recorded and later transcribed. All respondents were promised anonymity and pseudonyms are therefore used to ensure that end. In addition to formal interviews, I spent an hour (approximately) with each participant being shown the big data applications that they were most familiar with. While this time was not tape-recorded copious fieldnotes were taken.

Farm Hack

Research for this case study began on June 2014 and continued until January 2016. I grew the sample population by first reaching out to respondents whom I knew. I then further grew the sample by utilizing a snowball sampling technique. My intent was not to study Farm Hack as an organization *per se* but to examine how digital and legal "locks" were used and understood within this community. Those interviewed included individuals with different levels of Farm Hack involvement, from the deeply committed to others who, in the words of one, "participate from a distance" – monitoring the website, attending an occasional meeting, etc. A total of 18 producers from around the United States were interviewed who reported varied degrees of involvement with Farm Hack. Six managed less than 100 acres; five managed between 101 and 500 acres; four managed between 501 and 1,000 acres; and three managed more than 1,000 acres. Interviews lasted between 50 and 70 minutes and were tape-recorded and transcribed. Pseudonyms are used to protect the identity of respondents. Finally, I logged more than 50 hours of participant observations with this group. Most of that time was spent being shown around each individual farm, though I did attend two Farm Hack events. Extensive field notes were taken when no formal interviews were being conducted.

According to the group's website, Farm Hack is "a worldwide community of farmers that build and modify our own tools. We share our hacks online and at meet ups because we become better farmers when we work together" (http://farm-hack.org/ 310app/). There is no official membership list. The community is intentionally porous. As I was told by one respondent, the "hacking ethos is resistant to solidifying a group with membership lists, because that explicitly makes people 'not one of us,' which is a step away from people thinking if-you-not-with-us-you're-against-us." My principle interest in looking at this community involved better understanding how these farmers respond to firms' utilization of digital and legal locks. As such, Farm Hack came to be understood as a call and response to a highly proprietary style of socio-techno-agri-food governance.

Good Citizens Across Different Worlds

The book *Keywords*, written by Raymond Williams in 1985, offers a unique inquiry that I have taken methodological inspiration from when engaging with the subject of how we *make* worlds and not just *reproduce* them.

Like eyes are to the soul, words, according to Williams, are a portal into the very assumptions that underlie our ideas about the world. When we adopt meanings to a term we, often unknowingly, adopt ideas about how the world ought to be, which makes this inquiry into our shared vocabulary of immense practical consequence.

Using word clouds, I show how respondents from each of the case studies held radically different understandings of the keyword "good citizen," pointing to divergent food imaginaries.

Word clouds were created using data gathered from the following instruction: "Select three terms describing what *good citizen* means to you?" Before answering, participants were shown a list of roughly fifty terms for each keyword. The terms on this list were defined, to ensure all participants were operating from a shared understanding of the concepts. Terms were generated from extensive past research looking at how farmers, eaters, and others think about food-related phenomena. The words elicited from this instrument were then plugged into word cloud generating software. Words yielding two or fewer response were not included to improve the "cloud's" readability.

I will now present the findings of the word cloud instrument, followed by a discussion of what those images mean – in terms of the world(s) they imply – by layering them with data from the qualitative interviews. This will allow for tentative conclusions to be made about *why* these groups viewed and enacted the world(s) they did due to their respective engagements with code and data.

The world clouds generated for the keyword "good citizen" can be found in Figures 22.1 and 22.2. To start this analysis, note the difference in the top six terms mentioned for each group. Among those with big data industry, those were as follows: voters, shopper, rule maker, innovator, capitalist, and rule follower. Meanwhile, the six most mentioned terms among Farm Hack participants were voter, activist, rule breaker, innovator, rule maker, and sharer.

To unpack these divergent images, it would be useful to first differentiate between actors of citizenship and those who hold the status of citizenship (Isin 2009). The latter category refers to citizenship a as bundle of legal rights and responsibilities, signifying membership to a state. It is something one *has*. To be an actor of citizenship is to realize that subjects who are not citizens, in the aforementioned socio-legal sense, still act as citizens and that some of those acts have the potential to engender articulations with questions of rights, equality, difference, justice, and democracy. Citizenship in this sense is something one *does*.

Yet we can take this concept – actors of citizenship – still further, by making the distinction between active and activist citizens (Carolan 2017c). Active citizens are

Figure 22.1 *"Good citizen" according to big data industry*

those who perform a fairly conventional style of citizenship, those who engage in such activities as voting, volunteering, writing letters to the editor, and signing petitions. Note how the articulation of being a "good citizen" in big data industry word cloud parallels this idea of being an active citizen, with its emphasis on being a "shopper," "voter," "capitalist," and on being willing to write a "letter to the editor." Moreover, these imaginaries where not divorced from how farmers thought about and practiced big data and precision agriculture.

The following exchange involves a forty-something sales rep for widely use big data platform – Jeff. "I'm doing my part to be a good citizen by helping famers be better farmer," he told me. After asking for additional clarification, he offered the following.

"I'm encourage innovation" – note the appearance of "innovator" in Figure 22.1. "By assisting farmers buy the latest equipment" – note "shopper" in Figure 22.1 – "I'm helping feed the world." This was a common theme among those from this group, the linking of being a good citizen with productivist agriculture. This should not be surprising, given productivism's links to a number of terms prominently displayed in the sample's word cloud, including, beyond this mentioned in prior sentences, "capitalism" (e.g., Rannikko and Salmi 2017; Wilson 2001).

Active citizenship is premised on acts that are routinized, supporting already-established habits, social norms, conventions, and ways-of-being. While important, from the standpoint of nurturing civic health (food-based or otherwise), active citizens have limited capabilities of making a difference in part because their routinization, which is to say they have limited ability to afford novelty and spaces of experimentation. Activist citizens, conversely, are interested in challenging routine, understandings, and practices, which makes theirs a political *project* versus politics as usual.

Following others (Arditi 2014; Clarke 2014; Isin 2009), it is important to emphasize that a radical politics must do more than seek to reorder the state and markets. In some ways, that view of social change is a step too far, recognizing that without the creation of citizens who feel the need to *want* social change that aforementioned radical politics can never occur (Carolan 2011). We must therefore also look for spaces and experiences that leave "traces" (Arditi 2014) that can be reinvested by future mobilizations. Activist citizens engage in practices that help do this, by making the unthought-of thinkable and the undoable routine (Carolan 2013). Vehicles for enacting activist citizens therefore tend to operate outside conventional channels, and may even explicitly question and circumvent those channels.

This brings me to the style of citizenship represented in the Farm Hack word cloud – Figure 22.2. Note the appearance of such terms as "activist," "rule breaker," and "critic" in the image. This seems to offer clear indication that the status of a good

Figure 22.2 *"Good citizen" according to Farm Hack activists*

citizen for this group is someone who does not simply legitimize the status quo. Rather, it is someone who actively challenges it.

Examples of this also came out repeatedly in interviews with this group while articulating their engagement with code, data, and computer hardware. One such representative exchange occurred when interviewing Noelle, a self-described "farmer who stumbled upon open source and can't seem to leave it be." When telling me about what it meant for her to be a good citizen, she kept coming back to what she called "organized activism." In her words:

> To be a citizen is to push boundaries, and to work with others when doing it. [...] You're taught that citizens shop – hell, that's what they told us to do after 9/11; go shopping, remember? Citizens vote. [...] But that's such a limited way of creating change, in part because it pits you and you alone as the agent of change, when we're so much stronger together.

The word clouds by themselves tell us little about *how* these worldviews came to be. This is where the qualitative data come into the picture. I will now connect what you saw in the figures with the experiences and practices engendered by specific sociotechnical assemblages of the respective case studies. To help make sense of these divergent imaginaries around being a "good citizen," it is necessary to delve further in the qualitative data and discuss the underlying ontologies that inform these views.

Respondents expressed differing ontologies when articulating visions of being a citizen. These ranged, at one end of the spectrum, with highly individualistic visions of the world, populated with "consumers" (Nick, software engineer) and competitors – e.g., "I'd say the best citizens are those that compete successfully in the marketplace, which is how we'll feed the world" (David, executive). Farm Hack participants, meanwhile, identified worldviews grounded in collectivistic imaginaries. Here "farmers collaborate" (Sarah, farm hacker).

Emery (2015: 47) refers to the former position as resting on an "ideology of individualism." What is especially problematic about this outlook is that it leads farmers to see their neighbours as competitors: those from whom independence must be sought – note David's comment in the previous paragraph. "This," in Emery's words, "has the effect of masking the structural dependencies which farmers face [...] and limits the alternatives available to them to realize a view of independence that is maintained, rather than opposed, by interdependent collective action" (p. 49). Those articulating this position discussed being a good citizen and the material benefits generated from these practices through a narrowly individualistic lens, in terms of what they, as sovereign individuals, do versus as a process contingent on collective action with other likeminded farmers and activists.

Earlier in the chapter I discussed the DMCA, passed in 1998 to prevent digital piracy, and the 2015 ruling by the Librarian of Congress that granted an exception to the Act allowing anyone who owned a tractor (or car, truck, etc.) to tinker with its code. The reason why the DMCA exemption is a hollow win for most farmers is because it emerges out of a (neoliberal) worldview that privileges individualism – the idea that if only the farmer, and she alone, could tinker with her tractor's code the access problem would be solved.

The DMCA exemption, by rigidly ascribing to the ideology of individualism, might let farmers, as sovereign individuals, tinker with their equipment's computer. But as most farmers lack specialized training in code and software, this narrowly granted exemption does nothing in practice to reduce their dependency on agrifood firms and "approved" technicians. The DMCA exemption is an example of how, to quote one Farm Hack respondent, "free access isn't necessary fair access."

These technological artifacts have therefore politics because they have publics. *What* each publicizes, however, varies. Precision agriculture techniques make public, for instance, climatological trends, soil fertility levels, productivist orientations, and the like. Conversely, the technologies and techniques to emerge from the hacker community seem more trained on publicizing possibilities, versus the certainty of needing to feed 9 billion people eating agro-industrial commodities.

To be clear, by "publicize" I am not referring to a mere unveiling – to make public in the sense of making visible. John Dewey (1946), the great American philosopher and an early developer of pragmatism, worried about the interests, beliefs, and ideologies of elites becoming "fixed" and assuming a taken for granted status within dominant political and social cultures. To combat this he prescribed the technique of "experimentalism," which essentially involves the recruiting of the broader public to constantly reflect upon and question conventional habits and beliefs. Dewey believed this constituted an important first step in breaking up imposed rules of order and action that is necessary if meaningful social change is to occur.

How publics do this, however, is where I part company with many contemporary pragmatists, as they tend to place too much faith in the power of talk. Habermas (1987), for instance, develops his pragmatic insights by way of the concept of communicative rationality. For him, a vibrant public sphere composed of people talking and actively listening has the power to break the stranglehold on rationality by elites. What these communication-centered arguments miss is that publics also involve a *material* coming together, not just a talking together. And that is where these two groups differed greatly, in *how* these technologies materially brought people together.

Big data industry respondents *never* talked about precision agriculture as a vehicle that brought people together, collectivist ontological sort of way at least. When it did bring individuals together it was as consumers or even competitors. A representative quote illustrating this outlook came from Julie, a service technician for a large farm equipment company. The following statement comes on the heels of her telling me about how "precision agriculture brings people, farmers in particular, together by making them compete, and there's something very democratic about that." Curious, and a bit confused, by this statement, I asked for clarification.

> "It's [farming] becoming less about experience and knowledge. Precision agriculture puts everyone on an even playing field, as they all have micro-level data. [...] Those [farmers] that remain [in business], in the end, deserve it. May the best man or women win."
> "And the democratic piece? How is this 'democratic'?" I asked.

"As we were discussing earlier" – we were discussing what it meant to be a good citizen earlier – "the best way a farmer can be a good citizen is to work hard, run an efficient operation, and remain standing in the end."

Again, note the celebratory undercurrent toward individualism and competition in Julie's comments; sentiments you would not find privileged in the Farm Hack group.

Conclusion

In an article titled 'Do Artifacts Have Politics?', Winner (1980: 122–127) tells a story of the famous early twentieth-century political entrepreneur Robert Moses and the low-hanging overpasses leading into the beaches of Long Island. As Winner explains, Moses had these overpasses built to a specific height, under which only automobiles could pass. Buses stood too high and could not pass through these overpasses and therefore could not service the beaches. That this was done intentionally because the typical bus-user at the time was African American. In doing this, these overpasses served political ends: they helped keep the beaches of Long Inland a space for the wealthy automobile-owning (read: white) residents of the area.

In discussing this example, Winner's point is this: material artifacts affect and are in effect a type of politics. In his own words:

> [T]he devices, techniques, and systems we adopt shed their tool-like qualities to become part of our very humanity. [...] In a trivial sense it is true, for example, that "You can always turn off your television set." But given how central television has become to the content of everyday life, how it has become the accustomed topic of conversation in workplaces, schools, and other social gatherings, it is apparent that television is a phenomenon that, in the larger sense, cannot be "turned off" at all. (Winner, 1986: 12)

In sum, when one looks at technology as actively shaping the social landscape – rather than as something dead, inert, and politically inconsequential – the "age-old political questions about membership, power, authority, order, freedom, and justice" repeatedly appear and beg to be addressed (Winner 1986: 47).

Big data and precision agriculture, like what was envisioned and practiced by the big data industry sample, have this power because they instill and normalize a food imaginary predicted on large-scale, capital intensive agriculture. This technology is not scale neutral, as it attached to large, heavy pieces of equipment that assume expansive fields and an infrastructure designed around large-scale agriculture – large gates to access fields, well-packed gravel roads or paved roads, bridges that can support many tons, etc. (Carolan 2017a, 2017b, 2018)

This is not to suggest, however, a wholesale rejection of technology itself – whatever *that* would mean, as what is and isn't technology is a giant exercise of boundary work (Gieryn 1983). Technology can be embedded within more collectivist (collaborative) ontologies, as evidenced by how Farm Hack participants *do* technology.

When talking about many so called modern technologies – from, for example, biotechnology to precision agriculture and big data – membership in the decision- and design-making processes is anything but open and democratic (Wynne 2005). As Winner (1986) explains, disappearing are definitions of membership in this decision-making process that hinge on broad understandings of "citizenship." Instead, participation is increasingly decided by one's level of technological expertise (Jasanoff 2003; Wynne 2005). Winner (1986) notes that technology must be understood not simply as "tools we use" but as "forms of life" (p. 17), which is to so they not only functionally inject themselves into almost all aspects of life but also play a considerable role in shaping who we are and how we live.

The technologies engendered by the two samples seem to be coming from, and in turn appear to be supporting, divergent conceptions about who we are and how we live. Scholars would do well to (continue to) understand not only how technology reduces "forms of life," for example by way disciplining and modes of governmentality (Foucault 2007; Rose 1999). We should also turn our attention to those forms that are more collectivist, and thus potentially emancipatory, in nature.

References

American Farm Bureau. 2015. The Voice of Agriculture. *American Farm Bureau*. Available online at www.fb.org/newsroom/news_article/178/, accessed November 16, 2017.

Andrejevic, M. 2014. Big Data, Big Questions: The Big Data Divide. *International Journal of Communication*, 8: 17–32.

Arditi, B. 2014. "Insurgencies Don't Have a Plan – They Are the Plan: Political Performatives and Vanishing Mediators in 2011." pp. 113–39 in *The Promise and Perils of Populism: Global Perspectives*, edited by C. de la Torre, Lexington, KY: University of Kentucky Press.

Bobkoff, D. 2015 Seed by Seed, Acre by Acre, Big Data Is Taking Over the Farm. *Business Insider* 15 September. Available online at www.businessinsider.com/big-data-and-farming-2015-8, accessed November 9, 2017.

Carolan, M. 2010. The Mutability of Biotechnology Patents: From Unwieldy Products of Nature to Independent 'Object/S'. *Theory, Culture & Society*, 27(1),110–129.

Carolan, M. 2011. *Embodied Food Politics*. Burlington, VT: Ashgate.

Carolan, M., 2013. The Wild Side of Agro-Food Studies: On Co-experimentation, Politics, Change, and Hope. *Sociologia Ruralis*, 53(4), 413–431.

Carolan, M. 2017a. Publicising Food: Big Data, Precision Agriculture, and Co-Experimental Techniques of Addition. *Sociologia Ruralis*, 57(2), 135–154.

Carolan, M. 2017b. Agro-Digital Governance and Life Itself: Food Politics at the Intersection of Code and Affect. *Sociologia Ruralis*, 57(S1), 816–835.

Carolan, M. 2017c. More-than-Active Food Citizens: A Longitudinal and Comparative Study of Alternative and Conventional Eaters. *Rural Sociology*, 82(2), 197–225.

Carolan, M. 2018. "Smart" Farming Techniques as Political Ontology: Access, Sovereignty, and the Performance of Neoliberal and Not-So-Neoliberal Worlds. *Sociologia Ruralis*, DOI:10.1111/soru.12202

Clark, J., K. Coll, E. Dagnino, and C. Neveu. 2014. *Disputing Citizenship*. Chicago, IL: Policy Press.

Dawson, A. 2016. Digital Agriculture the Next Big Thing, Says Monsanto Official, *Manitoba Cooperator* May 3. Available online at www.manitobacooperator.ca/crops/digital-agriculture-the-next-big-thing-says-monsanto-official/, last accessed November 26, 2017.

Dewey, J. 1946. *The Public and Its Problems: An Essay in Political Inquiry*. Chicago, IL: Gateway.

Emery, S. B. 2015. Independence and Individualism: Conflated Values In Farmer Cooperation? *Agriculture and Human Values*, 32(1), 47–61.

Foucault, M. 2007. *Security, Territory and Population. Lectures at the College de France 1977–1978*. London: Palgrave,

Gieryn, T. 1983. Boundary-Work and the Demarcation of Science from Non-Science: Strains and Interests in Professional Ideologies of Scientists. *American Sociological Review*, 48(6), 781–795.

Habermas, J. 1987. *The Theory of Communicative Action, Volume 2: Lifeworld And System*. Cambridge: Polity.

Higgins, V., Bryant, M., Howell, A. and Battersby, J., 2017. Ordering Adoption: Materiality, Knowledge and Farmer Engagement with Precision Agriculture Technologies. *Journal of Rural Studies*, 55, 193–202.

Isin, E. 2009. Citizenship in Flux: The Figure of the Activist Citizen. *Subjectivity*, 29, 367–388.

Jasanoff, S. 2003. Technologies of Humility: Citizen Participation in Governing Science, Minerva, 41, 223–244.

Laudan, R. 2015. A Plea for Culinary Modernism. *Jacobin* May 22. Available online at www.jacobinmag.com/2015/05/slow-food-artisanal-natural-preservatives/ accessed November 9, 2017.

Michalopoulos, S. 2015. Europe Entering the Era of 'Precision Agriculture.' EurActiv.com October 23. Available online at www.euractiv.com/sections/innovation-feeding-world/europe-entering-era-precision-agriculture-318794, accessed November 20, 2017.

Rannikko, P. and Salmi, P., 2017. Towards Neo-Productivism? Finnish Paths in the Use Of Forest And Sea. *Sociologia Ruralis*. DOI:10.1111/soru.12195/

Rose, N. 1999. *Powers of Freedom: Reframing Political Thought*. Cambridge: Cambridge University Press.

USDA 2015. Crop Production Practices for Corn. United States Department of Agriculture, Washington, DC, https://data.ers.usda.gov/reports.aspx?ID=46941, accessed March 29, 2017.

Wiens, K. 2015. We Can't Let John Deere Destroy the very Idea of Ownership. *Wired* April 21. Available online at www.wired.com/2015/04/dmca-ownership-john-deere/.

Williams, R. 1985. *Keywords: A Vocabulary of Culture and Society*. Oxford: Oxford University Press.

Wilson, G., 2001. From Productivism to Post-Productivism ... and Back Again? Exploring the (Un) Changed Natural and Mental Landscapes of European Agriculture. *Transactions of the institute of British Geographers*, 26(1), 77–102.

Winner, L. 1980. Do Artifacts Have Politics? *Daedalus*, 109(1), 121–136.

Winner, L. 1986. *The Whale and the Reactor: A Search for Limits in an Age of High Technology*. Chicago, IL: University of Chicago Press.

Wynne, B. 2005. Reflexing Complexity: Post-genomic Knowledge and Reductionist Returns in Public Science. *Theory, Culture & Society*, 22, 67–94.

Zarco-Tejada, P., N. Hubbard and P. Loudjani 2014, Precision Agriculture: An Opportunity for EU farmers – Potential Support with the CAP, 2014–2020, Joint Research Centre (JRC) of the European Commission; Monitoring Agriculture ResourceS (MARS) Unit H04 Available online at www.europarl.europa.eu/RegData/etudes/note/join/2014/529049/IPOLAGRI_NT%282014%29529049_EN.pdf, accessed November 20, 2017.

Zoomers, A., A. Gekker, and M. Schäfer 2016. Between Two Hypes: Will "Big Data" Help Unravel Blind Spots in Understanding the "Global Land Rush? *Geoforum*, 69, 147–159.

23 Eating Our Way to a Sustainable Future?

Josée Johnston and Anelyse M. Weiler

1 Introduction

Eating can feel like a deeply personal, individual affair. We all have our own unique taste preferences and aversions. Eating is an activity through which we build our individual identity, a phenomenon that has become more pronounced with the rise of a foodie culture that celebrates "authentic" food choices (Johnston & Baumann, 2015). The current foodscape praises the creativity of celebrity chefs (Johnston & Goodman, 2015), while the world of social media gives individuals a platform to show off personal culinary creations and establish oneself as a "domestic goddess" (Rodney et al., 2017). Responsibility for food decisions is typically laid on the plates of individual eaters. A "good" food choice indicates a sophisticated palate, but a "bad" choice can trigger social judgement around body size, health, and status (Bowen, Elliott, & Brenton, 2014; Guthman & Dupuis, 2006).

While the dominant food culture is deeply individualistic, environmental sociology teaches us that our daily diets are intertwined with other eaters, other species and broader ecosystems. When we eat something, we ingest the non-human parts of the natural world – whether that is a mouthful of plant matter like lettuce, or a bite of animal flesh, like steak. Various plants, animals, bacteria and fungi allow us to survive, and to enjoy them, we indirectly consume the soil, water, sunshine, fertilizers (and fossil fuels) they require to grow. With each meal, we take in and absorb various bits of nature, incorporating plants and animals, and then return food nutrients back into the environment through compost, sewage, and landfill waste. These connections are essential to human life, but not always apparent – especially since more than half the people on the planet now live in cities. The backstory of food is mostly unknown to urban consumers in Global North countries, hidden from view by complex commodity chains, long-distance market connections, and layers of packaging.

Having peered behind the curtain to examine the backstory of food, many food activists and researchers are worried about how our collective food habits can be sustained into the future. Food scholars David Pimentel and Michael Burgess (2015) provide a dire warning: "Considering the declining availability of natural resources that support agriculture and food production, it is evident that both the quality of human life and humanity's survival are being threatened" (p. 182). Some argue our eating habits affect the environment more than almost any other human activity (Sage, 2011). Ecosystems are already under tremendous stress from twenty-first-century capitalism, and the main ways we nourish ourselves appear to be eroding many of the Earth's natural mechanisms for withstanding further human shocks and pressures (Steffen et al., 2015).

In this chapter, we describe some of the major environmental, social and ethical problems with the food system and how consumers are trying to address them through their everyday food choices. Our goal is to firmly establish the connection between what we eat and the natural world, and to outline some of the pressing problems that result from that connection. Clearly, consumers are not the sole actors responsible for creating an unsustainable food system; our analysis shows how consumer responsibility is just one current within the vast ocean of the food system. In the second section, we outline the dominant discourse of ethical eating. Popular marketing campaigns urge consumers to achieve a more sustainable, socially just food system by purchasing foods like local lettuce, organic tomatoes, and grass-fed beef. This approach is encapsulated in the rallying cry, "vote with your fork!" Lifestyle transformation, a close cousin of the idea of "shopping for social change," encourages eaters not only to change what they buy, but how they live – to grow their own food, make jam, live simply, and so on. While such approaches have an intuitive appeal and can offer a range of food pleasures, it is worth questioning how effective they are, and who they exclude. We conclude with a brief discussion of food democracy, a concept that challenges eaters and environmental sociologists to "think big" about how to create a food system that supports all the world's human (and non-human) inhabitants.

2 Food: A Bridge Between Humans and the Rest of the Natural World

How does food consumption fit in a broader environmental context? As we note above, eating represents one of the most concrete and intimate acts that link our bodies with social systems and the non-human natural world. To make sense of these complex linkages, social scientists use the concept of the *food system*. A food system involves the material factors (e.g., soil bacteria, water, animals) humans use to produce, distribute, consume and dispose of food, as well as social factors like history, culture, the economy, political institutions, and power relations. Food systems are dynamic and relational. They are shaped by the relationships between humans, between humans and an ever-changing environment, and by state and market forces that structure the global food economy. Despite the popularity of local food, most people in affluent countries get the mainstay of their diets through a globalized food system that involves a transnational flow of goods funnelled by powerful food corporations. The global food system can be dizzying in its complexity. Appreciating some of its major contours helps us understand why individual consumers can be motivated, but also deeply constrained in their ability to change the food system.

To identify the interrelated drivers of environmental harms in the food system, it can be useful to start with a single food. For example, by tracing the life of a single beef calf from its birthplace in a field to a burger joint, we can learn the food-system backstory of a person's hamburger habit. On a cultural level, hamburgers are considered a hallmark American institution; they exist in multiple formats, from cheap fast-food burgers to gourmet varieties, and are eaten weekly by almost half of Americans (Caldwell, 2014). The fat marbling, tender taste, and widespread availability of feedlot-fattened beef has suffused America's culinary imagination. By following the flow of money from calf to burger, we learn how an American cow's early diet of grass is replaced with government-subsidized corn, which is the main ingredient eaten by feedlot cattle in the United States. Cattle are brought to a feedlot so they can put on a maximum amount of weight before they are slaughtered, and a grain-based diet is the most efficient way to fatten them up. Compared to grass, a grain-based diet is difficult for cattle to digest, so sick cows are often fed antibiotics to keep them healthy. Many US farmers also give antibiotics to healthy animals because this promotes livestock growth.[1] The beef commodity chain has an inescapable economic logic: feedlots are an efficient way to mass-produce meat, especially given the low price of feed corn.

On both a regional and global scale, the beef commodity chain incurs serious environmental consequences. Growing mountains of grains, soybeans, and oilseed to feed cattle takes a large volume of fossil fuels and fertilizers, which can contaminate water sources and produce greenhouse gas (GHG) emissions (Weis, 2013, pp. 74–75). Feedlots generate harmful environmental by-products like lagoons of toxic manure, and cattle generate large volumes of methane, a GHG that is 25 times stronger than carbon dioxide. Indeed, an estimated 1/5 of global GHGs are now attributed to livestock (Weis, 2013, p. 78). While the model of industrialized animal agriculture was pioneered in the United States, it is now a worldwide phenomenon, with feedlots expanding alongside grain production for animals instead of humans. Political ecologist Tony Weis (2013) uses the term global "meatification" to refer to the "dramatic shift of animal flesh and derivatives from the periphery of human food consumption patterns, where it was for most of the history of agriculture, to the centre" (p. 67). Not only does meatification threaten the health of the biosphere and exacerbate climate change, but as Weis emphasizes, it also contributes to global inequality. A significant percentage of the world's grain sources are fed to livestock to produce meat for relatively privileged eaters.[2]

Thinking about how a hamburger fits within a larger food system – a system riddled with pollution, waste, ill health, and inequality – can help us better appreciate how a food product may only be cheap because the long-term environmental consequences haven't been factored into the price. Unfortunately, the hamburger

---

1 The issue of antibiotics in livestock production has become more prominent in public health debates. For example, in 2017, the Food and Drug Administration moved to eliminate the use of antibiotics that are medically important to humans for livestock growth promotion (Ferry & Benjamin, 2016).
2 Together, cattle, pigs, and chickens eat approximately "half of the world's wheat, 90 percent of the world's corn, 93 percent of the world's soybeans, and close to all of the world's barley" (Carolan, 2012, p. 182). Of course, not all of the farmland used to produce these grains and legumes would be suitable for growing crops that meet human preferences.

story is not unique, and many global food commodities share its basic features. So, how did we end up with a relatively toxic, energy-intensive, and wasteful food system? Answering this question requires more than finger-pointing at people chowing down on hamburgers. It demands a brief foray into the evolution of industrial agriculture, asking critical questions about the limits of technological innovations, the sufficiency of the food supply, the problem of waste, and the tendency towards corporate control.

Today, most of the plant-based and animal food products eaten by consumers in the Global North come from large-scale industrial agriculture.[3] Some environmental historians argue that first major rupture between humans and the rest of the world occurred with the dawn of agriculture that began some 10,000 years ago, and that this paved the way for future environmental degradation (Duncan, 1996; cf. Foster, 1999; Moore, 2000). Putting aside the question of agriculture's predisposition to denude the natural environment, we know that farming became much more ecologically disruptive in the post-World War II period. At this point, the industrial agriculture package of monocultures, synthetic agrochemicals, and high-yielding seed varieties became the norm, particularly in affluent countries (Gliessman, 2015). With the Green Revolution, this highly productive, chemical-intensive mode of food production was exported to Asia and Latin America. New agricultural techniques increased the yields of basic cereal crops like corn, wheat, and rice. The Green Revolution shifted the foundation of agricultural systems from subsistence polycultures (i.e., growing multiple crops together to protect soil fertility) towards energy-intensive, water-dependent, food monocultures reliant on fertilizers, pesticides, and irrigation.

While the Green Revolution is often discussed in a historical sense, these agricultural technologies are employed around the world today. The idea that technology alone can fix our problems remains a powerful belief undergirding the contemporary food system. Farmers are under constant pressure to stay on the "technological treadmill" by adopting new technologies as they struggle to stay one step ahead of pests and weeds that become resistant to current farming techniques (Carolan, 2012, pp. 18–21). New technologies are sold to farmers as a way to stay competitive, but they are also framed as a way to increase yields and address environmental problems in the food system. From the mid-1990s onward, one of the most influential – and controversial – agricultural technologies has been genetic engineering (GE). GE technology modifies the genetic structure of the plant or animal itself and transfers genes across species, producing *transgenic* crops. Some have dubbed the rapid expansion of transgenic crops in the United States, Canada, Brazil, Argentina, and India a "Gene Revolution."

Debates about the ecological implications of transgenic crops are complex and contentious, with critics charging that most transgenic crops are profit-driven interventions that have untested long-term implications for ecosystems and human

---

3 Non-agricultural or semi-agricultural modes of food provisioning remain culturally significant for communities around the world. Among Indigenous peoples of North America, these have included practices such as hunting, cultivating clam gardens, and introducing fire to promote the growth of food plants (Lepofsky & Caldwell, 2013; Turner, 2014).

health.[4] Although we can't resolve these debates here, we flag the issue of GE because it shows how difficult it is for a consumer to fully understand, let alone *fix* food system sustainability issues. When a consumer walks into the average American grocery store, they face shelves filled with transgenic foods – mainly from the corn and soy found in most processed foods. Consumer Reports (2014) found that "nearly all" of the processed foods they tested containing corn and soy also contained GE ingredients. The same study found that 92 percent of American consumers want GE foods to be labelled, but corporations have spent millions of dollars working to defeat consumer-ballot labelling initiatives. Even if GE foods are clearly labelled in the future, it would be difficult for such labels to reflect the complexity of the scientific issues at stake. For example, both profit-motivated corporations and publicly funded scientists use GE technology (Evanega & Lynas, 2015; Harriss & Stewart, 2015, p. 48). A consumer standing in a grocery store would find it difficult to distinguish a product engineered by publicly funded research to maximize a plant's drought resistance, from another corporate product engineered to increase herbicide-resistance and sell more herbicides.

Agricultural technology innovations like genetic engineering are frequently framed as a way to "feed the world." Given projections that the global population will reach nine billion humans by 2050, we might ask, have industrial agriculture technologies helped the global food system produce enough food to meet everyone's nutritional needs? Historically, the huge productivity gains from Green Revolution technologies allowed food production to outpace global population expansion. American agronomist Norman Borlaug, widely considered the "father" of the Green Revolution, was awarded the Nobel Peace Prize in 1970. He has been celebrated for introducing new high-yield crop technology that "may have prevented a billion deaths" from hunger (Easterbrook, 1997). Nonetheless, in 2017 the United Nations Food and Agriculture Organization declared world hunger was worsening and affected 11 percent of the global population (FAO, 2017). Although some argue that further technological innovation will end world hunger, others point out that a lack of access to healthy food occurs mainly because of distribution issues and economic inequality (Fraser et al., 2016). Many people simply do not have enough purchasing power to buy the food they need, and the production of more food doesn't guarantee hungry people will be able to feed themselves.

While the productivity gains of Green Revolution technologies have been impressive, the capacity of the global food system to continue nourishing people in the future is more uncertain. Some scholars argue that over the long term, industrial agriculture erodes the very ecological foundations that underpin food production (Gliessman, 2015). The planetary reach of agriculture is already immense: farming carpets 38 percent of the Earth's terrestrial surface, drinks up 70 percent of global freshwater for irrigation, and expels 19–35 percent of global

---

4 Champions of biotechnology claim transgenic crops will decrease the need for pesticides, among other purported environmental benefits. Others are skeptical of such claims and point out, for instance, that the corn rootworm has developed resistance to insecticides produced by transgenic Bt corn (Gassmann et al., 2014). More generally, critics contend that simply modifying single gene traits will not meaningfully resolve complex agricultural challenges (Fraser et al., 2016).

GHG emissions (Foley et al., 2011; Vermeulen et al., 2012). Both proponents and skeptics of new agricultural technologies are asking tough questions about the ability of the food system to keep delivering high yields as populations expand, weather patterns become more erratic with climate change, and the ecological contradictions of industrialized agriculture emerge (Carolan, 2012, p. 181; McHughen, 2015). Before the era of cheap fossil fuels, farming typically produced more energy than it used up. Today, one calorie of food energy in the United States requires about 7.3 to 10 calories of energy inputs (Heller & Keoleian, 2003; Neff et al., 2011; Pimentel & Giampietro, 1994). Access to inexpensive oil has historically enabled a relatively cheap and abundant food supply, but fossil fuel reserves are non-renewable and will diminish at some point in the future (Abas, Kalair, & Khan, 2015). Simultaneously, industrial agriculture methods are associated with loss of genetic diversity, nitrogen and phosphorus pollution, and changes to the landscape such as deforestation (Steffen et al., 2015; Pimentel & Burgess, 2015). Predictions about future demands for food have also sparked a debate on the trade-offs between using land to produce food or to conserve wildlife (aka "land sharing vs. land sparing") (Fischer et al., 2014).

Still others have questioned the assumption that we need to massively increase yields to feed future populations. The argument that 'the world already has enough food' (Fraser et al., 2016 p. 9) points out that the world already produces enough calories to meet people's needs, but that vast quantities of these calories are wasted. Indeed, a notorious feature of the industrial food system is how much valuable nutrition gets frittered away throughout the entire food chain (Spiker et al., 2017). In the last quarter of the twentieth century, we witnessed a troubling paradox: a swelling in the volume of calories available per person in North American supermarkets (Nestle, 2002) along with a dramatic increase in the volume of food wasted (Hall et al., 2009). The most widely cited estimate is that 30 to 40 percent of the world's food in rich and poor countries alike is wasted (Godfray et al., 2010, but see Bellemare et al., 2017). Carolan (2012) writes that the "food waste generated in the United States alone constitutes sufficient nourishment to pull approximately 200 million people out of hunger" (p. 235).

Although researchers debate what counts as "waste," we know that much of the energy, land, water, and other resources expended to produce food is squandered. In Global North countries, most wasted food is dumped in landfills, where it rots and turns into methane (Hall et al., 2009). In wealthy countries like the United States, the largest proportion of wasted food comes from households (Godfray et al., 2010; Parfitt et al., 2010). Eaters may individually try to reduce food waste through practices such as buying "ugly" vegetables and eating leftovers or, more problematically, by funnelling surplus unwanted food from wealthier to poorer eaters through the lens of charity (Soma, 2017). Still, government and industry policies play a major role in establishing norms and practices around wasted food. Food retailers influence how food is packaged, which food is purchased from growers, and which food is left to rot in the field (Carolan, 2012, p. 237). The culture of food waste is shaped by structural factors that transcend individual consumer control – factors like the cost of food, the presence of highly packaged processed foods, marketing strategies that encourage over-buying (e.g., buy-one-get-one-free deals), and confusing expiration date labels. Individuals can try to make a change to their everyday practices, but cultural and government regulatory strategies are both required to effectively deal with the problem of wasted food.[5]

Ensuring enough food for all, while simultaneously respecting the earth's natural limits, is made even more daunting by another feature of the global food system: the high degree of corporate concentration (Howard, 2016). If we imagine the food system in the shape of an hourglass, we can get a sense of how it is owned and controlled by oligopolies, meaning that a small number of firms hold a huge amount of power (e.g., Carolan 2012, pp. 41, 45; Howard, 2016, p. 13). At the top of the hourglass are the world's farmers (about 1.5 billion people), and at the bottom of the hourglass are the world's consumers (7.6 billion people). In the middle of the hourglass we find a handful of oligopolies at multiple stages of the food chain: agricultural input suppliers (e.g., DuPont), traders and processors (e.g., Cargill), food and beverage corporations (e.g., Coca Cola) and retail / supermarket giants (e.g., Walmart) (Burch & Lawrence, 2007; Clapp & Fuchs, 2009).

Corporate concentration impacts many of the food issues we have mentioned here. For example, it is estimated that after acquiring Monsanto in 2018, a single company, Bayer, now sells 24 percent of the world's pesticides and 29 percent of its seeds (Plumer, 2016). Just four firms control roughly 70 percent of the world's grains and oilseeds trade (Clapp, 2016, p. 105). Walmart controls one third of the US grocery retail market (Howard, 2016, p. 2). Because of the vertical and horizontal concentration of agribusiness power across the entire food system, consumers – and farmers – find it difficult to make substantial change on pressing environmental issues or on reversing longstanding inequalities (Holt-Giménez & Shattuck, 2011; Howard, 2016). We turn to this issue of ownership and control in our conclusion, where we discuss the possibilities for democratizing the food system so that there is greater transparency, and so that everyday people have meaningful decision-making power over the food-related issues that affect them. But first, we describe efforts to promote a more sustainable food system by encouraging consumers to change their diets.

3 Change Your Diet, Save The Planet?

Among relatively privileged eaters in the Global North, the dominant way people have responded to ecological crises in the food system has been through popular calls to shift to "green" lifestyles and consumption patterns. Eat-local initiatives urge consumers to reduce their food miles by buying products directly from local, and preferably organic, farmers (Smith & MacKinnon, 2007). Non-profit organizations have encouraged eaters to do their part, fighting climate change through Meatless Mondays and reducing household food waste (Evans, 2011; Laestadius et al., 2013). Consumers striving to make a difference can visit

---

5 One UK-based organization, WRAP, is organizing on waste issues to change individual consumer behavior as well as corporate practices through sector-wide voluntary agreements. See www.wrap.org.uk/food-waste-reduction

a growing number of farmers' markets, purchase organic products, and replace fast food with "slow food" that has been crafted to maximize biological integrity, tradition, and taste (Sassatelli & Davolio, 2010).

We can think of these various projects to change consumer food habits as fitting under the umbrella of "ethical consumption" or "ethical eating." Putting aside debates on what is *truly* philosophically "ethical" in the realm of consumption, we can recognize that the discourse of ethical eating proposes to save the earth, one forkful at a time. The various projects associated with ethical eating are united around the belief that a focus on consumers' food choices – voting with your fork – is the principal way to push the food system towards greater sustainability. There is an agreement on the overall principle of "eating for change," but the various projects within ethical consumer discourse can contradict each other. Vegans swap recipes for black bean burgers to reduce the planetary impact of meat production. Meanwhile, self-described conscientious carnivores grill grass-fed beef patties. As ethical consumers fill their carts with organic bone broth at Whole Foods Market, anti-capitalist freegans dig through the dumpster to recover granola bars past their best-before date. Although people disagree on how best to achieve a more sustainable food system, ethical consumption discourse coheres around the idea that consumers should eat, shop and live strategically to protect people and the planet.

The "vote with your fork" message has a lot of intuitive appeal and has been broadly taken up by food activists, small food producers and consumers. Meanwhile, food corporations selectively incorporate and rework these messages in their marketing efforts (Schleifer and DeSoucey, 2015). Buying honey from a local beekeeper offers a feel-good reward, but Starbucks also offers a feel-good message through its marketing materials about ethical sourcing and environmental sustainability. While it is understandable that consumers want to feel good about their shopping choices, it can be difficult for shoppers to separate corporate "greenwashing" from more meaningfully sustainable production practices (Johnston, Biro, & MacKendrick, 2009; Jones, 2015). Given the complexity of the environmental problems in the food system outlined in the previous section, one might question the feasibility of expecting consumers to simply eat and drink their way to a greener food system. Scholars have explored the myriad contradictions within the vote with your fork message (see Alkon, 2012; Johnston, 2008; Johnston & Cairns, 2012; Parker, 2013). Here, we hone in on why market fixes are appealing, but sometimes unhelpful or even counterproductive in dealing with ecological degradation in the food system.

To be clear: we are not suggesting that for-profit green businesses *never* improve the health of the environment. There is a long history of environmentally oriented social movements intersecting with markets, such as the natural foods movement in the United States, and these intersections can produce outcomes that include greener products and alternative technologies (see Haenfler et al., 2012; Hess, 2007; Miller, 2017). As Miller (2017) writes in her history of the American natural foods movement, "it is not always the case that private enterprise stands in opposition to movements for social change" (p. 4). A critical question to consider in such instances is the relative emphasis on consumers as agents of change versus enhanced market regulation, state policy, and a more equitable ownership of the means of production.

A focus on individual commodity solutions as the *primary* strategy for change can reinforce a neoliberal ideology, which suggests that environmental issues are best resolved through voluntary market efforts instead of state intervention in the marketplace (Brown & Getz, 2008; Lukacs, 2017). This is not a problem unique to food or environmental issues but reflects a longstanding "movements versus markets" tension for those trying to bring about social change. What are the limits of relying on neoliberal capitalist markets to realize the progressive social changes that movements want? Below, we outline a few major limitations: niche markets, externalities, market access, the "inverted quarantine" problem, and justice for hired workers and their families.

The case of organic foods illustrates why niche markets have a limited potential to enact social change, along with the movement-market tension endemic to ethical eating discourse. The market for organic foods developed alongside social movement efforts to improve soil health and promote sustainable farming practices (Guthman, 2014). The environmental benefits of organic agriculture involve many shades of grey based on where and what one is measuring, such as bird biodiversity, water use, soil erosion, or other ecological indicators (Seufert & Ramankutty, 2017). While organics represents only 1 percent of the area farmed globally, it is one of the mostly rapidly growing dimensions of agriculture worldwide (Seufert, Ramankutty, & Mayerhofer, 2017). Given the swift expansion of organics, many more people can find organic products in their grocery store, and millions of tons of toxic agrochemicals are kept out of the food system because of organic production methods. The presence of organics has also worked to expand public consciousness about sustainable farming practices, and it has even influenced conventional farms to adopt organic techniques such as cover-cropping (Seufert & Ramankutty, 2017).

While the expansion of organic food has many positive dimensions, this market fix has some significant limitations. The market growth of organics has largely been achieved via corporate firms. Many of these firms follow the letter of the law when it comes to organic farming standards but have been criticized for putting profits and growth ahead of meaningful sustainability practices (e.g., trucking in organic fertilizer to grow vast fields of monoculture crops rather than practicing diverse closed-loop farming) (Guthman, 2014; Haedicke, 2016; Obach, 2015, p. 160). Research suggests that organic consumers mainly buy organic food because they believe it is healthier, and not necessarily because of environmental motivations (Baumann et al., 2017, p. 71). Following consumers' preferences, organic regulations have shifted away from early organic farming's holistic ecological principles toward a shallower emphasis on "chemical-free" farming (Seufert et al., 2017). Rather than blame organic practitioners for "selling out," Obach (2015) argues that the "basic failure of the organic movement to achieve its transformative potential can be tied to its essential orientation towards the market" (p. 220). Why? Because the organic marketplace depends on price premiums (i.e., selling organic foods for a higher price-tag). As more market-oriented actors enter the organic sector, the price premium contracts (i.e., cheaper organic food becomes available). This diminishes producers' motivation to become organic and makes it more difficult for movement-oriented actors to compete with the efficiencies of large-scale corporate organics.

Consequently, there is a structural limit on the expansion of organics. Although organic growers may dream of a day when all farms are organic, Obach (2015) argues that organics will "likely remain a niche market indefinitely" (p. 234), and that *voluntary* environmental practices are an inadequate response to food system problems. Efforts to achieve sustainability would be more fruitfully geared toward government policies that set mandatory environmental practices for all actors, including non-organic producers.

Part of the reason why rarified eco-friendly foods are destined to remain niche goods is because of a core problem with capitalist markets: externalities. Economists use the concept of negative externalities to explain how markets fail to incorporate the full costs of a good. To explain the relevance of externalities, let's reconsider the example of a hamburger, a commonly used example of an underpriced food (e.g., Patel, 2011). Food writer Mark Bittman (2014) estimates that even by conservative estimates, the average hamburger would be vastly more expensive if it reflected the *true* public health and environmental costs of cattle production. Society bears huge costs from the heart disease associated with fast-food consumption, brutal employment conditions for many slaughterhouse workers, and the potential for feedlots to spread antibiotic-resistant superbugs (Anomaly, 2015). Instead of being reflected in the price of a burger, the environmental costs become someone else's probabilistic problem; they are diffused onto distant communities and future generations. Human costs such as illness or death are offloaded onto individuals and the health care system. While cattle incur the harms of systemic ill-treatment and death, eaters don't have to bear these ethical costs directly. Bittman (2014) sums it up, "[I]f those externalities were borne by their producers rather than by consumers and society at large, the [hamburger] industry would be a highly unprofitable, even a silly one. It would either cease to exist or be forced to raise its prices significantly."

The problem of externalities is not limited to hamburgers, but extends to many more foods and commodities. Environmental historian Jason Moore (2015) argues that capitalism has been made possible by treating nature as though it is external to humanity and therefore free or cheap. However, we seem to be encountering limits to exploiting nature in this way. The problem of externalities makes clear that while some consumers may benefit in the short run from capitalist markets (e.g., by having access to cheap hamburgers), market mechanisms on their own do not effectively deliver longer-term goals like environmental sustainability and human health.

Let's assume, then, that a concerned consumer could internalize all the costs of meat production by paying top-dollar for an organic grass-fed beef burger (cf. Capper, 2012). Clearly, asserting one's politics through ethical food consumption is a pay-to-play endeavor. A person's "votes" are determined by their financial resources – an idea that is anathema to the democratic principle of one person, one vote. While consumer income doesn't straightforwardly predict ethical consumption practices, studies find that ethical consumption practices may operate as a form of cultural capital that signals a person's social status to the rest of the world. An "eco-habitus" involves demonstrating one's care for the environment through a taste for high-status goods, along with practices such as swapping home-preserved food or signing up for a DIY butchery workshop (Carfagna et al., 2014). This eco-habitus can be more prevalent in professional classes and in groups with higher education (Baumann, Engman, & Johnston, 2015). Knowing the "right" eco-friendly lifestyle practices and consumption patterns can make eaters feel like they are drawing a symbolic boundary between themselves and those who do not belong to the same moral club. It can also reproduce disregard for the diverse pressures and life factors that make it structurally difficult for many groups to engage in ethical consumption practices (Johnston & Baumann, 2015).

Beyond the tendency to exclude certain eaters based on their social status, the problem with ethical consumption is also political. By implying that environmental protection can be achieved by composting, buying heirloom tomatoes or other individualized practices, it depoliticizes a collective problem. Food scholar Andrew Szasz (2007) uses the term "inverted quarantine" to refer to the idea of assembling a "personal commodity bubble for one's body" (2007, p. 97). Unlike a conventional quarantine that aims to protect an entire population, the "inverted" quarantine is focussed on *individual*-level commodity solutions. The eater's home and body act as a kind of personal sanitarium where they seek protection from toxic chemicals by purchasing goods like organic foods and distilled water. Where does that leave those living outside of the bubble?

It is understandable why consumers might buy organic food to seek protection from pesticides. As Szasz (2007) points out, however, this strategy is necessarily partial and induces a kind of "political anaesthesia." Even if consumers could afford to purchase strictly organic foods, the air we breathe, the electronic goods we use for entertainment, and other dimensions of our everyday lives are inescapably brimming with toxins. Standards for "good" middle-class mothering now include pressure to raise an "organic child" by purchasing organic products (Cairns, Johnston, & MacKendrick, 2013), but infant bodies contain an average of 200 toxic chemicals from the moment they are born (Szasz, 2007, p. 101; see also MacKendrick, 2018). Not only is the inverted quarantine impractical, but it can stifle people's will for political change. If society's most educated, privileged, and influential consumers feel like they can escape the ill-effects of a contaminated food system through careful purchases, then they may be less likely to prioritize cross-class struggles for state regulation and infrastructure that promotes long-term sustainability and human health for everyone (Kennedy, Parkins, & Johnston, 2016; MacKendrick, 2010, 2018). As we discuss later on, however, ethical consumption projects sometimes go hand-in-hand with modes of political engagement beyond the market (Baumann et al., 2015; Willis & Schor, 2012).

While efforts to achieve environmental well-being through shopping and food lifestyle practices are fraught with problems, addressing labor and social justice issues through the market is even more of a hornet's nest. Theoretically, ethical consumerism supports a three-legged stool of economic, ecological, and social sustainability. In practice, ethical food discourse in the Global North rarely foregrounds equity issues like immigration reform, poverty and hunger. For example, the US standards for organic production do not contain *any* labor criteria (Obach, 2015, p. 177). In fact, the organic movement in California has actively fought against regulatory initiatives that would improve farm working conditions (Getz, Brown, &

Shreck, 2008). In a study of gourmet food discourse, researchers found that "eating green" (organic, local) and animal welfare were prominent themes, while little or no attention was paid to issues of "social justice, hunger, food security, national food sovereignty and labour exploitation" (Johnston & Baumann, 2015, p. 142). While environmental issues in the food system can catalyze consumer concerns about personal health, social justice does not command the same consumer interest or price premium (Alkon, 2012). Employers in the food system typically view efforts to strengthen the power of the working class, such as unionization and movements for living wages, as a threat to profit. There are, of course, exceptions. For instance, the US-wide Restaurant Opportunities Center United has engaged with diners, restaurant employers, and employees to concretely advance worker pay and benefits (Jayaraman, 2016).

Fair trade is another exception to the relative neglect of labor and social justice issues in ethical eating discourse. Fair-trade certification programs are designed to enable producers of commodities like tea, coffee and chocolate to charge a price premium that generates a higher income for producers in the Global South (Brown, 2013). Fair-trade premiums are also intended to encourage democratic labor organization and environmental conservation efforts. In practice, fair trade schemes have a mixed record of fulfilling their promises. Fair trade-certified products often sell a simple story of what social justice looks like, and universalized fair-trade standards may clash with workers' visions for justice and ecological stewardship (Besky, 2014).

While the growth of fair-trade markets has helped improve the lives of many small farmers throughout the Global South, Cole and Brown (2014) argue that it is "neither a perfect system nor a solution to global poverty" (p. 52). In the case of coffee, insufficient consumer demand means that only 20 percent of the global supply of fair-trade coffee is sold at the fair-trade minimum price – the rest is dumped back into the regular market (Cole & Brown, 2014, p. 53). For coffee and tea grown on large plantations, the benefits of fair-trade premiums for workers and the environment are highly uncertain (Besky, 2014; Cole & Brown, 2014). A cup of fair-trade coffee could genuinely support a producer-owned cooperative in southern Mexico that is sustaining songbird habitat by growing coffee plants under a canopy of native trees. By contrast, a cup of fair-trade Darjeeling tea may paradoxically encourage businesses to cut down forests to make room for more tea fields, with the premiums flowing primarily into the pocketbook of a tea plantation owner (Besky, 2014). The fair-trade sector continues to struggle over the tension between movement activists and profit-oriented actors, and there is no simple way for consumers to know whether buying fair-trade goods will produce optimal social or environmental outcomes (Jaffee, 2007).

In the case of "domestic fair trade" certification schemes in Global North countries, researchers argue voluntary third-party regulation that fails to involve workers in setting and monitoring standards can both depoliticize and de-democratize labor issues (Brown & Getz, 2008). By contrast, farm worker-led initiatives such as the Coalition of Immokalee Workers' (CIW) Fair Food Program function as part of a more expansive movement to strengthen worker power in the food system

(Minkoff-Zern, 2017). The Fair Food Program is a workplace monitoring initiative driven by workers themselves to ensure fair wages and labor conditions. Simultaneously, the CIW leverages consumer power to pressure fast-food restaurant chains, grocery stores, and other large food purchasers to sign onto the Program and pay higher prices for goods such as tomatoes. Market-focused initiatives to strengthen the power of racialized workers exist alongside labor organizing struggles such as Familias Unidas por la Justicia (FUJ), a union led by Indigenous Oaxacan farm workers in Washington state (Bacon, 2017). Founded in 2013, FUJ is reportedly the first union in the state led by Indigenous workers. In 2017, it successfully signed a union contract with one of the state's biggest berry growers that includes an average wage of $15 per hour for all workers (Bacon, 2016; FUJ, 2017). Rather than simply offering consumers more product variation, contemporary initiatives for farm worker justice exemplify diverse approaches to engaging with the market and pushing for the redistribution of wealth and power. The cases of labor and fair trade also make clear that food production never involves a neat separation of humans from the environment. Numerous studies have shown how the fabric of the food system is braided together with systemic racism, colonialism, class exploitation, and ecological degradation (e.g., Gray, 2014; Holmes, 2013; Huseman & Short, 2012). In 2015, pollution caused an estimated 16 percent of premature deaths worldwide, with a disproportionately deadly impact on poor and marginalized communities (Landrigan et al., 2017). Scholars and activists use the concept of *environmental justice* to ask questions like the following: Who gets access to environmental goods like nutritious, chemical-free food, and who gets saddled with environmental harms like pesticide exposure (MacKendrick, 2018, p. 155)? Orchards, slaughterhouses, and fast-food counters are frequently staffed by low-wage workers, who are often (im)migrants and people of color (Bucklaschuk, 2016; Polanco, 2016). Ironically, the very farm workers who milk cows and harvest cherries experience higher-than-average rates of food insecurity and hunger (Minkoff-Zern, 2014; Weiler, McLaughlin & Cole, 2017). The bodies of people on the front lines of the industrial food system are also disproportionately poisoned by agrochemical toxins like chloropicrin, which was initially used as a poison gas in World War I and is now applied as a soil fumigant for crops like strawberries (Guthman & Brown, 2016). Consumers often assume that organic foods grown on small-scale, local family farms must mean better conditions for people who produce the food (Alkon, 2013; Gray, 2014). The evidence supporting this assumption is shaky at best (Cross et al., 2009; Dumont & Baret, 2017; Harrison & Getz, 2014; Weiler, Otero, & Wittman, 2016).

In a well-known TV episode of *Portlandia*, a young, white locavore couple interrogate a restaurant server for intimate details about "Colin," the chicken featured on the menu. Exactly how local is he? Did he have solid friendships with other chickens? The server obligingly brings them a photograph of Colin (who is, in fact, a hen) along with paperwork describing his life. Before they finish ordering their meal, however, the couple decamps from the restaurant so they can size up the ethical credentials of Colin's previous farm with their own eyes. This example from popular culture shows that many elements of ethical food discourse are now being poked fun at for being precious, privileged, and pretentious.

Given such depictions, it would be easy (and perhaps tempting) to dismiss ethical consumption as a frivolous exercise by foodies with time, money, cultural capital, and racial privilege to fuss over the provenance of what they put on their plates. However, as food scholars, we reject that approach as simplistic and reductionist. Many people in the industrialized North have become aware of the limits of shopping our way out of our environmental problems and the ambiguity of market projects for environmental change. We do not intend to paint a picture of naïve food consumers who blithely assume that every purchase will generate environmental transformation and social harmony. As we have outlined, market mechanisms have a slew of limitations, but sometimes they *are* linked to better environmental outcomes. The challenge for researchers is to parse out the intersections between movement and market, seeking to better understand when and how market forces work to reduce, co-opt and dilute social movement aspirations (e.g., Jaffee & Howard, 2010).

Finally, it is important to recognize the diverse and unexpected ways people are mobilizing through the politics of food consumption. Ethical eating can operate as a form of everyday feminist politics, offering women a way to express their political concerns through family foodwork – even as it may also reproduce the gendered burden of carework (Cairns & Johnston, 2015). Some consumers are throwing their weight behind what Alkon and Guthman (2017) call "the new food activism" – projects that may use market-based strategies as a starting point, but work toward harder-hitting and more imaginative forms of social transformation. Their book includes examples of how foodies and food activists can practice solidarity in useful ways through initiatives led by the communities most directly harmed by the industrial food system. Case studies highlight consumers allying with bakery workers in campaigns for good jobs, joining food cooperatives to democratize the ownership of shared resources, and advocating for state regulation against noxious pesticides. By using cooperative, oppositional and confrontational strategies to hold the state and capitalists accountable, new food activists are "work[ing] collectively for good jobs, healthy workplaces, affordable healthy food, land, and collective ownership of the means of production" (p. 15).

4 Conclusion: Building Food Democracy

We began this chapter by identifying how food choices are embedded in the natural world. While individual eating practices may feel disconnected from broader ecosystems, Euro-North American consumer habits are deeply implicated in a food system that systemically degrades the natural world, generates widespread animal suffering, and undermines human dignity. To address some of these problems, consumers are commonly encouraged to eat foods that are local, organic, fair trade, or humanely raised in order to advance human and environmental well-being. Ethical eating discourse urges consumers to "make a difference" by purchasing foods that promise benefits for individual health and the environment (e.g., substituting organic apples for conventionally grown fruit). Because everyone from small-scale producers to large corporations has tapped into the appeal of ethical eating, it is difficult for the average consumer to ascertain the differences between various production practices and the promises they actually deliver. A key tension within ethical eating involves the often-competing priorities of social change actors and market forces which, by definition, are oriented around profit. We argue this movement-market tension produces uncertain outcomes that should not be summarily dismissed, but should instead by critically interrogated. We identified several problems that result from a reliance on shopping for social and ecological change – the limited scope of niche markets, the unacknowledged costs of negative externalities, inequitable market access, the "inverted quarantine" phenomenon, and labor exploitation.

While consumers cannot simply eat their way to a more sustainable food system, various "eat for change" projects intersect with collective struggles for environmental justice and serve as a gateway for debating food policy issues in the public sphere. Chefs, bloggers, academics, and other food system actors are working to bring pressing issues to the forefront of consumers' food consciousness – issues like climate change, soil erosion, fertilizer over-use, genetic diversity, and workers' rights. Consumer-driven social change is often underpinned by the neoliberal belief that individual consumer choices and so-called free markets can solve the world's ecological woes. But under the banner of "food democracy" (alongside related concepts like "food justice" and "food sovereignty"), some threads of the food movement are demanding more. What do we mean by food democracy? The concept of food democracy sees food not simply as a commodity, but as a life good over which all eaters should have decision-making power (Johnston et al., 2009). Food democracy allows us to address the social values and power struggles at stake in food systems. At its root, food democracy involves the "idea that people can and should be actively participating in shaping the food system, rather than remaining passive spectators" (Hassanein, 2003, p. 79).

Of course, food democracy in practice is messy and complex. The actually-existing democratic institutions in which consumers can participate in food-related decisions are often built on a historical legacy that continues to oppress many people. For example, decision-making power over food and the environment is integrally tied to control of land and territory; many liberal democracies today are complicit in the ongoing dispossession of Indigenous lands through settler-colonialism or neo-colonialism (Coulthard, 2014). Complex issues like land ownership remain fundamental to realizing a more just and sustainable food system. Rather than offering a one-size-fits-all recipe to such challenges, the concept of food democracy involves identifying power inequalities in the food system and "making the social relations of food production, distribution, and consumption transparent and open to political contestation and transformation" (Johnston et al., 2009, p. 526).

Beyond merely voting with one's fork, food democracy calls on us to imagine what the food system might look like if we had meaningful choices over the issues that affect the land and our bodies. As with classical political democracy, food democracy moves beyond hard and fast binaries; it's not about *whether* the food system (or political system) is democratic. The issue concerns the *degree* of democracy present in the food system. Using a normative ideal like food democracy gives us a valuable vantage point from which to evaluate various "eating for change" projects and encourages pointed questions: Are eaters passive spectators, or can they meaningfully shape their own food choices? What are the trade-offs of using markets to implement social change? Who can influence local, national and global food policies, and which groups are systematically marginalized from decision-making? Is economic power concentrated at the top of national bureaucracies and transnational corporations, or is it horizontally dispersed and democratically structured? When we think in food democracy terms, eating becomes less of an individual pursuit and more of a collective challenge: the challenge is to ensure every eater, present and future, can access healthy, delicious meals while simultaneously valuing all other members of the natural world.

References

Abas, N., Kalair, A., & Khan, N. (2015). Review of fossil fuels and future energy technologies. *Futures*, *69*, 31–49.

Alkon, A. H. (2012). *Black, white, and green: Farmers markets, race, and the green economy*, Athens, GA: University of Georgia Press.

Alkon, A. H. (2013). The socio-nature of local organic food. *Antipode*, *45*(3), 663–680.

Alkon, A. H., & Guthman, J., eds. (2017). *The new food activism: Opposition, cooperation, and collective action*, Berkeley and Los Angeles, CA: University of California Press.

Anomaly, J. (2015). What's wrong with factory farming? *Public Health Ethics*, *8*(3), 246–254.

Bacon, D. (2016, October 3). Why these farm workers went on strike – and why it matters. *The Nation*, www.thenation.com/article/why-these-farm-workers-went-on-strike-and-why-it-matters/

Bacon, D. (2017, June 26). A new farm worker union is born. *The American Prospect*, http://prospect.org/article/new-farm-worker-union-born

Baumann, S., Engman, A., Huddart-Kennedy, E., & Johnston, J. (2017). Organic vs. local: Comparing individualist and collectivist motivations for "ethical" food consumption. *Canadian Food Studies/La Revue Canadienne des Études sur l'Alimentation*, *4*(1), 68–86.

Baumann, S., Engman, A., & Johnston, J. (2015). Political consumption, conventional politics, and high cultural capital. *International Journal of Consumer Studies*, *39*(5), 413–421.

Bellemare, M. F., Çakir, M., Peterson, H. H., Novak, L., & Rudi, J. (2017). On the measurement of food waste. *American Journal of Agricultural Economics*, *99*(5), 1148–1158.

Besky, S. (2014). *The Darjeeling distinction: Labor and justice on fair-trade tea plantations in India*, Berkeley and Los Angeles, CA: University of California Press.

Bittman, M. (2014). The true cost of a burger. *New York Times*, www.nytimes.com/2014/07/16/opinion/the-true-cost-of-a-burger.html

Bowen, S., Elliott, S., & Brenton, J. (2014). The joy of cooking? *Contexts*, *13*(3), 20–25.

Brown, K. (2013). *Buying into fair trade: Culture, morality and consumption*, New York, NY: NYU Press.

Brown, S., & Getz, C. (2008). Privatizing farm worker justice: Regulating labor through voluntary certification and labeling. *Geoforum*, *39*(3), 1184–1196.

Bucklaschuk, J. (2016). A temporary program for permanent gains? Considering the workplace experiences of temporary foreign workers in Manitoba's hog-processing industry. In S. A. McDonald & B. Barnetson, eds., *Farm workers in Western Canada: Injustices and activism*. Edmonton, AB: University of Alberta Press, pp. 101–119.

Burch, D., & Lawrence, G. (2007). *Supermarkets and agri-food supply chains: transformations in the production and consumption of foods*, Cheltenham, UK: Edward Elgar Publishing.

Cairns, K. & Johnston, J. (2015). *Food and femininity*, New York. Bloomsbury Press.

Cairns, K., Johnston, J., & MacKendrick, N. (2013). Feeding the 'organic child': Mothering through ethical consumption. *Journal of Consumer Culture*, *13*(2), 97–118.

Caldwell, M. (2014). The rise of the gourmet hamburger. *Contexts*, *13*(3), 72–74.

Capper, J. L. (2012). Is the grass always greener? Comparing the environmental impact of conventional, natural and grass-fed beef production systems. *Animals*, *2*(2), 127–143.

Carolan, M. (2012). *Sociology of food and agriculture*, New York: Earthscan Routledge.

Carfagna, L. B., Dubois, E. A., Fitzmaurice, C., et al. (2014). An emerging eco-habitus: The reconfiguration of high cultural capital practices among ethical consumers. *Journal of Consumer Culture*, *14*(2), 158–178.

Cross, P., Edwards, R. T., Opondo, M., Nyeko, P., & Edwards-Jones, G. (2009). Does farm worker health vary between localised and globalised food supply systems? *Environment International*, *35*(7), 1004–1014.

Clapp, J. (2016). *Food*, 2nd ed., Cambridge, UK: Polity Press.

Clapp, J., & Fuchs, D. A. (eds.) (2009). *Corporate power in global agrifood governance*, Cambridge, MA and London: MIT Press.

Cole, N. L., & Brown, K. (2014). The problem with fair trade coffee. *Contexts*, *13*(1), 50–55.

Consumer Reports (2014). Where GMOs hide in your food. *Consumer Reports*: www.consumerreports.org/cro/2014/10/where-gmos-hide-in-your-food/index.htm

Coulthard, G. S. (2014). *Red skin, white masks: Rejecting the colonial politics of recognition*, Minneapolis, MN: University of Minnesota Press.

Dumont, A. M., & Baret, P. V. (2017). Why working conditions are a key issue of sustainability in agriculture? A comparison between agroecological, organic and conventional vegetable systems. *Journal of Rural Studies*, *56*, 53–64.

Duncan, C. A. M. (1996). *The centrality of agriculture: Between humankind and the rest of nature*, Buffalo, NY: McGill-Queen's University Press.

Easterbrook, G. (1997, January). Forgotten benefactor of humanity. *The Atlantic*: www.theatlantic.com/magazine/archive/1997/01/forgotten-benefactor-of-humanity/306101/

Evans, D. (2011). Blaming the consumer–once again: the social and material contexts of everyday food waste practices in some English households. *Critical Public Health*, *21*(4), 429–440.

FAO. (2017). The state of food security and nutrition in the world 2017. *Food and Agriculture Organization of the United Nations*: www.fao.org/3/a-i7695e.pdf

Fischer, J., Abson, D. J., Butsic, V., et al. (2014). Land sparing versus land sharing: Moving forward. *Conservation Letters*, *7*(3), 149–157.

Foley, J. A., Ramankutty, N., Brauman, K. A., et al. (2011). Solutions for a cultivated planet. *Nature, 478*, 337–342.

Ferry, B., & Benjamin, M. (2016). Don't wait, be ready! New antibiotic rules for 2017. http://msue.anr.msu.edu/news/dont_wait_be_ready_new_antibiotic_rules_for_2017

Foster, J. B. (1999). Marx's theory of metabolic rift: Classical foundations for environmental sociology. *American Journal of Sociology, 105*(2), 366–405.

Fraser, E., Legwegoh, A., Krishna, K. C., et al. (2016). Biotechnology or organic? Extensive or intensive? Global or local? A critical review of potential pathways to resolve the global food crisis. *Trends in Food Science & Technology, 48*, 78–87.

Familias Unidas por la Justicia (FUJ) (2017, June 17). Historic union contract signed by FUJ and Sakuma Bros. Berry Farm: http://familiasunidasjusticia.org/en/2017/06/17/historic-union-contract-ratified-by-members-of-familias-unidas-por-la-justicia/

Gassmann, A. J., Petzold-Maxwell, J. L., Clifton, E. H., et al. (2014). Field-evolved resistance by western corn rootworm to multiple Bacillus thuringiensis toxins in transgenic maize. *Proceedings of the National Academy of Sciences, 111*(14), 5141–5146.

Getz, C., Brown, S., & Shreck, A. (2008). Class politics and agricultural exceptionalism in California's organic agriculture movement. *Politics and Society, 36*(4), 478–507.

Gliessman, S. R. (2015). *Agroecology: The ecology of sustainable food systems*, 3rd ed., Boca Raton, FL: CRC Press.

Godfray, H. C. J., Beddington, J. R., Crute, I. R., et al. (2010). Food security: The challenge of feeding 9 billion people. *Science, 327*(5967), 812–818.

Gray, M. (2014). *Labor and the locavore: The making of a comprehensive food ethic*, Berkeley and Los Angeles, CA: University of California Press.

Guthman, J. (2014). *Agrarian dreams: The paradox of organic farming in California*, Berkeley, CA: University of California Press.

Guthman, J., & Brown, S. (2016). Whose life counts: biopolitics and the "bright line" of chloropicrin mitigation in California's strawberry industry. *Science, Technology & Human Values, 41*(3), 461–482.

Guthman, J., & DuPuis, M. (2006). Embodying neoliberalism: Economy, culture, and the politics of fat. *Environment and Planning D: Society and Space, 24*(3), 427–448.

Haedicke, M. (2016). *Organizing organic: Conflict and compromise in an emerging market*, Stanford, CA: Stanford University Press.

Haenfler, R., Johnson, B., & Jones, E. (2012). Lifestyle movements: Exploring the intersection of lifestyle and social movements. *Social Movement Studies, 11*(1), 1–20.

Harriss, J., & Stewart, D. (2015). Science, Politics, and the Framing of Modern Agricultural Technologies. In R. J. Herring, ed., *The Oxford Handbook of Food, Politics, and Society*. New York: Oxford University Press, pp. 43–64.

Hassanein, N. (2003). Practicing food democracy: A pragmatic politics of transformation. *Journal of Rural Studies, 19*(1), 77–86.

Hall, K. D., Guo, J., Dore, M., & Chow, C. C. (2009). The progressive increase of food waste in America and its environmental impact. *PLoS ONE, 4*(11), e7940–6.

Harrison, J. L., & Getz, C. (2014). Farm size and job quality: Mixed-methods studies of hired farm work in California and Wisconsin. *Agriculture and Human Values, 32*(4), 617–634.

Heller, M. C., & Keoleian, G. A. (2003). Assessing the sustainability of the US food system: a life cycle perspective. *Agricultural Systems, 76*(3), 1007–1041.

Hess, D. J. (2007). *Alternative pathways in science and industry: Activism, innovation, and the environment in an era of globalization*, Cambridge, MA: MIT Press.

Holmes, S. (2013). *Fresh fruit, broken bodies: Migrant farmworkers in the United States*, Berkeley and Los Angeles, CA: University of California Press.

Holt-Giménez, E., & Shattuck, A. (2011). Food crises, food regimes and food movements: rumblings of reform or tides of transformation? *The Journal of Peasant Studies, 38* (1), 109–144.

Howard, P. H. (2016). *Concentration and power in the food system: Who controls what we eat?* London, UK and New York: Bloomsbury Publishing.

Huddart Kennedy, E. H., Parkins, J. R., & Johnston, J. (2016). Food activists, consumer strategies, and the democratic imagination: Insights from eat-local movements. *Journal of Consumer Culture, 18*(1), 149–168.

Huseman, J., & Short, D. (2012). "A slow industrial genocide": Tar sands and the indigenous peoples of northern Alberta. *The International Journal of Human Rights, 16*(1), 216–237.

Jaffee, D. (2007). *Brewing Justice: Fair trade coffee, sustainability and survival*, Berkeley, CA: University of California Press.

Jaffee, D., & Howard, P. H. (2010). Corporate cooptation of organic and Fair Trade standards. *Agriculture and Human Values, 27*(4), 387–399.

Jayaraman, S. (2016). *Forked: A new standard for American dining*, New York: Oxford University Press.

Johnston, J. (2008). The citizen-consumer hybrid: Ideological tensions and the case of Whole Foods Market. *Theory and Society, 37*(3), 229–270.

Johnston, J., & Baumann, S. (2015). *Foodies: Democracy and distinction in the gourmet foodscape*, 2nd ed., New York: Routledge

Johnston, J. & Cairns, C. (2012). Eating for change. In S. Banet-Wiser & R. Mukherji, eds., *Commodity activism: Cultural resistance in neoliberal times*. New York: New York University Press, pp. 219–239.

Johnston, J., & Goodman, M. K. (2015). Spectacular foodscapes: Food celebrities and the politics of lifestyle mediation in an age of inequality. *Food, Culture & Society, 18* (2), 205–222.

Johnston, J., Biro, A., & MacKendrick, N. (2009). Lost in the supermarket: The corporate-organic foodscape and the struggle for food democracy. *Antipode, 41*(3), 509–532.

Jones, E. (2015). Socially Responsible Marketing (SRM). In D.T. Cook & J. M. Ryan, eds., *The Wiley Blackwell Encyclopedia of Consumption and Consumer Studies*. Chichester, West Sussex: John Wiley & Sons, pp. 523–524.

Laestadius, L. I., Neff, R. A., Barry, C. L., & Frattaroli, S. (2013). Meat consumption and climate change: The role of non-governmental organizations. *Climatic Change, 120* (1–2), 25–38.

Landrigan, P. J., Fuller, R., Acosta, N. J., et al. (2017). The Lancet Commission on pollution and health. *The Lancet*, Online.

Lepofsky, D., & Caldwell, M. (2013). Indigenous marine resource management on the Northwest Coast of North America. *Ecological Processes, 2*(12), 1–12.

Lukacs, M. (2017, July 17). Neoliberalism has conned us into fighting climate change as individuals. *The Guardian*: www.theguardian.com/environment/true-north/2017/jul/17/neoliberalism-has-conned-us-into-fighting-climate-change-as-individuals

Lynas, M., & Evanega, S. D. (2015). The dialectic of pro-poor papaya. In R. J. Herring, ed., *The Oxford Handbook of Food, Politics, and Society*. New York: Oxford University Press, pp. 755–771.

MacKendrick, N. A. (2010). Media framing of body burdens: Precautionary consumption and the individualization of risk. *Sociological Inquiry*, *80*(1), 126–149.

MacKendrick, N. (2018). *Better safe than sorry: How consumers navigate exposure to everyday toxics*, Berkeley, CA: University of California Press.

McHughen, A. (2015). Fighting mother nature with biotechnology. In R. Herring, ed., *The Oxford Handbook of Food, Politics, and Society*. New York, NY: Oxford, pp. 431–452.

Miller, L. (2017). *Building nature's market: The business and politics of natural foods*, Chicago, IL: University of Chicago Press.

Minkoff-Zern, L.-A. (2014). Hunger amidst plenty: Farmworker food insecurity and coping strategies in California. *Local Environment*, *19*(2), 204–219.

Minkoff-Zern, L. A. (2017). Farmworker-led food movements then and now. In A. H. Alkon & J. Guthman, eds., *The new food activism: Opposition, cooperation, and collective action*. Berkeley and Los Angeles, CA: University of California Press, pp. 157–178.

Moore, J. W. (2000). Environmental crises and the metabolic rift in world-historical perspective. *Organization & Environment*, *13*(2), 123–157.

Moore, J. W. (2015). *Capitalism in the web of life: Ecology and the accumulation of capital*, London and Brooklyn: Verso.

Murphy, S. (2008). Globalization and corporate concentration in the food and agriculture sector. *Development*, *51*(4), 527–533.

Neff, R. A., & Parker, C. L. (2011). Peak oil, food systems, and public health. *American Journal of Public Health*, *101*(9), 1587–1597.

Nestle, M. (2002). *Food Politics*, Berkeley, CA: University of California Press.

Obach, B. (2015). *Organic struggle: The movement for sustainable agriculture in the United States*, Cambridge, MA: MIT Press.

Parfitt, J., Barthel, M., & Macnaughton, S. (2010). Food waste within food supply chains: Quantification and potential for change to 2050. *Philosophical Transactions of the Royal Society B: Biological Sciences*, *365*(1554), 3065–3081.

Parker, C. (2013). Voting with your fork? Industrial free-range eggs and the regulatory construction of consumer choice. *The Annals of the American Academy of Political and Social Science*, *649*(1), 52–73.

Patel, R. (2011). *The value of nothing: How to reshape market society and redefine democracy*, New York: HarperCollins.

Pimentel, D., & Burgess, M. (2015). Biofuels: Competition for cropland, water and energy resources. In R. Herring, ed., *The Oxford Handbook of Food, Politics, and Society*. New York: Oxford, pp. 181–201.

Pimentel, D., & Giampietro, M. (1994). Food, land, population and the U.S. economy. *Dieoff*, www.dieoff.com/page40.htm

Plumer, B. (2016, September 15). Why Bayer's massive deal to buy Monsanto is so worrisome. *Vox*, www.vox.com/2016/9/14/12916344/monsanto-bayer-merger

Polanco, G. (2016). Consent behind the counter: Aspiring citizens and labour control under precarious (im)migration schemes. *Third World Quarterly*, *37*(8), 1332–1350.

Rodney, A., Cappeliez, S., Oleschuk, M., & Johnston, J. (2017). The online domestic goddess: An analysis of food blog femininities. *Food, Culture & Society*, *20*(4), 685–707.

Sage, C. (2011). *Environment and Food*, London and New York: Routledge.

Sassatelli, R., & Davolio, F. (2010). Consumption, pleasure and politics: Slow Food and the politico-aesthetic problematization of food. *Journal of Consumer Culture*, *10*(2), 202–232.

Schleifer, D and DeSoucey, M. (2015). What your consumer wants: Business-to-business advertising as a mechanism of market change. *Journal of Cultural Economy*, *8*(2), 218–34.

Seufert, V., & Ramankutty, N. (2017). Many shades of gray – The context-dependent performance of organic agriculture. *Science Advances*, *3*(3), e1602638.

Seufert, V., Ramankutty, N., & Mayerhofer, T. (2017). What is this thing called organic?–How organic farming is codified in regulations. *Food Policy*, *68*, 10–20.

Smith, A., & MacKinnon, J. B. (2007). *The 100-mile diet: A year of local eating*, Toronto: Vintage Canada.

Soma, T. (2017). Gifting, ridding and the "everyday mundane": The role of class and privilege in food waste generation in Indonesia. *Local Environment*, *22*(12), 1444–1460.

Spiker, M. L., Hiza, H. A. B., Siddiqi, S. M., & Neff, R. A. (2017). Wasted food, wasted nutrients: Nutrient loss from wasted food in the United States and comparison to gaps in dietary intake. *Journal of the Academy of Nutrition and Dietetics*, *117*(7), 1031–1040.

Steffen, W., Richardson, K., Rockstrom, J., et al. (2015). Planetary boundaries: Guiding human development on a changing planet. *Science*, *347*(6223), 1259855-1–1259855-10.

Szasz, A. (2007). *Shopping our way to safety: How we changed from protecting the environment to protecting ourselves*, Minneapolis, MN: University of Minnesota Press.

Turner, N. (2014). *Ancient pathways, ancestral knowledge: Ethnobotany and ecological wisdom of Indigenous peoples of Northwestern North America*, Montreal & Kingston: McGill-Queen's University Press.

Vermeulen, S. J., Campbell, B. M., & Ingram, J. S. I. (2012). Climate change and food systems. *Annual Review of Environment and Resources*, *37*(1),195–222.

Weiler, A. M., McLaughlin, J., & Cole, D. C. (2017). Food security at whose expense? A critique of the Canadian temporary farm labour migration regime and proposals for change. *International Migration*, *55*(4), 48–63.

Weiler, A. M., Otero, G., & Wittman, H. (2016). Rock stars and bad apples: Moral economies of alternative food networks and precarious farm work regimes. *Antipode*, *48*(4), 1–23.

Weis, T. (2013). The meat of the global food crisis. *The Journal of Peasant Studies*, *40*(1), 65–85.

Willis, M. M., & Schor, J. B. (2012). Does changing a light bulb lead to changing the world? Political action and the conscious consumer. *The Annals of the American Academy of Political and Social Science*, *644*(1), 160–190.

24 Neoliberal Globalization and Beyond: Food, Farming, and the Environment

Geoffrey Lawrence and Kiah Smith

Introduction

Mitigating the environmental impacts of a neoliberalized agri-food system – including climate change – presents one of the greatest challenges facing the planet. This chapter investigates the social, political, and environmental characteristics and impacts of food and farming in the current era of neoliberal globalization. Both 'neoliberalism' and 'globalization' are broad and contested concepts, with many authors identifying plural, or hybrid, 'neoliberalisms' and 'globalizations' in recognition of the varieties of forms that have occurred in the course of history (Peiterse, 1996; Carolan, 2012).

Combining insights from critical political economy and political ecology, this chapter outlines the basic forms and features of the current era of neoliberal globalization that comprises multiple strands of actions, driven by a series of different processes, and resulting in a plethora of outcomes. Rather than seeing neoliberal globalization and its associated patterns of capital accumulation as 'totalizing' (Higgins and Larner, 2017, p. 3) it is viewed as being 'messy' and contested (Lewis et al., 2017). In this chapter a focus on capital, labour and land provides a framework to distil the main ways key actors have shaped food and farming (and environmental outcomes) within the current period of neoliberal globalization. As Newell (2013, pp. 3, 5) reminds us:

> For good or for bad ... the fate of the planet's ecology is increasingly bound up with the fate of contemporary capitalism. ... [T]he contradictions that are intrinsic to capitalism become ever more apparent in the ecological and social systems with which the global economy interacts, upon which it is based and which 'sustain' it ... this makes globalization first and foremost a political process.

As problems with capital, labour and land intersect with ecological constraints (such as climate change and declining fossil fuels) they have generated significant economic and ecological struggles (Tilzey, 2018) that are intrinsically political (McMichael, 2013a). The chapter examines the contemporary meanings of 'neoliberalism' and 'globalization', before turning to their impacts upon food and farming. The latter part of the chapter assesses the rise of 'alter-globalization', whereby the growth of neoliberal globalization (largely, the expansion of self-regulating markets) and the destructive nature of the present food regime, are being met with strong resistance.

Neoliberal Globalization and the Environment

Neoliberalism is both an ideology and a practice. In the post-World War II period, economists stressed the importance of individual freedom, private property rights, entrepreneurialism, market exchange, and removal of the 'heavy hand' of the state (Dumenil and Levy, 2004; Goldstein, 2012). The ideology of 'markets good/ government bad' was readily accepted in a period of 'stagflation' in the 1980s driven by policies to stimulate capital accumulation and to contain or reduce the wages of workers – largely through policies aimed at redistributing wealth to the capitalist class (Harvey, 2005; Heynen et al., 2007). The so-called 'world car' (with Transnational Corporations (TNCs) carefully planning the spatial production and assembly of vehicle components to maximize profits) is viewed as the quintessential example of production in a globalized world, as is its agri-food equivalent – the 'world steer' (global trade in hamburger meat) (see McMichael, 2000). In terms of the economic governance of this emerging global system, organizations such as the World Trade Organization (WTO), World Bank and the International Monetary Fund (IMF) have been actively involved in liberalizing trade, providing a platform for the settlement of disputes, and ensuring compliance to trade regulations (Oosterveer and Sonnenfeld, 2012). TNCs also play a pivotal role in global governance through quality assurance schemes, certification, and labelling (Kalfagianni and Fuchs, 2015). The policies and practices of 'economic rationalism' (Stilwell, 2002) associated with neoliberal ideology include: deregulation, privatization, trade liberalization, financial liberalization, the extension of private property rights, fiscal discipline, tax reductions (usually favouring the corporate sector), and cutbacks in welfare (redistributive) spending (Gray and Lawrence, 2001; Heynen et al., 2007; Bonanno, 2014).

Globalization is considered to be the 'widening, deepening and speeding up of worldwide interconnectedness' in most areas of contemporary life (Held et al., 1999, p. 1). Space and time are viewed as being compressed, allowing actions in one location to have immediate effects in other, distant, locations (Martell, 2010) – implying the emergence of a gradually tightening bond between the 'global' and the 'local' (Robertson, 1992). Phrases such as the 'global village' and 'borderless world' have arisen to recognize the intensification of cultural, social, economic, and political interactions among individuals, groups, and nations, but also the shared challenges of living in a world where terrorism, food insecurity, biodiversity loss and climate change are not nation-state specific, but reach across the globe (Martell, 2010). Global patterns of production, trade and flows of finance, and their governance, interface with environmental change as 'both a manifestation and a cause of much globalising activity' (Newell, 2013, p.7). For example, both liberal – and neoliberal – globalization have relied upon the use of fossil fuels, forests, water, land, and other environmental resources to fuel the growth of industrialization, unfettered production and consumption, and the geographical expansion of markets. This has had significant local- and global-scale impacts, including increasing greenhouse gas (GHG) emissions to unsustainable levels; indeed, global agriculture is responsible for between a quarter and a third of greenhouse gas emissions (McMichael, 2014). Yet, on a positive note, globalization has also facilitated the rise of global environmental governance and the increasing global recognition of shared responsibilities to protect environmental resources for future generations.

In relation to the environmental consequences of neoliberal globalization, there are differing interpretations based upon competing 'worldviews'. According to Clapp and Dauvergne (2011), for supporters of market neoliberalism the drivers of environmental degradation are lack of economic growth, poverty, market distortion and market failure, restrictive trade and investment policies, poorly defined property rights, and inappropriate state interventions (see Clapp and Dauvergne, 2011). An alternative perspective is that neoliberal globalization is based upon the overconsumption of commodities to satisfy capitalism's 'treadmill of production' (Tilzey, 2018). Here, neoliberal globalization – and its growth imperative – is seen as being responsible for unsustainable consumption of natural resources, more stress on waste sinks, and the collapse of global ecosystems, alongside growing social consequences such as poverty, inequality, and exposure to environmental harms (Clapp and Dauvergne, 2011). This critique seems well-founded, considering the following characteristics of global food security, agricultural production-consumption patterns, and environmental impacts:

- Worldwide, the number of undernourished people increased from 777 million in 2015 to 820 million in 2019
- Climate change compromises the ability to meet global food demand, predicted to require a doubling of food production by 2050 to feed a population of some 9 billion people
- In line with a doubling of food production by 2050, it could be expected that food waste will also double in that period, placing a huge burden on the environment. (Food loss and waste currently generates some 8 per cent of GHG emissions each year.)
- Between 2005 and 2015, some 26 per cent of the total damage and loss caused by climate-related disasters in developing countries was attributed to agriculture
- Up to 83 per cent of economic impact of drought falls on agriculture
- Agriculture is the largest water user worldwide, accounting for between 70 and 95 per cent of total freshwater withdrawals, and is a major source of water pollution from nutrients, pesticides, and other contaminants
- Livestock supply chains account for 14.5 per cent of total global GHG emissions, or 7.1 billion tonnes of CO_2-equivalent/year, while deforestation accounts for 10–11 per cent (Carolan, 2012; FAO, 2017a, 2017b).

According to many observers, accelerated economic growth in China, India, and the global South – combined with the move to a more western-style diet – is likely to exacerbate the environmental degradation associated with neoliberal globalization (see discussions in Weis, 2013, 2016; Smith, 2016). In the next section, we examine the consolidation of the neoliberalized form of food and farming that has occurred since the 1980s, and its environmental, social, and political consequences.

Neoliberal Food

Many aspects of food production, distribution, and consumption have been fostered by, and have become entrenched under, neoliberal globalization since the 1980s. In particular, 'supermarketization' and 'financialization' have played a key role in the demise of local food systems, the promotion of 'obesogenic' diets, and the creation of food waste, with implications for both the natural environment and for deteriorating conditions for labour in the food industry.

First, neoliberal policies aimed at providing a platform for the expansion of transnational capital have led to an increasingly dominant role played by supermarkets in global food provisioning (Dixon and Banwell, 2016) and the reconfiguring of agri-food chains globally (McMichael and Friedmann, 2007). While consumers welcome the convenience of supermarket shopping (along with lower prices and increased choice) (Gardner, 2013), this transformation is affecting food procurement, food marketing, food transportation, and culinary cultures. In the global South, supermarkets displace small retailers, wet-market traders, and small-scale wholesalers (Timmer, 2008; Gardner, 2013) and promote a western-style diet (Dixon, 2016; Dixon and Banwell, 2016). This diet is high in sugars, salts, saturated fats, and processed grains and is energy-rich but nutrient poor (Friel and Lichacz, 2010). It is also being 'meatified', which has major health and environmental (pollution and global warming) implications (Weis, 2016). Pechlaner and Otero (2015) have labelled this the 'neoliberal diet' in recognition of its corporate origins and global reach. It is a diet linked to growing obesity in both the global North and South, and at least partly explained by supermarkets' policy of selling highly processed foods and to the 'supersize me' marketing strategies of the fast food industry (Dixon and Broom, 2007; Delpeuch et al., 2009).

As supermarket concentration and power has increased so, too, has the growth in food waste. Through contracts with growers, supermarkets can reject fruits and vegetables that do not meet strict aesthetic (cosmetic) standards, with growers having little choice but to destroy the crop or feed it to livestock (Gille, 2013; Carolan, 2018). This is not only a cost borne by farmers: it represents a waste of energy, water, soil, and human labour – a cost borne by the environment and by society (Gille, 2013, p. 34). Supermarkets also dump products whose use by dates are close to expiring (Lawrence and Dixon, 2015), or sign 'forecast orders' that encourage food processors to over-order supplies which are wasted if the supermarkets order less than was originally forecast (Carolan, 2012, p. 130). As a consequence, it is estimated that between 30 and 50 per cent of all food grown and sold is wasted (Watson and Meah, 2013, p. 103).

Finance and agribusiness firms have, likewise, increased their power and influence within a neoliberalized global food system. Four firms dominate the trading and processing of grains such as soybeans, corn, wheat, and sugar – Archer Daniels Midland (ADM), Bunge, Cargill and Louis Dreyfus, the so-called 'ABCD' firms (Howard, 2016). Through mergers and buyouts they have significantly reduced industry competition. The 'concentration ratio' of the top four firms in any industry (or the CR4) is a predictor of market competition (and distortion). For example, with a CR4 of 20 per cent a market is competitive but concentrated, at 40 per cent it is highly concentrated and at 60 per cent it is considered to be 'significantly distorted' (Carolan, 2012, p. 41). Yet, the CR4 for the ABCDs is calculated to be some 80 per cent (Howard, 2016, p. 74), providing those firms with significant purchasing power and economically disadvantaging those who sell agricultural produce in a situation of many suppliers/few buyers (so-called 'monopsony') (Howard, 2016). Firms can buy cheaply from suppliers and sell higher-priced foods to purchasers along the food chain (Carolan, 2012; Howard, 2016). Importantly, heavy concentration is also evident in global seed firms (58 per cent) and agri-chemical firms (62 per cent) (Howard, 2016, p. 107). Three firms control 50 per cent of the global cocoa trade, seven firms 85 per cent of the tea trade, five firms control 75 per cent of the banana trade, while 65 per cent of the global chocolate market is controlled by five firms (Clapp, 2016a, p. 107). The level of concentration of the top four supermarkets is well above the 'significantly distorted' level of 60 per cent. It sits at 99 per cent in Australia, 91 in Sweden, 71 in France and the UK, and 60 in the USA (Carolan, 2013, pp. 102, 112). According to Clapp (2016a) and Howard (2016), these levels of market concentration produce negative impacts experienced by communities, workers, and the environment.

In Bonanno's (2015) assessment, neoliberal globalization has encouraged the 'flexibilization' of agri-food labour – especially in terms of work time, work activities, and employment conditions (including contracts). Such flexibility provides considerable benefits to transnational capital while ensuring workers remain marginalized and their wages compressed (Bonanno, 2015). According to Carolan (2012, pp. 107–10) those in the USA involved in food processing, transportation, packing, restaurant and fast food services receive wages below the national standard and this is particularly so for Asian, black and Latino workers. Wages have deteriorated significantly in meatpacking and in poultry processing as corporations have sourced cheaper products and/or have moved their operations to lower-cost regions – with minimum labour standards – around the world (Carolan, 2012; Bonanno, 2015; Michie, 2017). Globally, the food processing industry exhibits a number of common features: 'low-wage, low-skill, under- or non-unionised employees, a reliance on temporary labour from agencies, mostly migrant labour and higher proportions of casualised female workers' (Bryant, 2015, p. 350). Neoliberal policies have led to the removal of regulations that have helped protect food manufacturing and food transport workers, exposing them to increased accident and health risk in the workplace (Bonanno, 2015). Abattoir and CAFO workers, in particular, have constant exposure to polluted air, faecal emissions and toxic pesticides (Weis, 2013).

Finally, the finance sector has also increased its influence in the global economy. New actors (hedge fund operators, sovereign wealth funds, state owned enterprises, managed investment funds, and so forth), and new products (derivatives, credit default swaps, and commodity index funds, among others) (Konzelmann et al., 2013), are the defining components of neoliberal globalization (Fairbairn, 2014a; Schmidt, 2016; Lawrence and Smith, 2018). Commodity index funds (CIFs), for example, have allowed the bundling of unlike commodities (such as gold, oil, pigs, and corn) into a single item that can be traded, allowing food to be treated like any other commodity (Breger Bush, 2012). The purchase, and asset stripping, of agri-food companies by private equity 'raiders' is another example of finance extending its influence in the food sector (Schmidt, 2016), as is the increased use of derivatives and swaps by the ABCD firms as they engage in both hedging and speculation (Murphy et al., 2012). Before the global financial crisis of 2007/8, investors gambled on CIFs rising, which they did (doubling to US$400 billion between 2005 and 2008). But food prices rose as a result, with people in the global South – reliant upon imported foods – particularly affected. Food-price riots occurred in over 30 countries from 2007 to 2008 (Pechlaner and Otero, 2010). As food prices rose so, too, did the pressure to increase production by intensifying industrial agriculture on existing farmland, placing further stress on ecosystems (Clapp and Helleiner, 2012). Many new forms of investment – including biofuel production – have the potential to undermine ecological sustainability and 'land sovereignty' (Desmarais et al., 2017).

Neoliberal Farming

Globally, farming has been 'neoliberalized' as transnational agri-food corporations have sought to re-work relations between food production and food consumption. If farmers were once food *producers* for the *nation*, they are now positioned as *entrepreneurs* producing for a *global* economy (see Tilzey, 2018, p. 166, emphasis in original). While it must be emphasized that small-scale and peasant farmers continue to provide food for around 70 per cent of the world's population (quoted in McMichael, 2013a, p. 157), since the 1980s large-scale, industrial farming has become entrenched in the global North and has spread, via the 'green revolution', to the global South (McMichael, 2013a). Industrial farming is characterized by product specialization, intensive methods of production, the reliance upon synthetic chemicals (fertilizers, pesticides) and veterinary pharmaceuticals, application of computerized machinery, and sophisticated systems of irrigation (Argent, 2002; Lawrence, 2016). Examples are the monocultural production of crops like corn, wheat, and canola and, for animals, the Concentrated Animal Feeding Operations (CAFOs) which house tens of thousands of animals in confined spaces (Weis, 2013; and see Lawrence, 2016).

There is growing evidence that large-scale, capital-intensive 'productivist' agriculture is unsustainable (Clunies-Ross and Hildyard, 2013; Weis, 2016). Soil erosion, soil compaction, and fertilizer and pesticide run-off into streams and rivers, are common outcomes of monocultural cropping, while CAFOs produces huge volumes of animal biowastes that seep into waterways and cause widespread pollution (Weis, 2013, 2016). Heavily dependent upon fossil fuels in powering machinery, via nitrogen fertilizer applications, and in transport and refrigeration, industrial agriculture is also a major contributor of GHG emissions and, therefore, to climate change. It is predicted that release of nitrous oxide (N_2O) into the atmosphere will increase from 35 to 60 per cent up to 2030 as a direct result of farm intensification (Taylor and Entwistle, 2015, p. 68). Intensive farming, itself, is reliant on only a few selected plant varieties and animal breeds, displacing traditional stock and threatening biodiversity (Gomiero, 2015).

Although the search by corporations for low-cost food production areas has opened up new opportunities for farmers and farm-workers in the global South, this has also been costly for farmers. According to Lewis et al. (2017, p. 166) one of the major 'triumphs' claimed by adherents to neoliberalism is in technologies of audit and 'responsibilization'. Farmers have been encouraged (effectively, forced) to accept the standards imposed by the food industry giants and to comply with those standards through certification schemes (Thompson and Lockie, 2013; Kalfagianni and Fuchs, 2015). According to Bonanno and Cavalcanti (2012) fulfilling certification scheme requirements has resulted in longer working hours and less pay for farm workers, the growth of temporary employment contracts, and greater control and surveillance by farm owners over the labour process. Conditions faced by farm employees are often exploitative, particularly when labour comprises women and children (Collins, 2000). Neoliberal policies have watered down or removed agri-environmental regulations on the use of agri-chemicals and other hazardous substances, placing farm workers' health in jeopardy (Harrison, 2014). And, as Bonanno and Cavalcanti (2012) have indicated, local workers often have little choice: in a global world, corporations are 'footloose' and threaten to relocate their operations if labour is non-compliant or if communities begin to challenge their operations. The aim of neoliberal policies is to make agricultural labour more 'abundant, affordable, flexible and docile' (Harrison, 2014, p. 106). In places where illegal or migrant labour is hired, such as in the USA and the Mediterranean, workers have next to no rights or protection and face conditions of exploitation, work insecurity, racism, chemical poisoning, and a high risk of accidents (Holmes, 2013; Gertel and Sippel, 2014; Harrison, 2014).

Under neoliberal globalization, it is necessary for new areas of accumulation to be constantly brought into production in order to redress cyclical crises of capitalism (Tilzey, 2018). In the realm of farming, this has meant the acquisition of large tracts of farmland by agri-development corporations and finance sector entities such as sovereign wealth funds (SWFs). A number of factors help explain the so-called 'land rush' (Davis et al., 2015). These include: the growing global demand for food and biofuels, carbon trading as a source of future profit, the increasing scarcity of productive farmland as vast areas become 'desertified' through unsustainable farming practices, speculative opportunities for investors from rises in land values, national food security, and climate change (McMichael, 2013a; Kaag and Zoomers, 2014; Davis et al., 2015; Sippel et al., 2017). With agricultural land an attractive investment, private Farm Investment Management Organizations (FIMOs) have been compiling farmland portfolios and offering them to institutional investors who look toward capital gains over a ten or more year period (Fairbairn, 2014a, 2014b). Such investments – occurring in both the global North and South – are legal transfers and, while there are often local concerns about 'foreign' money encroaching upon farming, the transactions are sanctioned by law and are binding (Sippel et al., 2017).

In contrast to the land investment strategies described above, so-called 'land grabs' occur when an acquisition is illegal, underhanded, or unfair, dispossession results, and few local-level benefits flow directly from projects (McMichael, 2013a; Riddell, 2013). Samranjit (2014) identifies the three main 'enablers' of land grabbing as globalization, the (neo)liberalization of land markets, and the worldwide expansion of foreign direct investment. Over past decades millions of hectares of farmland, peatland, and forest have been purchased by SWFs and converted to monocultures of sugarcane, palm oil, and soybean, displacing smallholder farmers and pastoralists as part of a process of 'depeasantization' (McMichael, 2013a, p. 75). According to the public database Land Matrix (2018), some 49.5 million hectares of foreign-initiated global land deals have been concluded in recent decades, with another 20.2 million hectares pending. Hundreds of thousands of small-scale producers have been removed from the land as a consequence of land-grabbing (see Hays, 2014). Researchers have also identified other motives of finance capital and have reported on 'water grabs', 'green grabs' and – in a more general sense – 'resource grabs' occurring in the global South (Borras et al., 2013). All are driven by a neoliberal agro-industrial food regime in crisis (Borras et al., 2013; McMichael, 2013a, 2013b; Otero, 2014; Sekine and Bonanno, 2016). Importantly, the environment is being negatively impacted via land clearing and the introduction of industrial-scale agro-development (Lazarus, 2014).

Contesting Neoliberal Globalization: Alter-Globalization

The term alter-globalization refers to a broad set of ideologies, social movements, and actions emerging from global civil society, which are in opposition to the negative effects of globalization. Drawing together critiques of corporate financial power, global trade and over-production, the globalization of technology and culture, unequal development, and environmental degradation, alter-globalization draws attention to the failure of neoliberal structural reforms to both reduce poverty and to increase economic growth for the majority of people in the developing world. In contrast to *anti*-globalization, alter-globalization movements seek alternative ways for the world 'to globalize economically' (Broad and Cavanagh 2009, p. 110). Alter-globalization has a strong normative dimension. It is based upon shared critiques of neoliberal globalization, strong commitments to civil society processes, and a shared set of fundamental principles and values, resulting in a wide variety of concrete, democratic, actions that may be considered part of a broader movement (Broad and Cavanagh, 2006).

Alter-globalization is distinguished from neoliberal globalization on the basis of the following aims:

- Redefinition of development to emphasize the fulfillment of basic social, economic, cultural, and political rights, with progress measured in terms of improved health and well-being of children, families, communities, democracy, and natural environments

- Redistribution of power and wealth, and increasing the power of governments and citizen groups over markets dominated by TNCs
- Solidarity between groups in society worst affected by neoliberal globalization. It is not just a critique of the Washington consensus-institutions leading to reform, but of the growing marginalization of the South (especially Africa) from processes of exchange and capital accumulation
- Consensus governance, participatory democracy, solidarity, community, and bottom-up organizing, facilitated through global networks and the active involvement of government, business, NGOs and people's organizations.

Many alter-globalization groups are proposing a paradigm shift to replace the consumption-oriented, high-growth model with a low-carbon, low-growth economy where a more equitable income distribution would allow for a rise in the living standards of the poor (Broad and Cavanagh, 2009, p. 104).

Applied to food and farming, alter-globalization adherents argue that 'the undermining of small-scale, diversified, self-reliant, community-based agricultural systems and their displacement by corporate-run, export-oriented monocultures has been the primary cause of landlessness, hunger and food insecurity' (International Forum on Globalization, 2002, p. 172). Food sovereignty is the over-arching challenge to neoliberal globalization focusing, as it does, upon social and environmental sustainability (Charlton, 2016; McMichael, 2016). Food sovereignty promotes food and access to land for self-reliant food production by rural peoples as a fundamental human and economic right. Within a broader critique of neoliberalism, industrial (and export-oriented) agriculture is rejected for its negative environmental impacts contributing to climate change and biodiversity decline, high levels of corporatization, monopolization and problematic corporate accountability, unfair terms of trade, and the exacerbation of hunger and poverty. Unequal consumption (and the disparities between values and practices of those who produce and those who consume) is increasingly challenged. Concerns have also emerged around the politics and impacts of transgenic crops and biofuels, migrant and women's labour issues, violence and oppression, and indigenous rights in the global food system. The lack of access to land, food, and water have a major bearing upon unemployment, malnutrition, and poverty (Rosset, Patel and Courville, 2006; Altieri, 2009).

According to Borras et al. (2008), these complex global-local processes have prompted agrarian movements to localize their struggles, to privatize their activities in order to substitute states' withdrawal from social service delivery, and to internationalize in response to global restructuring. This has resulted in movements or networks of solidarity that are transnational (e.g. International Planning Committee for Food Sovereignty), regional (e.g. Network of Peasants and Producers Organizations of West Africa), sectoral (e.g. World Forum of Fisher Peoples), or local (e.g. Via Campesina includes mostly poor peasants and rural farmers mobilized at the grassroots, but on a global scale). Some movements are involved in dramatic anti-corporate actions, while others negotiate with the Food and Agriculture Organization of the United Nations (FAO), International Fund for Agricultural Development (IFAD), and The United Nations High Commissioner for Refugees

(UNHCR). The leading example of the latter is La Via Campesina's influence on the civil society mechanism within the FAO (see McKeon, 2017).

In response to concerns about land grabbing, in particular, but also pertaining to large-scale land acquisitions, more generally, four reform strategies have been suggested. These are: reformed trade rules to ensure transparency in exports and food security policy making; publicly managed grain reserves to dampen supply shocks, such as those increasingly linked to climate change; accessible finance for the poorest food importers when prices in international food markets spike; and, stronger national and international laws to govern land investments (Murphy, 2013). Of the latter, there is a lack of coherent legal frameworks and most national regulations are inadequate for regulating foreign investment. The overwhelming pattern is for multilateral and bilateral trade agreements to grant power to private companies to prioritise commercial over public interests, including public health and the environment, as well as legal rights of private corporations over governments (Murphy, 2013).

Since 2010, voluntary governance initiatives have emerged to promote more 'responsible' investment into agriculture and farmland. These include UNCTAD's Principles for Responsible Agricultural Investment (PRAI) and Principles for Responsible Investment in Farmland (PRI), FAO-led Voluntary Guidelines on the Responsible Governance of Tenure of Land, Fisheries and Forests in the Context of National Food Security (or, the Voluntary Guidelines), and Principles for Responsible Investment in Agriculture and Food Systems (PRIAFS) (Clapp, 2016b). PRAI's Principle 7 seeks to ensure environmental impacts of agricultural investment are minimized and mitigated, and that sustainable resource use is encouraged. However, the PRAI has been heavily criticized for not including civil society, for legitimating foreign investment in land, and for being too general to be practical. The Voluntary Guidelines and PRIAFS, by contrast, resulted from extensive consultation with the civil society mechanism, and the Committee on Food Security, within the FAO. These principles, guidelines and soft laws provide means for governments and civil society to explore public policy around land and land investment in particular, and to embed the right to food, food sovereignty, and food security into national policy-making (Murphy, 2013). A key challenge remaining is to shift from voluntary guidelines to binding laws. As Clapp (2016b) has cautioned, while voluntary responsible investment initiatives for agriculture can shift discourse, they are often vague and difficult to enforce, have low participation rates and are confusing due to multiple competing initiatives.

Third-party certification schemes and multi-stakeholder standard-setting mechanisms – such as fair and ethical trade – have also grown in an attempt to redress the negative labour impacts of transnational food supply networks. Originally designed to supply small-scale coffee farmers in the global South, fair trade has today expanded to include 20 different commodities in an industry worth over $US6 billion (Raynolds, 2014). Fair and ethical trade standards are largely based on globally agreed minimum labour standards (such as ILO labour codes) around health and wellbeing, workplace conditions, child labour, worker rights and empowerment, with additional requirements for good environmental practices, timely payments, gender equity, and community development (see Smith, 2014). Fair trade in particular aims to ensure that smallholders are paid fairer prices for their labour, and so represents a market-civil society mechanism for redistribution, for reconnecting producers and consumers, and for empowering producers. Whereas in the past, fair trade focused on recreating production-consumption networks where family farmers could earn a decent living from farming, sell and trade on local markets, and achieve self-sufficiency, today it has been expanded to apply to workers on plantations and in factories. This has been controversial, with local and global tensions emerging around the place of large-scale, commercial agriculture in fair trade, and failures to match global rhetoric with local realities for producers (Raynolds and Murray, 2007; Raynolds, 2014), especially women (see Smith, 2016).

From the perspective of alter-globalization, the capacity for market-based fair and ethical trade regulations to improve smallholder livelihoods and ecological sustainability is closely related to redistributive land reform that returns land to smallholder farmers for local/regional food production. Ghimire (2005, p. 15) defines land reform as primarily 'a political process as it involves interventions in local power relations'. However, a major limitation to pro-poor land reform is the capacity of the poor to benefit from institutions that were initially controlled by the resource-rich for their own benefit. Examples of movements for land reform seeking to address these concerns are the Brazilian Landless Movement (MST) and La Via Campesina.

The MST focuses on land occupation for achieving agrarian reform in Brazil. Landless day labourers, the urban homeless, people with substance abuse problems, unemployed rural slum dwellers, and peasant farmers who have lost their land, are mobilized to occupy unproductive land and plantations. They are also involved in the 'symbolic appropriation' of landed property, government buildings, and public thoroughfares. Since the 1970s, the MST has shifted from a marginal group of squatters to one of the most powerful and well-organized social movements in Brazil. It has encouraged credit co-ops, clinics, adult literacy programmes, and agro-ecological approaches to food growing (Vergara-Camus, 2009). Existing constitutional structure and legal frameworks further reinforce agro-ecology at state and federal levels (Holt-Giménez et al., 2009), with positive benefits for local and regional environments. That said, its promotion of land occupation has led to violent conflict with the authorities (Rosset et al., 2006). MST also influences global rural movements through its membership in Via Campesina – an international peasant federation, serving much like an umbrella organization for peasant movements around the world. Beginning formally in 1993, La Via Campesina has grown to facilitate networks of national farmers' movements, including over 164 (sub) national rural social movement organizations based in some 73 countries in Latin America, Caribbean, North America, Western Europe, South and East Asia, and Africa (La Via Campesina, 2015). Its main objectives are to:

> develop solidarity and unity among small farmer organizations in order to promote gender parity and social justice in fair economic relations, the preservation of land, water, seeds and other natural resources, food sovereignty, sustainable agricultural production based on small and medium-sized producers. (www.viacampesina.org).

La Via Campesina participates in international level policy-making forums, such as with the FAO and IFAD, has addressed the UN on the right to food, and is also a major force at global farmers' forums and the World Social Forum. For McKeon (2017), La Via Campesina – through inclusion in the FAO's Civil Society Mechanism – provides the world's foremost challenge to the present system of agrifood governance. It has been highly successful in reframing debates around land reform and land policy in reaction to the market-led model (McMichael, 2013a; 2016), with food sovereignty forming the basis of its alternative strategy to neoliberalism.

A cautionary note on the impact of the new social movements is, however, made by Woods (2016). While agreeing that there is widespread opposition to the industrial food system, he acknowledges a powerful 'hidden hand' of neoliberal globalization which is promoting a corporatized, agro-chemically based, 'business-as-usual' model of global food production. He observes that oppositional movements are both dispersed, and structurally and ideologically disconnected, thus limiting their capacity to challenge the present unsustainable trajectory (Woods, 2016). In other words, their ability to bring about significant change is currently quite limited.

Conclusion

As described above, the contradictions of neoliberal, globalized, food and agricultural systems are many. While the food and agricultural industry certainly produces abundant and relatively inexpensive food, it also creates vast social and ecological problems. These include biodiversity loss, environmental degradation, climate change (resulting from externalizing environmental costs and reliance on fossil fuels), poor labour conditions and low wages, tensions over land use, foods low in nutritional quality, and poor health outcomes for significant segments of the world's population. Yet, there are alternatives to the neoliberal globalization of food and farming, with many sectors of society providing critique and resistance.

This second part of this chapter has focused upon alter-globalization as one significant oppositional approach to neoliberal globalization. Alter-globalization challenges the very notion of capitalist growth. It is based upon recognition of finite natural resources and limits to consumption, and there is often a strong critique or rejection of market solutions to environmental problems. While concrete proposals vary greatly, alter-globalization movements generally argue for a very different form of rural development than is occurring within the current, neoliberal, trajectory. As described above, the movement is based upon the right to food and 'food sovereignty', self-sufficiency and local autonomy, relocalization and regionalization, redistributive land reform, fair trade rather than 'free trade', smallholder and family farming, de-privatization and de-corporatization, cooperatives, agro-ecology and organics, reversing the rules on intellectual property and patenting, and improved democracy and participation. Alter-globalizationists assume that while the market remains the central force of capitalist production and trade, it will produce unsustainable outcomes. Yet, the extent to which alter-globalization is able to challenge the hegemony of neoliberal globalization is questionable. While there have been signs of success (such as the Brazilian Landless Movement and La Via Campesina), the industrial system of agriculture, and food production and distribution, remains largely intact and is expanding in line with the interests and practices of transnational agri-food capital.

Acknowledgements

Emeritus Professor Lawrence is grateful for financial support from the Australian Research Council (DP 160101318), the Ministry of Education of the Republic of Korea and the National Research Foundation of Korea (NRF-2016S1A3A2924243), and the Norwegian Research Council (FORFOOD No. 220691).

References

Altieri, M. (2009). Ecological impacts of industrial agriculture and the possibilities for truly sustainable farming, *Monthly Review*, **50**(3), 60–71.
Argent, N. (2002). From pillar to post? In search of the post-productivist countryside in Australia, *Australian Geographer*, **33**(1), 97–114.
Bonanno, A. (2014). The legitimation crisis of neoliberal globalization: Instances from agriculture and food. In S. Wolf and A. Bonanno eds., *The Neoliberal Regime in the Agri-Food Sector: Crisis, Resilience and Restructuring*, London: Routledge, pp. 13–31.
Bonanno, A. (2015). The political economy of labor relations in agriculture and food. In A. Bonanno and L. Busch, eds., *Handbook of the International Political Economy of Agriculture and Food*, Cheltenham: Edward Elgar, pp. 249–63.
Bonanno, A. and Cavalcanti, J. (2012). Globalization, food quality and labor: The case of grape production in North-Eastern Brazil, *International Journal of Sociology of Agriculture and Food*, **19**(1), 37–55.
Borras, S., Franco, J. and Wang, C. (2013). The challenge of global governance of land grabbing: Changing international agricultural context and competing political views and strategies, *Globalizations*, **10**(1), 161–80.
Borras, S., Edelman, M. and Kay, C. (2008). Transnational Agrarian Movements: Origins and politics, campaigns and impact, *Journal of Agrarian Change*, **8**(2/3), 169–204.
Breger Bush, S. (2012). *Derivatives and Development: A Political Economy of Finance, Farming and Poverty*, New York: Palgrave Macmillan.
Broad, R. and Cavanagh, J. (2009). *Development Redefined: How the Market Met its Match*, Boulder, CO and London: Paradigm.
Bryant, L. (2015). Inequality regimes in food processing industries. In G. Robinson and D. Carson, eds., *Handbook on the Globalization of Agriculture*, Cheltenham: Edward Elgar, pp. 350–67.
Carolan, M. (2012). *The Sociology of Food and Agriculture*, London: Routledge.
Carolan, M. (2013). *Reclaiming Food Security*, London: Routledge.

Carolan, M. (2018). *The Real Cost of Cheap Food*, 2nd ed., London: Routledge.
Charlton, K. (2016). Food security, food systems and food sovereignty in the 21st century: A new paradigm to meet Sustainable Development Goals, *Nutrition and Dietetics*, **73**(1), 3–12.
Clapp, J. (2016a). *Food*, 2nd ed., Cambridge: Polity Press.
Clapp, J. (2016b). Responsibility to the rescue? Governing private financial investment in global agriculture, *Agriculture and Human Values*, **34**(1), 223–35.
Clapp, J. and Dauvergne, P. (2011). *Paths to a Green World: The Political Economy of the Global Environment*, 2nd ed., Massachusetts, MQ: MIT Press.
Clapp, J. and Helleiner, E. (2012). Troubled futures? The global food crisis and the politics of agricultural derivatives regulation, *Review of International Political Economy*, **19**, 181–207.
Clunies-Ross, T. and Hildyard, N. (2013). *The Politics of Industrial Agriculture*, London: Routledge.
Collins, J. (2000). Tracing social relations in commodity chains: The case of grapes in Brazil. In A. Haugerud, M. Stone, and P. Little, eds., *Commodities and Globalization: Anthropological Perspectives*, New York: Rowman and Littlefield, pp.97–109.
Davis, K., Rulli, M. and D'Odorico, P. (2015). The global land rush and climate change. Available at: http://onlinelibrary.wiley.com/doi/10.1002/2014EF000281/full, accessed 22 January 2018.
Delpeuch, F., Maire, B., Monnier, E. and Holdsworth, M. (2009) *Globesity: A Planet Out of Control?* London: Earthscan.
Desmarais, A., Qualman, D., Magnan, A. and Wiebe, N. (2017). Investor ownership or social investment? Changing farmland ownership in Saskatchewan, Canada, *Agriculture and Human Values*, **34**(1), 149–66
Dixon, J. (2016). The socio-economic and socio-cultural determinants of food and nutrition security in developed countries. In B. Pritchard, R. Ortiz and M. Shekar, eds., *Routledge Handbook of Food and Nutrition Security*, London: Routledge, pp. 379–90.
Dixon, J. and Banwell, C. (2016). Supermarketization and rural society futures. In M. Shucksmith and D. Brown, eds., *Routledge International Handbook of Rural Studies*, London: Routledge, pp. 227–49.
Dixon, J. and Broom, D. (eds) (2007). *The 7 Deadly Sins of Obesity: How the Modern World is Making us Fat*, Sydney: University of NSW Press.
Dumenil, G. and Levy, D. (2004). *Capital Resurgent: Roots of the Neoliberal Revolution*, Massachusetts, MA: Harvard University Press.
Fairbairn, M. (2014a). 'Just another asset class'? Neoliberalism, finance and the construction of farmland investment. In S. Wolf and A. Bonanno, eds., *The Neoliberal Regime in the Agri-food Sector: Crisis, Resilience and Restructuring*, London: Routledge, pp. 245–62.
Fairbairn, M. (2014b). 'Like gold with yield': Evolving intersections between farmland and finance, *Journal of Peasant Studies*, **41**(5), 777–795.
FAO. (2017a). *Water for Sustainable Food and Agriculture: A Report Prepared for the G20 Presidency of Germany*, Rome: FAO.
FAO. (2017b). *FAO's Work on Climate Change*, Rome: FAO.
Friel, S. and Lichacz, W. (2010). Unequal food systems, unhealthy diets. In G. Lawrence, K. Lyons, and T. Wallington, eds., *Food Security, Nutrition and Sustainability*, London: Earthscan, pp. 115–29.

Gardner, B. (2013) *Global Food Futures: Feeding the World in 2050*, London: Bloomsbury.

Gertel, J. and Sippel, S. R. (eds) (2014). *Seasonal Workers in Mediterranean Agriculture: The Social Costs of Eating*, London: Routledge.

Gertel, J. and Sippel, S. R. (2016). The financialization of food and agriculture. In M. Shucksmith and D. Brown, eds., *Routledge International Handbook of Rural Studies*, London: Routledge, pp. 215–26.

Ghimire, K. (2005). The contemporary global social movements: Emergent proposals, connectivity and development implications, *UNRISD Civil Society and Social Movements Programme Paper*, no. 19. Geneva: UNRISD.

Gille, Z. (2013). From risk to waste: Global food waste regimes. In D. Evans, H. Campbell and A. Murcott eds., *Waste Matters: New Perspectives on Food and Society*, UK: John Wiley and Sons, pp. 27–46.

Goldstein, N. (2012). *Globalization and Free Trade*, 2nd ed., New York: Facts on File.

Gomiero, T. (2015). Effects of agricultural activities on biodiversity and ecosystems: Organic versus conventional farming. In G. Robinson and D. Carson, eds., *Handbook on the Globalization of Agriculture*, UK: Edward Elgar, pp. 77–105.

Gray, I. and Lawrence, G. (2001). *A Future for Regional Australia: Escaping Global Misfortune*, Cambridge: Cambridge University Press.

Harrison, J. (2014). Situating neoliberalization: Unpacking the construction of racially segregated workplaces. In S. Wolf and A. Bonanno, eds., *The Neoliberal Regime in the Agri-food Sector: Crisis, Resilience and Restructuring*, London, Routledge, pp. 91–111.

Harvey, D. (2005). *A Brief History of Neoliberalism*, Oxford: Oxford University Press.

Hays, J. (2014). Land grabs, forced evictions and land seizures in Cambodia. Available at: http://factsanddetails.com/southeast-asia/Cambodia/sub5_2d/entry-2909.html, accessed 22 January 2018.

Held, D., McGrew, A. Goldblatt, A. and Perraton, J. (1999). *Global Transformations: Politics, Economics and Culture*, Cambridge: Polity.

Heynen, N., McCarthy, J., Prudham, S. and Robbins, P. (2007). Introduction: False promises. In N. Heynen, J. McCarthy, S. Prudham and P. Robbins, eds., *Neoliberal Environments: False Promises and Unnatural Consequences*, London: Routledge, pp. 1–21.

Higgins, V. and Larner, W. (2017). Introduction: Assembling neoliberalism. In V. Higgins and W. Larner, eds., *Assembling Neoliberalism: Expertise, Practice, Subjects*, New York: Palgrave Macmillan, pp. 1–19.

Holmes, S. (2013). *Fresh Fruit, Broken Bodies: Migrant Farmworkers in the United States*, Berkeley, CA: University of California Press.

Holt-Giménez, E., Patel, R. and Shattuck, A. (2009). *Food Rebellions! Crisis and the Hunger for Justice*, California: Food First Press.

Howard, P. (2016). *Concentration and Power in the Food System: Who Controls what we Eat?* London: Bloomsbury.

International Forum on Globalization. (2002). *Alternatives to Economic Globalization: A Better World is Possible*, San Francisco, CA: Berrett-Koehler Publishers.

Kaag, M. and Zoomers, A. (2014). *The Global Land Grab: Beyond the Hype*, London: Zed Books.

Kalfagianni, A. and Fuchs, D. (2015). Private agri-food governance and the challenges for sustainability. In G. Robinson and D. Carson, eds., *Handbook on the Globalization of Agriculture*, UK: Edward Elgar, pp. 274–90.

Konzelmann, S., Fovargue-Davies, M. and Wilkinson, F. (2013). The return of 'financialized' liberal capitalism. In S. Konzelmann and M. Fovargue-Davies, eds., *Banking Systems in the Crisis: The Faces of Liberal Capitalism*, London: Routledge, pp. 32–56.

La Via Campesina. (2015). We are La Via Campesina. Available at: www.cadtm.org/spip.php?page=imprimer&id_article=12142, accessed 30 March 2018.

Land Matrix. (2018). Online public database on land deals. Available at: www.landmatrix.org/en/, accessed 22 January 2018.

Lawrence, G. (2016). Food systems and land: Connections and contradictions. In M. Shucksmith and D. Brown, eds., *Routledge International Handbook of Rural Studies*, London: Routledge, pp. 183–91.

Lawrence, G. and Dixon, J. (2015). The political economy of agri-food: Supermarkets. In A. Bonanno and L. Busch, eds., *Handbook of the International Political Economy of Agriculture and Food*, Cheltenham: Edward Elgar, pp. 213–31.

Lawrence, G. and Smith, K. (2018). The concept of 'financialization': Criticisms and insights. In H. Bjørkhaug, A. Magnan and G. Lawrence, eds., *Financialization of Agri-food Systems: Contested Transformations*, London: Routledge, pp. 23–41.

Lazarus, E. (2014). Land grabbing as a driver of environmental change, *Area*, **46**(1), 74–82.

Lewis, N., Le Heron, R. and Campbell, H. (2017). The mouse that died: Stabilizing economic practices in free trade space. In V. Higgins and W. Larner, eds., *Assembling Neoliberalism: Expertise, Practice, Subjects*, New York: Palgrave Macmillan, pp. 151–70.

McKeon, N. (2017). Are equity and sustainability a likely outcome when foxes and chickens share the same coop? Critiquing the concept of multistakeholder governance of food security, *Globalizations*, **14**(3), 379–89.

McMichael, P. (2000). *Development and Social Change: A Global Perspective*, 2nd ed., California: Thousand Oaks Press.

McMichael, P. (2009). A food regime geneology, *Journal of Peasant Studies*, **36**(1), 139–69.

McMichael, P. (2013a). *Food Regimes and Agrarian Questions*, Halifax: Fernwood Publishing.

McMichael, P. (2013b). Land grabbing and security mercantilism in international relations, *Globalizations*, **10**(1), 47–64.

McMichael, P. (2014). Historicizing food sovereignty, *Journal of Peasant Studies*, **41**(6), 933–57.

McMichael, P. (2016). Food sovereignty. In B. Pritchard, R. Ortiz and M. Shekar, eds., *Routledge Handbook of Food and Nutrition Security*, London: Routledge, pp.335–48.

McMichael, P. and Friedmann, H. (2007) Situating the 'retailing revolution'. In D. Burch and G. Lawrence, eds., *Supermarkets and Agri-food Supply Chains: Transformations in the Production and Consumption of Foods*, Cheltenham: Edward Elgar, pp. 291–319.

Martell, L. (2010). *The Sociology of Globalization*, Cambridge: Polity.

Michie, J. (2017). *Advanced Introduction to Globalization*, Cheltenham: Edward Elgar.

Murphy, S. (2013). *Land Grabs and Fragile Food Systems: The Role of Globalization*, Institute for Agriculture and Trade Policy, Minnesota, MN: IATP.

Murphy, S., Burch, D. and Clapp, J. (2012). *Cereal Secrets: The World's Largest Grain Traders and Global Agriculture*, London: Oxfam.

Newell, P. (2013) *Globalization and the Environment: Capitalism, Ecology and Power*, Cambridge: Polity Press.

Oosterveer, P. and Sonnenfeld, D. (2012). *Food, Globalization and Sustainability*, London: Earthscan.

Otero, G. (2014) The neoliberal food regime and its crisis: State, agribusiness transnational corporations, and biotechnology. In S. Wolf and A. Bonanno, eds., *The Neoliberal Regime in the Agri-food Sector: Crisis, Resilience and Restructuring*, London, Routledge, pp. 225–44.

Pechlaner, G. and Otero, G. (2010). The neoliberal food regime: Neoregulation and the new division of labor in North America, *Rural Sociology*, **75**(2), 179–208.

Pechlaner, G. and Otero, G. (2015). The political economy of agriculture and food in North America: Toward convergence or divergence? In A. Bonanno, and L. Busch, eds., *Handbook of the International Political Economy of Agriculture and Food*, Cheltenham: Edward Elgar, pp. 131–55.

Pieterse, J. (1996). Globalization as hybridization. In M. Featherstone, S. Lash, and R. Robertson, eds., *Global Modernities*, London: Sage, pp. 45–68.

Raynolds, L. (2014). Fairtrade, certification, and labour: global and local tensions in improving conditions for agricultural workers, *Agriculture and Human Values*, **31**, 499–511.

Raynolds, L. and Murray, D. (2007). Fair Trade: Contemporary challenges and future prospects. In L. Raynolds, D. Murray and J. Wilkinson, eds., *Fair Trade: The Challenges of Transforming Globalization*, London: Routledge, pp. 223–32.

Riddell, P. (2013). 'Land grabs' and alternative modalities for agricultural investments in emerging markets. In T. Allan, M. Keulertz, S. Sojamo and J. Warner, eds., *Handbook of Land and Water Grabs in Africa*, London: Routledge, pp. 160–77.

Robertson, R. (1992). *Globalization: Social Theory and Global Culture*, London: Sage.

Rosset, P., Patel, R. and Courville, M. (eds) (2006). *Promised Land: Competing Visions of Agrarian Reform*, California: Food First Books.

Samranjit, P. (ed.) (2014). Land grabbing and impacts to small scale farmers in Southeast Asia sub-region. Available at: /www.iss.nl/sites/corporate/files/CMCP_60-Samranjit.pdf, accessed 22 January 2018.

Schmidt, T. (2016). *The Political Economy of Food and Finance*, London: Routledge.

Sekine, K. and Bonanno, A. (2016). *The Contradictions of Neoliberal Agri-food: Corporations, Resistance and Disasters in Japan*, West Virginia: West Virginia University Press.

Sippel, S. R., Larder, N. and Lawrence, G. (2017). Grounding the financialization of farmland: Perspectives on financial actors as new land owners in rural Australia, *Agriculture and Human Values*, **34**, 251–65.

Smith, K. (2014). *Ethical Trade, Gender and Sustainable Livelihoods*, London: Earthscan.

Smith, K. (2016). Food systems failure: Can we avert future crises? In M. Shucksmith and D. Brown, eds., *Routledge International Handbook of Rural Studies*, London: Routledge, pp. 250–61.

Stilwell, F. (2002). *Political Economy: The Contest of Economic Ideas*, South Melbourne: Oxford University Press.

Taylor, R. and Entwistle, J. (2015). Agriculture and environment: Fundamentals and future prospects. In G. Robinson and D. Carson, eds., *Handbook on the Globalization of Agriculture*, Cheltenham: Edward Elgar, pp. 31–76.

Thompson, L. and Lockie, S. (2013). Private standards, grower networks, and power in food supply systems, *Agriculture and Human Values*, **30**, 379–88.

Tilzey, M. (2018). *Political Ecology, Food Regimes and Food Sovereignty: Crisis, Resistance and Resilience*, London: Palgrave Macmillan.

Timmer, C. (2008). Food policy in the era of supermarkets: What's different? In E. McCullough, P. Pingali and K. Stamoulis, eds., *The Transformation of Agrifood Systems: Globalization, Supply Chains and Smallholder Farmers*, London: Earthscan, pp. 67–86.

Vergara-Camus, L. (2009). The politics of the MST: Autonomous rural communities, the state, and electoral politics, *Latin American Perspectives*, **36**, 178–191.

Watson, M. and Meah, A. (2013). Food, waste and safety: Negotiating conflicting social anxieties into the practices of domestic provisioning. In D. Evans, H. Campbell and A. Murcott eds., *Waste Matters: New Perspectives on Food and Society*, Chichester: John Wiley and Sons, pp. 102–20.

Weis, T. (2013). *The Ecological Hoofprint: The Global Burden of Industrial Livestock*, London: Zed Books.

Weis, T. (2016). Industrial livestock and the ecological hoofprint: Inequality, degradation and violence. In M. Shucksmith and D. Brown, eds., *Routledge International Handbook of Rural Studies*, London: Routledge, pp. 205–14.

Woods, M. (2016). Confronting globalisation? Rural protest, resistance and social movements. In M. Shucksmith and D. Brown (eds) *Routledge International Handbook of Rural Studies*. London: Routledge, pp. 626–37.

25 The Sociology of Environmental Morality: Examples from Agri-Food

Paul V. Stock

1 Introduction

Environmental sociology lays claim to introducing the importance of "the environment" to social relationships. We are both shaping the environment and being shaped by it in ways that are both positive and negative. At the heart of many environmental conflicts lie disagreements about what is good and good for persons and their community. In this chapter, I'd like to outline a subfield within environmental sociology that marries environmental sociology with the sociology of morality. While tackling some issues of the relationship between the environment and persons, the sociology of morality most often does this from a sociology of culture perspective. By incorporating some environmental sociology (and rural sociology) we can develop a sociology of environmental morality (with a nod to some previous attempts by Bell, 1994; 2018 and Stock, 2007; 2009; 2015) that will allow us to ask some different questions about what's good, and for whom, related to the environment. This chapter then will continue with an engagement with environmental ethics and the idea of the moral economy. From there we'll examine Michael Bell's concepts of natural conscience and multiple natures that shape identity and vice versa. Then we'll explore Justin Farrell's concepts of moral orders and Gabriel Abend's moral repertoires. We'll also examine the contributions of Richard Stivers and Jacques Ellul related to technological morality before offering some examples from agri-food studies.

2 Critique of Environmental Ethics and Moral Economy

Why not just focus on integrating environmental ethics or expanding the scope of the moral economy or environmental justice? We might ask, if sociology should be concerned with conflicts over what's environmentally good or bad, why shouldn't we just read environmental ethics? To a certain extent, we should. We should read widely to help us both comprehend what's outside of sociology, but also to help inject sociology where it is sorely needed. Heyd (2003) argues that environmental ethics is often too concerned with the creation of complete ethical systems to be of much use to real problems. Briefly then, environmental ethics, while an important field in and of itself, limits itself to the judgments and pronouncements of what is right or wrong towards or with the environment without offering us many tools to understand how or why people act the way they do toward or with the environment.

What about studies of the moral economy? Doesn't that offer insights about people's motivations of right and wrong towards other people and a concern (possibly) for the planet? In some ways, yes. The original concept explained the moral practical knowledge involved in structuring economic relationships embedded with cultural and interpersonal meaning necessary for the stability of the group (Thomson, 1971). One critique is then: What is the economy if it's not moral? Is it immoral? What is the good being emphasized in the moral economy then? For some it is an emphasis on understanding "moral" choices of production (e.g., higher wages, environmental consciousness) or those of consumers. The quick answer as to why the work on moral economy is not enough for building a successful sense of morality in environmental sociology is here. Developing a sociology of environmental morality can more easily put these moral (or immoral) designations in conversation with one another. Further, in each of these areas of literature the starting point of the analysis is that something is right or wrong, good or bad, just or unjust. The development of a sociology of environmental morality offers us some analytic tools to avoid such moralization if one so chooses.

It's also important to recognize the paucity and neglect of sociology's wrestling with morality, especially around the environment, of indigenous formulations of any kind. The inroads of indigenous ontologies and methods elsewhere seem not to have been engaged explicitly in environmental sociology until very recently.

3 Sociology of Morality and the Environment

3.1 Bell's Natural Conscience and the Good

Michael Bell, following George Herbert Mead and Erving Goffman, develops a sociology of environmental morality at the intersection of the self, social relationships, and how one experiences nature. First developed in *Childerley* (Bell, 1994), his ethnography of an English village experiencing a rapid influx of more urban dwellers after the completion of a major highway, Bell's concept of the natural conscience helps us puzzle together the role nature plays in our own self conceptions and group identity. Bell (1994) wrote that "Childerleyans find in nature a kind of moral preserve in a landscape of materialist desire, an alternative region of moral thinking I will call the natural conscience" (p. 138). Being "real country people" (or their sense of their "natural me" [p. 148]) serves as "the moral foundation for freedom" (149) for them. Bell discusses this self-conception of being a "real country person" as developed in opposition to the non-country people, a generalized other.

> In nature, they find a moral domain clearly free from the pollution of social motives and intrigues. This is the moral contrast that the villagers are ultimately trying to draw through their conceptions of nature.
> Nature's innocence of social intrigue gives it a further quality for many villagers. The contrast removes the things of nature from all moral criticism. Childerlayans therefore see nature as somehow more real, more true, more authentic – as something they can trust that is free of society's fickleness and back-stabbing.
> (Bell, 1994: 146)

This formulation of nature then develops as a natural conscience that reinforces their behavior towards and with nature: hunting is good or evil, these farming practices that cause pollution are necessary or harmful (see Lowe et al., 1997). Bell (1994) argues that, "The natural conscience is the imagination of something – some realm, some agent – apart from the collective conscience that can serve as a disinterested basis for values and the self." (151).

Bell expanded on these ideas in *The City of the Good* (2018) that explores the emergence of the idea of what is good over an expansive period of time from various faith backgrounds (see also Brewster and Bell, 2009). Bell describes this through the tension between the pagan and the bourgeoisie.[1] This tension between the pagan (or more country-oriented) and the bourgeois (more city-oriented), Bell argues, provides fundamentally different orientations to nature that are then used to build our social infrastructure. He documents this through a set of interrelated concepts. We collectively treat nature in three ways. First nature –"nature before nature" – is "how something is before and apart from any later manipulation or disguise" (Bell, 2018: 49). We still refer to such and such a thing's nature. Second nature is where "nature is a moral good." It's second nature that the Childerlayans used to develop their natural conscience. Further, that sense that nature is morally good helps develop the natural we or that "sense of nonpolitical community derived from a natural other" (Bell, 2018: 138).

The formation of a natural me (thus forming a natural we with other similar people) necessitates an other, or what Bell (2018: 225) calls a natural them. Because nature is beyond reproach, all things grounded in nature are, therefore, good; things from nature are then non-political. This also leads to conflict when different natural consciences come in contact with one other. Thus, we have the formulation of third nature where we see in others' conceptions of nature that are different than ours, then we see their idea of "nature as a moral bad, as the seat of desire and thus politics" (Bell, 2018: 141).

> One of the great questions of moral sociology must be how people can construct the bad as the good – as well as the good as the bad. The nonpolitical politics of the

---

1 "By 'bourgeois' – a word derived from the Latin for a fortified town – I mean the concerns over the justice of desire and the vicissitudes of wealth that originally arose in the city, but are no longer so confined. By 'pagan' – a word derived from the Latin for a country dweller – I mean the concerns over the troubles of disloyalty and the vicissitudes of agriculture and ecology that originally arose in the countryside, but are also not confined. They are not so confined because our contexts are not so pure. (And I should stress that by 'pagan' I do not mean New Age. I mean the ancient and living traditions that descend from the concerns of rural context the world over.)" (Bell, 2018: 12).

natural conscience are a common means of that construction, whether on the part of so much other social mischief in this difficult and beautiful world. (Bell, 2018: 224)

When one natural we comes in conflict with another natural we (with their opposite being a natural them, naturally) then the conflict over the competing idea of the second nature turns one natural we and their second nature into the other natural we's third nature. The competing ideas of The Good makes the conflict seem intractable. As Bell (2018) argues in the conclusion, "Morality isn't so easy. It's a tangled affair – tangled because morality is entangled with the world, not apart from it" (p. 273).

Where Bell (2018) tries to wring an understanding of the good out of the world's religious traditions, Farrell is less concerned with where ideas of what constitutes the good come from in the long term and more concerned about what happens when those traditions of what is good come in conflict with one another.

3.2 Farrell's Moral Orders

In wrestling with morality, the environment, and their related conflicts, Justin Farrell's (2015) work on moral orders, with only a passing referencing to Bell and the natural conscience, offers a good framework to complement Bell's work. Without being in direct dialogue with Bell's work, Farrell's moral orders, which draw heavily on the sociology of culture, resemble the natural conscience and the natural them, in practice. In *The Battle for Yellowstone*, Farrell outlines competing moral orders over the meaning of living a moral life in the US West. Farrell defines a moral order as "an interpersonally and institutionally shared structure of moral beliefs, desires, feelings and boundaries that are derived from larger narratives and rituals" (p. 10). He clarifies that "by 'moral' I mean an orientation toward what is right and wrong, good and bad, worthy and unworthy, just and unjust" (Farrell, 2015: 10). Most importantly these moral commitments or moral orders "shape basic practices [actions] ranging from how to treat friends, family, and coworkers, to public and political practices" (p. 10). Just as Hitlin and Vaisey (2010b: 10) describe "moral meanings" that help "create narrative coherence around an *individual's* life," Farrell contends that, "A fundamental motivation for human behavior is the struggle to enact and sustain moral order" (p. 11). Conflicts such as those related to the environment, landscapes, and climate change are "intense and [seemingly] unending because the deeper narratives and moral commitments at the heart of the conflict cannot be resolved by recourse to some external, objective, equally applicable standard. They are more like faith commitments that are 'true' only within their larger frameworks of cultural belief and practice" (p.11). Hence, in Bell's terminology, there exist competing natural consciences.

In the Greater Yellowstone Ecosystem, or GYE (an area that includes but is also much larger than Yellowstone National Park) moral conflicts rage over the reintroduction of wolves and how to handle bison, especially in relationship to ranching. Each of these conflicts pit competing moral orders against one another. On the one hand, the moral order of the Old West values emphasized physical labor (e.g., ranching, farming, and mining), private property, family ownership, rugged individualism, autonomy, and a sense of dominion over nature influenced by Manifest Destiny – what are often conflated as the physical and psychological attributes of American conservatism. For a variety of reasons that include demographic changes and new, emerging philosophical orientations to nature, such as Gaia, a New West moral order now revolves around scientific ecology and lifestyle amenities that treat the West as background landscape rather than a place to be worked and used for survival. It's a shift from land as provider to land as playground with echoes of Richard White's (1996) essay, "Are you an environmentalist or do you work for a living?" as well as Bell's Childerley.

Farrell concludes that these moral orders of the Old West and the New West compete such that those enacting a New West moral order morally devalue those enacting an Old West moral order. These competing repertoires (as Abend, below, would describe) of what is morally good are now in conflict; these different moral orders help explain the (environmental) social conflict over the reintroduction of wolves, the place of genetically pure bison, and the possibilities for expanding fracking. Not only is ranching, as many believe it to be practiced, causing desertification and erosion and contributing to global warming, but you, the rancher (because of these practices) are a bad person. The New West tries to "transform their opponent's sacred stories and core intrinsic values [like hard work and private property] from right to wrong, good to bad" (Farrell, 2015: 16). Thus the New West moral order undermines "what Stets and Carter (2012) call a 'moral identity' or theory of self" (Farrell, 2015: 90). That moral identity is "Constructed from both reality [practices and experience] and mythic nostalgia, this moral identity attempted to simplify life to its most basic and real elements, where what it means to be a good person was straightforward, and right and wrong were clearly defined" (90).

The New West moral order challenges and devalues the Old West moral order and those that hold fast to it. Conflicts over something like the reintroduction of the wolf to the GYE are not readily solved for the conflict is not simply a battle over the question of should we reintroduce the wolf? Yes or No?, but a matter of deeply held values that emerge from "deeper narratives that orient stakeholders' lives within an identifiable shared moral order informing what they want, what is 'good,' and why it all matters" (Farrell, 2015: 216).

Bell's natural conscience and Farrell's moral order and the consequent moral devaluation offer a great starting point for a sociology of environmental morality. From here we need to extend and open of the sociology of morality a little more as well as welcome back some sidelined and ignored work to bolster this theoretical framework we're calling the sociology of environmental morality.

3.3 Abend's Moral Backgrounds and Repertoires

Riding the resurgence of the sociology of morality (keeping in mind Durkheim's sincere hope that sociology would become a discipline with morality at its core [see Hitlin and Vaisey, 2010a]) is Gabriel Abend's work focused on the history of business ethics. Abend's work addresses what Hitlin and Vaisey (2010a: 9) lay out as the three themes in pursuing a resurgent sociology of morality: "(1) attention to social structures, resources, and power; (2) a focus on historically and socially patterned complexes of meaning; and (3) an emphasis on studying moral judgment, action, and discourse in ecologically valid contexts." You may notice the absence of anything related to the natural world. "Ecologically" in this context relates more to a social ecology than anything related to nature.

Before developing a sociology of environmental morality further, we need to incorporate Abend's (2014) significant combination of theorists and concepts to develop his moral background ideas including "variations on the concept of knowledge," "concepts that stress the body and embodied properties," and ideas about shared concepts what we might widely refer to as culture (Abend, 2014: 57). He argues that "all members of the extended [theoretical] family make a distinction between two realms or levels, and pay special attention to the one that lies underneath or behind the other" (Abend, 2014: 58).

For our purposes, Abend offers us an updated and novel framework related to Bell's and Farrell's work that help us make sense of moral actions in relationship to the environment. Abend (2014: 53) distinguishes between first order morality and the moral background that "facilitates, supports, or enables morality." First order morality focuses on judgments of whether something is right or wrong, good or appropriate – what others might refer to as ethics or ethical judgments. "The moral background facilitates, supports, or enables morality" (Abend, 2014: 53) such that it makes possible moral actions and practices as well as the judgements thereof. The moral background, specifically, is "a particular collection of para-moral elements" (Abend, 2014: 67). These para-moral elements include 1) Grounding; 2) Conceptual repertoires; 3) Objective of evaluation; 4) Method and argument; 5) Metaethical objectivity; and 6) Metaphysics. These various parts of the moral background shape a person's values and decisions related to how to relate to others, what is able to be morally evaluated or should be morally evaluated, what kinds of arguments are marshaled to justify such and such a behavior as moral as well as metaphysical justifications. By comparing two types of business styles, Abend illustrates that the Christian merchant type utilizes an ethics based on the Bible and a belief in God whereas the "Standards of Practice" type might justify a similar decision based on scientific naturalism with beliefs in empirical data or what we might now call best practices. On the other hand, the "Standards of Practice" type business ethics might justify a similar decision based on scientific naturalism with beliefs in empirical data or what we might now call best practices. Like Bell and Farrell, these competing types draw on moral backgrounds or orders that can often leave conflicts intractable. What is valuable from Abend's argument for our purposes though is not the conflict itself but the underlying existence of two playing fields such as it is two sets of rules – that classify good, better, or winning differently.

"Morality," Abend argues, "is underlain by background elements, whether actors realize it or not, whether they like it or not" (p. 364). What Farrell refers to as moral orders, again remembering his arguments come from cultural sociology, Abend writes about as moral repertoires. These "conceptual repertoires" have "two kinds of moral concepts."

Thin moral concepts include right and wrong, good and bad, appropriate and inappropriate, permissible and impermissible, and ought and ought not. Thick moral concepts include dignity, decency, integrity, piety, responsibility, tolerance, moderation, fanaticism, extremism, despotism, chauvinism, rudeness, uptightness, misery, exploitation, oppression, materialism, humanness, hospitality, courage, cruelty, chastity, perversion, obscenity, lewdness, civility, clemency, and friendship. One key difference between thin and thick moral concepts is this. You can apply a thin concept to anything you wish. . . . By contrast, the application of thick concepts is constrained by what the world is like. (Abend, 2014: 37)

4 Developing a Sociology of Environmental Morality

Bell, Farrell and Abend offer us the ideas of natural consciences, different natures, moral orders, the moral background, moral repertoires, and moral devaluation, that help us strengthen a relationship between the sociology of morality and environmental sociology. There are other examples, I'm sure, and this chapter should be seen as an invitation to this subfield in development. Further, each of us will privilege certain literatures, authors, and analyses of morality. I think, though, we can sketch out what the subfield could accomplish. Let's go through some common environmental sociology examples and apply some sociology of environmental morality to them.

Environmental justice (EJ) is a key area of research, not just for environmental sociologists, that comes from a specific moral orientation about how the world should look and pointing out that it can and should be otherwise. In fact, much of sociology might be described thusly (Smith, 2014). In terms of moral orders you might, for example, describe the moral orders at play in one of the defining battles for environmental justice in Warren County, North Carolina. In 1982, following the siting of a toxic waste incinerator in their predominantly African-American and impoverished community, citizens of Warren County and environmental allies protested the siting of the incinerator as well as the potential health and ecosystem effects of any potential incineration. These protests are often written about as the beginning of the environmental justice movement in the United States. You might describe the moral order or natural conscience of the local government, police force, and incinerator company as simply locating a place with cheaper real estate prices and potentially ineffective local government. On the other hand, the moral order or natural conscience of the former Civil Rights activists, local Native American, African-American, and poor residents, and other activists identified community health and well-being as more important than a new incinerator and that their ideas align with nature (Bell's second nature) and thus have more moral weight (McGurty, 2000). Warren County was also the location of the integrated experiment community called Soul City in the 1970s, which might help explain some of the protest's moral background (Strain, 2004). It's not as if environmental justice needs to be explained through a lens of environmental morality, but it can help us comprehend why these conflicts are often so protracted. The moral orders of business development (and often government) is one concerned with present financial concerns, not the historical legacy of what contributes to declining or suppressed real estate prices where others document the long history of racism and real estate geography, neighborhood formation or disintegration (Coates, 2014). EJ offers a great example of competing moral orders and repertoires as well as a way to understand much of the moral devaluation (intended and unintended) during conflicts. In this way, EJ offers a great example of studying the sociology of environmental morality where the places experiencing conflict are a huge part of what constitutes a community's identity and what its own well-being should be like.

Another common environmental sociology concept is disasters. Erikson's (1976) *Everything in its Path* describes not just the complicity of the coal company's dereliction of maintaining the integrity of the Buffalo Creek, West Virginia dam and the physical and ecosystem damage, but that the broken dam and resulting flood destroyed the fabric of the community and the collective morality (Erikson, 1976: 186–245). By destroying the "nature" in this place, residents' natural selves and, thus, their natural we were disrupted. Even the formation of a natural them (the coal company and the government) was not enough to retain the moral fabric of the community. Relatedly, the obvious racial differences in the outcomes for African-Americans (who were far more likely to have been displaced from their homes and city) following Hurricane Katrina made it one of the most studied disasters ever. In one instance the media coverage offered a moral judgment asserting the equivalence of looting, anarchy, and African-Americans, that in the overwhelming majority of cases was non-existent (Stock, 2007b).

Climate change offers a great example of competing moral orders in a few ways. Kari Norgaard's (2006, 2011) work on the management of emotions in response to the reality of climate change could also be illuminated by competing visions of what the world should look like and how the climate could be behaving. Many of these emotions come from a gap between the expected and the real. Additionally, competing moral orders of the role of science in shaping policies, the economy, and social life have become increasingly politicized (York, 2015). On one hand, like Farrell's New West, ecology and climatology are enrolled as justifications for state and supranational policies and action plans while on the other hand a moral order that emphasizes that not science, but business and the economy should dictate the response or denial of climate change's reality and forecasted impacts. These merchants of doubt are often tied into specific moral orders and repertoires connected to certain ideologies (McCright and Dunlap, 2011; Oreskes and Conway, 2010).

The last example from environmental sociology is the examination of population (Bates, 2015; Ehrlich, 1968). Since Malthus' essay in 1798, the competing moral orders over population divide into the broad categories of there are/will be too many people, therefore, we have to find a way to rein it in, versus how do we pragmatically take care of this many people in a dignified way, as in the arguments in the UN Declaration of Human Rights. Various versions of these arguments make proclamations of how we will provide enough food, or that we won't.

In all these cases the moral order or natural conscience one or a group ascribes to powerfully influences the kinds of analyses one prefers and even wants or writes about. It also has significant influence on the ability to reach pragmatic accords that are often advocated for in much of environmental sociology.

4.1 Exploring a Technological Morality

The idea of a sociology of morality is something I've tackled in various ways in my previous work, but in isolation (Stock, 2007a; 2009; 2015). At various times the focus has been on articulating relationships of identity and morality and at other times on developing arguments around care and farming practices. From here I'd like to pursue a certain strain of the sociology of morality and better apply it to an environmental example related to agriculture.

In, as Hitlin and Vaisey (2010a: 3) point out sometimes happens, a somewhat ignored part of the scholarship in the sociology of morality, Richard Stivers offers (as an extension of the work of French theoretician Jacques Ellul) that our lived morality is a technological morality. A technological morality is "a morality geared entirely toward efficiency" (Stivers, 1994: 8). "The new American morality...is exclusively one of power. But a morality of power is simultaneously a morality without meaning" (Stivers, 1994: x). Thus, our contemporary society lives by a morality that prizes competition, winning, and power above all other kinds of relationships with one another or the planet.

The ideological differences highlighted by Bell and Farrell – these competing natural "thems" and moral orders – are unified in the new American morality in "the common belief in technological progress and increased consumption" (Stivers, 1994: 3) that provides the path toward success, survival, happiness, and health (Stivers, 1994: 9). You could argue that, while scientifically informed, the New West morality is a different pursuit of power. This lack of meaning, more importantly by seeking (actively or unknowingly) to demean people of a different moral order exhibits such a technological morality. These moral orders are similar in form and content to what Stivers describes as a lived or social morality (Stivers, 1994). A "lived or social morality is an indirect expression of what is experience as sacred.... Lived morality is not primarily a matter of individual ethical choice or even the practice of virtue, but rather an expression of collective belief and allegiance" (Stivers, 1994: 5). "Whatever is most important to the continued existence of a society is spontaneously regarded as sacred" (Stivers, 1994: 5). Stivers, following Jacques Ellul, maintains that what is held as sacred in the contemporary era is technique, or the search for efficiency in all aspect of life.

Stivers develops his argument following Ellul's concept of technique. This is often translated as technology – that is, both physical technology (like computers, cell phones, electrical infrastructures), but also immaterial and psychological (like time clocks, educational testing). By technological morality the sole normative goal is one of radical efficiency in all things without regard to context or situation.

> [Technique] destroys, eliminates, or subordinates the natural world, and does not allow this world to restore itself or even to enter into a symbiotic relation with it. The two worlds obey different imperatives, different directives, and different laws which have nothing in common. Just as hydroelectric installations take waterfalls and lead them into conduits, so the technical milieu absorbs the natural. We are rapidly approaching the time when there will be no longer any natural environment at all. When we succeed in producing artificial *aurorae boreales*, night will disappear and perpetual day will reign over the plant. (Ellul, 1964: 79)

Ellul and Stivers claim that technique, not capitalism, is the defining feature of the contemporary world. Efficiency, through the pursuit of winning and power – which become a value, has become sacred and, thus, the lodestar of the good. The pursuit of power demarcates one as a good person. Those not pursuing power are, in Farrell's terms, morally devalued, which often falls on the rural, women, minorities, immigrants. For example, we might analyze that the reason the gender pay gap is presented as a personal deficiency because women are not winning or demanding the right to appropriate pay. What is deemed good (efficiency and power) is then held to be sacred and those that do not succeed (by getting the pay they deserve) are at fault. What should be emphasized is that there is an "inverse relationship between power and values" (Ellul, 2015). In our contemporary world, we have examples of what Stivers and Ellul describe as technological morality including life in the suburb (Baumgartner 1988), the office (Jackall, 1988), and parenting (Karlsson et al., 2013). It's no coincidence that these situations are also ripe for satirizing in situational and standup comedy (Dern, 2015). The Holocaust is also an example of technological morality (Todorov, 1996).

4.2 Examples from Agriculture

In the United States, ecology and these discussions of morality in farming are often kept at the level of ethics. More to the point, these ethical discussions are kept at the level of the prescriptive, or what Stivers would call theoretical morality (e.g., Thompson, 1995). Now that I've laid out some foundations and justifications for a sociology of environmental morality as well as some theoretical ideas not typically included in the mainstream sociology of morality, I'd like to offer a sociology of environmental morality analysis of two examples from the agri-food literature. The first is the growing body of work around the concept of the "good farmer." The second tackles work on animal agriculture in the United States.

4.2.1 The Good Farmer

People don't often say, "I'm a good farmer." Most often it's used to describe someone else. Further, in agriculture the technological morality and efficiency, relates to an ideology to feed the world (Rosin, 2013) or that good farmers maintain tidy farms (Burton, 2004b) – the efficiency of tidiness is affiliated with a normative assumption. The literature around the idea of someone being a "good farmer" has its theoretical origins in the UK (Burton, 2004). Burton (2004a; Burton and Wilson and 2006) also extended work on farmer identities. Further work on the "good farmer" has extended a Bourdieusian and social capital analysis (Sutherland, 2013; Sutherland and Burton, 2011; Sutherland and Darnhofer, 2012). In a departure, Stock (2007a) essentially analyzed the moral orders and natural consciences of organic farmers who had either converted from conventional agriculture or entered organic farming from a different profession. Michael Bell's (2004) work that examined the transition of farmer identities in Iowa resonates here as well. Other work in this vein includes Forney and Stock (2014) and emerging work on autonomy and care (Stock et al., 2014; Stock and Forney, 2014; Stock and Brickell, 2013). Literature classified by country and type of farming endeavor has emerged over recent years as well (Haggerty et al., 2009; Hunt, 2010; Huttenen et al., 2015; Riley, 2015; Silvasti, 2003). In all these cases, the work is wrestling with tensions between self, identity, natural conscience and the moral order, repertoires, and moral devaluation. In the case of farmers that converted away from conventional agriculture, Stock (2007a) documents (like Bell, 2004) the othering of those farmers by the "abandoned" natural we. By not considering the organic ones farmers, the moral orders and third nature of agriculture, at least in central Illinois in the mid-2000s, is on display. So in the case of good farmers, we have multiple instances where farmers that were too small, too different, too new feeling as if they were treated as non-farmers and for some that equates to being treated as non-persons which is felt most acutely by women, farmers of color, and/or non-heteronormative sexual identities (Leslie, 2017). This moral devaluation of one group with a certain moral order prevents a dialogue between two poles about the kinds of agriculture that will be necessary to address the major ecological and social stresses involved in doing an agriculture in step with the planet and the needs of the persons living upon it. As Carolan (2015) writes about understanding our roles as citizens related to new food projects, we might need to emphasize a wilding of our thinking. Here "wild" intimates an undisciplined, maybe even unexpected way of thinking about and experimenting with food system. This means wilding not only what and how we're willing to talk (where many pragmatists place their emphasis) about food, but also in what we do. Carolan (2015) emphasizes that, "to know something differently and see things in a new light we have to *do* something different" (p. 133). As Bell (2004) documents in the Iowa farmers involved in Practical Farmers of Iowa (PFI), the shift in their natural consciences came from shifting their farming practices which thus reinforced that shift in their natural me. And PFI in turn developed into a natural we with BigAg serving as a natural them. Thus, what it meant to be a good farmer took both talk and doing, together, to change from one natural conscience/moral order to another because they are, by design, pretty resilient.

4.2.2 Animal Agriculture

Ellis (2013) offers us a specific example of farmers, ranchers in this instance, promoting and proclaiming a moral order/natural conscience of a kind of animal production. "When stewardship is done properly, when it is in balance with the environment, it is seen as good for everyone and everything: the ranchers, the animals, and the environment" (Ellis, 2013: 435). As Ellis argues, the characterization of one's self as a good farmer/rancher, while contributing to negative ecosystem effects is about hiding power. Ellis refers to this as symbiotic (Burton might argue symbolic) ideology. Ellis goes on to argue that

> stewardship and husbandry are closely related and that both describe a process of interaction. These values are part of a narrative of balance and co-constitution that constructs ranching work as part of a natural system. I term this narrative the *symbiotic ideology* because it obscures our view of the cultural values that allow people to use nonhuman animal bodies and the environment for their own ends. That is to say, it obscures our view of dominion and the power relationships inherent in animal agriculture. (p. 429)

Stivers (1994) further argues, "that technical and bureaucratic rules are the 'morality' of technology" (p. ix) which defines our current era. This helps explain the overlap of a self-proclaimed anarchist farmer in Kansas (newfarmeersproject.com) and Joel Salatin's (one of the most well-known public advocates for sustainable agriculture) Christian-inspired rant against government bureaucracy (Pollan, 2006: 226–238).

Ellis offers us both a description of the ranchers' natural conscience and moral orders, what Farrell would most certainly characterize as Old West. We also see strains of the moral custom outlined by Stivers. Most importantly, the New West values demean the ranchers' self-identity as stewards and caretakers of the land. And yet, the emphasis on the efficiency and husbandry that produces meat in the long run that results in ecological destruction and the turning of living bodies into commodities is based on a system of domination, according to Ellis (2013: 445), which also reveals a moral stance on the author's part. This is not a critique of that, but merely to point out that, as environmental sociologists, there are these moral orders, repertoires, and identities at play for us as sociological observers as well. In his documentation of the chicken tournament, Leonard (2014), describes the competition that chicken companies put their farmers into that creates not only a playful tournament in which someone is trying to win (the most or heaviest or fastest birds), but turns their livelihoods into a competition. The bad farmers livelihoods, reputations, and sometimes lives are damaged as a result. The "losers" in this rigged system are thus morally devalued. In this system they are turned into a natural them. Stivers would argue that a technological morality dominates both such that the lives of the animals and the farmers are secondary to the pursuit of power and technique. What we can glean from these examples of animal agriculture, though, is that there exist competing conceptions of which nature is good and by extension which people are good depending on to which natural conscience or moral order they ascribe.

5 Conclusions

We've seen nuanced and specific iterations of how we understand the good in good farmers. My examples here are not meant to instantiate or claim one is more correct, but merely to highlight the importance of exploring the assumptions, moral backgrounds, and motivations that both flow from understandings of goodness, but also contribute to in a dialectical and dialogical (what Bell, 2018: 275 refers to as multilogical) dialogue of place, ecosystems, cultural tradition, education, status, gender, geography, type of operations, economic standing, policy, moral backgrounds, repertoires, and technique. We could spend hours debating these various sides, but that is not the point.

What Bell's natural conscience and multiple natures and Farrell's moral orders point to is an opportunity to expand our study of the role of morality, lived morality, and moral custom in the relationship between our human actions and the planet, its resources, other species, and our biosphere. An expanded understanding of these, often in competition, relationships leads to better understanding, contributes to better meaning making about our relationship to the non-human world and with each other, and lastly, can help us potentially find some places of agreement from which to work on problems collectively. The sociology of environmental morality affords us the tools to both continue to critically evaluate relationships, while also recognizing the values of those that think differently.

References

Abend, G. (2014). *The Moral Background: An Inquiry into the History of Business Ethics*. Princeton, NJ: Princeton University Press.

Bates, D. C. (2015). Population, Demography, and the Environment. In K. Gould & T. L. Lewis, eds., *Twenty Lessons in Environmental Sociology*, 2nd ed., Oxford, UK: Oxford University Press, pp. 118–136.

Bell, M. (1994). *Childerley: Nature and Morality in a Country Village*, Chicago, IL: University of Chicago Press.

Bell, M. (2004). *Farming for Us All: Practical Agriculture & the Cultivation of Sustainability*. University Park, PA: Penn State Press.

Bell, M. (2018). *City of the Good: Nature, Religion, and the Ancient Search for What Is Right*, Princeton, NJ: Princeton University Press.

Brewster, B. H., & Bell, M. M. (2009). The environmental Goffman: Toward an environmental sociology of everyday life. *Society & Natural Resources*, **23**(1), 45–57.

Burton, R. J. (2004a). Reconceptualising the "behavioural approach" in agricultural studies: a socio-psychological perspective. *Journal of Rural Studies*, **20**(3), 359–371.

Burton, R. J. F. (2004b). Seeing through the 'good farmer's' eyes: Towards developing an understanding of the social symbolic value of 'productivist' behaviour. *Sociologia Ruralis*, **44**(2), 195–215.

Burton, R. J. F., & Wilson, G. A. (2006). Injecting social psychology theory into conceptualisations of agricultural agency: Towards a post-productivist farmer self-identity? *Journal of Rural Studies*, **22**(1), 95–115.

Carolan, M. (2015). Re-Wilding Food Systems: Visceralities, Utopias, Pragmatism, and Practice. In P. Stock, M. Carolan, & Rosin, Christopher, eds., *Food Utopias: Reconsidering Citizenship, Ethics and Community*, London: Routledge, pp. 126–139.

Coates, T. (2014). The case for reparations. *The Atlantic*.

Dern, N. (2015). The comedy of Sociology. *Footnotes*, **43**(4), 7.

DeVault, M. L. (1994). *Feeding the Family: The Social Organization of Caring as Gendered Work*, Chicago, IL: University of Chicago Press.

Ehrlich, P. (1968). *The Population Bomb*, New York: Sierra Club/Ballantine Books.
Ellis, C. (2013). The Symbiotic Ideology: Stewardship, Husbandry, and Dominion in Beef Production. *Rural Sociology* 78(4):429–449.
Ellul, J. (1964). *The Technological Society*, New York: Vintage.
Ellul, J. (2015). *The Political Illusion*, Eugene, OR: Wipf and Stock Publishers.
Erikson, K. T. (1976). *Everything in Its Path*, New York: Simon and Schuster.
Farrell, J. (2015). *Battle for Yellowstone: Morality and the Sacred Roots of Environmental Conflict*, Princeton, NJ: Princeton University Press.
Forney, J., & Stock, P. V. (2014). Conversion of family farms and resilience in Southland, New Zealand. *International Journal of Sociology of Agriculture and Food*, 21(1), 7–29.
Gould, K., & Lewis, T. L. (eds.). (2015). *Twenty Lessons in Environmental Sociology*, 2nd ed., Oxford: Oxford University Press.
Haggerty, J., Campbell, H., and Morris, C. (2009). Keeping the stress off the Sheep? Agricultural intensification, neoliberalism, and "good" farming. *Geoforum* 40(5), 767–777.
Heyd, T. (2003). The case for environmental morality. *Environmental Ethics*, 25(1), 5–24.
Hitlin, S., & Vaisey, S. (2010a). Back to the future. In S. Hitlin & S. Vaisey, eds., *Handbook of the Sociology of Morality*, New York: Springer.
Hitlin, S., & Vaisey, S. (eds.). (2010b). *Handbook of the Sociology of Morality*, New York: Springer.
Hunt, L. (2010). Interpreting orchardists' talk about their orchards: The good orchardists. *Agriculture and Human Values* 27(4), 415–426.
Huttunen, S., Mela, H., and Hildén, M. (2015). Good farmers, good adapters? How a cultural understanding of good farming affects the adaptive capacity of farmers. In A. Paloviita and M. Järvelä, eds., *Climate Change Adaptation and Food Supply Chain Management*. London: Routledge, pp. 107–118.
Jackall, R. (1988). *Moral Mazes: The World of Corporate Managers*, New York: Oxford University Press.
Karlsson, M., Löfdahl, A., & Prieto, H. P. (2013). Morality in parents' stories of preschool choice: narrating identity positions of good parenting. *British Journal of Sociology of Education*, 34(2), 208–224.
Kessler, A., Parkins, J. R., & Huddart Kennedy, E. (2016). Environmental harm and "the good farmer": conceptualizing discourses of environmental sustainability in the beef industry. *Rural Sociology*, 81(2), 172–193.
King, L., & McCarthy, D. (Eds.). (2009). *Environmental Sociology*, 2nd ed., Lanham, MD: Rowman & Littlefield Publishers.
Leonard, C. (2014, October 1). The Chicken Competition. *Guernica: A Magazine of Art and Politics*.
Leslie, I. S. (2017). Queer farmers: Sexuality and the transition to sustainable agriculture. *Rural Sociology*, 82(4), 747–771.
Lowe, P., Clark, J., Seymour, S., & Ward, N. (1997). *Moralizing the Environment: Countryside Change, Farming and Pollution*, London: UCL Press Limited.
McCright, A. M., & Dunlap, R. E. (2011). The politicization of climate change and polarization in the American public's views of global warming, 2001–2010. *The Sociological Quarterly*, 52(2), 155–194.
McGurty, E. M. (2000). Warren County, NC, and the emergence of the environmental justice movement: Unlikely coalitions and shared meanings in local collective action. *Society & Natural Resources*, 13(4), 373–387.

Nelson, J., & Stock, P. (2016). Repeasantisation in The United States. *Sociologia Ruralis*, **58**(1), 83–103.

Norgaard, K. M. (2006). "People want to protect themselves a little bit": Emotions, denial, and social movement nonparticipation. *Sociological Inquiry*, **76**(3), 372–396.

Norgaard, K. M. (2011). *Living in Denial: Climate Change, Emotions, and Everyday Life*, Cambridge, MA: MIT Press.

Oreskes, N., & Conway, E. M. (2010). *Merchants of Doubt*, New York: Bloomsbury Press.

Pollan, M. (2006). *The Omnivore's Dilemma: A Natural History of Four Meals*, New York: Penguin.

Riley, M. (2016). Still being the "Good Farmer": (Non-)retirement and the preservation of farming identities in older age. *Sociologia Ruralis*, **56**(1): 96–115.

Rosin, C. (2013). Food security and the justification of productivism in New Zealand. *Journal of Rural Studies*, **29**, 50–58.

Smith, C. (2014). *The Sacred Project of American Sociology*, New York: Oxford University Press.

Silvasti, T. (2003). The cultural model of "the Good Farmer" and the environmental question in Finland. *Agriculture and Human Values*, 20(2):143–150.

Stets, J. E., & Carter, M. J. (2012). A theory of the self for the sociology of morality. *American Sociological Review*, **77**(1), 120–140.

Stivers, Richard. (1994). *The Culture of Cynicism: American Morality in Decline*. Oxford: Blackwell.

Stock, P. V. (2007a). "Good Farmers" as reflexive producers: An examination of family organic farmers in the US Midwest. *Sociologia Ruralis*, **47**(2), 83–102.

Stock, P. V. (2007b). Katrina and anarchy: A content analysis of a new disaster myth. *Sociological Spectrum*, **27**(6), 705–726.

Stock, P. V. (2009). *The Original Green Revolution: The Catholic Worker Farms and Environmental Morality*, Fort Collins, CO: Colorado State University.

Stock, P. V. (2015). Contradictions in hope and care. In *Food Utopias: Reimagining Citizenship, Ethics and Community*, Oxon, UK: Routledge/Earthscan, pp. 171–194.

Stock, P. V., & Brickell, C. (2013). Nature's good for you: Sir Truby King, Seacliff Asylum and the greening of health care in New Zealand, 1889–1922. *Health & Place*, **22**, 107–114.

Stock, P. V., & Carolan, M. (2011). A utopian perspective on global food security. In C. J. Rosin, P. V. Stock, & H. Campbell, eds., *Food Systems Failure: The Global Food Crisis and the Future of Agriculture*, London: Earthscan, 114–128.

Stock, P. V., & Forney, J. (2014). Farmer autonomy and the farming self. *Journal of Rural Studies*, **36**(0), 160–171.

Stock, P. V., Forney, J., Emery, S. B., & Wittman, H. (2014). Neoliberal natures on the farm: Farmer autonomy and cooperation in comparative perspective. *Journal of Rural Studies*, **36**: 411–422.

Strain, C. (2004). Soul City, North Carolina: Black power, utopia, and the African American dream. *The Journal of African American History*, **89**(1), 57–74.

Stivers, Richard. (1994). *The Culture of Cynicism: American Morality in Decline*. Oxford, UK: Blackwell.

Sutherland, L.-A., & Burton, R. J. F. (2011). Good Farmers, good neighbours? The role of cultural capital in social capital development in a Scottish farming community. *Sociologia Ruralis*, **51**(3), 238–255.

Sutherland, L.-A. (2013). Can organic farmers be "good farmers"? Adding the "taste of necessity" to the conventionalization debate. *Agriculture and Human Values*, **30**(3), 429–441.

Sutherland, L.-A., & Darnhofer, I. (2012). Of organic farmers and "good farmers": Changing habitus in rural England. *Journal of Rural Studies*, **28**(3), 232–240.

Thompson, E. P. (1971). The moral economy of the English crowd in the eighteenth century. *Past and Present*, **50**, 76–136.

Thompson, P. B. (1995). *The Spirit of the Soil: Agriculture and Environmental Ethics*, New York: Routledge.

Todorov, Tzvetan. (1996). *Facing the Extreme: Moral Life in the Concentration Camps*. New York: Metropolitan.

White, R. (1996). 'Are You an Environmentalist or Do You Work for a Living?': Work and Nature. In W. Cronon, ed., *Uncommon Ground: Rethinking the Human Place in Nature*, New York: Norton, pp. 171–185.

York, R. (2015). The Science of Nature and the Nature of Science. In K. Gould & T. L. Lewis, eds., *Twenty Lessons in Environmental Sociology*, 2nd ed., Oxford, UK: Oxford University Press, pp. 95–104.

PART VII

Social Movements

26 Alternative Technologies and Emancipatory Environmental Practice

Chelsea Schelly

Introduction: Sociology's Historical Relationship with Technology

When Emile Durkheim claimed that social facts could only be explained by other social facts, he arguably set the course for an artificially constructed disciplinary silo that considered humans as embedded in cultures and social institutions that are somehow divorced from their physical and material compositions (Catton and Dunlap, 1980; Benton et al., 2001). Yet for decades, scholars working from multiple perspectives have challenged the "social determinism" of sociology's earliest foundations to demonstrate how human beings are embedded, co-constituted, and even produced by the material worlds around them (White, Rudy, and Gareau, 2015). Environmental sociologists continue to grapple with the ways in which social contexts are inextricably linked to material contexts in the form of both "natural" and "built" environments (Dunlap and Catton, 1983), including the material reality of the technological infrastructures that support human life.

Human beings are arguably defined by their use of technology to meet their needs and comforts, albeit in a wide diversity of ways across both time and space. "Technology" in the sense used here refers to the constellation of materials and material systems that humans design and employ for their usefulness. Technologies are often organized into technological systems or infrastructures, comprised of constellations of various and sometimes overlapping technological networks to provide humans with, for example, food to eat and the means to cook and store it; electricity in their homes; cars and the systems of roads, maintenance providers, and service technologies that support them; water at the turn of a tap; commercial air travel, and the Internet.

The technologies that humans employ and the infrastructural systems that humans utilize to organize technologies clearly impact the organization of other aspects of society. Technologies are sometime viewed as being the determinative base on which economic systems rely (Smith and Marx, 1994). Karl Marx has long been accused of technological determinism, with evidence provided in the form of claims that "the hand-mill gives you society with the feudal lord; the steam-mill society with the industrial capitalist" (1847). In scholarship from the field of science and technology studies, others describe technology as "co-productive" of the social order (Jasanoff,

2004), while still others describe the social construction of technology itself (Bijker, Hughes, and Pinch, 1987).

Technology is also understood as shaping both human behavior and human cognition. Martin Heidegger argued that technology is "enframing" (see Heidegger, 1977), meaning that technological systems create a frame through which humans view, perceive, interpret, and understand the world. Michel Foucault argued that technological systems operate as strategies of power to produce humans and their social groups (in the sense Foucault meant by power as productive, that power actively shapes human beings to be who they are, and to do what they do, and to think how they think, see Foucault 1980; see also Burchell, Gordon, and Miller, 1991). These understandings of technology relate to the etymological and conceptual territory of technology as technique, as a strategy of rational pursuit, suggesting that technology as technique captures both how humans organize their use of their natural world and how they perceptively approach that organization (Horkheimer and Adorno, 1972; see also Schatzberg, 2006). Emile Durkheim's nephew Marcell Mauss understood that the use of material things is socially shaped, in that even our physical, bodily, corporeal engagement with technologies, as tools, is learned through and defined by our social interactions (1973). These perspectives on technology offer depth, richness, and diversity, but they also all suggest that technological systems shape both the organization of society and the organization of the cognitive categories humans use to make sense of daily life.

Contemporary Dominant Technologies as "Monotechnics"

With the advent and then established dominance of technological systems based on fossil fuel resources and centralization with widespread distribution, technological systems arguably congealed around a few key attributes and common social consequences. Lewis Mumford characterized the technological systems that dominate in contemporary US contexts as "monotechnics," sharing authoritarian tendencies, based on domination by expertise and exclusion (Mumford, 1934). These systems are rigid, inflexible, and oriented toward concentration of power. Modern technological infrastructures like the systems of energy generation, distribution, and provision are perhaps the clearest example of monotechnics. These systems isolate users so that they are not in any meaningful way connected to the technologies that support life itself, while simultaneously making users dependent on a largely invisible sociotechnical network including the technologies themselves as well as the technical and political experts and written and unwritten rules of engagement that maintain them (Schelly, 2017).

The technologies that dominate the organization of human communities in many contexts across much of the world today can be described as examples of monotechnics; these systems are oriented toward ever-expansive growth and rely on continuous supplies of inputs from nonrenewable resources as well as technical expertise and dominance. The systems of monotechnics that provide for the very means of life, subsistence needs and comforts like shelter, food, water, heat, and light, are rigid, locked-in, and devastating for both the planet and human communities. The current means of organizing and engaging technological infrastructures to meet human needs and comforts is literally dismantling the basic climatic balance of the earth system necessary to maintain human life. The "externalities" of these systems poison water, air, soil, and humans, destroying the prospects of human health in entire communities. The organization of these systems concentrates both wealth and power into the hands of a very few technical, economic, and political elites, contributing to the imbalanced power relations that perpetuate environmental and social harms for the sake of profit. These systems are often highly inefficient, involving substantial loss of the very energetic resources employed to produce and maintain those very systems. These systems require coordination among massive political networks across geopolitical scales to ensure the subsidies, standards, and bureaucratic specializations required to maintain them. The consequences of dominant technologies can be questioned and criticized from multiple perspectives.

Alternative Technologies as "Polytechnics"

Given the wide-ranging concerns posed by these dominant technological systems, alternatives are in some cases and at some times conscientiously pursued. In contrast to monotechnics, according to Mumford, polytechnics are characterized as "broadly life-oriented" (1967, p. 9); they are flexible, adaptive, suitable to local conditions, and focused on providing for the needs of life. Adoption of alternative technologies – systems characterized as polytechnics given their flexibility, local suitability, adaptability, and above all their orientation to providing for the needs of life rather than denying the needs of life to many at for the benefit of few – comes in many forms. What is "alternative" about alternative technologies is arguably not only their scale (distributed or localized) or source of their power (often renewable energy resources) but also the orientation to life, politics, and practice that they promote.

Some social groups are at least partially defined by their stance towards technology. The Amish, for example, eschew modern technologies because of the undesirable social consequences perceived as associated with them. Other groups of less explicitly defined social actors have also articulated arguments in favor of alternative technologies as having alternative and more desirable political consequences for the organization of broader society. These alternative technologies may be used to localize, decentralize, and democratize access to resources such as energy (Winner, 1986), food (Goodman et al., 2012), and transportation (Bardhi and Eckhardt, 2012).

Dubbed the appropriate technology movement, there arose in the 1960s a coalition of voices expressing another vision for the possibilities of technological development, one focused on shifting from monotechnics to polytechnics. The appropriate technology movement took aim at development policies that limited technological assistance to monotechnical forms, such as industrialization of agriculture and extraction of fossil fuel resources. The movement argued that technologies were inherently political (see the work of Langdon Winner on the inherent politics of technologies, Winner, 1980; 1986) and that decentralized and distributed technologies based on renewable energy resources and local mechanisms of ownership and control offered more desirable political, social, and environmental consequences than the dominating monotechnics that rich nations attempted to pawn off for profit in the guise of development assistance. Also referred to with concepts like the "soft energy path" (see Amory Lovins, 1976; 1977; 1978) and "tools for conviviality" (see Ivan Illich, 1973), proponents of appropriate technology argued that small is beautiful (Schumacher, 1973), that humans can indeed flourish by sharing spaceship earth (Fuller, 1969), and that alternative technologies were associated with alternative and more desirable sociopolitical systems for the organization of human life in ways that promoted democratization, community resiliency, and environmental sustainability.

Without much success in influencing development policy, the appropriate technology movement also conjointly manifested an influence on the US counterculture movement of the 1960s. Ideas about moving "back to the land" were well suited for and congruent with promoting the use of small-scale, decentralized, and distributed technologies. These technologies most often rely on resources available at the site of use, including renewable resources like water, wind, and sunlight. The appropriate technology movement may not have had substantial influence on policies guiding international development and technology transfer, the original ambition of scholars and practitioners promoting the ideas of appropriate technology (Pursell, 1993). Yet the movement did help to establish alternative technologies as offering a clear, intended, and visual alternative to the centralized and largely fossil fuel based dominant technologies, an alternative option in both form and sociopolitical consequence.

The move to alternative technologies as promoted by the appropriate technology and back-to-the-land movements has been criticized as nothing more than a move to "building a better mousetrap" (Winner, 1986). In other words, some argue that alternative technologies reinforce the same sociopolitical consequences they seek to avoid by individualizing responsibility to avoid environmental harm (rather than building collective social movements, see Maniates, 2001) and by suggesting that technologies themselves can somehow save society from itself. However, these critiques are challenged in the face of new ideas from environmental social theory suggesting that human engagement with technology must be viewed in terms of its consequences for shaping cognitive categories and internalized values via new forms of social practice. Thus, the forms of engagement involved in using alternative technologies to support human needs and comforts may facilitate alternative ways of thinking, and alternative modes of thought may themselves precipitate emancipatory social change.

Alternative Technologies as Emancipatory Practice

Alternative technologies based on decentralized forms of organization and distribution are often viewed as environmentally responsible technologies, given that they often involve use of renewable resources and result in lessened environmental damages when compared to dominant, conventional, fossil fuel-based options. These kinds of technologies, given that they are not dominant but are in fact alternatives that have to be more actively, consciously sought, are often presumed to be used by people who are motivated by environmental concerns and who have heightened environmental values. A vast body of research in environmental sociology has been dedicated to examining the relationship between environmental values and environmentally responsible behavior, without much success in actually predicting or promoting environmentally responsible behaviors (Heberlein, 2012). Despite a range of theoretical models and presumed predictors (see, for example, Schwartz, 1973; Bandura, 1977; Ajzen, 1991; Bamberg and Möser, 2007), these perspectives share a common foundational presumption that attitudes and values predicate and can be used to predict action. This presumption, while extremely influential for shaping scholarship and policy, is arguably fundamentally flawed (Shove, 2010).

Instead of viewing human behavior as premeditated and determined by human values, theories of social practice instead argue that behavior is best understood as largely habitual, unthinking, and contextualized based on existing social and material conditions (Reckwitz, 2002). Social practices involve commonly shared patterns of behavior that also unconsciously reinforce behavioral norms; in other words, we humans use technology in regularized and patterned ways, these patterns are common and shared among social groups, and these patterns involve both shared action and shared thought about normal and expected action (Spaargaren, 2003; 2011).

Based on the scholarship of thinkers like Michel Foucault and Pierre Bourdieu (1977) and arguably consistent with a pragmatist theory of action (see Dewey, 1910; 1929), theories of social practice argue that conscious thought (about beliefs, values, or attitudes) does not necessarily predicate action and that action can most accurately be understood as patterned, habitual, and embedded in social and material contexts that constrain both what humans do and how humans think about what they do. Scholarship on social practice offers new insight into what limits engagement with more environmentally sustainable technological arrangements (Kennedy, Cohen, and Krogman, 2016) and how existing material systems constrain the contexts in which humans can consider the possibilities for creative, emancipatory action.

Based on a practices theory perspective, alternative technologies that involve decentralized and distributed systems and that rely on renewable resources for meeting human needs and comforts have consequences beyond their impacts on environmental or economic systems. Alternative technologies also have the capacity to reshape social thought as a result of reorganizing patterns of social practice. Because alternative technologies change how humans engage with the material world, they can also change how humans think about engagement with both ecological and social systems. It is because of this capacity to change behavior and as a consequence change thought that alternative technologies can be viewed as radically emancipatory.

From this perspective, behavior is viewed as more typically resulting from the unconscious enactment of habits that illustrate and reinforce rather that being the result of values. Further, behavior is both constrained and enabled by built environments, policy contexts, and lifestyle patterns as well as any stated value preference. This perspective requires a fundamental shift in how sociologists, including environmental sociologists, conceptualize the link between thought and action. In a field that has dedicated substantial time and energy to linking existing attitudes, values, and beliefs to rationally considered action as well as to understanding how to change attitudes in order to change behaviors, this shift is fundamental. It requires foundational changes to how scholars theorize and study behavior (Schelly, 2016).

This perspective also suggests that alternative technologies may be most accurately defined as those that require alternative or changed practice. Using Howard Becker's trick to utilize the case to define the concept (Becker, 1998), alternative technology can be defined based on how it reorients practice relative to the practice consequences of currently dominating technological systems. Dominant technological systems encourage users to adopt relatively unthinking patterns of engagement in which they are dependent on experts and monetary payment for service provision; these technologies suggest that users are consumers but not decision makers and certainly not creators. Alternative technological systems, on the other hand, facilitate engaged practice in which users are actively involved in the creation and maintenance of the technological systems that support everyday life and are physically and ideologically engaged as citizens rather than merely consumers (Meyer, 2015).

Consider, for example, the technological infrastructures that comprise electricity generation and distribution in many social contexts, including the contemporary United States. Electric energy is generated using predominantly fossil fuel-based resources, is generated in large scale and centralized facilities, and is then distributed via a complex system of transmission. The infrastructures involved are often owned by large corporate entities with very little direct or personal connection to the end users of the resource.

This system can and does integrate renewable energy resources into the generation mix, without requiring any fundamental change to the scales or forms of distribution, transmission, or ownership. While utility scale installations of solar PV and wind powered electricity generation are increasing in prevalence, they do not in themselves challenge the dominant model of electrical energy service provision via corporate and technical elites nor do they require changes in electricity use practice on behalf of the user. Thus, while solar PV systems and wind turbines are certainly examples of renewable energy technology and are certainly valuable contributions to the existing electrical energy technological infrastructure, they are arguably not examples of alternative technologies, as defined herein.

Alternative technologies, in contrast, facilitate alternative practice. The dominant technologies used for electrical energy provision in the United States require economic dependence, in that users must pay a monthly bill in exchange for use of electricity energy services, every month for the rest of their lives. Yet they do not require attentiveness to energy resources or energy supply; users can flip a switch and expect the lights to come on regardless of the weather or the time of day. Alternative technologies are those that change how people act, and in turn, how they think about that action. Thus, alternative energy technologies are those that require awareness of energy resources and energy usage and that may require changes in energy use practice, in exchange for changes in the forms of dependence experienced by users (who are now dependent on the availability of electricity rather than the availability of money to pay for electricity).

Residential solar technology adopters, who install solar electric technology on their individual homes, do not necessarily need to change their energy use practices; if they remain connected to the electric utility grid, they will continue to be provided with an uninterrupted supply of energy regardless of whether the sun is shining (they also will lose access to electricity if the larger grid infrastructure fails, despite having the technology to produce electricity at the site of use). Yet residential solar technology adopters do change their energy use practices after installation; many discuss incessantly checking their power production via their meter or an Internet application that allows them to track electricity production and consumption in the home. Many also discuss becoming more interested in energy efficiency updates in their home only after installing their solar system and having heightened engagement with their home's energy use (Schelly, 2014a; 2014b). A model of human cognition based on a presumption of rational calculation taking place prior to decisive action cannot explain the reality of changed practice and changed priorities only after technological change occurs.

This conceptualization of alternative technology as changing social practice and as having the potential to create emancipatory practice outcomes applies to myriad examples. One of the largest challenges in localizing food systems so that more people are growing, buying, and eating fresh, seasonal produce is the simple fact that most people have very little experience with preparing or eating such foods. In many a local food movement group, members explain that it is only after exposure to fresh broccoli or Brussels sprouts that people who grew up eating the frozen foodstuffs version of these foods realize they like them. It sometimes takes direct exposure to the technological alternatives to realize the value of their pursuit. This suggests a real and significant direction of policy making, which can be used to either limit or to maximize human experiences with alternative technologies designed for life and liberation.

The kinds of alternative technologies that shift how users engage with technologies and how they think about technological engagement, often involve changing both organizational scale and forms of ownership. Technologies that are decentralized and based on smaller scales of organization and control are arguably more congruent with alternative but more desirable forms of social organization and more democratic sociopolitical consequences (Lovins, 1976). These alternative politics, in addition to the alternative forms of practice, are arguably essential in defining alternative technologies and identifying their potential.

Yet this does not mean that all technologies should be owned, controlled, and maintained at the individual level, which those committed to technological efficiency narrowly defined would recognize as highly inefficient while those committed to challenging the ideology and consequences of neoliberalism would recognize as problematically individualized. In fact, many forms of alternative technology are most effectively organized at the level of a community collective

(Schelly, 2017), suggesting that the most promising alternative technologies are not those pursued in individual isolation but rather those that encourage, facilitate, or even require cooperation and communication, other forms of alternative practice that are stifled by currently dominant technological systems.

Alternative Technologies as Emancipatory Social Movements

Thus, alternative technologies are those that challenge the sociopolitical consequences of as well as the dominant patterns of engagement with dominant technological systems. If the dominant technological systems in contemporary societies are characterized by sociopolitical consequences of dependence and isolation (Schelly, 2017), alternative technologies are arguably characterized by congruence and connection; they require that users live in congruence with the limits of the natural world but also in connection with it and with other users. Alternative technologies also require users to shift technological practice so that they are more actively engaged with the ecological systems that ultimately support human existence.

It is in this shift that alternative technologies have the most emancipatory potential. By changing how humans engage with the world, they also have the potential to change how humans think about engagement. Because all technology involves bodily engagement, and categories of thought are reshaped through this corporeal experience, alternative technologies that involve different physical patterns of practice also create different ideas about the value and potential of engaged, emancipatory social practice (Schelly, 2018). Thus, the very process of adopting emancipatory practices via alternative technology use opens up opportunities for future adoption of new technologies, new practices, and new categories of thought about what is humanly desirable and possible. Alternative technologies also change the scales and forms of technological organization in ways that challenge political, economic, and ideological domination and encourage collective cooperation.

Adoption of alternative technologies can be dismissed as involving an apolitical and individualized focus on a technological fix to social problems. Certainly, some arguments regarding the role of technology in society are narrowly and erroneously defined (for example, see Asafu-Adjaye et al., 2015). Yet decades of quality scholarship in environmental sociology and related fields has convincingly illustrated that the consequences of technological systems are not limited to their impacts of environmental resources or ecological systems. Technologies are indeed political, in that they are highly consequential for the distribution of resources, including economic and monetary resources but also knowledge resources and the distribution of power and control.

The technological systems that dominate in contemporary societies are destroying environmental resources from individual species to the entire global climate while simultaneously exacerbating corporate domination, extreme poverty, and lack of meaningful social engagement with the very systems that support human life on earth. Alternative technological systems are those that challenge these sociopolitical consequences. They do so by changing the political consequences of technology through decentralization and descaled systems of ownership and control. They also do so by changing how humans physically engage with the technological systems that provide for human needs and comforts, because these forms of engagement have the potential to consequentially change how humans think about their relationship to technologies, the environment, and one another. Thus, adoption of alternative technologies as form of emancipatory mobilization should not be dismissed as individual, apolitical, or narrowly oriented to a technical fix. Rather, alternative technologies support alternative practices and alternative politics, both of which involve emancipatory consequences for human freedoms and for human care, including care for the natural systems that support life on earth as well as care for our fellow species.

Arguments for alternative technologies that operate in congruence with more democratic, participatory, and emancipatory politics certainly are not new, having been articulated by Lewis Mumford starting in the 1930s and popularized in the appropriate technology movement of the 1960s. These articulations recognize that social systems are fundamentally constituted by the ways humans use and organize technological systems to meet their needs and comforts. Further, scholars have argued for decades that currently dominating forms of technology are exploitative of both humans and nature while simultaneously suggesting that the only meaningful role for most humans when it comes to technological engagement is as a consumer, paying for services over which one has very little influence or control.

Alternative technologies have the possibility of opening up alternative consequences for both sociopolitical arrangement and the constitution of social practice. In order for this to occur, environmental sociologists must at the very least take the material organization of society seriously as an analytically important explanatory context for understanding the social world. Scholarship must also contend with the very real challenge provoked by theories of social practices and reshape their questions and their methods to examine the extent to which human behavior involves linked processes of action and cognition rather than rationally calculated motivations based on predetermined or set value systems. Finally, environmental sociology as a field can contribute to the development of emancipatory environmental practice by defining alternative technologies in terms of their emancipatory potential and by interrogating social realities that contribute to or inhibit the realization of that potential.

While there is a robust literature on social movements aimed at shaping technological choices (Hess, 2005; 2007), it is imperative for scholarship to also recognize engagement with alternative technologies as a form of social movement participation (Rubin, 2018). Alternative technological systems that localize control of resources needed to sustain human life may even play a significant role in social movements aiming to confront recent surges in rural authoritarian populism (Scoones et al., 2018). Yet there is still much work to be done to connect technological change and emancipatory practice to sites of resistance and productive change for societal transformation (Temper et al., 2018).

By engaging with alternative technologies as a form of social movement participation, scholars and activists alike are arguably engaging in making a claim, a claim about the emancipatory potential of alternative technologies (Rose, 2012; Schelly, 2014c; Rose, 2018). Perhaps most importantly, scholars and practitioners must also themselves engage with alternatives in order to fully appreciate the role of technological systems in shaping the emancipatory potential of environmental social practice. Through seemingly simple acts like growing a garden or riding a bike in lieu of using a car, or through more committed acts like living with renewable electricity generation or shifting towards more communal forms of technological ownership and use, sociologists can both better understand the role of technology in shaping cognition and practice as well as contributing to the changes necessary to utilize technology for emancipatory social practice.

References

Ajzen, I. (1991). The theory of planned behavior. *Organizational Behavior and Human Decision Processes* 50, 179–211.

Asafu-Adjaye, J., Blomquist, L., Brand, S. et al. (2015). An ecomodernist manifesto. *The Breakthrough Institute*. www.ecomodernism.org/ accessed January 1, 2018.

Bamberg, S., and G. Möser. (2007). Twenty years after Hines, Hungerford, and Tomera: A new meta-analysis of psycho-social determinants of pro-environmental behavior. *Journal of Environmental Psychology* 27, 14–25.

Bandura, A. (1977). Self-efficacy: Toward a unifying theory of behavioral change. *Psychological Review* 84, 191–215.

Bardhi, F., & Eckhardt, G. M. (2012). Access-based consumption: The case of car sharing. *Journal of Consumer Research*, 39(4), 881–898.

Becker, H. S. (1998). *Tricks of the Trade: How to Think about your Research while You're Doing it*. Chicago, IL: University of Chicago Press.

Benton, T., Buttel, F., Catton Jr., W.R., (2001). *Sociological Theory and the Environment: Classical Foundations, Contemporary Insights*. Lanham, MD: Rowman & Littlefield Publishers.

Bijker, W. E., Hughes, T. P. and Pinch, T. J., eds. (1987). *The Social Construction of Technological Systems: New Directions in the Sociology and History of Technology*. Cambridge, MA: MIT Press.

Bourdieu, Pierre. (1977). *Outline of a Theory of Practice*. Cambridge, MA: Cambridge University Press.

Burchell, G., Gordon, C., and Miller, P. eds. (1991). *The Foucault Effect: Studies in Governmentality*. Chicago, IL: University of Chicago Press.

Catton Jr., W. R., & Dunlap, R. E. (1980). A new ecological paradigm for post-exuberant sociology. *American Behavioral Scientist*, 24(1), 15–47

Dewey, J. (1910). *How we Think*. London: Heath & Co.

Dewey, J. (1929). *The Quest for Certainty: A Study of the Relation of Knowledge and Action*. New York: Minton, Balch & Company.

Dunlap, R. E. and Catton Jr., W. R., (1983). What environmental sociologists have in common (whether concerned with "built" or "natural" environments). *Sociological Inquiry*, 53(2–3), 113–135.

Foucault, M. (1980). *Power/Knowledge: Selected Interviews and Other Writings, 1972–1977*, edited by Colin Gordon. New York: Pantheon Books.
Fuller, B. R. (1969). *Operating Manual for Spaceship Earth*. Carbondale, IL: Southern Illinois University Press.
Goodman, D., DuPuis, E. M., & Goodman, M. K. (2012). *Alternative Food Networks: Knowledge, Practice, and Politics*. London: Routledge.
Heberlein, Thomas A. (2012). *Navigating Environmental Attitudes*. New York: Oxford University Press.
Heidegger, Martin. (1977). *Questions Concerning Technology and Other Essays*. New York: Harper & Row.
Hess, D. J. (2005). Technology-and product-oriented movements: Approximating social movement studies and science and technology studies. *Science, Technology, & Human Values*, 30(4), 515–535.
Hess, D. J. (2007). *Alternative Pathways in Science and Industry: Activism, Innovation, and the Environment in an Era of Globalization*. Cambridge, MA: MIT Press
Horkheimer, M., and Adorno, T. W. (1972). *The Dialectic of Enlightenment*. New York: Herder and Herder.
Illich, I.. (1973). *Tools for Conviviality.* New York: Harper & Row.
Jasanoff, S., ed. (2004). *States of Knowledge: The Co-Production of Science and Social Order*. London: Routledge.
Kennedy, E. H., Cohen, M. J., and Krogman, N. (2016). *Putting Sustainability into Practice: Advances and Applications of Social Practice Theories*, Cheltenham, UK: Edward Elgar.
Lovins, Amory. (1976). Energy strategy: The road not taken? *Foreign Affairs* 55, 65–96.
Lovins, Amory. (1977). *Soft Energy Paths: Toward a Durable Peace*. New York: Harper & Row.
Lovins, Amory. (1978). Soft energy technologies. *Annual Review of Energy* 3, 477–517.
Maniates, M. (2001). Individualization: Plant a tree, buy a bike, save the world? *Global Environmental Politics* 1, 31–52.
Marx, K. (1847). *The Poverty of Philosophy*. London, Martin Lawrence Limited.
Mauss, M. (1973). Techniques of the body. *Economy & Society* 2, 70–88.
Meyer, J. (2015). *Engaging the Everyday: Environmental Social Criticism and the Resonance Dilemma*. MIT Press.
Mumford, Lewis. (1934). *Technics and Civilization*. New York: Harcourt, Brace and Company.
Mumford, Lewis. (1967). *The Myth of the Machine*. New York: Harcourt Brace & World.
Pursell, Carroll. (1993). The rise and fall of the appropriate technology movement in the United States, 1965–1985. *Technology and Culture* 34, 629–637.
Reckwitz, Andreas. (2002). Toward a theory of social practice: A development in culturalist Thinking. *European Journal of Social Theory* 5, 243–263.
Rose, M. (2012). Dwelling as marking and claiming. *Environment and Planning D: Society and Space* 30(5), 757–771.
Rose, M. (2018). Consciousness as claiming: Practice and habit in an enigmatic world. *Environment and Planning D: Society and Space* Online First. DOI:10.1177/0263775818784754
Rubin, Z. (2018). My year pooping in a bucket: Lifestyle, cultural, and social movements in the "Node" at Dancing Rabbit Ecovillage. PhD, Sociology, University of Missouri, Columbia, Missouri.

Schatzberg, E. (2006). "Technik" Comes to America: Changing meanings of "technology" before 1930. *Technology and Culture*, 47(3), 486–512.

Schelly, C. (2014a). Residential solar electricity adoption: What motivates, and what matters? A Case Study of Early Adopters. *Energy Research and Social Science* 2, 183–191.

Schelly, C. (2014b). Transitioning to Renewable Sources of Electricity: Motivations, Policy, and Potential. Pages 62–72 in *Controversies in Science and Technology, Volume 4*. Edited by Daniel Lee Kleinman, Karen Cloud-Hansen, and Jo Handelsman. New York: Oxford University Press.

Schelly, C. (2014c). Are residential dwellers marking and claiming? Applying the concepts to humans who dwell differently. *Environment and Planning D: Society and Space, 32* (4), 672–688.

Schelly, C. (2016). Understanding energy practices. *Society & Natural Resources* 29(6), 744–749. http://dx.doi.org/10.1080/08941920.2015.1089613

Schelly, C. (2017). *Dwelling in Resistance: Living with Alternative Technologies in America*. New Brunswick, NJ: Rutgers University Press.

Schelly, C. (2018). Bringing the body into environmental behavior: The corporeal element of social practice and behavioral change. *Human Ecology Review* 24 (1) 137–154.

Schumacher, E.F. (1973). *Small is Beautiful: Economics as if People Mattered*. New York: Perennial.

Schwartz, Shalom H. (1973). Normative explanations of helping behavior: A critique, proposal and empirical test. *Journal of Experimental Social Psychology* 9, 349–364.

Scoones, I., Edelman, M., Borras Jr., S. M. et al. (2018). Emancipatory rural politics: confronting authoritarian populism. *The Journal of Peasant Studies*, 45(1), 1–20.

Shove, Elizabeth. (2010). Beyond the ABC: Climate change policy and theories of social change. *Environment and Planning A* 42, 1273–1285.

Smith, M.R. and Marx, L., eds. (1994). *Does Technology Drive History? The Dilemmas of Technological Determinism*. Cambridge, MA: MIT Press.

Spaargaren, Gert. (2003). Sustainable consumption: A theoretical and environmental policy perspective. *Society & Natural Resources* 16, 687–701.

Spaargaren, Gert. (2011). Theories of practices: Agency, technology, and culture: Exploring the relevance of practice theories for the governance of sustainable consumption practices in the new world-order. *Global Environmental Change* 21, 813–822.

Temper, L., Walter, M., Rodriguez, I., Kothari, A., & Turhan, E. (2018). A perspective on radical transformations to sustainability: resistances, movements and alternatives. *Sustainability Science*, 13(3), 747–764

White, D., Rudy, A., & Gareau, B. (2015). *Environments, Natures and Social Theory: Towards a Critical Hybridity*. New York: Palgrave Macmillan.

Winner, Langdon. (1980). Do artifacts have politics? *Daedalus* 109, 121–136.

Winner, Langdon. (1986). *The Whale and the Reactor: A Search for Limits in the Age of High Technology*. Chicago, IL: University of Chicago Press.

27 The Global Fair Trade Movement: For Whom, By Whom, How, and What Next

Elizabeth A. Bennett

Acronyms

AJP – Agricultural Justice Project
ATO – Alternative Trade Organization
CIW – Coalition of Immokalee Workers
DFTA – Domestic Fair Trade Association
FI – Fairtrade International
FTAO – Fair Trade Advocacy Organization
FTF – Fair Trade Federation
FTIS – Fair Trade International Symposium
NGO – Non-Governmental Organization
SPP – Símbolo de los Pequeños Productores (Small Producers' Symbol)
WFTO – World Fair Trade Organization
WDSR – Worker-Driven Social Responsibility

1 Introduction

"Fair trade" is a vision for a world in which "justice and sustainable development are at the heart of trade structures and practices." The goal is for "everyone, through their work, to maintain a decent and dignified livelihood and develop their full human potential" (FI & WFTO 2009). The global fair trade movement is comprised of non-governmental organizations (NGOs), businesses, social enterprises, communities, faith groups, consumer advocacy networks, and others committed to this idea. Fair trade advocates focus on the ways in which market transactions impact groups that are vulnerable to marginalization (exclusion from the market), oppression (limited opportunities and/or freedoms), and/or exploitation (over-work and/or under-compensation) (Raynolds and Bennett, 2015).

This chapter aims to provide a snapshot of the current fair trade movement by addressing four provocative questions: 1) Fair trade for whom? 2) Fair trade by whom? 3) How may fair trade labeling and certification support these goals (or not)?

and 4) What next for fair trade's approach to capitalism and the state? For new readers, the chapter provides an introduction to fair trade that focuses primarily on the current moment. For readers more familiar with fair trade, it offers an updated summary of key debates in the field, drawing heavily on literature published in the last five years.[1] This chapter also highlights linkages between fair trade and the environment. It describes the ways in which fair trade can support environmental conservation, raises questions about climate justice, and shows how certification programs can pit people against the planet, rather than supporting both. The following section provides basic background information on the fair trade movement.

2 Historic and Contemporary Fair Trade

Consumers have long used their purchasing power as a tool for social change (Boström, Micheletti, & Oosterveer 2019). In the 1820s, for example, US Quakers and free Black abolitionists spearheaded the "free produce" movement and promoted cotton, clothing, sugar without slave labor to support the abolition movement (Brown 2015). In the mid-twentieth century, several initiatives emerged in the US and Europe to empower economically disadvantaged groups though direct trade relationships based on trust and equity. For example, in the 1940s, an American Mennonite nun imported and sold Puerto Rican artisans' handicrafts without keeping a share of the profits. Likewise, European groups imported and sold handicrafts from communities affected by war or experiencing poverty (Anderson 2015; Brown 2015). These initiatives are often identified as the origins of the contemporary fair trade movement (van Dam 2015). In the 1950s, 60s, and 70s, these small-scale practices expanded and extended to new regions and products. "Alternative trade organizations" (ATOs) such as Ten Thousand Villages published mail-order catalogues and opened brick-and mortar "world shops" to bolster sales of fairly traded goods (Bennett 2012a).

In the late 1980s a group of fair trade organizers decided to increase fair trade sales by collaborating with conventional brands and retailers, a practice called "mainstreaming." Working with businesses *not* oriented around fairness was (and still is) very contentious within the movement. Supporters of this tactic developed a label that could be used by mainstream brands (e.g., well known chocolatiers or coffee roasters) to market a line of their products as "fair trade." These products would then be sold in conventional supermarkets, as opposed to world shops. By the late 1990s several fair trade labels had emerged, and in 1997 they united to form Fairtrade International (FI, formerly Fairtrade Labelling Organization, or FLO). Today, FI is the largest fair trade certification program in the world, though several others have

---

1 This chapter synthesizes analysis from my own scholarship and academic engagement in the field, my co-edited *Handbook of Research on Fair Trade* (Raynolds & Bennett 2015), a review (in early 2018) of nearly 700 books and articles on fair trade, most published in the previous five years, and participation in the two most recent Fair Trade International Symposium (FTIS) meetings. Those interested in learning more may wish to explore the Fair Trade Institute website (/www.fairtrade-institute.org) or participate in the next Fair Trade International Symposium (FTIS) (www.fairtradeinternationalsymposium.org/ftis2018).

emerged (Bennett, 2012a). Fair trade certification programs include not only social criteria but also environmental standards, as discussed in the following section.

Today, numerous and diverse market actors identify their work as "fair trade." Some fair trade businesses commit to fully integrating fairness into their business model, while others simply include fair trade certified products in their supply chain. Some fair trade consumers contribute to the movement by educating their communities through the Fairtrade Towns programs, while others simply "vote with their dollar" (Samuel, Peattie, & Doherty 2018). Surveys and market studies show that consumers are willing to pay more for fairly traded products, but the amounts vary by consumer demographic and type of product (GlobeScan, 2016; Marconi, Hooker & DiMarcello 2017). The global fair trade movement is diverse, decentralized, contentious, and rapidly expanding. The following sections highlight today's most salient debates: For whom? By whom? How? and What next?

3 Fair Trade for Whom?

Ethical consumerism, as a tool, can be leveraged to benefit any community or group.[2] From the 1940s to 1990s, "fair trade" and "alternative trade" typically referred to market interventions and consumer practices in the Global North[3] that supported people in the Global South (or refugees in the North) who were experiencing economic adversity and related challenges (such as lack of access to education or health care).[4] Fair trade advocates often educated themselves and their communities about the ways in which international trade agreements benefitted the Global North and the legacies of colonialism, more broadly. They used the expression "trade not aid" to express the idea that trade justice would be more successful in empowering citizens of the Global South than charity, foreign aid, or other types of development programs. In the 1990s, as globalization transformed labor dynamics worldwide, the concept of "fair labor" emerged to recognize that issues of fairness are not limited to international trade, but also occur within domestic markets.

Today, the question "Fair trade for whom?" is highly contentious and often contended. This section highlights four debates within this conversation. First, fair trade was initiated to support smallholder farmers and artisan cooperatives – should it expand to support workers on large plantations? Second, fair trade was initiated as a way for groups in wealthier countries to support communities vulnerable to exploitation, oppression, marginalization, or poverty in poorer countries – should it expand to support struggling populations in the Global North? Third, can the fair trade concept be expanded to support the world's most vulnerable populations, such

---

2 As Stolle and Huissoud (2019), Micheletti and Oral (2019), and Lekakis (2019) illustrate, ethical consumerism can be used to advance anti-democratic, racist, or nationalist causes just as easily as the progressive agenda with which it is typically associated.
3 Note that here "North" refers to developed, wealthy, high income, or "First World" countries, as opposed to the geographic region of the Northern Hemisphere.
4 See Naylor (2014), p. 275, on the concept of "Global North" and "Global South" in the context of fair trade.

as migrants, women, homeworkers, and refugees? Finally, to what extent can and should fair trade be used to pursue environmental objectives?

3.1 Which Groups in the Global South?

In the twentieth century, fair and alternative trade efforts typically targeted smallholder farmers (farmers who own and farm their own small plot of land) and small-scale artisans. Fair trade advocates sought out existing coops and helped form new cooperative associations. The idea was that by strengthening grassroots organizations, supporting resistance against land-grabs, and promoting political solidarity networks, fair trade could not only provide market access but also support broader social change. Although fair trade has historically aimed to support smallholder farmers, in some contexts it has been extended to people without land who are working on large plantations, such as tea pluckers in India, who have participated in fair trade since the 1980s (see Besky 2015).

As fair trade expands to new products one debate is whether the movement should target smallholder farmers or (landless) workers hired to work on large estates. Supporters of estate certification argue that fair trade should target the most vulnerable populations (e.g., landless laborers), that some corporations will only buy fair trade from estates (so certifying them does not create competition with smallholders), and that including workers allows the benefits of fair trade to reach more people. Those who oppose certifying estates argue that Fairtrade certification may not be as effective or transformative for hired laborers as it is for smallholder producers, which has been the case in tea, for example. They also argue that supporting smallholders can catalyze rural social movements, slow urban migration, and support sustainable agriculture. The debate about whether or not fair trade should be extended to coffee estates became heated in 2012. Ultimately, FI decided that for coffee it would only certify smallholder farmer cooperatives – not estates. In response, one board member withdrew from the organization and established a new fair trade label, FairTrade USA, committed to supporting hired laborers on coffee plantations (see Cole & Brown 2014; Raynolds 2014; Bennett 2016; Linton & Rosty 2015; Jaffee & Howard 2016; Valiente-Riedl 2016).

3.2 Vulnerable Groups in the Global North?

As unionists and social justice advocates have long pointed out, exploitation, oppression, and marginalization are not limited to the Global South. In the 1990s globalization and emerging organic certification programs increased attention to labor exploitation in the Global North. Simultaneously, Fairtrade International's certification program gained worldwide recognition. In this context, several "fair labor" and domestic fair trade groups have emerged in the North. Supporters of fair trade production in the Global North argue it is appropriate to leverage public awareness of international fair trade to draw attention to labor issues in all countries. Opponents argue that fair trade should continue in its tradition of highlighting *international* drivers of *global* inequalities and addressing colonial legacies. Some also suggest that a North/South model of fair trade may wrongly reify, stereotype, or normalize differences between North and South instead of closing the opportunity gap between them (Leissle 2012; Naylor 2014). Further complicating this debate is the emergence of South–South fair trade, which fair trade items produced in the South are also sold in the South (and not exclusively to international tourist populations) (Doherty, Smith, & Parker 2015). Today, much of the movement remains focused on the South, but in solidarity with efforts to benefit people in the North (see Naylor 2014; Brown & Getz 2015; Howard & Allen 2016). In 2009, the WFTO and Fairtrade International issued a joint statement that emphasized this point:

> Fair Trade is a trading partnership, based on dialogue, transparency and respect, that seeks greater equity **in international trade**. It contributes to sustainable development by offering better trading conditions to, and securing the rights of, marginalized producers and workers – **especially in the South**. (FI & WFTO 2009, emphasis added)

In the United States, the Domestic Fair Trade Association (DFTA), Agricultural Justice Project (AJP), and Worker-Driven Social Responsibility (WDSR) Network are committed to extending fair trade concepts to vulnerable workers within the country.

3.3 Excluded Populations

Studies suggest that fair trade initiatives, because of both which groups they have targeted and how they have been organized, have excluded some of the most marginalized populations including migrant workers, unorganized smallholder farmers, home workers, and processors of agricultural products (e.g., Cramer et al. 2014; Loconto & Dankers 2014; Pinto et al. 2014). Although some fair trade initiatives specifically aim to support women – with varying levels of success – such as Café Femenino (Alegría 2016), others have failed to address barriers to women's leadership and participation (Smith, Kuruganti & Gema 2015). Similarly, despite fair trade's historic and ongoing work to prevent the necessity of economic migration (Lewis & Runsten 2008) and support refugees (Blanchard & Mackey 2018), few initiatives focus on undocumented workers without asylum status.

3.4 Fair Trade and the Environment

The question of whether and how fair trade could, should, or must address environmental issues is also contentious. Although transportation, processing, packaging, and retail space can generate adverse environmental consequences, the fair trade movement has largely focused on environmental issues at the farm level. Some argue fair trade producers should be *required* to engage in environmentally conscientious practices, such as organic farming, as doing so may promote long-term sustainability, increase income, and attract consumers. Opponents, however, interpret this as a form paternalism, hypocrisy, and neocolonial control. They highlight the ways in which overconsumption and industrialization in the North have generated environmental problems and suggest marginalized groups in the South should not be required to address the consequences. This debate is further complicated by climate change, however, as commodity producers face increased vulnerability and risk (see Noble 2017; Jaffee & Howard 2010). Fairtrade International's environmental standards prohibit many agricultural inputs and provide price incentives to comply with organic standards. It is challenging to parse out whether and how these standards affect producers' practices and participation in fair trade. Additionally, the relationship between fair trade and environmental outcomes may differ by region, year, and sector. Yet, a study of coffee from Latin America and the Caribbean may be illustrative, as coffee is the greatest fair trade product (by sales) and 80 percent of the world's fair trade coffee comes from the region. The findings suggest that fair trade can *directly* benefit the environment by reducing agrichemical use and increasing adoption of organic methods. *Indirectly*, fair trade can support biodiversity, water conservation, nutrient cycling, erosion control, pollination, and pest control (Bacon, Rice, & Maryanski 2015). Despite these potential benefits, some argue that the fair trade model – even with strict environmental requirements – fails to address environmental issues related to monocrop farming, especially when applied to large cash crop estates (Jaffee & Howard 2010). While fair trade initiatives generally aim to prioritize their commitments to fairness with environmental concerns, eco-labels and sustainability certifications tend to put environment first. This tension (between fair trade and sustainability labeling) can pit people against the environment, instead of finding ways to support both (Bennett 2018).

This section highlighted several debates about "who counts" as a legitimate beneficiary of fair trade. They point to the challenge of whether and how to privilege one group over another (Besky 2015; Jaffee & Howard 2016). At its best, these conversations are about collaboration among actors with different allegiances. At worst, it might be considered a form of biopolitical sorting, suggesting that some bodies count more than others, and are thus more worthy of empowerment and economic justice (Guthman & Brown 2016).

4 Fair Trade by Whom?

From the 1940s to the late 1980s, fair and alternative trade initiatives were typically organized as partnerships between 100 percent committed companies or NGOs (in the North) and producers of fair trade goods (in the South). Organizers in the North often described their role as "empowering" producers to increase their agency and shape their own lives and livelihoods. For many, this is what distinguished fair trade from charity. Since the 1990s, other types of actors have been facilitating fair trade in new and different ways. Both then and now, the politics of *who* decides *what* is fair are not only fraught with controversy but also of serious consequence – empowerment and development programs are much more effective when the target beneficiaries are in charge (Bacon 2010; Koenig-Archibugi 2017; Raynolds 2017). This section examines who is organizing fair trade and to what extent they have succeeded at including producers in leadership – a form of empowerment in itself.

4.1 Fair Trade Authorities

Fairtrade International and the World Fair Trade Organization (WFTO, formerly IFAT) are widely recognized as the fair trade movement's most powerful organizations, though their perspectives are not necessarily representative of the broader movement (Bennett 2012b). Fairtrade International, the world's largest certification organization, is led by a Board of Directors and Membership Assembly. Votes are evenly shared between the producers of fair trade goods and the NGOs responsible for marketing them (Bennett 2016).[5] Some critics suggest that this model still limits producer empowerment because corporations can influence standards (both through their purchasing power and direct participation in the standards-setting process) (Jaffee 2014), producers lack sufficient resources to facilitate democratic leadership among themselves (Sutton 2013), and dialogue with unions remains limited (Stevis 2015). Compared to other large, global standards-setting organizations, however, it is much more inclusive of the target beneficiaries (Bennett 2017).

The WFTO is a membership for seriously committed fair trade businesses. It self-identifies as the "global authority on" and "guardian of" fair trade values and principles (WFTO 2018). WFTO's president identifies wealth distribution as the most critical problem in the world and argues that conventional business models are hardwired to exacerbate inequality (Dalvai 2018). Its 400 members, located in seventy-five countries, are divided into five world regions (Pacific/North America, Asia/Africa, Europe, Latin America, and Africa/Middle East), each of which has its own governance system and is represented on the board of directors. Membership is contingent on passing through a process of self-assessment (every 2–3 years), monitoring audits (every 2–6 years), and peer visits (every 2–6 years). An online participatory monitoring mechanism invites both the members and the public to submit statements of concern about members' practices (Davenport & Low 2013).

4.2 Companies and Brands

Today, many companies, brands, and social enterprises identify as "100% fair trade." Fair trade principles are at the heart of their business models. Divine Chocolate, for example, is a UK-based chocolatier that is 50 percent owned by the Ghanaian cocoa producer organization Kuapa Kokoo. Equal Exchange, a US-based worker-owned cooperative, fosters fair trade relationships with the producers of the products it sources. Individuals who develop fair trade businesses are often driven by values, altruism, and an impactful experience relevant to fair trade (Brown 2015). The alternative trade models they create can go further toward empowering producers than simply trading certified products (Doherty & Huybrechts 2013; Cater, Collins & Beal 2017). Fair trade-oriented companies can apply to join the WFTO. In the United States and Canada,

---

[5] Although it was not always this way – producers were sometimes excluded from leadership because the marketing organizations thought it would increase the certification's credibility with consumers (see Bennett 2016).

companies are also eligible to join the Fair Trade Federation, a membership organization of over 250 fair trade companies that:

1) Create opportunities for economically and socially marginalized producers;
2) Develop transparent and accountable relationships;
3) Build capacity;
4) Promote fair trade;
5) Pay promptly and fairly;
6) Support safe and empowering working conditions;
7) Ensure the rights of children;
8) Cultivate environmental stewardship; and
9) Respect cultural identity. (FTF 2018)

Companies that are not fully oriented around fair trade (and would thus not be eligible for Fair Trade Federation membership) also engage in fair trade by buying fair trade certified products and/or developing company codes of conduct that include fair trade principles. Research on these types of companies show that there is great diversity in the extent to which these activities contribute to fair trade outcomes. Advocates of this type of engagement often note that large multinational companies can have profound impact on global supply chains. They can purchase large quantities of fair trade certified products, shift norms of "acceptable" labor standards, and invest profits in research, product development, and educational programs that can empower producers. Critics suggest that companies that engage with fair trade in superficial ways may be more interested in marketing, protecting economic interests, and mitigating brand risks than empowering producers. "Fair washing" (making trade *appear* fair, even it is not) can confuse consumers about which initiatives to believe in and support, and may thus prevent a challenge to the fair trade movement (Doherty, Davies, & Tranchell 2013; Elder, Lister & Dauvergne 2014; Barrientos 2016).

4.3 Multi-Stakeholder Initiatives

Around the turn of the twenty-first century, companies, NGOs, and businesses began creating "multi-stakeholder initiatives" (MSIs) – organizations, typically non-profit, charged with developing voluntary standards and certification programs to facilitate sustainability and fair labor certifications. As the name suggests, they are developed and governed by diverse stakeholders. Advocates suggest that this gives traditionally marginalized groups a seat at the decision-making table and facilitates cooperation and collective action among corporations accustomed to competing with one another. Critics note that MSIs' governance structures, as described in their constitutions, often exclude hired workers and smallholder farmers, or give them few votes, compared to traditionally powerful groups (Bennett 2017). They also note that even when traditionally marginalized groups are (formally) included, their perspectives may be dismissed, ignored, or bureaucratically limited from impacting policy outcomes (Fransen 2012; Kohne 2014; Cheyns 2014; Cheyns & Riisgaard 2014).

4.4 New Models of Empowerment

In the last decade, the producers of fair trade products have initiated several high-profile fair trade ventures. The Small Producer Symbol (or Símbolo de los Pequeños Productores, SPP) is a fair trade certification system launched *by* Latin American smallholder producers, *for* smallholder producers in 2011. Its standards reflect the challenges that producers have faced in participating in other fair trade certification programs. For example, buyers (e.g., coffee roasters in the Global North) must commit to purchasing at least 5 percent of their overall volume from SPP producers in order to use the logo (Renard & Loconto 2012). In the United States, the Worker-Driven Social Responsibility (WDSR) network includes several producer-initiated fair labor programs. One is the Fair Food Program, an initiative of the Coalition of Immokalee Workers (CIW). The CIW is a worker-based human rights organization in southwest Florida. Its Fair Food Program is a worker-driven farm certification program that recruits major fast food franchises (such as McDonald's) to sign a legally binding contract to only purchase produce from Fair Food certified farms (Brudney 2016; CIW 2018). Similarly, the Milk with Dignity campaign, launched in 2014, requires participating companies to sign a legally enforceable contract in which they commit to sourcing all dairy from farms that adopt a farmworker-authored code of conduct, educate farm workers about their rights, and permit third party compliance audits. Companies must also contribute to economic justice though premiums paid directly to farmers and farm workers. In 2017, Ben & Jerry's signed the first Milk with Dignity Agreement. Due to the ice cream company's large purchasing volume, the majority of Vermont's dairy farmers are now employed on Milk with Dignity farms (Milk with Dignity 2018).[6] The SPP, Fair Food, and Milk with Dignity are all driven by farmers and farm workers and emerged in explicit response to – and as a critique of – top-down, elite-led, corporate-run fair trade initiatives and social responsibility programs that can reify status quo (unfair) social and economic relationships (Gordon 2017).

5 Why (Not) Labeling and Certification?

Today, there are several fair trade certification programs as well as many sustainability certification programs that include standards related to the principles of fair trade (Bennett, 2017).[7] While every program is different, typically a non-profit organization (e.g., Fairtrade International) brings stakeholders (companies, farmers, environmentalists, etc.) together to develop standards for what "counts" as fair trade and which types of groups are eligible for certification. An eligible group (e.g., a cocoa cooperative) pays a third party auditor (e.g., SCS Global Services) to conduct a site visit to collect data, and evaluate whether the organization is in compliance with the standards. The standards-setting organization then determines whether the

---

6 For more on labor justice in the US dairy sector, see Keller 2019.
7 For a consumer's guide to fair trade labels, see the Fair World Project's "Reference Guide" (2019, updated version forthcoming).

audited entity can be certified. Brands (e.g., Green & Black, a chocolatier) purchase the certified products and pay the standards-setting organization (e.g., Fairtrade International) a licensing fee to use the logo on their packaging and in their marketing materials.

There are many benefits of using certification as a tool to facilitate fair trade. First, collective standards-setting, publicly available standards, and third party verification can lend credibility to fair trade claims and elicit higher prices for producers (Distelhorst & Locke 2018). Second, certified goods can be incorporated into conventional (non-fair trade) supply chains, increasing the fair trade market size (Raynolds 2012). Third, environmental labels may aim to increase their appeal by adding in some social standards – improving labor conditions within environmental programs (Brown & Getz 2008).

Certification programs also generate many challenges, risks, and negative externalities. First, the standards-setting process can privilege some types of knowledge over others (Loconto & Hatanaka 2017), marginalize some types of groups, and lack transparency (Ponte and Cheyns 2013). Second, standards and auditing criteria may not capture the complexity of reality (Loconto 2015). Third, producers are required to pay an auditing fee but are not guaranteed to receive a higher price (Ortiz-Miranda & Moragues-Faus 2015). Fourth, certifications typically target some aspects of the supply chain but not others. For example, Fair Trade USA's textiles must be cut and sewn at a certified factory, but earlier stages of preparing the textile may occur at uncertified processors. Fifth, the message that consumers can promote trade justice by simply "looking for the label" conceals the complexities of trade justice and (paradoxically) can contribute to "commodity fetishism" – a cultural process in which the products of labor are given social value, and the people producing them are reduced to factors of production (Gunderson 2013, drawing on Marx 1977 [1867]). Finally, comparative studies show that certification programs can be a weak substitute for more collaborative management approaches (e.g., Muller, Vermeulen & Glasbergen 2012).

5.1 Competing Labels: Race to the Bottom or Race to the Top?

The proliferation of labels has generated an additional set of questions. On the consumer side, there are concerns about "label fatigue" – consumers feeling overwhelmed by the task of comparing and evaluating labels, and keeping up with rapid changes in labeling over time. As a result, consumers may stop asking questions and simply trust all labels – allowing weak labels to flourish – or reject the value of certifications all together (Walske & Tyson 2015; Castka & Corbett 2016). For producers, there is concern that, as the market becomes saturated with certified goods, prices may decline or they may have to price fair trade goods as conventional products in order to make a sale (Dragusanu, Giovannucci & Nunn 2014).

One of the most contentious debates, as certifications compete with one another for market share, is whether they are racing to the top (who can deliver the most impact?) or the bottom (who can attract the most companies with low standards?) (Reinecke, Manning & von Hagen 2012). Many scholars point to the ways in which certifications have diluted standards and impact to producers in order to gain favor with large corporations able to award large contracts (Jaffee 2012). Sustainability labels, for example, typically have very weak social standards (compared to fair trade programs). For example, many reinforce the legal minimum wage and basic labor laws instead of bolstering them (Bennett 2018). On the other hand, standards-setting organizations have competed with one another to be seen as the most credible and producer-oriented organizations (Bennett 2016), and there have been several collaborations across certifications to solve common problems, such as how to calculate and implement a living wage. For example, in 2003 several labeling programs collaboratively established the ISEAL Alliance, an accreditation organization for social and environmental standards-setting organization. In 2013, several member organizations launched the Living Wage Coalition to collectively address the challenges of calculating and implementing living wage standards. Studies show that certifications sponsored by social movement organizations (as opposed to companies or industry associations) and organized as non-profit (as opposed to for-profit) are more likely to comply with best practices for standards-setting organizations, though other factors also play a role (Cashore & Stone 2014; van der Ven 2015; Li & van't Veld 2015). Overall, it is clear that while labeling and certification can increase sales of fair trade products, this approach generates risks, challenges, and negative externalities. Fair trade advocates are divided on whether these shortcomings can (or will) be mitigated and whether the benefits outweigh the limitations (Bezençon & Blili 2009; Wilson & Mutersbaugh 2015).

6 What Next for Fair Trade, Capitalism, and the State?

6.1 Fair trade and Capitalism

The global fair trade movement has always been in tension with market-based capitalism. On one hand, the objective of fair trade is to give marginalized groups access to the market, to empower them in leveraging capitalism to their advantage, and to help them navigate international trade. On the other hand, the vulnerability that many groups face has been perpetuated or exacerbated by market logic, capitalism, and globalization. The question of how to manage the tension of simultaneously working within and against the market place has long challenged the movement (Raynolds, Murray, & Wilkinson 2007). Here, I describe three approaches to capitalism currently at work in the movement: reform, revolution, and rejection. These are ideal types – few fair trade advocates or initiatives would conform exactly to just one – and are intended to illustrate the diverse types of engagement that occur within the movement.

Reformers see the movement's goals as shaping capitalism into an international economic order that sustains and supports workers, smallholder farmers, and other marginalized groups. They see fair trade as a tool for reembedding capitalism and reigniting the social contract (Fridell 2007, 2013). Instead of fighting against capitalism, they aim to bolster regulations that protect the people who work with in it. The Fair Trade Advocacy Organization (FTAO), for example, conducts research and policy advocacy around trade law, anti-trust law, and state procurement policies, reforming current trade institutions to better support fair trade outcomes (FTAO 2018). Opponents of this *reformist* perspective argue that these are solutions are inadequate for fixing a system that is fundamentally broken. They point to increasing income inequality, uneven accumulation of wealth, and elite capture of political institutions as evidence that the capitalist system requires a more comprehensive transformation. Within this group, some fair trade advocates are *revolutionaries* – they aim to replace capitalism with more communal or pro-social economic models. Equal Exchange, for example, was founded by three individuals – an Anarchist, a Leninist, and a Marxist – who were inspired by the Nicaraguan Revolution and driven to support grassroots political change (Bennett 2012a). Other advocates *reject* all engagement with the capitalist system. Instead of *reforming* or *revolutionizing* the system, they promote alternative, opt-in models of exchange, such as time banks, communal living, subsistence farming, and clothing swaps. These methods of "lifestyle politics" contribute to social change by challenging norms, developing alternatives, recruiting participants, and facilitating the reimagining of political and economic life (Haenfler, Johnson, & Jones 2012).

6.2 Private Regulation and the State

The debate of how fair trade should approach capitalism and the state is part of a broader discussion of whether private regulation can or should replace the regulatory role of the state. Private regulation, or non-state actors creating regulations around business practices, emerged in the late 1990s in part to respond to the ways in which globalization challenged states' sovereignty over economic and social policy-making. Private regulation takes many forms, including corporate codes of conduct, multi-stakeholder initiatives, alternative business models (such as social enterprise and the "triple bottom line"), and fair trade (Auld, Renkens, & Cashore 2015).

There is clear consensus that private regulation, alone, is inadequate for protecting vulnerable groups and empowering them within globalized supply chains (Mayer & Gereffi 2010). Many scholars advocate "layering" private and public regulations, suggesting that they may be mutually reinforcing (Overdevest & Zeitlin 2014). Private regulation is more effective in states that participate actively in the International Labour Organization (ILO), adopt stringent labor laws, and protect freedom of the press (Toffel, Short, & Ouellet 2015). This means that for initiatives like fair trade to achieve their potential, states must invest in producers (Jena & Grote 2017), protect workers (Lyon 2015), and close legal loopholes that enable exploitation (Bloomfield 2014). International organizations, transnational diplomacy, and foreign policy also have an important role to play in constraining the damage that investment firms, multinational corporations, and global supply chain actors can impose by chasing weak regulation (Verbruggen 2013).

7 The Future of Fair Trade

Over the last seventy-five years – and, in particular, the past twenty-five – the global fair trade movement has proliferated in several ways: its market share has increased; more consumers are aware, interested, and willing to pay; new types of products bear a fair trade label; an increasingly diverse set of companies are making fair trade claims; and advocates continue to create new organizations to facilitate research, education, and advocacy. As fair trade expands in these ways, debates are emerging and evolving around whether the new activities are scaling up the most important aspects of fair trade or diluting its efficacy. Additionally, as climate change and climate justice come to the fore of public concern, there is increasing attention to the question of how fair trade initiatives can engage the environment whilst focusing on their commitment to fairness and social justice. In the context of these debates about who should benefit, how to govern, whether to certify, and when to engage capitalism and the state, fair trade scholars seem to converge on a few conclusions.

First, authentic fair traders must push back against the most egregious forms of fair-washing. The objective of fair trade is to empower vulnerable populations to take charge of their own lives and livelihoods and to resist market relationships that perpetuate economic marginalization, labor exploitation, and all forms of oppression. Ventures that eschew producer empowerment or obscure systems of exploitation cannot be considered part of the movement. Furthermore, given the movement's mission of empowerment, initiatives must not limit fair trade producers from exercising agency (e.g., by excluding them from positions of power and leadership). They also must not reify, entrench, or obscure the economic disparities and power dynamics that divide fair trade producers from fair trade consumers or advocates.

Second, while certification programs can contribute to fair trade, they are not (on their own) sufficient tools for reforming, transforming, or building alternatives to the contemporary capitalist system. Certification programs must be accompanied by more transformative approaches that truly challenge the inequitable results of the current market system. Some emerging fair trade initiatives – such as Worker-Driven Social Responsibility (WDSR) initiatives and the Small Producer Symbol (SPP) certification program – seem quite aligned with these commitments to transformative change. Others – such as certification programs simply enforcing the legal minimum wage – are not only superficial, but also harmful. These fair trade facades not only confine producers to a broken system, but also confuse consumers and complicate fair trade in earnest.

Third, whether fair trade advocates aim to reform capitalism or create alternatives to it, they must in some way attend to the important role that public policies play in making the market more or less fair. The state may not be the only actor shaping the global economy, but it is an important one. The Fair Trade Advocacy Organization (FTAO) has offered leadership in this area. Its trade justice campaigns aim to reform international trade agreements and government procurement policies in ways that promote North-South equality, reflect fair trade principles, and improve the livelihoods of marginalized producers and workers in the South.

Authentic fair trade puts smallholder farmers, workers, and marginalized groups in positions of leadership and power. It both empowers them to take control over their own lives and works to dismantle the structures of oppression that perpetuate income, class, wealth, and other forms of inequality. Credible fair trade initiatives not only support people in being responsible stewards of the environment, but also address issues of climate justice by placing the onus for change on the most wealthy, powerful, and super-consumptive groups and institutions. Those who identify with the fair trade movement may not agree on every issue, but they are united in these commitments to empowerment, structural equality, and environmental justice.

Acknowledgments

Thank you to Chloe Safar for research assistance and editing; Nicole Godbout for editing; Julie C. Keller for excellent feedback; Stefano Ponte, Tamara Trownsell, and others who provided helpful comments at the 2018 International Studies Association (ISA) Annual Meeting in San Francisco; the Center for Fair & Alternative Trade (CFAT) at Colorado State University and the Fair Trade International Symposium (FTIS) for igniting and nourishing a vibrant community of fair trade scholars; and to the students, faculty, and administration of Lewis & Clark College for providing a supportive and provocative intellectual home.

References

Alegría, D. C. (2016). The case of Café Femenino: The limitations of gender-conscious fair trade. *World Development Perspectives*, **1**, 1–3.

Anderson, M. (2015). *A History of Fair Trade in Contemporary Britain: From Civil Society Campaigns to Corporate Compliance*. London: Springer.

Auld, G., Renckens, S., & Cashore, B. (2015). Transnational private governance between the logics of empowerment and control. *Regulation & Governance*, **9**(2), 108–124.

Bacon, C. M. (2010). Who decides what is fair in fair trade? The agri-environmental governance of standards, access, and price. *Journal of Peasant Studies*, **37**(1), 111–147.

Bacon, C. M., Rice, R. A., & Maryanski, H. (2015). Fair trade coffee and environmental sustainability in Latin America. In L. T. Raynolds & E. A. Bennett, eds., *The Handbook of Research on Fair Trade*. London: Edward Elgar, pp.388–404.

Barrientos, S. (2016). Beyond fair trade: Why are mainstream chocolate companies pursuing social and economic sustainability in cocoa sourcing? In M. P. Squicciarini & J. Swinnen, eds., *The Economics of Chocolate*. Oxford: Oxford University Press, pp. 213–227.

Bennett, E. A. (2012a). A short history of Fairtrade certification governance. In B. Granville and J. Dine, eds., *The Processes and Practices of Fair Trade: Trust, Ethics, and Governance*. London: Routledge, pp. 43–78.

Bennett, E. A. (2012b). Global social movements in global governance. *Globalizations*, **9**(6), 799–813.

Bennett, E. A. (2016). Governance, legitimacy, and stakeholder balance: Lessons from Fairtrade International. *Social Enterprise Journal*, **12**(3), 322–346.

Bennett, E. A. (2017). Who governs socially-oriented voluntary sustainability standards? Not the producers of certified products. *World Development*, **91**, 53–69.

Bennett, E. A. (2018). Voluntary sustainability standards: A squandered opportunity to improve workers' wages. *Sustainable Development*, **26**(1), 65–82.

Besky, S. (2015). Agricultural justice, abnormal justice? An analysis of fair trade's plantation problem. *Antipode*, **47**(5), 1141–1160.

Bezençon, V., & Blili, S. (2017). Fair trade managerial practices: Strategy, organisation and engagement, *Journal of Business Ethics*, **90**(1), 95–113.

Blanchard, S., and Mackey. E. (2018). *Standing with Refugees*. Baltimore: Catholic Relief Services. Retrieved from https://ethicaltrade.crs.org/community-stories/standing-refugees-fair-trade-tradition/

Bloomfield, M. J. (2014). Shame campaigns and environmental justice: Corporate shaming as activist strategy. *Environmental Politics*, **23**(2), 263–281.

Boström, M., Micheletti, M., and Oosterveer, P. 2019. *The Oxford Handbook of Political Consumerism*. Oxford: Oxford University Press.

Brown, K. (2015). Consumer politics, political consumption and fair trade. In L. T. Raynolds & E. A. Bennett, eds., *The Handbook of Research on Fair Trade*. London: Edward Elgar, pp. 157–173.

Brown, S., & Getz, C. (2008). Towards domestic fair trade? Farm labor, food localism, and the "family scale" farm. *GeoJournal*, **73**(1), 11–22.

Brudney, J. (2016). Decent labour standards in corporate supply chains. In J. Howe & R. Owens, eds., *Temporary Labour Migration in the Global Era: The Regulatory Challenges*. Oxford and Portland, OR: Hart Publishing, pp. 351–376.

Cashore, B., & Stone, M. W. (2014). Does California need Delaware? Explaining Indonesian, Chinese, and United States support for legality compliance of internationally traded products. *Regulation and Governance*, **8**(1), 49–73.

Castka, P., & Corbett, C. J. (2016). Governance of eco-labels: Expert opinion and media coverage. *Journal of Business Ethics*, **135**(2), 309–326.

Cater, J. J., Collins, L. A., & Beal, B. D. (2017). Ethics, faith, and profit: Exploring the motives of the U.S. fair trade social entrepreneurs. *Journal of Business Ethics*, **146**, 185–201.

Cheyns, E. (2014). Making "minority voices" heard in transnational roundtables: The role of local NGOs in reintroducing justice and attachments. *Agriculture and Human Values*, **31**(3), 439–453.

Cheyns, E., & Riisgaard, L. (2014). Introduction to the symposium: The exercise of power through multi-stakeholder initiatives for sustainable agriculture and its inclusion and exclusion outcomes. *Agriculture and Human Values*, **31**(3), 409–423.

CIW. (2018). http://ciw-online.org/.

Cole, N. L., & Brown, K. (2014). The problem with fair trade coffee. *Contexts*, **13**(1), 50–55.

Cramer, C., Johnston, D., Oya, C., & Sender, J. (2014). *Fairtrade, Employment and Poverty Reduction in Ethiopia and Uganda*. London: DFiD. Retrieved from http://ftepr.org/publications/#publication-563.

Dalvai, R. (2018). Embedding Fairness in Sustainable Development. Remarks at the Fair Trade International Symposium, University of Portsmouth, Portsmouth, UK, June 26.

Davenport, E., & Low, W. (2013). From trust to compliance: Accountability in the fair trade movement. *Social Enterprise Journal*, **9**(1), 88–101.

Distelhorst, G., & Locke, R. M. (2018). Does compliance pay? Firm-level trade and social institutions. *American Journal of Political Science* **62** (3), 695–711.

Doherty, B., Davies, I. A., & Tranchell, S. (2013). Where now for fair trade? *Business History*, **55**(2), 161–189.

Doherty, B., & Huybrechts, B. (2013). Connecting producers and consumers through fair and sustainable value chains. *Social Enterprise Journal*, **9**(1), 4–10.

Doherty, B., Smith, A., & Parker, S. (2015). Fair trade market creation and marketing in the Global South. *Geoforum*, **67**, 158–171.

Dragusanu, R., Giovannucci, D., & Nunn, N. (2014). The economics of fair trade. *Journal of Economic Perspectives*, **28**(3), 217–236.

Elder, S. D., Lister, J., & Dauvergne, P. (2014). Big retail and sustainable coffee: A new development studies research agenda. *Progress in Development Studies*, **14**(1), 77–90.

Fairtrade International (FI) & World Fair Trade Organization (WFTO). (2009). A charter of fair trade principles. https://wfto.com/sites/default/files/Charter-of-Fair-Trade-Principles-Final%20(EN).PDF

Fair World Project (Commerce Equitable France, Fair World Project, Fairness Research Network on Fair Trade, and Forum Fairer Handel) (2019). Reference Guide. https://fairworldproject.org/wp-content/uploads/2019/12/international-Guide-to-Fair-Trade-Labels-2020-Edition.pdf

Fransen, L. (2012). Multi-stakeholder governance and voluntary programme interactions: Legitimation politics in the institutional design of corporate social responsibility. *Socio-Economic Review*, **10**(1), 163–192.

Fridell, G. (2007). *Fair Trade Coffee: The Prospects and Pitfalls of Market-Driven Social Justice*. Toronto: University of Toronto Press.

Fridell, G. (2013). *Alternative Trade: Legacies for the Future*. Halifax: Fernwood Publishing.

FTF. (2018). www.fairtradefederation.org/.

FTAO. (2018). www.fairtrade-advocacy.org/.

GlobeScan. (2016). Assessing Public Support for Regulation for Fairer Trading Practices. London.

Gordon, J. (2017). The Problem with Corporate Social Responsibility (unpublished paper). Fordham University School of Law.

Gunderson, R. (2013). Problems with the defetishization thesis: Ethical consumerism, alternative food systems, and commodity fetishism. *Agriculture and Human Values*, **31**(1), 109–117.

Guthman, J., & Brown, S. (2016). I will never eat another strawberry again: The biopolitics of consumer-citizenship in the fight against methyl iodide in California. *Agriculture and Human Values*, **33**(3), 575–585.

Haenfler, R., Johnson, B., & Jones, E. (2012). Lifestyle movements: Exploring the intersection of lifestyle and social movements. *Social Movement Studies*, **11**(1), 1–20.

Howard, P. H., & Allen, P. (2016). Consumer willingness to pay for domestic 'fair trade': Evidence from the United States. *Renewable Agriculture and Food Systems*, **23**(3), 235–242.

Jaffee, D. (2012). Weak coffee: Certification and co-optation in the fair trade movement. *Social Problems*, **59**(1), 94–116.

Jaffee, D. (2014). *Brewing Justice: Fair Trade Coffee, Sustainability, and Survival (updated edition)*. Berkeley, CA: University of California Press.

Jaffee, D., & Howard, P. H. (2010). Corporate cooptation of organic and fair trade standards. *Agriculture and Human Values*, **27**(4), 387–399.

Jaffee, D., & Howard, P. H. (2016). Who's the fairest of them all? The fractured landscape of U.S. fair trade certification. *Agriculture and Human Values*, **33**(4), 813–826.

Jena, P. R. & Grote, U. (2017). Fairtrade certification and livelihood impacts on small-scale coffee producers in a tribal community of India. *Applied Economic Perspectives and Policy*, **39**(1), 87–110.

Keller, J.C. (2019). *Milking in the Shadows: Migrants and Mobility in America's Dairyland*. Rutgers, NJ: Rutgers University Press.

Koenig-Archibugi, M. (2017). How to diagnose democratic deficits in global politics: The use of the "all-affected principle." *International Theory*, **9**(2), 171–202.

Kohne, M. (2014). Multi-stakeholder initiative governance as assemblage: Roundtable on Sustainable Palm Oil as a political resource in land conflicts related to oil palm plantations. *Agriculture and Human Values*, **30**(3), 469–480.

Leissle, Kristy. (2012). Cosmopolitan cocoa farmers: Refashioning Africa in divine chocolate advertisements. *Journal of African Cultural Studies*, **24**(2), 37–41.

Lekakis, E. (2019). Nationalist struggles in Europe. In M. Boström, M. Micheletti, and P. Oosterveer, eds., *The Oxford Handbook of Political Consumerism*. Oxford: Oxford University Press, pp. 663–680.

Lewis, J., & Runsten, D. (2008). Is fair trade organic coffee sustainable in the face of migration? Evidence from a Oaxacan community. *Globalizations*, **5**(2), 275–290.

Li, Y., & van't Veld, K. (2015). Green, greener, greenest: Eco-label gradation and competition. *Journal of Environmental Economics and Management*, **72**, 164–176.

Linton, A. and Rosty, C. (2015). The US market and fair trade certified. In L.T. Raynolds, and E. A. Bennett, eds., *The Handbook of Research on Fair Trade*. London: Edward Elgar, pp. 333–353.

Loconto, A. (2015). Assembling governance: The role of standards in the Tanzanian tea industry. *Journal of Cleaner Production*, **107**, 64–73.

Loconto, A., & Dankers, C. (2014). *Impact of International Voluntary Standards on Smallholder Market Participation in Developing Countries*. Rome: FAO.

Loconto, A., & Hatanaka, M. (2017). Participatory guarantee systems: Alternative ways of defining, measuring, and assessing "sustainability." *Sociologia Ruralis*, **58**(2), 412–432.

Lyon, S. (2015). The hidden labor of fair trade. *Labor Studies in Working-Class History of the Americas*, **12**(2), 159–176.

Marconi, N. G., Hooker, N. H., & DiMarcello III N. (2017). What's in a name? The impact of fair trade claims on product price. *Agribusiness*, **33**(2), 160–174.

Marx, K. (1977 [1867]). *Capital*, vol. 1. New York, NY: Vintage.

Mayer, F., & Gereffi, G. (2010). Regulation and economic globalization: Prospects and limits of private governance. *Business and Politics*, **12**(3), 1–26.

Micheletti, M., & Oral, D. (2019). Problematic political consumerism: Confusions and moral dilemmas in boycott activism. In M. Boström, M. Micheletti, and P. Oosterveer, eds., *The Oxford Handbook of Political Consumerism*. Oxford: Oxford University Press, pp. 699–720.

Milk with Dignity. (2018). https://migrantjustice.net/milk-with-dignity.

Muller, C., Vermeulen, W. J. V., & Glasbergen, P. (2012). Pushing or sharing as value-driven strategies for societal change in global supply chains: Two case studies in the British–South African fresh fruit supply chain. *Business Strategy and the Environment*, **21**(2), 127–140.

Naylor, L. (2014). "Some are more fair than others": Fair trade certification, development, and North-South subjects. *Agriculture and Human Values*, **31**(2), 273–284.

Noble, M. D. (2017). Chocolate and the consumption of forests: A cross-national examination of ecologically unequal exchange in Cocoa exports. *Journal of World Systems Research*, **23**(2), 236–268.

Ortiz-Miranda, D., & Moragues-Faus, A. M. (2015). Governing fair trade coffee supply: dynamics and challenges in small farmers' organizations. *Sustainable Development*, **23**(1), 41–54.

Overdevest, C., & Zeitlin, J. (2014). Assembling an experimentalist regime: Transnational governance interactions in the forest sector. *Regulation & Governance*, **8**(1), 22–48.

Pinto, L. F. G., Gardner, T., McDermott, C. L., & Ayub, K. O. L. (2014). Group certification supports an increase in the diversity of sustainable agriculture network-rainforest alliance certified coffee producers in Brazil. *Ecological Economics*, **107** (November), 59–64.

Ponte, S., & Cheyns, E. (2013). Voluntary standards, expert knowledge and the governance of sustainability networks. *Global Networks*, **13**(4), 459–477.

Raynolds, L. T. (2012). Fair trade Flowers: Global certification, environmental sustainability, and labor standards. *Rural Sociology*, **77**(4), 1–27.

Raynolds, L. T. (2014). Fairtrade, certification, and labor: Global and local tensions in improving conditions for agricultural workers. *Agriculture and Human Values*, **31**, 499–511.

Raynolds, L. T., & Bennett, E. A. (2015). *The Handbook of Research on Fair Trade*. London: Edward Elgar. http://doi.org/10.4337/9781783474622.

Raynolds, L. T. (2017). Fairtrade labour certification: the contested incorporation of plantations and workers. *Third World Quarterly*, **38**(7), 1473–1492.

Raynolds, L. T., Murray, D. L., & Wilkinson, J. (eds.). (2007). *Fair Trade: The Challenges of Transforming Globalization*. New York: Routledge.

Reinecke, J., Manning, S., & von Hagen, O. (2012). The emergence of a standards market: Multiplicity of sustainability standards in the global coffee industry. *Organization Studies*, **33**(5–6), 791–814.

Renard, M.-C., & Loconto, A. (2012). Competing logics in the further standardization of fair trade: ISEAL and the símbolo de pequeños productores. *International Journal of Sociology of Agriculture and Food*, **20**(1), 51–68.

Samuel, A., Peattie, K., & Doherty, B. (2018). Expanding the boundaries of brand communities: The case of Fairtrade Towns. *European Journal of Marketing*, **52**(3/4), 758–782.

Smith, S., Kuruganti, K., & Gema, J. (2015). *Equal harvest: Removing the barriers to women's participation in smallholder agriculture*. London: FairTrade Foundation.

Stevis, D. (2015). Global labor politics and fair trade. In L.T. Raynolds and E. A. Bennett, eds., *The Handbook of Research on Fair Trade*. London: Edward Elgar, pp. 102–119.

Stolle, D., & Huissoud, L. (2019). Contemporary examples of undemocratic political consumerism. In M. Boström, M. Micheletti, and P. Oosterveer, eds., *The Oxford Handbook of Political Consumerism*. Oxford: Oxford University Press, pp. 625–642.

Sutton, S. (2013). Fairtrade governance and producer voices: stronger or silent? *Social Enterprise Journal*, **9**(1), 73–87.

Toffel, M. W., Short, J. L., & Ouellet, M. (2015). Codes in context: How states, markets, and civil society shape adherence to global labor standards. *Regulation & Governance*, **9**, 205–223.

Valiente-Riedl, E. (2016). To be free and fair? Debating fair trade's shifting response to global inequality. *Journal of Australian Political Economy*, **78**, 159–185.

van Dam, P. (2015). The limits of a success story: fair trade and the history of postcolonial globalization. *Comparativ: Zeitschrift für Globalgeschichte und vergleichende Gesellschaftsforschung*, **25**(1), 62–77.

van der Ven, H. (2015). Correlates of rigorous and credible transnational governance: A cross-sectoral analysis of best practice compliance in eco-labeling. *Regulation & Governance*, **9**(3), 276–293.

Verbruggen, P. (2013). Gorillas in the closet? Public and private actors in the enforcement of transnational private regulation. *Regulation and Governance*, **7**(4), 512–532.

Walske, J., & Tyson, L. D. (2015). Fair Trade USA: Scaling for impact. *California Management Review*, **58**(1), 123–144.

WFTO. (2018). https://wfto.com/.

Wilson, B. R., & Mutersbaugh, T. (2015). Fair trade certification, performance and practice. In L. T. Raynolds & E. A. Bennett (eds.), *The Handbook of Research on Fair Trade*. London: Edward Elgar, pp.281–297.

28 Possibilities for Degrowth: A Radical Alternative to the Neoliberal Restructuring of Growth-Societies

Barbara Muraca

1 The Re-Emergence of the Degrowth Discourse

Critique of economic growth is not new on the arena of political debate and social movements. Its first appearance dates back to the 1970s in occasion of the publication of the report to the Club of Rome "Limits to Growth" (Meadows et al., 1972) and was mainly fueled by the threatening scarcity of key natural resources, such as oil. The analysis of biophysical limits for economic production, inspired i.a. by the findings of what later would be called Ecological Economics, depicted the impossibility of continuing on a growth-path indefinitely. The French term *décroissance* (later translated back into English as degrowth) appeared in 1979 as a translation of Georgescu-Roegen's concept of "declining state" in a collection under the title "La décroissance. Entropie – Écologie – Économie" (Georgescu-Roegen, 1995) and established itself as an alternative to zero-growth or steady-state. From its very beginning in the French discussion, degrowth emerged as the knot between two parallel lineages: the rising awareness of the devastating impact of further economic growth on the environment with its consequences on the sphere of production and the promise of liberation from the socio-cultural logic of growth in terms of the productivism diktat that colonizes alternative imaginaries for a good life (Duverger, 2011; Muraca, 2013).

During the restructuring of the global economy in the 1980s and 1990s, the critique of growth almost disappeared under the new paradigm of a "sustainable development," which more adequately mirrored the new managerial approach of neoliberalism (D'Alisa et al., 2014; Muraca & Döring, 2018).

Less than twenty years later, the spreading awareness of global warming brought growth critique back on the agenda. The degrowth discourse resurfaced through a detour: critics of globalization in the Global North built a strong alliance with new social movements emerging in the Global South, including indigenous and peasants movements, to denounce the devastating face of the Western model of development and to call for resistance and alternatives to its pervasive hegemony (Muraca, 2013; Muraca & Döring, 2018).

The global financial crisis of 2008 fueled the fear that the growth path might come to its end, not only due to environmental constraints, but also to endogenous factors in the economic structure of modern societies. Since then, long-term shrinking scenarios, loaded with the prospect of stagnation, social inequality, and pauperization, have become plausible and intersect with degrowth narratives (Miegel, 2010; Muraca, 2014). Degrowth activists feel compelled to clarify that degrowth is not the mere reverse of GDP growth, all other things being equal. The slogan "Your recession is not our degrowth" articulates the key difference between shrinking scenarios under neoliberal austerity and degrowth as a vision for a radical transformation of societies. While both perspectives share the analysis that a shrinking economy is impending as a consequence of the (capitalist) mode of production in industrialized societies that jeopardize their very biophysical conditions of reproduction, the shrinking scenario remains caught within the business-as-usual neoliberal narrative and calls for adaptation to the allegedly inevitable inequality and pauperization (Muraca, 2017). The degrowth proposal experiments instead with possibilities for an ecologically sustainable and socially equitable path beyond growth. Although the degrowth vision does call for a reduction of the material size of the economy, it goes well beyond advocating a society with a smaller metabolism under unchanged conditions and envisions "a society with a metabolism which has a different structure and serves new functions" (D'Alisa et al., 2014, p. 4). Degrowth radically questions the logic of growth as the central goal of political decisions and aims at liberating social forces towards a truly democratic self-determination of the basic conditions for common living. As such, degrowth plays a major role in repoliticizing the sustainability discourse, i.a. by challenging the managerial approach to environmental problems and demanding a democratization of societal relations with nature (Asara et al., 2015; Muraca & Döring, 2018).

As I show in this chapter, the degrowth movement strikes with its critique and modes of resistance the very core of the neoliberal *stealth revolution* (Brown, 2015). Far from representing an eminently "culturalistic" critique[1] against economic growth, degrowth embodies forms or resistance against the disciplining modes of operation of the pervasive neoliberal rationality both in terms of its *substantial goals* and of its *modus operandi*. It embodies a concrete utopia aimed at a radical transformation of the *Social Imaginary* (Muraca, 2014).[2]

2 From the Crisis of Growth-Societies to Neoliberal Restructuring

2.1 Growth-Societies and their Crisis

Modern capitalist societies, especially under the Fordist regime, owe their relative institutional stability to the capacity of dynamically stabilizing themselves over and

---

1 For Martinez-Alier et al. degrowth *à la Française* is characterized, alongside the tradition of political ecology, by what they call a "culturalist" perspective aimed at a critique of developmentalism (2010).
2 According to Castoriadis, every society constitutes itself as more or less coherent whole of institutions, by creating a comprehensive universe of meaning, which is not determined by historical necessity (Castoriadis, 2010). The Social Imaginary refers to a deep, collective self-understanding that legitimizes and confers sense to institutions and practices.

over again via growth, intended as further material expansion of the economic sphere, technological augmentation, and cultural innovation (Rosa et al., 2017). Far from being merely one contingent factor among others, growth is structurally essential to overcome their recurrent crises. This is even more evident under the so-called social-democratic pact of Fordism (Dardot & Laval, 2017). The prospect of increasing growth rates was key in supporting a historically rather unique period of comparably low inequality, combined with substantial public expenditures for and the corresponding decommodification of education, health, and social security (Piketty, 2014).[3] Given the essential role of economic growth, its main measurement in terms of GDP, which was established during WWII as a way of managing the war economy (Schmelzer, 2015), became progressively the key indicator for the "health" of a national economy and ultimately the capital goal of political interventions (Brown, 2015).

Economic growth came thus to be *explicitly*[4] the decisive factor for the stabilization of post-war societies, by guaranteeing fiscal stability, employment, and the social and political integration of capitalist democracies (Offe, 1983). The social-welfare state – Habermas argues – "may not impair the conditions of stability and the requirements of mobility of capitalist growth" among other reasons because "adjustments to the pattern of distribution of social compensations trigger reactions on the part of privileged groups unless they can be covered by increases in the social product and thus do not affect the propertied classes; when this is not the case, such measures cannot fulfill the function of containing and mitigating class conflict" (1984, 348). The Fordist pact among social classes and the corresponding stabilization of Western democracies are rooted in the ongoing promise of social mobility, social security for a large majority of people, and a relatively moderate inequality.

On the verge of the 1970s the post-war felicitous, dynamic stability (at least for countries of the Global North) came to a first, significant stall. Particular political-economic conditions created a fertile soil for the renewal of neoliberal ideology and for the hegemonic position that it would achieve in the decades to come. On the one hand, the Fordist model of growth ran against its self contradictions[5] and endogenous limits (Dardot & Laval, 2017, pp. 152ff.): profit of enterprises begun to decline with the increasing saturation of domestic markets, while successful struggles of the working class reduced the capitalist margin of profit. The relative stability of the international monetary system started staggering and a new monetarist policy slowly prevailed, driven inter alia by the US need of money for the Vietnam War. The Keynesian model gradually fell into disgrace and left the stage to the re-emerging neoliberal ideology.

---

[3] It would be inappropriate to attribute the relative stability of Welfare States in the post-war period exclusively to the structural factor of economic growth. Of course, successful social struggles as well as the hegemonic role played by Keynesianism contributed to justify the specific regime called Fordism. However, economic growth constituted a fundamental basis for this success.

[4] Structurally, growth has always been linked to capitalist accumulation – but it is with the formalization of GDP as its official measurement at the level of a national economy that it becomes an explicit political goal (Schmelzer, 2015).

[5] "The Keynesian welfare state is a victim of its success. By (partly) eliminating and smoothening crises, it has inhibited the positive function that crises used to perform in the capitalist process of 'creative destruction'" (Offe, 1983, p. 241).

While it slowly became a commonplace to blame the central bureaucracy of the – increasingly indebted – Welfare States for the crisis, a growing support for alternative paradigms emerged across the political spectrum. It is not only the explicit hostility of the economic elite against redistribution policies, high taxation, and labor market regulation, that fueled neoliberal sympathies. In a sense, a certain weariness of the centralized bureaucratic State and the surfacing recalcitrance against the dull system of production that generates alienation and suffocates creativity inspired new forms of social critique of capitalism more inclined to welcome a turn towards narratives of self-realization and the diversification of work.[6]

Against this background, the ecological crisis of the 1970s and the report to the Club of Rome represented a major element of destabilization in an already increasingly fragile situation. The raising awareness that ecological conditions might jeopardize the foundations of industrialized societies reshuffled the cards of the game. The vulgate that the dark prognoses of the Club of Rome did not come true and that economic growth nonetheless continued undisturbed in the following decades, as weak sustainability scholars and advocates of win-win scenarios for green growth still hold, misses a central point: it was precisely *because* the warnings were taken very seriously, especially by the economic elite, that forces joined for a radical restructuring of societies in order to unleash new possibilities for growth and profit accumulation.

2.2 Neoliberalism as Answer to the (Ecological) Crisis

The neoliberal restructuring of society from the late 1970s ongoing cannot be dissociated from the threat represented by the ecological crisis. It can be understood as a systematic and forward-looking answer to *Limits to Growth*. To summarize it with degrowth scholar and activist Francois Schneider, the "capacity to exploit" of the economy had to be increased and liberated from constraints in all dimensions – material, human, temporal, spatial – (Schneider, 2008). The neoliberal restructuring reclaimed growth no longer as a support for redistribution policies, but as a driver for increasing (private) accumulation: "Rather than undermining the growth paradigm, what has been called the monetarist 'counter-revolution,' the 'marketization' of economics, or the rise of 'neoliberalism' merely re-articulated growthmanship in a new guise" (Schmelzer, 2015, p. 267). With some notable exceptions, like French non-orthodox Marxists involved in a profound critique of productivism (Gorz, Castoriadis, Illich), forces of the political Left, normatively embedded in the partial successes of Fordism, kept holding onto the traditional model of economic growth and did not anticipate the historical and planetary significance of the ecological crisis.

---

6 According to Boltanski and Chiapello capitalism generates ideological support by incorporating at least in part the claims articulates by its critics. Accordingly, the 'artistic critique' is led by "an ideal of liberation and/or of individual autonomy, singularity and authenticity" (2005, p. 176). It offers a somehow fertile soil to the new narrative of neoliberal entrepreneurial risk. As I will show later, however, this does not mean that the artistic critique is entirely coopted by the New Spirit of Capitalism. The degrowth movement embodies the submersed lineage of a more subversive artistic critique.

Against both neoliberal and social-democratic growthism, the re-emerging degrowth discourse explicitly reclaims that very antiproductivist tradition of French political ecology, along with a radical critique of globalization and development (Muraca, 2013; Muraca & Schmelzer, 2017; Demaria et al., 2013).

The complexity of historical transformations, which led to a new *accumulation regime*[7] beyond Fordism, forbids a superficial analysis and the rather one-sided interpretation of a clear design plotted by a small political-economic elite. Although the so-called *neoliberal strategy* is characterized by a series of explicit political objectives, such as dismantling the social state, privatizing public enterprises, deregulating the financial sector etc., I follow Dardot and Laval in considering it, with Foucault, a *strategy without a strategist*, i.e. without a specific and clearly locatable intentionality, that operates through the generalization and multiplication of practices and techniques of power (Dardot & Laval, 2017).

As mentioned earlier, neoliberalism is characterized by a renewed drive to enable the further expansion of the capitalist mode of production threatened i.a. by the ecological crisis. In doing that, it paradoxically embraces the challenge launched by Ecological Economists to focus on *life* as the core productive process that generates novelty (Georgescu-Roegen, 1971). Neoliberal strategies are "no longer only based on the exploitation of labor in the industrial sense, but also on that of knowledge, life, health, leisure, culture etc" (Lazzarato 2004, p. 205; see also Foucault, 1980). They aim at the capitalist valorization of life in all its forms and tap into life's *autopoietic* power, embodied i.a. in intellectual creativity, emotional attachment, desires, and the generation of novelty. Accordingly, neoliberal exploitation involves "cognitive, cultural affective and communicative resources (the life of individuals) as much as territories, genetic heritage (plants, animals, and humans), the resources necessary to the survival of the species and the planet (water, air, etc.). It is about putting life to work" (Lazzarato, 2004, p. 205). Exploiting life's creativity requires a different approach, because living beings are self-creative processes that do not react mechanically to stimuli. All attempts at controlling and managing life, from genetic engineering to synthetic biology, have to come to terms with the recalcitrance of living processes (Köchy, 2014). Valorizing life's productivity requires the *activation*[8] of and cooperation with processes of self-organization that cannot be controlled by targeted manipulation. The intervention is more successful when it operates on the *environmental situation*, thus generating a framework of conditions under which specific actions are triggered by positive or negative selection. Neoliberal governmentality implies precisely an *environmental type of intervention*, in which "action is brought to bear on the rules of the game rather than on the players" (Foucault 2008, p. 260; Oksala, 2016).

---

7 The term refers to the way in which in a specific historical period production and consumption are framed and how their reproduction over time is guaranteed and stabilized by a corresponding mode of regulation. The term Fordism addresses the specific mode of accumulation and regulation of the postwar period (Aglietta, 2000).

8 No wonder that the term is used both in synthetic biology to address the activation of genes and in sociology to address the new role of the State under neoliberalism (Lessenich, 2008).

Generally, no mode of production can reproduce itself over time without successfully establishing a shared imaginary that legitimizes it and is embodied by individuals in their desires, forms of living, intentions, or, in other words, *subjectivity* (Read, 2002; Bloch, 1976).[9] The mere imposition by violence would not be effective without a phase of "normalization," in which the new imaginary is embraced and habitualized until it becomes not only self-evident, but also "natural" (Read, 2002, p. 45). Accordingly, modes of production do not just happen within the economic sphere in a strict sense, but imply and are implied by *modes of subjection*,[10] i.e. the constitution of subjectivities that support and reproduce that very mode of production. Industrial capitalism operated by transforming "people's bodies and their time" into "labor power and labor time" (Foucault quoted in Read, 2002) via techniques of power that discipline subjects more or less directly.[11] Submission to power is a necessary condition for existing and being "seen" in the realm of social relations. At the same time, however, the subject is itself produced as a *subject*, i.e. as a possibility for agency. *Subjection* is therefore at the same time also *subjectivation*, a type of relation to oneself that is not merely produced by submission to an external norm, but that is also constituted and re-constituted by a process of self-transformation triggered by the external norm, but not reducible to it (Dardot 2011, p. 189). As I will show in section 3, this is the leverage point of resistance that degrowth can mobilize.

While the *mode* of subjectivation of industrial capitalism is predominantly the employment of disciplinary techniques of power, its *substance* (goals, contents of norms, and social ontology) is to form subjects according to the rationality of utility-maximizing self-interested individuals, free and willing to enter into voluntary contracts and subordinated to an *apparatus of efficiency* (Dardot & Laval, 2017, pp. 257ff). The economic man of the nineteenth century is expected to be a calculating individual – a man of exchange (Foucault, 2008) – and an efficient worker (Dardot & Laval, 2017). However, the logic of nineteenth-century *homo oeconomicus* was neither pervasive nor totalizing. At least as a normative ideal, modern man was divided between the citizen, the *homo politicus* as holder of rights and concerned with the common good, and the self-interested economic man (Dardot & Laval, 2017; Brown, 2015). The pervasive economization of neoliberalism evacuates instead even the desires for democracy (Brown, 2015).

On the one hand, neoliberalism is characterized by a continuity with the traditional capitalist mode of production and subjection represented by disciplinary and disciplining interventions directed to the "players," their bodies and minds. This is particularly (albeit not exclusively) true in the form of new colonialism triggered by

---

9 Following Balibar (and Althusser), the imaginary represents ideology as the *material production of subjectivity* (Read, 2002), or mode of subjection that is inseparable from the mode of production: "[I]n any historical conjuncture, the effects of the imaginary can only appear through and by means of the real, and the effects of the real through and by means of the imaginary; in other words, the structural law of causality in history is the detour through and by means of the other scene" (Balibar 1995, p. 160).
10 Mode of production and mode of subjection are linked in a reciprocal causation and constitute each other (Read, 2002).
11 Industrial capitalism would have not been possible without the forcible creation of new subjectivities that would readily identify with the normative framework of wage labour. The destitution of peasants and their exodus into the cities as beggars were not enough to turn them into factory workers (Read, 2002).

globalization in countries of the Global South that Naomi Klein has dubbed *the shock doctrine* (Klein, 2007). Here processes of primary accumulation (Read, 2002) recall the violent interventions of nineteenth-century England.[12] Expanding the capacity to exploit takes here the form of aggressive neo-extractivism, landgrabbing, and privatization of the still existing commons. On the other hand, and especially in the Global North, neoliberal governmentality matures as a new mode of governing life that operate *with* and not *against* life's power.

With respect to its *mode of operation*, governmentality intervenes on the environment rather than on the single actors in order to create a *situation* that triggers action neither with recourse to violence nor to contractual agreement (Dardot, 2011). This is what Foucault famously calls "conduct of conduct," where the word "conduct" means at the same time guiding others and behaving, i.e. guiding oneself. Accordingly, governmentality pre-arranges (*aménager par avance*) a space of possibilities, within which individuals will then have to inscribe their own actions (Dardot, 2011, p. 195, own translation). The disciplinary power of neoliberalism no longer "fabricates docile bodies (in the factory, the madhouse, or the prison)" but "structurates the action field (*champ d'action*) of individuals" (Dardot, 2011, p. 195). This evokes the theoretical framework of evolutionary theory applied to life, according to which no overarching teleology presides the development of organisms, but only the internal drive to survive regulated by natural selection. Neoliberal theory is heavily indebted to Spencer's rendering of evolution as applied to social systems (Dardot & Laval, 2011). Unleashed competition and the survival of the fittest are the guiding principle that justified in the nineteenth century the rejection of State intervention to regulate social and economic interactions. Later on, Hayek (1993) introduced the term *catallaxis* to address the "spontaneous order" of the market that neither conforms to a manufactured system according to a clear design (*taxis*) nor reflects a natural order that operates independently of human will and has its drive in itself (*cosmos*). Rather than a natural, self-regulating system tending to equilibrium – as neoclassical economics would say – the market is a dynamic, spontaneous system, within which individuals struggle for survival and improvement, by adapting to the selection mechanisms of the environment, in which they are embedded. In analogy to natural evolution, the market is independent of any particular goal or design and market actors operate creatively in reaction to general rules. However, other than natural evolution, for Hayek its rules have to be instituted by humans.

The mode of subjectivation of neoliberal governmentality structurates the field of action that renders livability possible. As Butler writes: "subjection is the paradoxical effect of a regime of power in which the very 'conditions of existence,' the possibility of continuing as a recognizable social being, requires the formation and maintenance of the subject in subordination" (1997, p. 27). Successful adaptation in the biological realm is not a simply passive reaction, but a self-creative and productive process that generates novelty and surplus value. This is why the mode of

---

12 By this I do not mean that this violence was not perpetrated already in the 19th Century and earlier in other parts of the world. But the form of early colonialism resembles more closely what Foucault calls *raison d'Etat* or the arbitrary rationality of the sovereign himself (2008) that governs by force rather than by disciplinary apparatuses.

subjectivation of neoliberalism operates by activating creativity and by triggering structures of desire. If it is at all true that for living beings desire is always and foremost *desire to persist in one's own being* (Spinoza quoted in Butler, 1997, p. 28), then the catallactic model (as Hayek, 1993 terms it) mobilizes desire as "the relay of apparatuses for steering conduct" (Dardot & Laval, 2017, p. 260). In its schizophrenic nature, neoliberalism liberates desire beyond the strict requirements of its mode of production, but also constantly tries to reincorporate it by tapping into its productivity (Deleuze & Guattari, 1983) – this is even more evident today with the functioning of social media orchestrated around flows of desires (for recognition, sociality, affection, appearance etc ...) and fears (of anonymity, social blaming, solitude, etc ...). The underlying logic is not only that of "more," but also of "better" (contacts, likes, reads, etc.).

With regard to its *substance*, neoliberal ideology is formed around the idea that unleashed competition is the only guarantor of freedom against totalitarianism and State control. Right after WWII, in a diagnosis that is diametrically opposed to that of Polanyi, ordoliberals[13] as well as Austrian neoliberals agreed that the market as spontaneous order driven by competition is the only force that can counteract the return of fascism. However, precisely because competition is not naturally given, it has to be made into the highest objective of governmental interventions (Brown, 2015). In order to foster the competition logic via governmentality, the whole of society has to be modeled according to the structure of the enterprise (Dardot & Laval, 2017). This is a totalizing logic that neither requires nor tolerates any parallel normativity. The differentiation between citizen and economic man is lost and the new unitary subject of neoliberalism is born as *entrepreneur of himself* (Dardot & Laval, 2017; Foucault, 2008). Under neoliberalism, workers should no longer consider themselves as labor, but as (human) capital. What counts is their *investment value*: the entrepreneurial man seeks to enhance his portfolio value in all domains of life through a constant logic of self-investment that attracts investors – no social or private sphere should escape the pervasiveness of the entrepreneurial logic (Brown, 2015).

The new work ethic calls for a strict self-management and the rationalization of desire (free from stigmatization or ascetic repression) in order to constantly improve one's own opportunities (Dardot & Laval, 2017). Life-long learning, constant training on and alongside the job, as well as investment in health, fitness, communication skills are all building blocks of the personal enterprise to enhance *resilience*[14] and *self-valorization*.

Self-determined working times, self-realization on the job, and freedom of choice are at the same time the seductive lure of the new paradigm and its most insidious trap: in the name of freedom of choice through "liberalization" a *situation* is created,

---

13 German School of thought based in Freiburg and aiming at renewing classical liberalism right after WWII.
14 How interesting that the term resilience is being increasingly used in social sciences to address active adaptability to environmental conditions (Walker & Cooper, 2011). The originally ecological term fits Hayek's catallactic model of the market-society well, as it refers to multi-dynamic stabilities rather than to a status of equilibrium.

in which choosing becomes compulsory and each choice implies personal responsibility for success or failure. The State becomes the manager and motivator of the entrepreneurial logic, by *activating* peripheral groups to become "subject-agents of their own employability, 'self-enterprising' beings taking themselves in hand" (Dardot & Laval, 2017, p. 174). The combination of the seductive discourse of individual self-management (qua portfolio of assets) operating via governmental modes of subjectivation and more direct disciplinary interventions against pockets of recalcitrance or inertia successfully restore the capitalist mode of production in a new fashion.

Resistance and possibilities for radical transformation are limited in the wake of the totalizing logic of competition and investment. Economic growth, no longer only in its literal sense of GDP nor as the sum of material and energy flows, but together with its symbolic twin brothers acceleration, intensification, and enhancement, solidifies into a *mental infrastructure* (Welzer, 2011) that seems to re-ingest any attempt to deviate from it.

Precisely in this context, degrowth can embody a radical challenge to the neoliberal logic, by operating at the same time on the *mode* of subjectivation and the *substantial* goals of neoliberal governmentality. Resistance is not simply antagonistic negation, but – with Foucault – a creative process: "to create and recreate, to transform the situation, to participate in the process, that is to resist" (1997, p. 168).

3 Degrowth as a Vision for a Radical Transformation of Society Beyond Neoliberalism

The degrowth discourse is heterogeneous and multifaceted. Depending on the geographical traditions (Muraca & Schmelzer, 2017) and the theoretical inspirations that flow into different practices, social experiments, and societal debates, it can take different forms. It is difficult to speak of degrowth as a social movement (Petridis, Muraca, & Kallis, 2015) because it lacks the homogeneity of a political program and a clearly identifiable organizational structure. The best way of representing it is by adopting a concept introduced in the title of the book *Degrowth in Movement(s)*: in the book germane movements, social experiments, and resistance groups are invited to articulate their relationship with the degrowth idea and to highlight overlapping grounds, differences, common strategies, and visions for transformation (Burkhart, Schmelzer, & Treu, 2020). The degrowth discourse is attractive precisely because of its horizontal, "spontaneous" organizing, less an umbrella than an archipelago, in which (at least so far) diverse approaches of resistance and social critique can find a distinct place while being connected by common seaways, without merging. Alliances with neighboring movements, both antagonistic and prefigurative (feminist and queer, commons, climate camp, transition towns, postdevelopment, indigenous, global environmental justice, altermondialiste etc.) contribute a rich and complex perspective of critique and of radical imagination for alternative futures.

With respect to the *substance* of neoliberal rationality, degrowth represents a radical rupture with the pervasive diktat of growth in all its dimensions (Muraca & Döring, 2018). It inherited both the feminist critique of *monetary* growth in terms of GDP and the environmentalist fight against the growth of *material* and energy flows. It quickly adopted the sociological analysis of the *structural* function of growth in stabilizing capitalist societies and the diagnosis of their crisis. Degrowth in movement(s) envisions a radical transformation of basis institutions, in order to render them independent of the growth mechanism, while guaranteeing equality, solidarity, and (real) democracy (Asara, 2016; Muraca, 2014). Together with the postdevelopment tradition, degrowth unmasks the violence behind the *development* narrative and builds alliances with *Buen Vivir* in Latin America and *Radical Ecological Democracy* in India (Escobar, 2015; Kothari, Demaria, & Acosta, 2014). Since the very beginning of the French degrowth discourse, growth has been furthermore identified as a hegemonic mode of signification that constitutes the social imaginary, by pervasively forming patterns of recognition, the articulation of needs, and modes of desiring (Latouche, 2009; Castoriadis, 1997; Petridis, Muraca, & Kallis, 2015). Accordingly, the growth logic has colonized the lifeworld and molded our very self-understanding as social agents.

In some interesting ways, the representation of society in the degrowth discourse is similar to Hayek's catallactic model and shares with it the critique of centralized design of social processes.[15] Society is envisioned as a spontaneous, policentric, dynamic, and self-creative order that does not need a central power to function. However, its fulcrum is not competition, as in Hayek's Spencerian (mis)interpretation of Darwin's evolutionary theory, but conviviality. Inspired by the M.A.U.S.S. school,[16] by French political ecology (Illich, Gorz, Castoriadis), by feminist theories, and postcolonial thinking, the social ontology of degrowth revolves around the idea that being human means being-in-relation (the original meaning of *inter-esse*) and aims at forms of *cum-vivere* (living together) based on solidarity and reciprocity. According to an alternative reading of Darwin's evolutionary theory that counters Spencer's interpretation, civilization is considered as characterized by the prevalence of "social instincts" and of sympathy that can neutralize the eliminatory aspects of natural selection (Dardot & Laval, 2017). Degrowth displays a strong anarchist root both in its core ideas and in its practices of resistance and struggles.

In alliance with the global commons movement, degrowth embodies the idea that cooperation and sharing rather than competition are the engine of creativity, novelty,

---

15 In the heterogeneity of degrowth discourse, "degrowth by design" plays some role in the literature (Victor), but is less influential in the movements and their self-understanding. A large part of degrowth activists tend to identify with forms of diffuse and subversive actions, including a significant vicinity to anarchism and the alternative practical left (Eversberg & Schmelzer, 2018; Victor, 2008).

16 *Movement Antiutilitariste dans le Sciences Sociales* is the acronym of a sociological school that takes inspiration from Marcel Mauss's anthropology of the gift. The "gift" is considered as the foundation of the social bond in its triadic non-obligatory obligation of giving, receiving, and giving back or giving further. The gift keeps traditional societies together and enables all other forms of exchange (contracts, barter etc.) (Caillé, 2004; Muraca, 2013).

and human flourishing. While capitalist appropriation superimposes arbitrarily a new sphere of controlled scarcity on the otherwise freely floating cooperation of minds, the new commons movement struggles to liberate the virtual power of creation from the constraints of capitalistically valorized "work" (Lazzarato, 2004). A form of class consciousness is emerging as the "commons" "has become one of the keywords in the global class struggle against neoliberal capitalism." (Yang & Howison, 2010). The battle around the new global commons, not intended as things, but as modes of relations struggling for spontaneity and collective self-determination, hits neoliberal rationality in its core. Commons refer to the activity that *institutes* a new way of dealing with things and ideas in terms of sharing, taking charge, and taking (commonly) care of them (Dardot & Laval, 2014).

Degrowth is learning from feminist ecological economics the concept of *(re)productivity* that refers to the irreversible processes – social as well as ecological – of generation of novelty and their regeneration (Biesecker & Hofmeister, 2010). (Re)productivity encompasses both production and reduction (i.e. death, loss, fading) at the complex intersection between ecological and social interactions. The flourishing and regeneration ability of living processes (including human life and health) requires time, favorable conditions, and is neither automatic nor ever circular. This is why the leading principle is not investment, but *care* in its bivalent and asymmetric modality of being taken care of (i.e. supported and regenerated) and caring for the fragility of living processes for the provision of what is necessary to live and flourish (Biesecker & Hofmeister, 2010). Care refers to "everything we do to maintain, contain, and repair our 'world' so that we can live in it as well as possible. That world includes our bodies, ourselves, and our environment, all of which we seek to interweave in a complex, life-sustaining web" (Tronto, 1993, p. 103). In a degrowth framework, care should neither be organized by the State nor by the market, but embedded in commons-based self-managed projects (Orozco, 2014).

An even more radical challenge to the neoliberal logic comes from the Mediterranean roots of degrowth that embody the indolent recalcitrance against a subject-centered narrative of self-realization, planning, and (the illusion of) control of future opportunities. The Mediterranean strategy of *absence* that combines mimicry and local forms of subsistence for the preservation of social reproduction, i.e. "the parasitic capture of resource flows from the colonizing powers (related to mimicry) and the small scale self-production" (Romano, 2012, p. 588), combined with the ancient tradition of destroying debts and evening out inequalities in community-based celebrations, remind of Bataille's analysis of *dépense inproductive* (unproductive expenditure) with respect to the potlatch (1933).[17] *Dépense inproductive* provocatively challenges the moral norm, rooted in the neoliberal rationality, of investing efficiently and successfully in one's own personal portfolio.

Finally, even the very term degrowth with its pessimistic undertone carries a message of liberation from the diktat of "better." Against the pervasive logic of positivity at any cost, Latouche claims that "better" is often the enemy of good

---

17 Gift-giving ceremony practiced in the Northwest Coast of North America and used by anthropologists to address similar ceremonies and systems of gift exchange in other parts of the world.

(2001). The goal of a sustainable and equitable degrowth path is to secure the commonly negotiated conditions for a good life for all (Muraca, 2014). In the dynamic process of life, "good" is not a dull, static condition, but a place for care, maintenance, attention, and convivial embodiment.

With respect to the *mode* of subjectivation, degrowth represents an even more radical alternative to neoliberalism. The success of the neoliberal logic is its attempt at totalizing the normative space of action, by extending the entrepreneurial logic to all dimensions of life (personal, social, and biological). However, the very structure of legitimation of neoliberalism runs against its own overarching goal.

First and foremost, as Hayek's idea of *catallaxis* expresses, the real is neither monolithic nor pre-determined, but dynamically open. Accordingly, there is indeed no "society" as a whole operating in unison, but the "'ensemble of the social relations' ..., not that already constituted and indissoluble organic totality that would be the "thesociety," in the singular, but the unstable complex of antagonistic forces, in the plural, whose conflicts, at each instant, make, unmake, and remake that which is nothing but a precarious resultant." (Macherey, 2012).

In the neoliberal restructuring of society, as Foucault has rightly acknowledged, a new possibility for a policentric civil society is given that – he hoped – might lead to a more tolerant space for diversities.[18] Against the central control of the State and its forms of normalization (including the patriarchal family model during the "glorious" Fordist period), a new anarchist understanding of society can emerge against the Spencerian distortion of social interactions in terms of competition and self-valorization. A shift towards a new radical social formation is anticipated and embodied by social experiments, like the global commons, the international network of solidarity economy, or the countless post-capitalistic experiments worldwide (Gibson-Graham, 2006).

Since even the strongest ideology depends on some form of legitimation that somehow has to address the quest for *de-alienation* and to embody a promise for a good life, the set of norms and values that legitimize current institutions, practices, and behavior always bear a *surplus of meaning* (Bloch, 1976) that goes beyond the mode in which they are actually implemented under specific historical and social conditions. Against the totalizing pervasiveness of neoliberal rationality, by employing the excess of meaning as a lever for change, degrowth can operate as a *concrete utopia* (Bloch, 1976; Muraca, 2014) that kindles fantasy and opens possibilities for alternatives imaginaries: "Fantasy is what allows us to imagine ourselves and others otherwise. Fantasy is what establishes the possible in excess of the real; it points, it

---

18 One can understand Foucault's attraction to neoliberal governmentality with reference to its modus operandi at least as the promise that it embodied at the beginning of the 1970s: "On the horizon of this analysis we see instead the image, idea, or theme-program of a society, in which the field is left open to fluctuating processes, in which minority individuals and practices are tolerated" (2008, pp. 259–260). The sentence continues with the famous definition of governmentality as environmental intervention. Indeed, while this possibility might be slumbering beneath the image, idea or theme-program of neoliberal governmentality, it is long gone with the implementation of the *substance* of neoliberal ideology. As Brown remarks, Foucault "was surprisingly unimaginative about the implications of the neoliberal refashioning of the subject as human capital" (2015, p. 78).

points elsewhere, and when it is embodied, it brings the elsewhere home" (Butler, 2004, p. 29).

Concrete utopias rest on already established values and subversively transform them by shifting their meaning, while re-interpreting and embodying them in alternative practices (Muraca, 2017). Other than abstract utopias, which are led by merely wishful thinking and operate as palliative against daily alienation, concrete utopias envision the *real-possible*, what is already slumbering in the meanders of the actual world, and with *militant optimism* identify and actively seize the potentials and tendencies for transformation in the present (Bloch, 1976). Concrete utopias operate as a kind of catalyst that corroborates hidden tendencies pointing towards alternative futures.

Social reality resembles a complex fabric made of several threads that are woven together to compose visible patterns (Muraca, 2014). Hegemonic rationalities lure attention to the dominant pattern, while concealing the processes that actualize it under the surface – this is the power of the neoliberal TINA-mantra (*There Is No Alternative*) famously articulated by Margaret Thatcher. Yet, in a dynamic field, the main pattern results from the threads that keep reconfiguring it, by being woven over and over again into a uniform arrangement. The same threads can be combined in different patterns by a minimal shift in each repetition mode.

By pulling apart the fabric, the alleged "naturalness" of the status quo and its legitimizing imaginary can be unmasked as an *instituting process* thus triggering a shift from heteronomy (legitimation lays in something "other" to society, such as the market or the hypostatization of individual freedom) to autonomy (when a society becomes aware of its self-instituting character) (Castoriadis, 2010).

Acting *outside* of the field of possibilities framed by the dominant rationality is impossible, because it would mean to remain socially unintelligible. Becoming a subject results from the subjection to the general imaginary that constitutes the basic conditions for livability. However, at the same time, qua "power *exerted* on a subject, subjection is nevertheless a power *assumed by* the subject" (Butler, 1997, p. 11).

Because in a pluralist and dynamic ontology nothing can persist in existence without being re-enacted over and over again, norms (including patterns of behavior and recognition) can only become operative by being constantly repeated and re-actualized in each process of subjectivation. No matter how little autonomy is accessible in repeating the norms, every re-enaction bears the possibility of a re-signification, a shift, even if only in the form of subversive parodizing:

> The subject is compelled to repeat the norms by which it is produced, but that repetition establishes a domain of risk, for if one fails to reinstate the norm "in the right way," one becomes subject to further sanction, one feels the prevailing conditions of existence threatened. And yet, without a repetition that risks life – in its current organization – how might we begin to imagine the contingency of that organization, and performatively reconfigure the contours of the conditions of life? (Butler, 1997, p. 29).

With reference to pervasive heteronormativity, Butler shows how powerful the parody of heterosexual norms by queer performances can be in challenging the obviousness of the norm and breaking open (risky and painful) spaces of livability amidst the impossibility of being "seen" and "read." Given the (attempted) totalizing efficacy of neoliberal rationality, it seems promising to employ Butler's heterosexual matrix as a point of reference. It is true that the heterosexual matrix has a transhistorical validity that crosses paths with patriarchy and gender-binarism throughout time and space – with few, important exceptions that have been obscured by dominant narratives. It is also true that heteronormativity has been exacerbated with the advent of capitalism to serve the social reproduction of capital (Federici, 2004). Paradoxically, neoliberalism seems to offer, instead, a space for liberation under the guise of a new normativity that operates on structures of desire and is open to – prima facie – any type of desire. However, despite all the differences, neoliberal governmentality presents some key similarities to the heterosexual matrix in its pervasive mode of functionalizing desire to the entrepreneurial logic, while marginalizing any alternative attempt at escaping the competitive race. Not aiming at improving one's own career, not participating in the orgiastic celebration of fitness and wellness self-investment, not enhancing performance, might not make individuals entirely unintelligible, but their space of livability is significantly restricted or limited to prefabricated, peripheral identities.

Degrowth can learn from alliances with queer subversive strategies when it comes to challenge the pervasive attempt at a "naturalization" of the neoliberal growth logic and to open spaces for radically alternative imaginaries and practices. Resistance to the neoliberal logic operates from within through the re-signification of established patterns via subversive experiments aiming at the detoxification of desire – the powerhouse of subjectivation processes – and the *decolonization* of the dominant imaginary (Latouche, 2009).[19]

Becoming – or being made into – a subject is not a once and for all event, but a continuous process of reiteration and re-enaction (Dardot, 2011). If there is "no doer behind the deeds" (Nietzsche quoted by Butler, 1990) it is because the doer is done in the process of doing – who we are results time and again from what we do and from the possibilities of doing around us: "If I am someone who cannot be without doing, then the conditions of my doing are, in part, the conditions of my existence" (Butler, 2004, p. 3). This is why common acting (*agire commun* – Dardot, 2011) in subversive social experiments can create conditions for alternative subjectivities, in which different and differing processes of self-transformation can take place.

Given the governmental framework of neoliberalism, strategies of resistance require interventions on the environment, the *situation* (Dardot, 2011), in order to frame niches within the folds of the given, in which – in analogy to an ecological system – selection operates differently and alternative life forms are rendered possible and livable. In the cracks of a multifaceted and open present, radical social experiments open protected spaces, in which activists not only ignite alternative imaginaries for a radical transformation of society, but also build *spaces of*

---

19 While there is value in addressing the coloniality of neoliberal imaginaries, the term "decolonization" in the degrowth discourse does not (yet) fully addresses the material and literal dimension of what decolonization actually calls for.

experience against the invasion of life by the neoliberal logic (Pleyers, 2011, p. 39). Accordingly, *prefigurative* activism considers the struggle "as a process of *creative experimentation* in which the values of 'another world' are put into practice within organizations" and in daily life (Pleyers, 2011, p. 38). Radical social experiments at the same time anticipate future possibilities and frame the space in which these possibilities can be experienced, lived and tested today. They operate thus as laboratories of liberation (Muraca, 2014), where alternatives are literally forged and where participants can find the power and the motivation for resisting, building alliances, and continue the transformation in other areas of life. Prefigurative activism considers the transformation of the relations to oneself and to others as an essential part of the struggle for "'escaping the spirit of competition and consumerism promoted by neoliberalism' (an activist during the 'Beyond the ESF' gathering, London, 2004)" (Pleyers, 2011, pp. 38–39).

An essential function of utopias is the 'education of desire' (Levitas, 2010). In the language of a horizontal, spontaneous, and dynamic order, this means that the multiple social experiments within and adjacent to the degrowth archipelago can enable collective learning processes, in which people commonly mobilize and re-appropriate desire in different ways, to foster livability and care instead of investment and competitive struggles. In these protected spaces individuals are supported by material structures of solidarity and alternative patterns of recognition. The concomitant care for political goals and for the daily embodiment of social interactions and modes of desiring creates environmental interventions that trigger alternative modes of subjectivation.

4 Conclusion

Through the degrowth discourse the critique of economic growth of the 1970s has re-emerged at the beginning of the millennium in a new and more radical form. Against the neoliberal *stealth revolution*, degrowth embodies a radical alternative project, both with respect to the *substantial* goals and to the *mode of operation* of neoliberal governmentality. In its heterogeneity, degrowth opens spaces for radical imaginaries, practices, and experiences that challenge the neoliberal, pervasive logic of growth and self-optimization, while experimenting possibilities for alternative subjectivities and new modes of being.

We are now probably faced with the emerging of a new phase of neoliberal rationality that presents elements of a more direct intervention on people's lives along with softer environmental interventions aimed at enhancing self-valorization. The promise of growth (in a wider sense) for personal and social advancement no longer holds and its force to trigger desire might soon be fading. The current resilience discourse in sociology can be a symptom of a shift of narrative from competition aimed at self-improvement to a harsh competition for survival under impelling crisis scenarios (see Graefe, 2019). Accordingly, investing in one's own skills would be no longer attached to the promise of social advancement, but a necessary condition for dynamic adaptation. In the context of a possible refeudalization scenario (Neckel,

2010), degrowth in movement(s) embodies an even more radical form of resistance for radical democracy, collective autonomy, and solidarity. It fosters an entirely different meaning of resilience in terms of a commonly determined, active, and manifold transformation of the conditions for common living beyond growth.

References

Aglietta, M. (2000). *A Theory of Capitalist Regulation: The US Experience*, London: Verso.
Asara, V. (2016). The Indignados as a Socio-Environmental Movement: Framing the Crisis and Democracy. *Environmental Policy and Governance*, 26, 527–542.
Asara, V., Otero, J., Demaria, F., & Corbera, E. (2015). Socially Sustainable Degrowth as a Social–Ecological Transformation: Repoliticizing Sustainability. *Sustainability Science*, 10 (3), 375–384.
Balibar, E. (1995). The Infinite Contradiction. In *Depositions: Althusser, Balibar, Macherey, and the Labor of Reading*. Yale French Studies, 88, 142–165.
Bataille, G. (1933). *La Notion de dépense*. Paris: Éditions Lignes.
Biesecker, A., & Hofmeister, S. (2010). (Re) Productivity: Sustainable Relations Both Between Society and Nature and Between The Genders. *Ecological Economics* 69 (8), 1703–1711.
Bloch, E. (1976). *Das Prinzip Hoffnung*, Frankfurt a.M.: Suhrkamp.
Boltanski, L., & Chiapello, E. (2005). The New Spirit of Capitalism. *International Journal of Politics, Culture and Society*, 18 (3–4), 161–188.
Brown, W. (2015). *Undoing the Demos. Neoliberalism's Stealth Revolution*. New York: Zone Books.
Burkhart, C., Schmelzer, M., & Treu N. (eds.) (2020). *Degrowth in Movement(s)*. Zero Books.
Butler, J. (1990). *Gender Trouble*. New York: Routledge.
Butler, J. (1997). *The Psychic Life of Power*. Stanford: Stanford University Press.
Butler, J. (2004). *Undoing Gender*, New York. Routledge.
Caillé, A. (2004). Marcel Mauss et le paradigme du don. *Sociologie et sociétés*, 36 (2), 141–176.
Castoriadis, C. (1997). *The Imaginary Institution of Society*. Cambridge: MIT Press.
Castoriadis, C. (2010). *A Society Adrift: Interviews and Debates, 1974–1997*. New York: Fordham University Press.
D'Alisa, G., Demaria, F., & Kallis, G. (eds.) (2014). *Degrowth: A Vocabulary for a New Era*. Oxford: Routledge.
Dardot, P. (2011). La subjectivation à l'épreuve de la partition indivuel-collectif. *Revue du Mauss*, 83, 187–210.
Dardot, P., & Laval, C. (2014). *Commun. Essai sur la révolution au XXIème siècle*. Paris: La Decouverte.
Dardot, P., & Laval, C. (2017). *The New Way of the World: On Neoliberal Society*. London: Verso.
Deleuze, G., & Guattari, F. (1983). *Anti-Oedipus. Capitalism and Schizophrenia*. Minneapolis: University of Minnesota Press.
Demaria, F., Schneider, F., Sekulova, F., & Martinez-Alier, J. (2013). What is Degrowth? From an Activist Slogan to a Social Movement. *Environmental Values*, 22(2), 191–215
Dörre, K., S. Lessenich, & Rosa, H. (2015). *Sociology, Capitalism, Critique*. London: Verso.

Duverger, T. (2011). De Meadows à Mansholt: L'invention du "zégisme." *Entropia*, 10:114–123.

Escobar, A. (2015). Degrowth, Postdevelopment, and Transitions: A Preliminary Conversation. *Sustainability Science*, 10(3), 451–462.

Eversberg, D., & Schmelzer, M. (2018). The Degrowth Spectrum: Convergence and Divergence within a Diverse and Conflictual Alliance. *Environmental Values 27*, 245–267.

Federici, S. (2004). *Caliban and the Witch*. New York: Autonomedia.

Foucault, M. (2008). *The Birth of Biopolitics: Lectures at the Collége de France, 1978–79*. New York: Palgrave Macmillan.

Foucault, M. (1997). Sex, Power, and the Politics of Identity. In Rabinow, P. (ed.), *Essential Works of Foucault: Ethics, Subjectivity, Truth*, New York: The New York Press, pp.163–173.

Foucault, M. (1980): The Politics of Health in the 18th Century. In Gordon, C. (ed.), *Power/Knowledge: Selected Interviews and Other Writings*. New York: Pantheon Books, pp. 113–127.

Georgescu-Roegen, N. (1971). *The Entropy Law and the Economic Process*, Cambridge: Harvard University Press.

Georgescu-Roegen, N. (1995). *La Décroissance. Entropie – Écologie – Économie*. Paris: Éditions Sang de la terre. Original edition, 1979.

Georgescu-Roegen, N. (2011). Quo Vadis Homo sapiens sapiens. In Bonaiuti, M. (ed.), *From Bioeconomics to Degrowth*. New York: Routledge, pp.158–170.

Gibson-Graham, J.-K. (2006). *A Postcapitalist Politics*. Minneapolis: Minnesota University Press.

Graefe, S. (2019). *Resilienz im Krisenkapitalismus. Wider das Lob der Anpassungsfähigkeit*. Bielefeld: Transcript Verlag.

Habermas, J. (1984). *The Theory of Communicative Action*. Boston: Beacon Press.

Hayek, F. A. (1993). *Law, Liberty, and Liberty*. London/ New York: Routledge.

Klein, N. (2007). *The Shock Doctrine*. New York: Penguin.

Kothari, A, Demaria, F., & Acosta, A. (2014). Buen Vivir, Degrowth and Ecological Swaraj: Alternatives to Sustainable Development and the Green Economy. *Development*, 57 (3), 362–375.

Köchy, K. (2014). Lebensbegriffe in den Handlungskontexten der Synthetischen Biologie, in *Jahrbuch für Wissenschaft und Ethik*, 18 (2013), 133–172.

Latouche, S. (2001). *La déraison de la raison économique*, Paris: A. Michel.

Latouche, S. (2009). *Farewell to Growth*, Cambridge: Polity Press.

Lazzarato, M. (2004). From Capital-Labour to Capital-Life. *Ephemera*, 4 (3), 187–208.

Lessenich, S. (2008). *Die Neuerfindung des Sozialen*. Bielefeld: transcript.

Levitas, R. (2010). *The Concept of Utopia*. Bern: Peter Lang.

Martinez-Alier, J., Pascual, U., Vivien, F.-D., & Zaccai, E. (2010). Sustainable De-Growth: Mapping the Context, Criticisms and Future Prospects of an Emergent Paradigm. *Ecological Economics*, 69 (9), 1741–1747.

Macherey, P. (2012). Judith Butler and the Althusserian Theory of Subjection. *Décalages*: 1 (2). Translated by Stephanie Bundy. Available at: http://scholar.oxy.edu/decalages/vol1/iss2/13. accessed on February 18, 2018.

Meadows, D. H., Meadows, D., Randers, J. & W. W. Behrens (1972). *The Limits to Growth; A Report for the Club of Rome's Project on the Predicament of Mankind*. New York: Universe Books.

Miegel, M. (2010). *Exit: Wohlstand ohne Wachstum*. Berlin: Propyläen.
Muraca, B. (2010). *Denken im Grenzgebiet: prozessphilosophische Grundlagen einer Theorie starker Nachhaltigkeit*. Freiburg/ München: Alber.
Muraca, B. (2013). Decroissance: A Project for a Radical Transformation of Society. *Environmental Values*, 22 (2), 147–169.
Muraca, B. (2014). *Eine Gesellschaft jenseits des Wachstums*. Berlin: Wagenbach.
Muraca, B. (2017). Against the Insanity of Growth: Degrowth as a Concrete Utopia. In Heinzekehr, J., & P. Clayton (eds.) *Socialism in Process: Ecology and Politics toward a Sustainable Future*. Anoka: Process Century Press, pp.145–168.
Muraca, B., & Döring, R. (2018). From (Strong) Sustainability to Degrowth. A Philosophical and Historical Reconstruction. In Caradonna, J. (ed.) *Routledge Handbook of the History of Sustainability*, London/ New York: Routledge, pp. 339–362.
Muraca, B. & Schmelzer, M. (2017). Degrowth: Historicizing the Critique of Economic Growth and the Search for Alternatives in Three Regions. In Borowy, I., & Schmelzer, M. (eds.) *History of the Future of Economic Growth*, Abingdon: Taylor & Francis, pp. 174–197.
Neckel, S. (2010). Refeudalisierung der Ökonomie. Zum Strukturwandel kapitalistischer Wirtschaft. *Neue Zeitschrift für Sozialforschung* 8 (1), 117–128.
Offe, C. (1983). Competitive Party Democracy and the Keynesian Welfare State: Factors of Stability And Disorganization. *Policy Sciences* 15 (3), 225–246.
Oksala, J. (2016). *Feminist Experiences*. Evanston: Northwestern University Press.
Orozco, A. P. (2014). *Subversión feminista de la economía. Aportes para un debate sobre el conflicto capital-vida*, Madrid: Traficantes de Sueños.
Petridis, P., Muraca, B., & Kallis, G. (2015). Degrowth: Between a Scientific Concept and a Slogan for a Social Movement. In Martinez-Alier, J., & Muradian, R. (eds.) *Handbook of Ecological Economics*, Cheltenham: Edward Elgar, pp. 176–200.
Piketty, T. (2014). *Capital in the Twenty-First Century*, Cambridge: Harvard University Press
Pleyers, G. (2011). *Alter-Globalization: Becoming Actors in a Global Age*, Cambridge: Polity Press.
Read, J. (2002). Primitive Accumulation: The Aleatory Foundation of Capitalism. *Rethinking Marxism*, 14 (2), 24–49.
Romano, O. (2012). How to Rebuild Democracy, Re-Thinking Degrowth. *Futures* 44, 582–589.
Rosa, H., Dörre, K., & Lessenich, S. (2017). Appropriation, Activation and Acceleration: The Escalatory Logics of Capitalist Modernity and the Crises of Dynamic Stabilization. *Theory, Culture & Society*, 34 (1), 53–57.
Schmelzer, M. (2015): The Growth Paradigm: History, Hegemony, and the Contested Making of Economic Growthmanship. *Ecological Economics*, 118, 262–272
Schneider, F. (2008). Macroscopic Rebound Effects as Argument for Economic Degrowth. In Schneider, F., & Flipo, F. (eds.) *Proceedings of the First International Conference on Economic Degrowth for Ecological Sustainability and Social Equity*, Paris: Research & Development, INT, 29–36.
Tronto, J. (1993). *Moral Boundaries: A Political Argument for an Ethic of Care*, New York: Routledge.
Victor, P. (2008). *Managing Without Growth. Slower by Design, not Disaster.*, Cheltenham: Edward Elgar.
Walker, J., & Cooper, M. (2011). Genealogies of Resilience: From Systems Ecology to the Political Economy of Crisis Adaptation. *Security Dialogue*, 42 (2), 143–160.

Welzer, H. (2011). *Mental Infrastructures: How Growth Entered the World and Our Souls*, Heinrich Böll Stiftung, Berlin.

Yang, M., Howison, D. (2010): Introduction to the Special Issue "Commons, Class Struggle, and the World." *Borderlands* 11(2). Retrieved from www.borderlands.net.au/vol11no2_2012/yanghowison_commons.pdf.

29 Achieving Environmental Justice: Lessons from the Global South

Pearly Wong

1 Environmental Justice Movement

The use of the term "Environmental Justice" (EJ) as a form of political activism is widely credited to movements in the US South in the 1980s by African American communities against the unequal environmental costs shouldered by minority and poor communities (Bullard, 1999; Taylor, 2000). The use of the language of EJ has since extended far beyond its origins. The term now applies to the resistance of transnational and global environmental inequalities (Roberts, 2007), which incorporates a range of issues such as land-grabbing, food sovereignty, biopiracy, water justice etc. (Martinez-Alier et.al., 2016). The EJ movement is viewed to be distinct from the larger mainstream environmental movement, for it has sought to redefine environmentalism as much more integrated with the social needs of human populations (Pellow & Brulle, 2005).

The literatures traditionally conclude that environmental movements tend to be disproportionately populated by a new middle class of highly educated individuals (Cotgrove & Duff, 1980; Cotgrove & Duff, 1981; Cotgrove, 1982). But empirical cases, especially those from the global South, challenge this argument. In fact, most of the EJ movements in the global South initiate in the rural, with access to resources and livelihood as the motivating factors (Haynes, 1999). Some start in the form of unions and resemble class struggles. Yet, they transcend class structure, as intellectuals, and at times elites, are parts of those movements. Rather than against the economic ruling class, they oppose the state and other political and socio-cultural institutions (Laclau, 1985, Evers, 1985), including international and multi-lateral organizations.

I make a distinction here between EJ movements in the global North and the global South. Due to the distrust in civil public spheres, and constant power struggle for inclusion in the post-colonial contexts (Williams and Mawdsley, 2006), many who reside within the territories of the Southern countries, especially those who maintain distinctive identities and way of lives, are still struggling for recognition by the state. In countries where education is not as widespread and poverty is prevalent, paternalistic attitudes are rampant and the poor and minorities are often dismissed as backward, in need of "help" from both modern nation-states and development projects (Wong, 2018). For instance, according to a study on declining swidden agriculture in Southeast Asia (Fox et al., 2009), swiddeners are classified as minorities in the country, who are viewed to be at a primitive "stage" in a social evolutionary perspective (Cramb et al., 2009; Sturgeon 2005). Policy documents, laws, practices, and attitudes by states have ranged from the tagging of swidden cultivators as "lower quality people" in Southwest China to "isolated backward populations" in Indonesia (Li, 1999). Thus, the power disparity is especially large among certain populations and the states in the global South.

However, due to economic dependency of many Southern countries to multilateral financial institutions such as the World Bank, as well as on international development aid and trade, many Southern movements find an alternative pathway towards contesting domestic policies, by relying on the "boomerang" effect (Keck & Sikkink, 1998). The process entails support by international allies in pressuring Northern governments, which in turn influence international norms and institutions, forcing the Southern domestic governments to comply. While it is recognized that EJ issues could be faced by the poor and minorities in the North, it is unlikely that they would externalize their protests to achieve domestic goals.

I analyze several classic Southern EJ movements which used the "boomerang" effect. These include the Rubber Tappers' movement in Brazil, the Narmada anti-dam movement in India, the Zapatista uprising in Mexico, and the Ogoni movement in Nigeria. I do not include, for instance, the Chipko movement in India, which did not rely on this strategy to succeed. The chosen movements cover a range of environmental issues: mega dam construction, land-grabbing, severe environmental pollution and indigenous sovereignty, as well as major continents of the global South – Asia, Africa, Central and South America. The period of these movements coincided with the rise of attention towards the social and environmental effects of internationally funded development projects. They have been among the most widely cited movements in the literature and by the global civil society. Brief information on each movement is given below, following the historical timeline:

1.1 The Rubber Tappers' (RT) Movement

The RT movement started in the form of workers' union in the 1970s, when ranchers from Southern Brazil began to purchase massive land areas and forcibly evict rubber tappers from their forest land in Acre, Brazil. The strategy employed by the RT unions was *"empates,"* which involved groups of activists confronting laborers hired to clear the forests and attempting to persuade them to abort their work (Keck, 1995; Schwartzman, 1991). However, during this period, the RT union leaders were continuously assassinated and the Democratic Ruralist Union (UDR) was formed by the ranchers to combat the RT unions' movements (Schwartzman, 1991). In 1985, the National Council of Rubber Tappers (CNS) was established by Chico Mendes and other key union leaders, and the First National Rubber Tappers Congress was organized. The congress produced a proposal for extractive forest reserves as a more sustainable solution to secure rubber tappers' livelihoods (Schwartzman, 1991). Chico Mendes was brought to the United States to present the proposal to the US Congress, the World Bank and

Inter-American Development Bank, which were involved in funding a road project in the region (Keck, 1995; Schwartzman, 1991). The proposal was endorsed, compelling the Brazilian government to establish extractive reserves in the Amazon. Mendes later was assassinated but became an iconic figure in protecting the Amazon (Simons, 1988). As of 2017, around seventy extractive reserves have been established in Brazil (Instituto Socioambiental, 2017).

1.2 Narmada Bacho Andolan (NBA)

The Narmada river in India flows through the states of Gujarat, Madhya Pradesh, and Maharashtra. One of the most controversial dams proposed on the river is the Sardar Sarovar Project (SSP), which could result in the displacement of hundreds of thousands of people and widespread environmental damage (Narula, 2008; Ekins, 1992; Dwivedi, 1997, 1998). In 1985, the World Bank agreed to finance the SSP despite the absence of a sound resettlement and rehabilitation plan. Local opponents, environmental activists, and academic, scientific, and cultural professionals formed the Narmada Bacho Andolan (NBA), or the "Save Narmada Movement," led by Medha Patkar, who organized protest marches, rallies, sit-in, hunger strikes, etc. against the dam construction (Narula, 2008). In 1989, Patkar testified in front of the US Congress on the effects of the dam (Narula, 2008; Dwivedi, 1997, 1998). Due to mounting international pressure, the World Bank was forced to form an Independent Review Mission for the project and eventually withdrew its support (Narula, 2008; Dwivedi, 1997, 1998). The struggle against the SSP marked the first time that a community-level opposition effectively redirected a World Bank policy (Narula, 2008). The movement stepped up its campaign calling for a moratorium of World Bank funding of large dam projects all over the world through the Manibeli Declaration (Dwivedi, 1998).

1.3 The Zapatista Movement

The Zapatista Army of National Liberation (EZLN), often referred to as the Zapatistas, is a left-wing revolutionary political and militant group based in Chiapas, Mexico. The EZLN guerrilla forces, in cooperation with indigenous peoples, took over San Cristobal de las Casas, Chiapas on January 1, 1994 (Harvey, 1998). Factors for the uprising included ecological crisis and unavailability of productive land for the indigenous communities (Harvey, 1998), due to long-term exploitation of the indigenous people (Rus, Castillo and Mattiac, 2001). Though the Mexican President, Zedillo, launched a military offence against the ELZN in 1995, the conditions were quickly restored for negotiation by the Secretary of the Interior, Moctezuma. In 1996, The San Andrés Accords were signed, with the terms of peace and constitutional change that guaranteed the rights of the indigenous peoples of Mexico. The movement was recognized as the world's first postmodern revolution (Golden, 2001; McGreal, 2006) and first to have proliferated using the internet. It is now part of the larger movement of anti-globalization.

1.4 The Ogoni Movement

The Ogoni movement was launched in 1990 in response to decades of environmental degradation caused by oil mining by the Royal Dutch Shell Company in the Niger Delta. The Movement for the Survival of the Ogoni People (MOSOP), founded by Ken Saro-Wiwa, submitted the Ogoni Bill of Rights to the government of Nigeria and the United Nations. The bill contains the Ogonis' demand for self-determination and fair representation in all Nigerian institutions, as well as the right to protect the environment from degradation (MOSOP, 1990). The lack of response by the Nigerian government led MOSOP to directly address Shell (Agbonifo, 2009). On January 4, 1993, MOSOP organized a massive peaceful protest. In response to the protest and Shell's withdrawal from the region, the unstable national government forcefully suppressed MOSOP's activities, through direct crackdown and staged ethnic violence (Agbonifo, 2006; Cayford, 1996; Bob, 2002; Osha, 2006). As repression within the country increased, international support grew. Greenpeace and Amnesty International led international campaigns for Saro-Wiwa's release during 1992 and 1993 when he was repeatedly arrested (Osha, 2006). On May 21, 1994, four Ogoni chiefs who had disagreed with Saro-Wiwa were murdered. Saro-Wiwa and eight other activists were arrested and accused of murder. They were publicly executed on November 10, 1994, after an unregulated trial widely recognized as a travesty (Agbonifo, 2009; Cayford, 1996; Bob, 2002; Osha, 2006), resulting in international outcry.

2 Actors in EJ Movements

I focus on actors involved in the chosen EJ movements as the means to illustrate processes of internationalization and operationalization of the "boomerang" effect. I classify actors into international actors, domestic actors, and movement actors. As I will demonstrate below, both international and local dimensions of a movement do not operate in a vacuum but are mutually dependent. Domestic and movement actors act in response to global political economy and international pressure, while international allies capitalize on grassroots mobilization and charismatic leaders at the local level to legitimize their support.

2.1 International Actors: International Institutions, Foreign Governments, and International Allies

Evidence of internationalism and internationalization is found in multilateral institutions such as the United Nations, the World Bank, and the World Trade Organization (WTO); in the growing intergovernmental relations (Slaughter 2004); in regional alliances like the EU; in networks of informal ties among nongovernmental organizations, and advocacy networks. Internationalism today is complex, horizontal, and vertical, giving a wide range of potential venues for contention (Tarrow, 2005). Three major types of international actors are identified in the selected movements:

international institutions (the World Bank and the InterAmerican Development Bank), foreign governments (United States, Finland, EU, Japan) and international NGOs and allies (the Environmental Defense fund, Greenpeace, etc.)

International institutions, i.e. the World Bank and Inter-American Development Bank (IDB), were the focal point for contention (Tarrow, 2005) in two out of the four selected movements: the Narmada Bacho Andolan (NBA) in India and the Rubber Tappers' (RT) movement in Brazil. At that time, the World Bank had agreed to finance the SSP on the Narmada river, while the IDB was supposed to fund the road project in Acre, an extension of the road project from Rondonia previously funded by the World Bank.

International NGOs and allies mediated the process of "externalizing" these movements' demands to the international arena. "Externalization" is defined as the vertical projection of domestic claims onto international institutions or foreign actors (Tarrow, 2005). For instance, Chico Mendes was brought by the Environmental Defense Fund (EDF) and the National Wildlife Federation to speak with members of the US Congress and with World Bank and IDB staff in Washington and Miami (Keck, 1995). The banks formally endorsed the RT's proposal for extractive forest reserves within the same trip (Schwartzman, 1991), compelling the Brazilian government to implement those proposals in projects funded by them. This chain of pressure cascading from movements and their international allies, to foreign governments, then international institution, and back to domestic government, resulting in a change of domestic policy, is the typical "boomerang" effect described by Keck and Sikkink (1998).

Likewise, for the NBA movement, EDF arranged for the NBA's leader, Medha Patkar, to testify in front of the US Congressional Sub-Committee on Natural Resources, Agricultural Resources and Environment, on the effects of the SSP. Other allies of NBA included the Narmada International Action Committee, Friends of the Earth, and Japan's Overseas Economic Cooperation Fund (Dwivedi,1997). Due to the advocacy efforts by these international allies, the US, Finnish, Japanese and EU governments consecutively pressured the bank to withdraw its funding. The World Bank was forced to establish an Independent Review Mission and found the evidence of the project to be flawed, eventually cancelling the funding agreement.

International alliances also existed for both the Zapatista movement in Mexico and the Ogoni movement in Nigeria. Within days of the Zapatista uprising, supporters began to circulate information through email lists (Shirley 2001). The mobilization facilitated by internet communities created an outpouring of support both on and off the Internet (Martinez-Torres 2001; Olesen 2004). The attention drawn to the Mexican government made it impossible for them to conduct large-scale repression (Castells 1997; Downing 2001).

Unlike Zapatista, MOSOP's network of international allies (including Greenpeace and Amnesty International), however, could not help the movement in evading repression. Though the repression did further strengthen the Ogoni's international support, Saro-Wiwa and eight other MOSOP leaders were eventually arrested, falsely charged with murders, and publicly executed. This outcome was especially shocking because in the weeks before the execution, the Ogoni network convinced important international politics figures, such as President Bill Clinton, Prime Minister John Major, and President Nelson Mandela, of the injustice being done by the Nigerian government. But the Abacha regime cared little about its reputation abroad. Moreover, the severest sanction, an international boycott of Nigerian oil, was never seriously considered. The Ogoni movement has never recovered from Abacha's depredations (Bob, 2002).

While international allies are essential factors mediating the transnational activism process, and contributing in garnering widespread attention and resources, the relatively effective movements (in terms of achieving goals) seem to be those with additional venues for contention, i.e. the banks. The presence of banks incurs the involvement of foreign governments, which effectively impact the decisions by the banks in favor of the movements. However, the involvement of international institutions and foreign governments are restricted to movements which are issue specific and targeting projects funded by them, precluding any resistance for larger societal transformation.

In addition, the "boomerang" effect is dependent on the extent to which domestic governments are reliant on external banks' funding. The Brazilian government has, since the movement, established extractive reserves in the country, but the Indian government continued the dam project with its own funds despite the withdrawal of the World Bank from the project. The dam was completed and inaugurated in September 2017 by Indian Prime Minister, Narendra Modi. (Express Web Desk, 2017). The NBA case demonstrates that the "boomerang" effect can sometimes fail. Though the strategy of targeting the World Bank was successful, it did not help to achieve the movement's goal of stopping the dam project.

2.2 Domestic Actors: State Authorities, Political Elites, and Countermovements

It is not straightforward to categorize domestic actors into state authorities, political elites, and countermovements, as these actors have overlapping identities and are not coherent entities by themselves. Their actions could simultaneously provide opportunities and constraints to the movements, as demonstrated by the selected cases.

In the context of the RT movement, the environmental movement grew significantly in Brazil during the transition from military to democratic rule in the late 1970s and 1980s. Repression did not occur directly on the RT but through the form of inaction when ranchers forcefully evicted the RT from their land and assassinated union leaders with impunity. Ranchers even formed the Democratic Ruralist Union (UDR) to combat the unions' movements. Mendes, during his time as a city councilor from 1977 to 1982, consistently criticized the state authorities in many of his speeches for not enforcing the law in favor of the have-nots, and serving only their own personal interests. (Rodrigues and Rabben, 2007). Nevertheless, the political liberalization that began in Brazil in 1974 enabled diverse forms of grass-roots organizing, many under the umbrella of the Catholic church (Keck, 1995). By 1985 there were a series of accomplishments by the RT movement – formation of rural unions in the state, demonstrations against deforestation and expulsion, popular education projects, successful mobilization against evictions elicited by the local landholding elites, etc. (Schwartzman, 1991). The movement could not have achieved the same outcomes if it had taken place a decade earlier.

In the case of the SSP on the Narmada river, there was an inter-state conflict regarding the shape of the project and distribution of benefits between Gujarat, Madya Pradesh (M.P.), and Maharashtra. Most of the submergence was to occur in M.P., including fertile agricultural land and land for government projects. As a result, the M.P. congress party, initially as an opposition party, supported the NBA. Upon winning the state election, the party continued tacit support for the NBA through the establishment of the Five Member Group (FMG), and two high-level committees to consider resettlement issues and tackle the problems of the Scheduled Tribes in the state (Dwivedi, 1997). On the contrary, the Gujarati government organized rallies, festivals and exhibitions throughout Gujarat to highlight the benefits of the SSP. In a short time, all major political parties, Gujarati NGOs, the chambers of commerce, farmers' associations and even Gandhian social activists in Gujarat extended support to the project (Dwivedi, 1997), presenting a strong countermovement to the NBA.

The Zapatista uprising also occurred in an interesting time when NAFTA was being negotiated. Both President Salinas and the US government ignored prior information of possible unrest by the Zapatistas to avoid any distraction (Collier and Quaratiello 1999; Ronfeldt 1998; Schulz 1998). In order to placate human rights and financial interests, the Mexican government reformed its military policy throughout the conflict in Chiapas, creating divisions between the political and military elites (Muñoz, 2008). The Zapatistas movement had been able to gain allies within the legislative branches of Mexico due to intra-party elite tension (Tarrow 1998). The Zapatistas also gained support from the Mexican public, by allowing people to participate in voting on the topics they would discuss in their negotiations with the Mexican government (Collier and Quaratiello 1999; Nash 2001). There were Counter-Zapatista groups among cattle ranchers, coffee producers, businessmen, Chiapas residents, and rural communities and peasant organizations affiliated with the authoritarian political party, the Revolutionary Institutional Party. This countermovement had a positive, reciprocal effect on the Zapatistas' activity (Inclan, 2012).

The Movement for the Survival of the Ogoni People in Nigeria took place in a time of extreme political instability, marked by repeated coups and seizure of power by different regimes, and annulation of election results. MOSOP was able to organize during this period and successfully launched mass protest. Following the suspension of production in the Ogoni land by the Royal Dutch Shell due to "breakdown of law and order," (Human Rights Watch, 1995) repressive measures against the Ogoni stepped up. Soldiers were asked to confront aggressive communities and to "shoot troublemakers if necessary." (Boele, 1995). After General Abacha took power in 1994, he brutally repressed the Ogoni movement. The River State Internal Security Task Force began a series of raid on Ogoni villages, shooting, beating, raping, and detaining Ogoni people. The MOSOP leaders were charged with having instigated the mob that killed pro-government Ogoni chiefs. Despite criticism from international organizations regarding the irregularities of the trial, Saro-Wiwa and eight other MOSOP leaders were publicly executed (Cayford, 1996).

From these examples, it is apparent that states are not coherent, single-standing entities which either repress or facilitate movements. The identities of authorities, political parties, movements, and countermovements are dynamic and overlapping. For instance, the Gujarati government itself presented the countermovement for NBA and a political party has been involved in counter-Zapatista activities. The M.P. Congress Party was part of the NBA movement as an opposition party, until it was elected and provided support as a state government instead.

States are also increasingly constrained by the larger political economy context, with the presence of multilateral institutions and agreements, such as the United Nations, the World Bank, NAFTA, etc., as shown by the case of Mexican government during the Zapatista uprising. Nevertheless, states still hold tremendous power as they could punish their internal dissidents regardless of international reactions, if they chose so, as observed in the Ogoni case.

2.3 Movement Actors: Movement Leaders

Nepstad and Bob (2006) highlighted three important capitals of movement leadership: cultural, social, and symbolic. Cultural capital refers to familiarity with local values and the ability to connect with a mass base (Barker, Johnson, and Lavalette 2001; Veltmeyer and Petras 2002), as well as knowledge of the cultural principles and political trends within the broader public they seek to engage. Such knowledge needs to be applied along with media skills, rhetorical abilities, and strategic practicality (Nepstad and Bob, 2006); Social capital is proficiency in strong face-to-face relations and weak distant connections to broader networks (Diani and McAdam 2003), through media, Internet and other communication methods (Nepstad and Bob, 2006); Symbolic capital refers to prestige, honor, and social recognition, the embodiment of ideals by leaders, for which people struggle and remain faithful to, in face of a difficult and dangerous cause (Nepstad and Bob, 2006). I argue that all the three capitals were demonstrated in the leadership of the selected movements, and are thereby crucial for organizing transnational activism.

For example, Chico Mendes, the leader of the RT movement, was from a rubber tapper's family himself and learnt how to read from his teacher, Euclides, who showed him the great exploitation of the rubber tappers (Rodrigues and Rabbin, 2007). His early days as a union leader coincided with the increasingly tense confrontation with the ranchers. This strengthened his social and cultural capitals, and trained him to be capable of negotiating on an equal footing with his enemies. Realizing the importance of allies, he sent written messages to NGOs asking for help to support the RT movement or denounce the death threats he was receiving. Eventually, he was invited to the United States and won international awards, which he dedicated to all rubber tappers (Rodrigues & Rabbin, 2007). His symbolic capital was apparent when the news of his death made the front page of the *New York Times* (Simons, 1988).

Marcos, the spokesperson of the Zapatista movement, though not an indigenous person himself, remained in the jungle for over a decade, learning about local needs, indigenous culture, and ethnically based discrimination (Nepstad and Bob, 2006). As a result, the EZLN's ideology, goals, and strategies gradually shifted to suit the agenda of the indigenous population (Nepstad and Bob, 2006). Marcos took advantage of the world's fascination with the NAFTA and anti-globalization themes, and steered the Zapatistas into pioneering and piggybacking on the emerging global justice/anti-globalization movement. Marcos's writings were the main channel through which the movement communicated with the national and international public and an essential mechanism in the creation of the Zapatista network identity (Russell, 2005). Marcos's persona became a central part of the Zapatista movement, with symbolic meanings conferred upon him by others. As Marcos himself stated, he symbolized "romantic, idealistic expectations, namely the white man in the indigenous world, akin to references in the collective unconscious, Robin Hood" (Scherer García, 2001). His own composite identity – intellectual, revolutionary, activist, artist, worker, feminist, Indian, Mexican, multilingual, cosmopolitan – encouraged others to go beyond traditional identity affiliation and to join the network and embrace the hybridity of their own identities. His figure granted outsiders permission to become involved in the struggle (Russell, 2005).

Saro-Wiwa, the main leader for the MOSOP, was heavily invested in his people and understood their plight. He noticed that the Ogoni people lacked a mass organization to press for their rights (Agbonifo, 2009). He conceived the idea of the MOSOP, united rival factions within the Ogoni community, mobilized both elitist and mass support, and wrote key documents including the group's manifesto – the Ogoni Bill of Rights. Saro-Wiwa's fluency in English, knowledge of the international cultural and political scene, and experience in media, helped him to understand the value of publicity (Saro-Wiwa, 1995). He also had unusual access to major environmental and human rights networks as a writer and president of the Association of Nigerian Authors, who had won several fellowships to study, lecture, and tour Europe, the United States, and the Soviet Union in the late 1980s (Nepstad and Bob, 2006). He successfully initiated the Ogoni Day March in January 1993 (Nepstad and Bob, 2006), and garnered international support in the process. After the execution of Saro-Wiwa, Nigeria was suspended from the Commonwealth. International attention continued until, in 2001, the African Commission on Human and Peoples' Rights reached a verdict against the government of Nigeria (ACHPR, 2002).

Overall, the successes and the reputations of these movements were closely linked to their iconic leaders. These leaders were often the main, even sole, contact point between movements and international audiences. Their knowledge of transnational actors reinforced the multiple reframing that made international support more likely (Bob, 2002). Apart from their strategic leadership in mobilizing resources and achieving movement goals, their lives and struggles were of interest to so many that biographies and memoirs about Saro-Wiwa, Marcos, and Chico Mendes have been published. Their works were recognized through international awards, bringing additional attention to their movements (Patkar and Saro-Wiwa received the Right Livelihood Award while Mendes received the UN Environment Program's Global

500 award). The status of such leaders as heroic figures could be one of the most effective factors for drawing and sustaining international support and attention.

3 Framing Process in EJ Movements

As mentioned in the previous section, leaders require social capital to appeal to local and wider audiences. The stories constructed using such capital, about who suffers, who is the villain and what are the solutions, constitute the movement's framing. In this regard, "local" movements are more valuable to wider audience if they frame their interests to align with internationally resonant issues and broader NGO goals (Bob, 2002). Framing alignment was defined as "the linkage of individual and social movement organizations' (SMOs) interpretive orientations, such that some set of individual interests, values, and beliefs and SMO activities, goals, and ideology are congruent and complimentary." (Snow et. al., 1986). In transnational setting, movements target not only individuals for frame alignment, but also international civil society and multilateral institutions. To take advantage of transnational advocacy network, one must appeal to both local masses for mobilization and foreign allies for strength and resources, through global framing of local issues. This global framing processes are observed in almost all selected movements.

The RT movement began as unions for agricultural workers. Their struggle was mostly to preserve traditional livelihoods against the encroachment of the ranchers who threatened to turn tropical forests into pastures. The movement's initial identity and goals were not environmentalist, but books and television documentaries made the struggle into a metaphor of "Saving the Brazilian Rain Forest" (Keck, 1995). Hence, the movement expanded its original concern on extractive and land rights and became part of a broader movement for social justice and eventually part of a global environmental struggle. The framing brought the rubber tappers access to new allies and new institutional arenas for activism (Keck, 1995). At a time when egalitarian values were increasingly labeled utopian, environmentalism was a new universal value on which many of the same claims could be grounded. It gave substance to environmentalists' claim that preserving tropical forests was a universal good (Keck, 1995).

Similarly, the NBA began in the 1980s as a struggle for just resettlement and rehabilitation (R&R) of people being displaced by the SSP Dam. However, doubts were raised about the government's capabilities and will to implement the policies, as well as the availability of large quantity of land required for resettlement. In June 1988, after two organizations, the EDF and the Friends of the Earth, testified on the inadequate environmental impact assessments and cost-benefit analysis undertaken by the project authorities (Ekins, 1992 citing Udall, 1989), the focus was shifted to an entire opposition of the dam, for preserving the environmental integrity and natural ecosystems of the entire valley (Karan, 1994). The use of the language of risks, uncertainties, and environmental impacts was instrumental to the movement success in the international arena (Dwivedi, 1997).

Saro-Wiwa was first rejected when he approached the Greenpeace International, the Friends of the Earth International, and other environmental NGOs in 1990–1991 (Bob, 2002). He framed the Ogoni's cause in terms similar to those long voiced in Nigeria: the neglect of the Ogoni ethnic rights by the Nigerian state. However, he quickly learnt to highlight the environmental dimensions of the Ogoni's plight, particularly their exploitation by the Royal Dutch Shell company. This change of framing successfully gained international NGOs' support and media attention (Nepstad and Bob, 2006).

With regard to tactics, local movements improve their chances of support if they employ approaches acceptable to prospective international backers (Bob, 2002). At the start of the revolt, the Zapatistas identified themselves as armed revolutionary forces in Mexico. But in the efforts of gaining international support, they quickly switched their way to "civil resistance" and adopted the slogan *"mandar obedeciendo"* ("leading by obeying" civil society) (Nepstad and Bob, 2006).

From the examples above, the process of global framing is reflective of the North–South power imbalance. Global framing differs considerably from local framing of movement issues, or is otherwise rejected. Yet, the resources and leverage offered by international support is perceived to be so valuable that the global framing process is observed in all selected movements. For the Southern movements, global framing is not a process which takes place in domestic ground, but mutually shaped by the movement and its international allies, as well as the global media, in the process of 'externalizing' their movements.

4 The Importance of 'Local'

Internationalization is one of the many protest tactics employed by movements. These movements drew upon historical traditions, and derived strategies targeting multiple actors, not just international ones. Participants in movements like the RT have a shared history of oppression, peasantry and union activism. The MOSOP and the ELZN were both fighting for place-based sovereignty. For larger movements like the NBA, the leaders were able to exploit the Gandhian ideology and the *Sarvodaya* movement, which was already widely appealing to Indian society. The ideology is characterized by *Satyagrahas* (political action based on non-violence), asceticism, and self-sacrificing for the greater good, demonstrated through protest tools such as *jal samarpan* (sacrificial drowning) and hunger strikes, combined with strategies of *rasta roko* (road blockades) and *gaonbandi* (refusing the entry of state officials into the village) (Dwivedi, 1997).

Thus my focus on the international dimension of Southern EJ movements by no means suggests that its importance supersedes internal mobilization. In fact, there are examples of successful movements which do not rely on international alliances, such as the famous Chipko movement (Shiva and Bandyopadhyay, 1986). The movement was based on an age-old concept of "Dhandak" (Guha, 1989). Women of Garahwali have always worked together in collecting firewood, fodder, and water for mutual company and to prevent attacks. Due to people's dependence on forests, the idea of conservation and sustainability of the resource base formed an integral part of the sociocultural ethos of society (Tewari, 1995). Through tree-hugging, or Chipko, as a means of non-violent protest, they repeatedly stopped contractor workers from logging activities. The movement spread quickly to other parts in the region (Sethi, 1993). Globally, Chipko movement was the first to demonstrate that environmentalism is not only an agenda of the rich, but concerns daily lives of the poor.

It is not only that the Northern movements are supporting the Southern ones, but the latter can inspire and affect changes around the world. For instance, the NBA was one of the first anti-dam cases in the world which successfully affected a World Bank policy (Narula, 2008), and inspired anti-dam movements globally, such as the Brazilian anti-dam movement (Rothman, 1993). Through NBA's calling for a moratorium of World Bank funding of large dam projects all over the world, leading to the Manibeli Declaration and later the Curitiba Declaration (Dwivedi, 1997), the World Commission on Dams was established in February 1998 to review the impacts of major dams in the world. The report of the Commission supported the shaping of global norms around dams (Scheumann, 2008), and was immediately utilized by anti-dam campaigns in Thailand and Mozambique (Sneddon and Fox, 2008).

The ELZN is another extraordinary example. The Zapatistas' online presence was unprecedented, and scholars used the term 'Zapatista effect' to describe the impact of the movement (Cleaver, 1998). The influence of the pro-Zapatista mobilization reached across at least five continents, and the EZLN regularly sent messages to others around the world (Cleaver, 1998). In January 1996, Zapatistas called for continental and intercontinental "encounters" to discuss, among other things, contemporary global neoliberal (capitalist) policies, methods of elaborating a global network of opposition and formulas for connecting various projects for alternatives. Grassroots activists from over 40 countries and five continents attended the meetings – 3,000 to Chiapas and 4,000 to Spain (Cleaver, 1998). In this case, Zapatista itself has become a global movement.

In short, international mobilization cannot substitute local ones, and support for EJ movements is not necessarily one way–from the Northern allies to Southern movements. Instead, Southern EJ movements, given support, could potentially affect global agenda. These should be kept in mind by the Northern EJ actors, while negotiating a support for their Southern counterparts.

5 Conclusion

While international NGOs and allies are essential for mediating the transnational processes of these movements, the more effective movements, in terms of achieving specific goals, are those which find international institutions as additional venues of contention and incur the involvement of foreign governments. But using such venues also precludes movements for larger societal transformation – such as those drastically challenging the power of international institutions themselves. It is apparent here that global networks and internationalization support claim-making of some Southern EJ movements. Yet, it is economic globalization which brings conflicts between global capital and affected populations in the first place. These reaffirm Tarrow's observation (2005) that globalization and internationalization present both threats and opportunities for these movements. States are increasingly restrained by the presence of multilateral institutions and agreements, but still held sovereign power to repress its domestic dissenters, albeit with a suffering of international reputation. The effectiveness of the "boomerang" effect rests on the strength of domestic governments and their dependence on international resources. The effect can fail in cases where domestic governments are able to seek alternative resources in funding their domestic projects. Thus, internationalization, though important, is not sufficient in itself in achieving movement goals.

The demand for environmental justice is a dynamic process, dependent on capital, and opportunities arising from interactions among different actors. This is illustrated by the contestation and change of framing for movements upon interaction with international allies and global media. Hence, the simplistic understanding of environmental justice as equitable distribution of environmental resources and costs is largely inaccurate. In fact, different versions of environmental justice advocated by the above movements, from extractive reserves to self-determination, reflect the diversity of values and agency, as well as the timing and contextual opportunities captured by those movements. These suggest the importance of understanding environmental justice contextually, which entails the recognition of people's realities and aspirations, and the respect of their identities and territories (Wong, 2018).

The case studies here provide an important lesson that the Southern EJ movements could and should take advantage of internationalization and develop corresponding strategy, in terms of framing their environmental causes globally, reaching out to international allies, as well as establishing a lasting impression (including an iconic figure) to garner sympathetic support from outsiders. The additional arena for contention at the international level is a unique opportunity for movements in the global South. For their Northern counterparts, whose governments are the major decision makers themselves in the multilateral financial institutions and international organizations, international support is thus unlikely to exert meaningful pressure. For the many Northern movements with a global agenda, this chapter should provide lessons on effective ways to support their Southern counterparts, such as aiding the global framing of Southern movements while being attentive to their major needs and concerns, bridging Southern leaders and Northern governments, targeting multilateral institutions, and providing resources and networks, etc.

References

Adebanwi, W. (2004). The Press and the Politics of Marginal Voices: Narratives of the Experiences of the Ogoni of Nigeria. *Media, Culture & Society, 26*(6), 763–783. http://doi.org/10.1177/0163443704045508

The African Commission on Human and Peoples' Rights, ACHPR (2002, May 27). *Communication 155/96: Social and Economic Rights Action Center (SERAC) and Center for Economic and Social Rights (CESR) / Nigeria.* 30th Ordinary Session,

13–27 October 2001. Banjul, Gambia. Retrieved from www.achpr.org/communications/decision/155.96/

Agbonifo, J. (2009). Oil, Insecurity, and Subversive Patriots in the Niger Delta: the Ogoni as Agent of Revolutionary Change. *Journal of Third World Studies*, 26(2), 71–106. Retrieved from http://connection.ebscohost.com

Barker, C., Johnson, A., & Lavalette, M. (2001). Leadership Matters: An Introduction. In C. Barker, A. Johnson, M. Lavalette (eds.) *Leadership in Social Movements* (pp. 1–23). Manchester: Manchester University Press.

Bob, C. (2002). Political Process Theory and Transnational Movements: Dialectics of Protest among Nigeria's Ogoni Minority. *Social Problems*, 49(3), 395–415. http://doi.org/10.1525/sp.2002.49.3.395

Boele, R. (1995). *Report of the UNPO mission to Investigate the Situation of the Ogoni of Nigeria*. The Hague: UNPO. Retrieved from http://unpo.org/images/reports/ogoni1995report.pdf

Bullard, R. D. (1999). Dismantling Environmental Justice in the USA. *Local Environment* 4(1), 5–20. https://doi.org/10.1080/13549839908725577

Castells, M. (1997). *The Power of Identity, Vol. II*. Oxford, UK: Blackwell Publisher

Cayford, S. (1996). The Ogoni Uprising: Oil, Human Rights, and a Democratic Alternative in Nigeria. *Africa Today*, 43(2), 183–198. Retrieved from www.jstor.org/stable/4187095

Cleaver, H. (1998). The Zapatistas and the Electronic Fabric of Struggle. In J. Holloway & E. Pelaez (eds.) *Zapatista! Reinventing Revolution in Mexico* (pp. 81–103). London: Pluto Press.

Collier, G., & Quaratiello, E. (1999). *Basta! Land & The Zapatista Rebellion in Chiapas*. Oakland, CA: Food First Books.

Cotgrove S. (1982). *Catastrophe or Cornucopia: The Environment, Politics, and the Future*. Chichester: Wiley

Cotgrove, S. & Duff, A. (1980). Environmentalism, Middle-Class Activism and Politics. *Sociological Review*, 28:333–351. http://doi.org/10.1111/j.1467-954X.1980.tb00368.x

Cotgrove, S. & Duff, A. (1981). Environmentalism, Values, and Social Change. *British Journal of Sociology 32* (1), 92–110. http://doi.org/10.2307/589765

Cramb, R. A., Colfer, C. J. P., Dressler, W. et al. (2009). Swidden Transformations and Rural Livelihoods in Southeast Asia. *Human Ecology*, 37(3): 323–346. http://dx.doi.org/10.1007/s10745-009-9241-6

Diani, M, & McAdam, D. (2003). *Social Movements and Networks: Relational Approaches to Collective Action*. Oxford: Oxford University Press.

Downing, J. (2001). *Radical Media: Rebellious Communication and Social Movements*. Thousand Oaks, CA: Sage Publications

Dwivedi, R. (1997). *People's Movements in Environmental Politics: A Critical Analysis of the Narmada Bachao Andolan in India*. The Hague: Institute of Social Studies.

Dwivedi, R. (1998). Resisting Dams and "Development": Contemporary Significance of the Campaign against the Narmada Projects in India. *The European Journal of Development Research 10* (2), 135–183. http://doi.org/10.1080/09578819808426721

Ekins, P. (1992). *A New World Order: Grassroots Movements for Global Change*. London, New York: Routledge

Evers, T. (1985). Identity: The Hidden Side of Movement in Latin America. In D. Slater (ed.) *New Social Movements and the State in Latin America* (pp. 43–71). Amsterdam: CEDLA Online Archive.

Express Web Desk (2017, September 17). PM Modi inaugurates Sardar Sarovar Narmada Dam: Top quotes. *The Indian Express*. Retrieved from http://indianexpress.com/article/india/pm-narendra-modi-speech-inaugurates-sar-sarovar-narmada-dam-top-quotes-gujarat-4847648/

Fox, J., Fujita, Y., Ngidang, D. et al. (2009). Policies, Political-Economy, and Swidden in Southeast Asia. *Human Ecology, 37*(3), 305–322. http://doi.org/10.1007/s10745-009–9240-7

Golden, T. (2001, April 8). Revolution Rocks. Thoughts of Mexico's first postmodern guerrilla commander. *The New York Times*. Retrieved from www.nytimes.com/books/01/04/08/reviews/010408.08goldent.html

Goldstone, J.A. (2004). More Social Movements or Fewer? Beyond Political Opportunity Structures to Relational Fields. *Theory and Society 33*(3–4): 333–365. https://doi.org/10.1023/B:RYSO.0000038611.01350.30

Guha, R. (1989). *The Unquiet Woods: Ecological Change and Peasant Resistance in the Himalaya*. Delhi: Oxford University Press.

Harvey, N. (1998). *The Chiapas Rebellion: The struggle for land and democracy*. Durham: Duke University Press.

Haynes, J. (1999). Power, Politics and Environmental Movements in the Third World. *Environmental Politics, 8*(1), 222–242. http://doi.org/10.1080/09644019908414445

Human Rights Watch. (1995). Nigeria: The Ogoni crisis: A case-study of military repression in southeastern Nigeria. Human Rights Watch/Africa Report (July) 7: 5. Retrieved from www.hrw.org/legacy/reports/1995/Nigeria.htm

Inclán, M. (2012). Zapatista and Counter-Zapatista Protests: A Test of Movement–Countermovement Dynamics. *Journal of Peace Research 49* (3), 459–472. http://doi.org/10.1177/0022343311434238

Instituto Socioambiental. Extractive Reserve. Retrieved 19 December 2017 from https://uc.socioambiental.org/en/uso-sustent%C3%A1vel/extractive-reserve

Karan, P. P. (1994). Environmental movements in India. *Geographical Review 84*(1), 32–41.

Keck, M. E (1995). Social Equity and Environmental Politics in Brazil: Lessons from the Rubber Tappers of Acre. *Comparative Politics, 27*(4), 409–424. Retrieved from www.jstor.org/stable/422227

Keck, M. E., & Sikkink, K.(1998). *Activists beyond Borders: Advocacy Networks in International Politics*. Ithaca: Cornell

Keck, M. E., & Sikkink, K. (1999). Transnational Advocacy Networks in International and Regional Politics. *International Social Science Journal, 51*(159), 89–101. http://doi.org/10.1111/1468–2451.00179

Laclau, E. (1985). New Social Movements and the Plurality of the Social. In D. Slater (ed.) *New Social Movements and the State in Latin America* (pp. 27–42). Amsterdam: CEDLA Online Archive. Retrieved from www.cedla.uva.nl/50_publications/pdf/OnlineArchive/29NewSocialMovements/pp-27–42(Laclau).pdf

Li, T. M. (1999). Marginality, Power and Production: Analyzing Upland Transformations. In Li, T. M. (ed.) *Transforming the Indonesian Uplands: Marginality, Power and Production* (pp. 1–59). Amsterdam: Harwood Academic.

Martinez-Alier, J., Temper, L., Del Bene, D., & Scheidel, A. (2016). Is There a Global Environmental Justice Movement? *Journal of Peasant Studies, 43* (3), 731–755. http://doi.org/10.1080/03066150.2016.1141198

Martinez-Torres, M. E. (2001). Civil Society, the Internet, and the Zapatistas. *Peace Review 13*, 347–355. https://doi.org/10.1080/13668800120079045

McGreal, S. (2006). The Zapatista Rebellion as Postmodern Revolution. *Journal of Critical Postmodern Organizational Science*, 5(1), 54–64. Retrieved from http://tamarajournal.com/index.php/tamara/article/viewFile/252/pdf_83

Mertig A.G., & Dunlap R.E. (2001). Environmentalism, New Social Movements, and the New Class: A Cross-National Investigation. *Rural Sociology 66*, 113–136. http://doi.org/10.1111/j.1549–0831.2001.tb00057.x

MOSOP. (1990). Ogoni Bill of Rights. Port Harcourt: Saros. Retrieved from www.bebor.org/wp-content/uploads/2012/09/Ogoni-Bill-of-Rights.pdf

Muñoz, J. A. (2008). Protest and Human Rights Networks: The Case of the Zapatista Movement. *Sociology Compass*, 2(3), 1045–1058. http://doi.org/10.1111/j.1751–9020.2008.00115.x

Narula, S. (2008). *The Story of Narmada Bachao Andolan: Human Rights in the Global Economy and the Struggle Against the World Bank*. New York University Public Law and Legal Theory Working Papers. Paper 106. Retrieved from http://lsr.nellco.org/nyu_plltwp/106

Nash, J. (2001). *Mayan Visions: The Quest for Autonomy in an Age of Globalization*. New York: Routledge.

Nepstad, S., & Bob, C. (2006). When do Leaders Matter? Hypotheses on Leadership Dynamics in Social Movements. *Mobilization*, 11(1), 1–22. Retrieved from http://mobilization.metapress.com/index/013313600164m727.pdf

Olesen, T. (2004). The Transnational Zapatista Solidarity Network: An Infrastructure Analysis. *Global Networks 4*, 89–107. http://dx.doi.org/10.1111/j.1471–0374.2004.00082.x

Osaghae, E. E. (1995). The Ogoni Uprising: Oil Politics, Minority Agitation and the Future of the Nigerian State. *African Affairs*, 94(376), 325–344. Retrieved from www.jstor.org/stable/723402

Osha, S. (2006). Birth of the Ogoni Protest Movement. *Journal of Asian and African Studies*, 41(1–2), 13–38. http://doi.org/10.1177/0021909606061746

Pellow, D. N., & Brulle. R. J. (2005). *Power, Justice, and the Environment: A Critical Appraisal of the Environmental Justice Movement*. Cambridge, MA: Massachusetts Institute of Technology Press.

Piven, F., and Cloward, R. (1977). *Poor People's Movement's: How They Succeed, Why They Fail*. New York: Pantheon Books

Roberts, J.T. (2007). Globalizing Environmental Justice: Trend and Imperative. In R. Sandler and P. Pezzullo (eds.) *Environmental Justice and Environmentalism: The Social Justice Challenge to the Environmental Movement* (pp. 285–308). Cambridge, MA: MIT Press

Rodrigues, G., & Rabben, L. (2007). *Walking the Forest with Chico Mendes: Struggle for Justice in the Amazon*. Austin: University of Texas Press. Retrieved December 20, 2017, from Project MUSE database, https://muse.jhu.edu/book/13916

Ronfeldt, D. (1998). *The Zapatista 'Social Netwar' in Mexico*. Santa Monica, CA: Rand.

Rothman, F. D. (1993). Political process and peasant opposition to large hydroelectric dams: The case of the Rio Uruguai Movement in southern Brazil, 1979 to 1992 (Doctoral Dissertation). University of Wisconsin-Madison.

Rus, J., Hernández Castillo, A., & Mattiace, S. L. (2001). The Indigenous People of Chiapas and the State in the Time of Zapatismo: Remaking Culture, Renegotiating Power. *Latin American Perspectives*, 28 (2), 7–19. Retrieved from www.jstor.org/stable/3184983

Russell, A. (2005). Myth and the Zapatista movement: Exploring a Network Identity. *New Media and Society*, 7 (4), 559–577. http://doi.org/10.1177/1461444805054119

Saro-Wiwa, K. (1995). *A Month and a Day: A Detention Diary*. Ibadan: Spectrum Books.

Scherer García, J. (2001). "*La entrevista insólita*," interview with Subcomandante Marcos, March 10, Proceso. Retrieved from www.youtube.com/watch?v=5C4no09zGtE

Scheumann W. (2008). How Global Norms for Large Dams Reach Decision-Makers. In W. Scheumann, S. Neubert, M. Kipping (eds.) *Water Politics and Development Cooperation* (pp. 55–80). Berlin: Springer. Retrieved from https://link.springer.com/chapter/10.1007/978-3-540-76707-7_3#citeas

Schulz, M. (1998). Collective Action across Borders: Opportunity Structures, Network Capacities and Communicative Praxis in the Age of Advanced Globalization. *Sociological Perspectives*, *41*: 587–611. https://doi.org/10.2307/1389565

Schwartzman, S. (1991). Deforestation and Popular Resistance in Acre: From Local Social Movement to Global Network. *The Centennial Review*, *35*(2), 397–422. Retrieved from www.jstor.org/stable/23739139

Sethi, H. (1993). Survival and Democracy: Ecological Struggles in India. In P. Wignaraja (ed.) *New Social Movements in the South: Empowering the People* (pp. 122–148). New Delhi: Vistaar.

Shirley, S. (2001, September). "*Zapatistas Organizing in Cyberspace: Winning Hearts and Minds?*" Paper presented at the Conference of the Latin American Studies Association, Washington, DC. Retrieved from http://lasa.international.pitt.edu/lasa2001/shirleysheryl.pdf

Shiva, V., & Bandyopadhyay, J. (1986). The Evolution, Structure, and Impact of the Chipko Movement. *Mountain Research and Development Mountain Research and Development*, *610446*(2), 133–142. Retrieved from www.jstor.org/stable/3673267

Simons, M. (1988, December 24). Brazilian Who Fought to Protect Amazon Is Killed. *The New York Times*. Retrieved from www.nytimes.com/1988/12/24/world/brazilian-who-fought-to-protect-amazon-is-killed.html

Slaughter, A. (2004). *A New World Order*. Princeton: Princeton University Press

Sneddon, C., & Fox, C. (2008). Struggles Over Dams as Struggles for Justice: The World Commission on Dams (WCD) and Anti-Dam Campaigns in Thailand and Mozambique. *Society & Natural Resources*, *21*(7), 625–640. http://doi.org/10.1080/08941920701744231

Snow, D., Rochford, E., Worden, S., & Benford, R. (1986). Frame Alignment Processes, Micromobilization, and Movement Participation. *American Sociological Review*, *51*(4), 464–481. Retrieved from www.jstor.org/stable/2095581

Sturgeon, J. C. (2005). *Border Landscapes: The Politics of Akha Land Use in China and Thailand*. Seattle: University of Washington Press

Tarrow, S. (2005).*The New Transnational Activism* (Cambridge Studies in Contentious Politics). Cambridge: Cambridge University Press. http://doi.org/10.1017/CBO9780511791055

Taylor, D. (2000). The Rise of the Environmental Justice Paradigm. *American Behavioural Scientist*, *43* (4), 508–580. https://doi.org/10.1177/0002764200043004003

Tewari, D. (1995). The Chipko: The Dialectics of Economics and Environment. *Dialectical Anthropology*, *20*(2), 133–168. Retrieved from www.jstor.org/stable/29790401

Udall, L. (1989). *Statement on behalf of EDF et al. concerning the environmental and social impacts of the World Bank financed Sardar Sarovar dam in India before the Sub-Committee on Natural Resources*, Agricultural Research and Environment, 24 October, EDF. Washington DC.

Veltmeyer, H., & James P. (2002). The Social Dynamics of Brazil's Rural Landless Workers' Movement: Ten Hypotheses on Successful Leadership. *Canadian Review of Sociology and Anthropology*, *39*(1), 79–97. http://doi.org/10.1111/j.1755-618X.2002.tb00612.x

Williams, G., & Mawdsley, E. (2006). Postcolonial Environmental Justice: Government and Governance in India. *Geoforum*, *37*(5), 660–670. http://doi.org/10.1016/j.geoforum.2005.08.003

Wong, P. (2018). *Recognizing Plurality, Heterogeneity and Agency – A Contextualized Approach Towards Environmental Justice*. Unpublished Manuscript.

30 Conclusion: Envisioning Futures with Environmental Sociology

Julie C. Keller, Michael M. Bell, Michael Carolan, and Katharine Legun

In the final chapter of this far-reaching handbook, we turn to the future of the field, giving special attention to the most exciting and promising developments in theory and practice. But first, we want to retrace our steps to remind the reader of our intentions in putting together this collection.

Environmental sociology has become quite broad and highly multifaceted in a short period of time. As one of us has noted elsewhere (Bell & Ashwood, 2016), in the late 1990s social scientists had just begun thinking about environmental questions. Now, however, the idea that the social sciences have something to offer to the study of environmental problems is unlikely to give one pause. As this collection demonstrates through its theoretical approaches and empirical work, environmental sociology brings together a diverse set of scholars and perspectives to analyze how environment and society are intertwined, examining the causes and consequences of environmental problems, and proposing solutions.

As this handbook makes clear, we strongly believe that a comprehensive study of environment and society requires the participation of thinkers from a wide swath of disciplinary perspectives. Sociologists are certainly not the only scholars working in environmental sociology. Though they may dominate in numbers, sociologists have found collaborators from a range of disciplines who are also tackling the connections between environment and social life. Political science, philosophy, geography, economics, anthropology, psychology, and history, as well as interdisciplinary fields such as public health and gender and women's studies, are all important starting points for developing a growing community of scholars in environmental sociology.

We have also been mindful in curating this collection to draw from different generational perspectives, incorporating the wisdom of more senior environmental sociologists who ushered the field through its infancy, as well as scholars with fewer rings in their trees. We see intergenerational conversations as a critical ingredient for imagining and implementing solutions to the complex socioecological problems that confront us. Envisioning and carrying out sustainable futures no doubt requires the guidance of young voices, older voices, and those in between.

As we mentioned in the introductory chapter, when envisioning this collection we wanted to include voices from a range of different places about those places. All four editors are white, English-speaking, and from a predominantly North American academic tradition. In selecting contributors, we wanted to avoid reinforcing these privileged social locations by bringing voices from the margins to the center. Important chapters in the handbook reflect this vision. To name just a few, these include Pearly Wong's overview of environmental justice movements in the Global South (Volume 2, Chapter 29), Henri Acselrad's analysis of capitalism and environmental inequalities in Brazil (Volume 1, Chapter 26), and Gustavo A. García-López's conceptualization of the commons and "communing" practices (Volume 1, Chapter 11).

We recognize, however, that most chapters in the handbook do not challenge the dominance of white, English-speaking, North American – or European – perspectives in environmental sociology. Our field is of course not alone in this dominance, as Raewyn Connell (2007) emphasized in her critique of sociology's exclusion of ideas from the Global South. In the discussion that follows, we are especially mindful of the risks of reproducing this exclusion and others. To help account for this, we pay particular attention below to exciting developments in environmental sociology from perspectives in the Global South, from non-white voices, and that concern topics related to racial or indigenous-based inequality and justice.

Promising Developments in Theory and Practice

A major theoretical thread found throughout this handbook consists of questions over what constitutes the environment, society, and relations between the two. Such questions were central to the birth of environmental sociology, when Catton and Dunlap (1978) called for abandoning The Human Exceptionalism Paradigm that defined sociology, and these inquiries continue to feature prominently in the work of environmental sociologists today.

One entry point for understanding the connection between environment and society is to interrogate the cultural assumptions that keep these two concepts separated. Connell (2007), for instance, describes the universalist assumptions of sociological theories born in northern contexts. One of these is the centrality of the individual in theories about the workings of society. The focus on individuals and their relationships to the structures that organize social life is indeed pervasive in sociology, and in fact constitutes what we call "the sociological imagination." This way of apprehending the social world is of course a core feature of sociology, and is often the very first lesson we teach to students in introductory courses to the discipline. Yet we must be able to recognize its limitations. Not only does this perspective ignore cultural differences in conceptualizing the social world, but it also presents serious barriers for understanding the interconnectedness of environment and society, and developing what Bell and Ashwood (2016) have termed "the environmental sociological imagination."

In García-López's chapter on commons, communing, and communality, he points out that Latin American scholars are lately focused on the concept of *autonomismo*, a concept that prioritizes "the communal over the individual, the connection to the Earth over the separation between humans and non-humans, and *buen vivir* (roughly:

the good life) over the economy" (Escobar, 2014, García-Lopez's translation). Not only does *autonomismo* encourage us to see the interconnectedness of environment and society, but it also challenges the separation of the rational against the emotional, says García-Lopez, urging us toward a *sentipensar la tierra*, or "feeling-thinking [with] the Earth."

Bell (Volume 1, Chapter 5) points to a deep and longstanding conflict at the heart of the shift to an increasingly separatist view: the material and ideological conflict of the pagan and the bourgeois that arose with the great expansion of cities, states, and empires in the late Iron Age, and that is still very much with us today. In the pagan traditions of people of the countryside, the human, the non-human, and the divine are all *entangled* together in an often quite riotous interactiveness that even extends to the sexual and political – a realistic view that is by no means understood as necessarily good. In the bourgeois traditions of the people of the city, there is what Bell calls an *ancient triangle* of separations between the human, nature, and the divine, a separateness that enables nature and the divine to be envisioned as non-political and thus a basis for the good and the claims for innocence that underlie what Bell terms *non-political politics*. As well, this claim of bourgeois goodness promotes attributions of pagan moral backwardness, helping legitimate the exploitation of rural peoples and their land and resources, leading to urban and class accumulation.

Another focus that challenges the meaning of environment and society through an emphasis on power relations is the quickly growing field of environmental justice (EJ) studies. As contributors to this collection note, the emergence of EJ as a concept can be traced to the important work of African Americans in the US South in the 1980s, protesting environmental harms that disproportionately affected their communities. A critical moment in the modern EJ movement was the First National People of Color Environmental Leadership Summit held in Washington, D.C. in 1991. The Principles of Environmental Justice that resulted from this meeting have inspired countless social movements around the world, along with establishing the urgency of EJ as an interdisciplinary academic field.

This collection includes a number of exciting contributions from EJ studies that push this area forward. David N. Pellow (Volume 1, Chapter 25), for instance, calls for sharpening the focus of EJ in a new approach that he calls Critical Environmental Justice. This includes strengthening intersectional foundations, adding multiscalar components, integrating a more comprehensive analysis of power and hierarchy, and finally, embracing the principle of indispensability. Pellow explains that CEJ offers additional categories of social inequality, beyond the more common focus on race and class, to add complexity to our understanding of how marginalized populations experience environmental injustices. Deepening the intersectional basis of justice questions to more fully include the effects of gender, sexuality, citizenship, ability, and even species, would enrich EJ studies.

This fairly recent articulation of intersectionality seems to come from movements themselves. Looking to the work of practitioners on the ground, we see evidence of intersectionality at work in a variety of movement-building spaces. For instance, the Total Liberation frame among recent ecological movements brings together several frames from other movements, including ecofeminism and environmental justice

(Pellow & Brehm, 2015). And, as Julie C. Keller discusses (Volume 2, Chapter 5), the formation of queer food spaces grounded in LGBTQ rights, environmentalism, and often anti-racist principles, demands an environmental justice framework that is driven by intersectional priorities. In *Farming While Black*, Leah Penniman (2018) describes a "matrix of intersectionality" in which she invites the reader to mark down their multiple identities, including race, class, and gender, but also sexual orientation, ability, and documentation, among others. On the farm and organizing space that Penniman co-founded, Soul Fire Farm in upstate New York, this matrix is used to name and challenge various sources of privilege so that they can more effectively dismantle injustices in the food system and beyond.

Intersectional approaches such as these also offer promise for increased cross-disciplinary conversations within EJ studies. Phillip Warsaw (Volume 1, Chapter 29) notes that economic approaches to environmental injustice tend to rely on concepts such as political and distributional inequity, but an analysis of the power structures that create and maintain those inequities is often lacking. Warsaw argues that intersectionality provides a way to strengthen that analysis, while also bringing environmental sociologists and ecological economists together. Social ecological economics, with its critical stance toward both neoclassical macro and microeconomic theory, is central to these emerging conversations.

Pearly Wong's (Volume 2, Chapter 29) review of EJ movements in the Global South ties in nicely to the theme of strengthening EJ's focus on power and hierarchy, particularly when it comes to the state. Some of these movements have seen success through forming international allies that can apply external political pressure to the state. Although successful in some cases, this strategy has not always resulted in the state recognizing their claims. Also writing about state power, Leslie King points out in her analysis of the environmental impacts of capitalism that an exclusive focus on government regulations and enforcement is unlikely to lead to environmental justice (Volume 1, Chapter 28). She calls for more work in environmental sociology on identifying the legal and financial dimensions of corporate power in order to promote change.

For Michael Löwy (Volume 1, Chapter 10), environmental justice must be achieved through a large-scale reorientation of our society towards ecosocialism. He proposes that ecosocialism is a radical alternative framework that aims to develop a society that is ecologically oriented and socially just. He proposes that this can be achieved through collective ownership of the means of production, democratic engagement and planning, and a new technological structure. In taking the ecosocialist approach, Löwy, along with colleagues, takes the incisive critique of capitalism offered by political economists and offers a socialist alternative, but intervenes in more traditional accounts of socialism by suggesting that production processes be altered, rather than simply their property status. Indeed, he suggests we need to revolutionize the productive apparatus itself through radical transformation, undertaken through democratic planning and control.

Questions around democracy and environmental justice are also elaborated in the context of national parks in Africa by Maano Ramutsindela, who describes how parks were part of a colonial project, both reflecting and institutionalising colonial relations (Volume 1, Chapter 14). By drawing on empirical examples and a rich body of international scholarship, Ramutsindela furthers work on the colonization of nature, describing how parks can be a form of enclosure, and their administration the assertion of colonial power relations.

Democracy and justice are also discussed by Henri Acselrad in his chapter on environmental action and environmental movements in Brazil. He describes how the environment became a particular kind of contested terrain as a result of the character of state interactions with environmental management through dictatorial approaches aimed at securing resources for exploitation, and later neoliberal approaches that further concentrated resource access and power. These have undermined the democratic capacity of environmental management in Brazil, and in a context of wide social inequalities, present particular strategic roadblocks to environmental activists and those dispossessed.

Others rethink theory and practice in novel ways, engaging with ideas bubbling up across the disciplinary landscape. Resilience thinking became a popular part of systems thinking in ecology in the early 2000s, but also captured the attention of most scholars working on topics related to the environment and society. The resilience approach analyzed the degree to which a social group and its related ecosystems were able to adapt to shocks and thrive. Soon, those working in the area saw the concept of resilience everywhere, inspiring scholars, the public, and policymakers alike. As Hale and Carolan (Volume 1, Chapter 24) suggest, "resilience thinking" is not without its shortcomings. It can be seen to over-emphasize the importance of a particular state of society or the environment, normalizing the status quo, for example. While often presented in apolitical terms, then, resilience thinking comes with its own politics that must be accounted for. To address these concerns, they propose we can take a more porous, fluid, and potentially radical resilience approach by considering "relational resilience," emphasizing the importance of diverse connections across space and time. Others have similarly developed the concept of resilience in a sociologically radical way attuned to issues of justice, and we can see some of that work in the edited collection *Resilience, Environmental Justice and the City* (Caniglia, Vallée, & Frank, 2017). These works transform what can be a rather limiting and often depoliticized concept and develop it into a much more nuanced approach to socio-ecological justice with radical possibilities.

Other work tackles ethics and the development of ethical economies. Like Bell's chapter considering the development and evolution of a nature equated with goodness, discussed earlier, Paul V. Stock delves into texts dealing with environment and morality to help us develop a more concerted and sustained engagement with environmental morality (Volume 2, Chapter 25). In the chapter he proposes a subfield that merges environmental sociology with a sociology of morality explicitly, and explores a technological morality.

We see questions of morality, goodness, and its practical economic application in work exploring diverse and community economies. In their text *Take Back the Economy: An Ethical Guide for Transforming our Communities*, Gibson-Graham,

Cameron, and Healy (2013) outline a theory and practical approach to strategically combating capitalist dynamics and instead enacting a more democratic, moral, and ecological alternative. This work is an extension of some of the early community economies ideas developed by Gibson-Graham, and put forward in their book *The End of Capitalism (As We Knew It)*. The Community Economies Research Network (CERN) is a blossoming global network of people engaged with the work of Gibson-Graham, and who have been elaborating the ideas in exciting new ways. Several of the chapters have drawn on the community economies work including Elizabeth S. Barron's chapter on "Emplacing Sustainability in a Post-Capitalist World" (Volume 1, Chapter 12), Gustavo A. García-López's chapter "Commons, Power and (Counter)Hegemony" (Volume 1, Chapter 11), and Damian White and J. Timmons Roberts's chapter "Post Carbon Transition Futuring: For a Reconstructive Turn in the Environmental Social Sciences?" (Volume 1, Chapter 15). Many other chapters touch on the themes around the practical enactment of moral economies through forms of commoning and communities gaining control of assets that allow them to thrive socially, ecologically, and culturally. Some of these have been articulated through ideas around community resource management, touched on through chapters like Bianca Ambrose-Oji's chapter on community forestry (Volume 2, Chapter 15) or Nathan Young's chapter on adaptive co-management (Volume 2, Chapter 21).

Promising Avenues Across Disciplines

While we hope we have assembled a diverse set of contributors from a range of disciplinary perspectives in this handbook, we do see potential for deeper conversations in a few different fields.

First, we see rich opportunities for bringing the work of environmental sociologists closer to the field of public health. Carrera and Brown (Volume 2, Chapter 7), as well as Manuel Vallée (Volume 2, Chapter 8) identified numerous ways of linking these two fields. The "One Health" framework (e.g., Chien, 2013) offers much promise for fruitful conversations between these two areas. An additional area of collaboration is the connection between access to green space and health outcomes (e.g., Bezold et al., 2018), particularly for marginalized populations.

Second, although this handbook makes an effort to integrate perspectives from the humanities, there is much more room for sustained engagement. Environmental historians, for instance, have much to offer to environmental sociology, and vice versa. In a comprehensive study of timber unions and environmentalism across the twentieth century, Erik Loomis (2016) untangles entrenched assumptions about the incompatibility of workers and environmental concerns. This research has much to offer to environmental sociologists who are interested in bringing a deeper focus on labor to the field. Moreover, we have much to learn from the arts, both in terms of embracing the full diversity of forms of human expression and communication, and embracing less rationalistic forms of thought and their tendency to unnecessarily bound our sense of the possible.

Third, we have yet as a field to take fully to heart the value of reaching beyond the disciplinary to the transdisciplinary, welcoming the non-academic knowledge of the practitioner into our "careful, committed conversations," as we phrased it in the introduction to this handbook. Scholarly knowledge is valuable and so is practitioner knowledge, but each tends to undervalue what the other has to offer, largely because of institutional barriers. We just don't have occasion to talk much with each other. Academic conferences are usually mostly just that: academic. Scholars are able to attend because their institutions support their travel (although less than they should!), as they recognize that the prestige of their institutions depends in part upon evaluation by academic peers. The practitioner from NGOs or from frontline communities has no such institutional need, and moreover is much less likely to have financial support for attending the wonderful conferences where the disciplines and interdisciplines discuss vital issues of environmental sustainability and justice. These are structural issues that will take concerted work to overcome, if we are to engage a broad and transformative conversation about the "environmental sociological imagination."

Conclusion

As we write this concluding chapter in 2019, the United Nations just released a report finding that over one million species are at risk of extinction due to the devastating ecological impacts of humans (Plumer, 2019). In the United States, the Trump administration continues rolling back environmental regulations, shrinking the power of the Environmental Protection Agency and relaxing standards in favor of economic interests (Gibbens, 2019). Across the Atlantic, France's Minister of the Environment announced his resignation in 2018, citing President Macron's unwillingness to make serious progress toward environmental goals as reason for his departure (Chrisafis, 2018).

We are not blind to the dark reality that is unfolding around us. Yet we do remain hopeful with each glimpse that society is ready to take serious steps to prevent the further destruction of our planet. As we write, the House of Commons in the United Kingdom declared a climate change emergency, with Labour Party leaders pushing for net-zero emissions by 2050 and demanding a zero waste economy (BBC, 2019). At the same time, students around the world, from New Delhi to Washington, D.C., are walking out of schools and holding demonstrations to protest inaction over climate change (Associated Press, 2019).

It is in these urgent times that we especially need guidance about the interconnectedness of humanity and environment. The chapters in this handbook lay out the most significant and promising trajectories of thought in the field of environmental sociology. On a practical level, these authors offer a wide range of tools for understanding today's socioecological problems and for developing solutions to address them. As we emphasized in the introduction to this handbook, the pressing challenges we face in this bleak moment will require strong communities and strong coalitions. The future of our precarious planet depends on a collective effort.

References

Associated Press. (2019, March 15). Students worldwide walk out of school to push for action on climate change. *Washington Post*. Retrieved from https://wapo.st/2TQRJ5u?tid=ss_tw&utm_term-.456b3ee65fde

BBC. (2019, May 1). UK Parliament declares climate change emergency. *BBC News*. Retrieved from www.bbc.com/news/uk-politics-48126677

Bell, M. M., & Ashwood, L. L. (2016). *An Invitation to Environmental Sociology* (5th ed.). Thousand Oaks, CA: Sage.

Bezold, C. P., Banay, R. F., Coull, B. A. et al. (2018). The Association Between Natural Environments and Depressive Symptoms in Adolescents Living in the United States. *Journal of Adolescent Health*, 62(4), 488–495. https://doi.org/10.1016/j.jadohealth.2017.10.008

Caniglia, B. S., Vallée, M., & Frank, B. (eds.). (2017). *Resilience, Environmental Justice and the City*. New York: Routledge.

Catton, W. R., & Dunlap, R. E. (1978). Environmental Sociology: A New Paradigm. *The American Sociologist*, 13, 41–49.

Chien, Y.-J. (2013). How Did International Agencies Perceive the Avian Influenza Problem? The Adoption and Manufacture of the 'One World, One Health' Framework. Sociology of Health & Illness, 35(2), 213–226. https://doi.org/10.1111/j.1467–9566.2012.01534.x

Chrisafis, A. (2018, August 28). French environment minister quits live on radio with anti-Macron broadside. *The Guardian*. Retrieved from www.theguardian.com/world/2018/aug/28/french-environment-minister-quits-live-on-radio-with-anti-macron-broadside

Connell, R. (2007). *Southern Theory: Social Science and the Global Dynamics of Knowledge*. New York: Polity.

Escobar, A. (2014) *Sentipensar con la tierra: Nuevas lecturas sobre desarrollo, territorio y diferencia*. Medellín: Ediciones UNAULA.

Gibbens, S. (2019, February 1). 15 ways the Trump administration has impacted the environment. National Geographic. Retrieved from www.nationalgeographic.com/environment/2019/02/15-ways-trump-administration-impacted-environment/

Gibson-Graham, J. K., Cameron, J., & Healy, S. (2013). *Take Back the Economy: An Ethical Guide for Transforming our Communities*. Minneapolis, MN: University of Minnesota Press.

Li, T. M. (2014). What is Land? Assembling a Resource for Global Investment. *Transactions of the Institute of British Geographers*, 39(4), 589–602.

Loomis, E. (2016). *Empire of Timber: Labor Unions and the Pacific Northwest Forests*. New York: Cambridge University Press.

Pellow, D. N., & Brehm, H. N. (2015). From the New Ecological Paradigm to Total Liberation: The Emergence of a Social Movement Frame. *Sociological Quarterly*, 56(1), 185–212. https://doi.org/10.1111/tsq.12084

Penniman, L. (2018). *Farming while Black: Soul Fire Farm's Practical Guide to Liberation on the Land*. White River Junction, VT: Chelsea Green Publishing.

Plumer, B. (2019, May 6). Humans Are Speeding Extinction and Altering the Natural World at an 'Unprecedented' Pace. *New York Times*. Retrieved from www.nytimes.com/2019/05/06/climate/biodiversity-extinction-united-nations.html

Index

ABC models of behavior, in green governance models, 225
Abend, Gabriel, 433–5
accumulation of resources, water, 305
accumulation regimes, 482
Achieving Results in Communities, 255–6
ACM. *See* adaptive co-management
active citizenship, 381–6
   environmental governance through, 262
   word clouds and, 382, 383
actor-network theory (ANT), 165–7, 173
   material semiotics and, 166
   'new materialisms' and, 169–70
   political economy systems and, 171–2
   power and, 166–7
acute myeloid leukemia (AML), 144, 149
adaptive co-management (ACM), of natural resources, 360–72
   academic literature on, 366–7
   adoption of, as policy ideal, 361–3, 371–2
   challenges with, 363–6
      epistemological, 364–5
      gulf between types of knowledge, 364–5
      procedural, 365
      relational, 365–6
      structural, 365
   environmental sociology and, 369–71, 372
   in governance, 362–3
   limitations of, 361
   origins of, 361–2
   in participation paradigm, 362
   in practice, 363–6
      critiques of, 368–9
      in deliberative democracies, 368–9
      in representative democracies, 368–9
adult learning theories, 34
advocacy biomonitoring, 129–30
AERs. *See* Alberta Energy Regulators
affordability crisis, for water, 311
Africa. *See* Sub-Saharan Africa; *specific countries*
African Americans, masculine identity among, 109
agencement, in assemblage theory, 167
aggregation mechanisms, in spatial data, 76–7
Agricultural Justice Project (AJP), 463

agricultural movements. *See also* sustainable agriculture movement
   Community Supported Agriculture, 251–2
   La Via Campesina, 421–2
   Worldwide Opportunities on Organic Farms movement, 216–17
Agricultural Resource Management Survey (ARMS), 377
agricultural technology, in 'Gene Revolution', 394
agriculture, sociology of morality and, 438–40
   animal agriculture, 439–40
   for farming, 438–9
AJP. *See* Agricultural Justice Project
Alberta Energy Regulators (AERs), 336
Alberta province (Canada), fracking in, 328–38
   under Climate Leadership Plan, 332
   natural gas production and, 330–2
      public need for, 332
      reserves for, 330–2
   oil production compared to, 330
   sociotechnical imaginaries and, 329–34
      public concern as pillar of, 335
      safety concerns as pillar of, 333–4
      silencing of public voices as pillar of, 335–8
alienation, 181
alter-globalisation, 418–22
   distinction from neoliberal globalisation, 418–19
   farming and, 419–20
   food systems and, 419–20
   land grabbing and, 420
   transnational food supply networks, 420–2
   voluntary governance initiatives, 420
alternative spatial scales, 69
alternative technologies, 449–56
   adoption of, 454
   as emancipatory practice, 450–4
   emancipatory social movements and, 454–6
   as polytechnics, 449–50
   practices theory perspective, 451–2
alternative trade organisations (ATOs), 460
AMA. *See* American Medical Association
amateur work, in degrowth movement, 215–17
American Medical Association (AMA), 149–53
American Sociological Association, xix
AML. *See* acute myeloid leukemia
animal agriculture, 439–40

animal tourism, 192–3
　on Okunoshima Island, 192–6
animals
　human–animal relations, 196–9
　　domesticatory practices, 196
　　future of, 199–200
　　human connections with animals, 188–9
　　human–animal studies, 188, 189
　　　conflict theory and, 189
　　　methodology of, 190
　　　multi-species ethnography and, 191, 199–200
　　　sociology in, 189–92, 199–200
　　　symbolic interactionism approach in, 189–90
　　rabbits, 196–9
　trait-based breeding, through genomic editing, 349–50
ANT. *See* actor-network theory
Anthropocene, 26
　climate change and, 19–20, 23–4
　imaginary
　　ecological feminism and, 25
　　nature in politics influenced by, 23
antibiotic use, in livestock production, 392
anti-green messaging, masculine identity and, 105, 106–9
　through feminine stereotyping, 108
　in gender theory, 108
　through masculine overcompensation, 106–7
　'white male' effect, 108–9
anti-toxics movement
　environmental breast cancer movement, 120–1
　environmental justice movement and, 120
　Environmental Protection Agency and, 118–19
　legacy of, 118–21
　under National Environmental Protection Act, 118–19
　scope of, 118–21
　women's involvement in, 120–1
　　for environmental reproductive justice, 120–1
applied sociology, CBR in, 50
aquaculture, ecological degradation and, 351
ARMS. *See* Agricultural Resource Management Survey
articulation. *See* logos
Asad, Talal, 190
assemblage theory, 167–9, 173
　agencement, 167
　governance and, 168
　'new materialisms' and, 169–70
　political economy systems and, 171–2
ATOs. *See* alternative trade organisations
autocorrelation, in spatial analysis, 67
autonomismo concept, 516–17

'barefoot' epidemiology, 118
Barry, Andrew, 170–1
The Battle for Yellowstone (Farrell), 432
Beck, Ulrich, 15
Bell, Michael M., 91, 430–1, 439
　Childerly, 430–1

The City of the Good, 431–2
　natural conscience, 429, 430–2
　　bourgeois and pagan, tensions between, 431
Bellah, Robert, 29
Bennett, Jane, 170, 173
big data industry, in precision technologies, 380, 385, 386
biodiversity, genomic science and, 343
biological determinism, 342
biological theories of criminality. *See* criminality
biomarkers, in environmental health effects, 122
biophysical conditions, in environmental sociology, disciplinary neglect of, xx
Blaen Brân Community Group, 255
Boal, Augusto, 47
Borlaug, Norman, 394
Bourdieu, Pierre, 451
　on DI approaches, 33–34, 36–7
　habitus concept, 238
Brazil
　Brazilian Landless Movement, 421
　climate change in, multi-level governance of
　　Brazilian Forum of Climate Change, 289–90
　　civil society organisations and, 289
　　under Climate Convention guidelines, 289
　　Conference of Parties and, 294
　　establishment of climate policies, 290–1
　　implementation of policies for, 295–6
　　through local climate policies, 294–6
　　methodological aspects of, 287
　　mitigation actions, 291–2
　　mobilisation phase on, 289
　　National Policy on Climate Change, 291–2
　　overview of, 296–7
　　political responses to, 289–96
　　as state level, 292–4
　　as subnational level, 292–6
　climate issues in, 287–9
　greenhouse gas emissions in, 286–7
　　by activity sector, 288–9
　　increase of, 287–9
　Landless Rural Workers Movement, 88–9
　water privatisation in, 309
Brazilian Forum of Climate Change, 289–90
Brazilian Landless Movement (MST), 421
Bridle, James, 163–4
Buen Vivir, in Latin America, 487
Bulbeck, Chilla, 193, 198
bureaucracy, Weber on, 35–6
Burgess, Michael, 390–1
Bush, George W., xxii, 88

Callaghan, Paul, 349
Callon, Michael, 165–6
Canada. *See also* Alberta province; fracking
　CBR in, for climate change, 51
　Climate Leadership Plan in, 332
　'cap and trade' systems, carbon markets in, 267–8

capitalism, 22
  climate, carbon markets and, 269
  colonialism and, 16
  degrowth movement and, 481, 483
  environment as externality and, 399
  evolution of, through modernisation and technology, 163–4
  global fair trade movement and, 469–70
  imperialism and, 16
  slavery and, 16
  as treadmill of production, 413
Capitalocene, 26
carbon economy, 269
carbon markets
  in 'cap and trade' systems, 267–8
  creation of, 269–73
    through corporate networks and organisations, 270–1
    under critical political economy approach, 269–70
    ecological modernisation theory and, 269
    through non-governmental organisations, 271
    through policy coalitions, 270–1
    through state involvement, 271–2
  definition of, 259
  'double movement' and, 277–8
  ecological modernisation theory and, 269, 274–5
  in economic sociology, 273
  embeddedness of, 277–8
  emerging, 279
  environmental pragmatism and, 279
  environmental sociology and, 276–9
  in EU
    ETS market, 271, 273–4, 275–6, 279
    origins of, 269
  as 'fictitious commodity', 277–8
  function and purpose of, 273–6
  future studies of, 277, 278–9
  global value of, 267
  greenhouse gas emissions mitigation, 273–4
  moral meaning of, 273
  offset programs and, criticisms of, 275, 276
  origins of, 267–8, 269–73
    in carbon economy, 269
    climate capitalism and, 269
    in EU, 269
  performativity approach to, 273
  in Polanyian economic sociology, 277
  potential effectiveness of, 275
  research on, 272–3
  in sociology, critical examination of, 279
  theoretical approach to, 268–9
  in U.S., 279
Carson, Rachel, 108, 118–19
Carver, George Washington, 98
catallaxis, 489
Catton, William, xix–xx
CBNRM. *See* community-based natural resource management

CBR. *See* community-based research
CCP. *See* Cultural Cognition Project
certification, in global fair trade movement, 467–9
charitable organisations, environmental governance with, 257–8
CHE. *See* Collaborative on Health and the Environment
Chicago School
  on human–environment interactions, 63
  privatisation of resources and, 306
  urban and community development approaches, 63
Childerly (Bell), 430–1
Chipko movement, 507–8
CIFs. *See* commodity index funds
citizenship. *See* active citizenship
The City of the Good (Bell), 431–2
civic science mechanisms, 130
civil society
  in Brazil, climate change approaches by, 289
  climate change and, 285
  environmental governance in, for forestry management, 250
  with civil society organisations, 258–9
CIW. *See* Coalition of Immokalee Workers
CL. *See* collaborative learning
climate capitalism, carbon markets and, 269
climate change
  Anthropocene and, 19–20, 23–4
  in Brazil. *See* Brazil
  CBR and
    in Canada, 51
    in environmental sociology, 51
    for food systems, 51–2
  civil society involvement in, 285
  consumption pattern strategies and, 286
  ecological feminism and, 24–5
  'economic man' concept and, 20–2
  ecological modernisation theory, 21, 22
  government responses to
    as stakeholders, 285
    at state and municipal levels, 285
    theoretical approaches to, 285–7
  metamorphosing concept for, 15
  moral orders and, 436
  as transformative, 15–16
  treadmill of production theory and, 22
Climate Convention, Brazil under, 289
Climate Leadership Plan, in Canada, 332
Coalition of Immokalee Workers (CIW), 467
collaboration, in CBR, 48
collaborative governance, of natural resources. *See* adaptive co-management
collaborative learning. *See also* environmental management policy; natural resource management
  adult learning theories, 34
  emergence of, 33–5
  experiential learning theories, 34
  foundations of, 34
  systems as part of, 33

collaborative learning (CL), 32–5, 43
Collaborative on Health and the Environment (CHE), 141
collective action, in environmental governance, 262
collective constitution, of plants, 179–80
colonialism, 16, 404, 483–4
　carbon colonialism, 275
commodification of resources
　of nature, 304–5
　water, 303, 304–5
　　'fictitious commodities', 304–5
　　as 'uncooperative commodity', 314–15
　　through water management, 304, 313–15
commodity index funds (CIFs), 415–16
communicative systems, in collaborative learning, 37–9
Community Learning Partnership, 55
Community Supported Agriculture (CSA), 251–2
community-based natural resource management (CBNRM), 52–3
　in communities in poverty, 52–3
community-based research (CBR)
　in applied sociology, 50
　challenges with, 48–50
　　of co-optation, 50
　　in process, 49–50
　　validity as, 48–9
　in collaborations between academics and activist-researchers, on environmental toxicity, 128–9
　as collaborative, 48
　Community Learning Partnership and, 55
　definition of, 47–8
　environmental justice through, 53–4
　　in epidemiological work, 53
　　in public health, 53
　　in urban areas, 54
　in environmental sociology, 50–2, 54–5
　　for climate change, 51
　　for food systems, 51–2
　future directions for, 54–5
　in Global South, 47
　in natural resource management (CBNRM), 52–3
　　in communities in poverty, 52–3
　origins of, 47
　project-based approach to, 49–50
　qualitative research approach, 49
　scope of, 47–8
　in sociology, 50
　sustainable behavior change from, 53
　terms for, 47–8
Confederation Paysanne, 88–9
Conference of Parties (COP), 294
confidentiality, in spatial analysis, 69–70
conflict theory, 189
Connell, Raewynn, 104
The Conservation of Races (DuBois), 342
consumerism. *See* ethical consumption
continuous spatial data, 65
COP. *See* Conference of Parties

Corbyn, Jeremy, 312
Crick, Francis, 345
criminality, biological theories of, 342
CRISPR technology, 347–8
critical construction tools, in sociology, 13
Critical Environmental Justice framework, for LGBTQ populations, 98
critical plant studies, 178
critical political economy approach, to carbon markets, 269–70
CSA. *See* Community Supported Agriculture
Cultural Cognition Project (CCP), 108–9
culture, definition of, 236–7. *See also* sustainability cultures
'cultures' framework, for sustainability cultures, 239–41
　actor-centered methodology for, 240–1
　applications of, 241
　collective components of, 240
Cyborg Manifesto (Haraway), 169

Daly, Herman, 221–2
data ownership, in genomic science, 344
decommodification of work, 216–17
'decroissance', 206–7, 478
degradation. *See* ecological degradation
degrowth by design, 487
degrowth movement, economic
　accumulation regimes, 482
　capitalism and, 481, 483
　critiques of growth, 208–9
　'decroissance' and, 206–7, 478
　definition of, 206
　democratic transitions to, 209–11
　'depense' and, 215
　descaling as distinct form, 217–18
　development of, 206–7
　ecological sustainability and, 205
　feminist ecological economics, 488
　global commons movement, 487–8
　in Global North, 207, 484
　neoliberalism and, 481–6
　　environmental types of intervention, 482
　　material production of subjectivity, 483
　　modes of production and, 483
　nonscalability as element of, 217–18
　politics of, 212–13
　practices of, 212–13
　　ecological objectives in, 212
　　institutional mechanisms of, 213
　　social justice objectives in, 212–13
　purpose of, 207–13
　re-emergence of, 478–9
　re-imagination of economy, 213–17
　　through amateur work, 215–17
　　decommodification of work, 216–17
　　frugal abundance concept, 215
　　'having enough' concept, 215
　repolitisation of the economy, 211–12

stealth revolution and, 479
theoretical approach to, 205–6
transformation of society as result of, 486–92
voluntary transitions to, 209–11
Deleuze, Gilles, 167
deliberative democracies, adaptive co-management in, 368–9
democracy. *See also* deliberative democracies; representative democracies
through food systems, 403–5
'depense', 215
depression, from environmental toxicity, 122–3
descaling, degrowth movement as distinct from, 217–18
determinism. *See* biological determinism
Dewey, John, 385
DFTA. *See* Domestic Fair Trade Association
DI approaches. *See* discourse-intensive approaches
Digital Millennium Copyright Act (DMCA), U.S. (1998), 379, 384–5
disciplines
  feminist sociology as, 13–14
  neglect of biophysical conditions, in environmental sociology, xx
discourse-intensive (DI) approaches
  Bourdieu on, 33–4, 36–7
  collaborative learning, 32–5, 43. *See also* environmental management policy; natural resource management
    adult learning theories, 34
    emergence of, 33–5
    experiential learning theories, 34
    foundations of, 34
    systems as part of, 33
  communicative systems in, 37–9
  designers of, 31
    Bourdieu on, 36–7
  to environmental management policy, 31–3
    complexity of, 32
    controversy over, 32–3
    'Solomon Trap' in, 33
    uncertainty of knowledge as element of, 33
  in-group/out-group competition and, 39–40
  macro variables in, 39
  Luhmann on, 33–4, 37–9
  methodological approach to, 29–30
  to natural resource management, 31–3
    bureaucratic features of, 35–6
    complexity of, 32
    controversy over, 32–3
    'Solomon Trap' in, 33
    uncertainty of knowledge as element of, 33
    Weber on, 35–6
  overview of, 40–3
  in practice, 33–34, 40
  scope of, 30–2
  Sherif on, 33–34, 39–40
  Weber on, 33–34
    on natural resource management, 35–6

on nature of bureaucracies, 35–6
discrete spatial data, 65
disease framing, in medical information coverage
  by American Medical Association, 149–53
  on environmental pollution, 138–40
    ideological opposition to preventive approaches, 152–3
    in online medical publishing sites, 140
    reductionist approaches to, 139–40
    socio-economic implications of, 139
  on WebMD, 149–53
    coverage of toxicants, as cause of leukemia, 149
    financial interests as factor in, 150–2
    political economy approach, 149–50
    public health interests in, prioritisation over, 150–2
diseases, environmentally induced, 118
diversionary reframing, of sociotechnical imaginaries, 326–7
DMCA. *See* Digital Millennium Copyright Act
Domestic Fair Trade Association (DFTA), 463
'double movement,' carbon markets and, 277–8
Douglas, Mary, 197
Dublin Principles, 303
DuBois, W. E. B., 342
Durkheim, Emile, 447, 448
dynamics of change, materials in, 163

Earth Day, xix
Earth Overshoot day, 223
eating. *See* food consumption and cultures; food systems
EBCM. *See* environmental breast cancer movement
eco-habitus, 399–400
ecological degradation
  genomic science and, 350–2
    of aquaculture, 351
    metabolism theories, 351
    'New Ecological Paradigm' and, 350
    of oceans, 350–2
  social oppression and, 103–4
ecological feminism
  Anthropocene imaginary and, 25
  climate change and, 24–5
  nature and, ontological politics of, 25
ecological modernisation theory
  carbon markets and, 269, 274–5
  climate change and, 21, 22
  'economic man' concept and, 21, 22
ecological sustainability, degrowth movement and, 205
ecomodern masculinity, 111
economic degrowth. *See* degrowth movement
economic growth, 480. *See also* growth-societies
'economic man' concept, climate change and, 20–2
  ecological modernisation theory, 21, 22
economic sociology
  carbon markets in, 273
  Polanyian, 277
eco-queer movements, 94–7
Edelman, Murray, 326

Edelstein, Michael, 124
education of desire, 492
EFMs. *See* emergency fiscal managers
Einstein, Albert, 43
Ellul, Jacques, 437–8
emancipatory practice, 450–4
emancipatory social movements, 454–6
embeddedness, of carbon markets, 277–8
emergency fiscal managers (EFMs), 311
Emergency Planning and Community Right-to-Know Act (EPCRA), U.S. (1986), 125
emerging carbon markets, 279
emission trading schemes, definition of, 259. *See also* carbon markets
Encyclopedia of DNA Elements (ENCODE), 346
endocrine disrupter hypothesis, 121–2
energy efficiency. *See* household energy efficiency
The Entropy Law and the Economic Process (Georgescu-Rogan), 206
Entwhistle, Barbara, 74–5
environmental breast cancer movement (EBCM), 120–1
environmental ethics, 429–30
environmental governance
  through active citizenship, 262
  through collective action, 262
  for forestry management
    in civil society, 250
    contested views on, 250–2
    overview of, 261–3
    polycentric systems in, 250–1
    through public ownership, 260–1
    in Scotland, 258, 259–60
    in transnational social movements, 251
    in U.K., with citizen and community forestry groups, 252, 253–4. *See also specific community groups*
    in Wales, 256–7, 259–60
  through hybrid co-governance arrangements, 260–1
  innovation with community groups, 255–61
    Achieving Results in Communities, 255–6
    Blaen Brân Community Group, 255
    Llais Y Goedwig community, 256–7
    Natural Resource Wales Action Plan, 257
  innovation with state or public sector, 259–61
    community benefits from, 260
    environmental benefits from, 260
    Friends of Tower Hamlet Cemetery Park, 260
    Friends of Westonbirt Arboretum, 260–1
    through volunteering, 259–61
  innovation with third sector, 257–9
    with charitable organisations, 257–8
    with civil society organisations, 258–9
    Loch Arkaig Community Forest, 258
    with non-governmental organisations, 251–2
    Woodlland Trust, 258
    Woodmatters, 258–9
  social innovation in, 262
environmental justice
  actors in, 500–2
    domestic, 502–4
    international, 500–2
  anti-toxics movement and, 120
  through CBR, 53–4
    in epidemiological work, 53
    in public health, 53
    in urban areas, 54
  framing process in, 506–7
  genomic science and, 352–4
  global development initiatives in, 73–4
    United Nations Sustainable Development Goals, 74
  in Global North, 508
  in Global South, 508–9
    Chipko movement, 507–8
    internationalisation of, 507–8
    Narmada Bacho Andolan, 499, 503, 506
    Ogoni Uprising, 500, 507
    Rubber Tappers' Movement, 498–9, 502–3, 506
    theoretical approach to, 497–8
    Zapatista Movement, 499, 503
  LGBTQ populations and, 97–9
    Critical Environmental Justice framework for, 98
    Food Justice Studies and, 99
    in queer intentional communities, 97–8
  movement leaders in, 504–6
  natural and organic foods movement and, 402
  Principles of, 353
  social oppression and, 103–4
  sociology of morality and, 435–6
  through spatial analyses, of geographic data, 76
  through spatial data, 76
  for water privatisation, 311
environmental management policy, DI approaches to, 31–3
  complexity of, 32
  controversy over, 32–3
  'Solomon Trap' in, 33
  uncertainty of knowledge as element of, 33
environmental pragmatism, 279
Environmental Protection Agency (EPA), U.S., 75, 118–19
environmental reproductive justice, 120–1
environmental sociology. *See also* community-based research
  academic literature on, xxi–xxii
  adaptive co-management and, 369–71, 372
  biophysical conditions and, disciplinary neglect of, xx
  carbon markets and, 276–9
  definition of, xix–xx
  disciplinary approach to, 520–1
  evolution of, xix–xxii
  future developments for, 516–20
  gender gap in, 106
  genomic science in, 344–5
  in Global North, 14–15
  post-modern perspectives on, xx–xxi
  as practice, future developments for, 516–20

public sociology and, xxii
social justice and, 26–7
spatial analyses for. *See* spatial analyses
spatial data on. *See* spatial data
theoretical approach to, 11–14
environmental sustainability, masculine identity and, 103
environmentally induced diseases, 118
EPA. *See* Environmental Protection Agency
EPCRA. *See* Emergency Planning and Community Right-to-Know Act
epidemiology, 117–18
'barefoot', 118
in collaborations between academics and activist-researchers, on environmental toxicity, 130–1
environmentally-induced diseases, 118
'popular', 118
epigenetics, 346–7
epistemophilic drive, 184
Erikson, Erik, 436
ESDA. *See* exploratory spatial data analysis
État de siège, 24
ethical consumption, 397, 400–1
fair trade markets and, 401–2
global fair trade movement and, 461–2
in popular culture, 402–3
ethics
environmental, 429–30
in genomic science, 352–4
ETS (carbon) market, 271, 273–4, 275–6, 279
EU. *See* European Union
eugenics, 342
Eurocentrism, sociology and, 16
European Union (EU)
carbon markets in
ETS market, 271, 273–4, 275–6, 279
origins of, 269
privatisation of resources in, water, 311–12
Everything In Its Path (Erikson), 436
experiential learning theories, 34
exploratory spatial data analysis (ESDA), 68
extinction risks, 521

Fair Trade Advocacy Organization (FTAO), 469–70
fair trade markets, for food systems. *See also* global fair trade movement
in Global North, 401–2
in Global South, 401–2
worker-led initiatives in, 401–2
fair trade movement. *See* global fair trade movement
Fairtrade International, 463, 464, 465
Farm Hack, 380–1
Farm Investment Management Organizations (FIMOs), 417
farming. *See also* precision agriculture technologies
alter-globalisation and, 419–20
neoliberal globalisation and, 416–18
Farm Investment Management Organizations and, 417

in Global North, 416
resource grabs, 417–18
sovereign wealth funds and, 417
sociology of morality for, 438–9
Farming While Black (Penniman), 518
Farrell, Justin, 432
on moral orders, 432–5
feminine identity
in anti-green messaging, stereotyping of, 108
in gender theory, 104
feminism, 13–15, 17, 18–19, 25, 169–70
ecological
Anthropocene imaginary and, 25
climate change and, 24–5
nature and, ontological politics of, 25
ecological economics, in degrowth movement, 488
relational agriculture, 96–7
feminist agrifood systems theory, 94
feminist sociology
as discipline, 13–14
gender concepts in, 23
scope of, 11
'fictitious commodities'
carbon markets as, 277–8
water as, 304–5
FIMOs. *See* Farm Investment Management Organizations
Flint Water Crisis, 124, 311
food activism, 390–1
food activists, 390–1
food consumption and cultures, 391–6
dietary changes and, 396–401
ethical consumption, 397, 400–1
fair trade markets and, 401–2
in popular culture, 402–3
Green Revolution and, 393
as individualistic, 390
through natural and organic foods movement, 397–9
Food Justice Studies, 99
food sovereignty, 419
food systems. *See also* agricultural movements; sustainable agriculture movement
activists and, 390–1
alter-globalisation and, 419–20
CBR for, 51–2
components of, 391–2
corporate concentration of, 396
democracy through, 403–5
dietary changes and, 396–401
eco-habitus, 399–400
environmental consequences of, 392–3
from antibiotic use, in livestock production, 392
in Global North, 393
fair trade markets
in Global North, 401–2
in Global South, 401–2
worker-led initiatives in, 401–2
feminist agrifood systems theory, 94
Food Justice Studies and, 99

food systems (cont.)
  food provisioning modes in, 393
  'Gene Revolution', 393, 394
    agricultural technology in, 394
    ecological implications of, 393–4
    in Global North, 396–401
      environmental consequences of, 393
      fair trade markets, 401–2
    in Global South, in fair trade markets, 401–2
  Green Revolution and, 393
    productivity gains of, 394–5
  natural and organic foods movement, 397–9
    environmental justice and, 402
    fair trade markets for, 401–2
  neoliberal globalisation and. *See* neoliberal globalisation
  transnational food supply networks, 420–2
  wasted resources in, 395–6
forestry management, 249–63
  community-based, 251–2
  environmental governance in. *See* environmental governance
  theoretical approach to, 249–50
  United Nations Conference on the Environment and Development (Rio Conference), 249
Foster, John, 138
Foucault, Michel, 448, 451
  on neoliberalism, 489
fracking (hydraulic fracturing), in Canada, 324–5, 327–38. *See also* Alberta province
  arguments for, 327–8
  citizen responses to, 328
  fluid composition of, 333
  The Hansards, 325
  manufacturing of consent on, 324–5
  opposition to, 328
  public silence on, 324–5
  social impacts on, 328
  sociotechnical imaginaries and, 325, 338
    conceptual background of, 325–7
    discourse as element of, 326
    diversionary reframing of, 326–7
France, *Confederation Paysanne* in, 88–9
freeganism, 226
Freire, Paulo, 47
Freud, Sigmund, 184
Friends of Tower Hamlet Cemetery Park, 260
Friends of Westonbirt Arboretum, 260–1
frugal abundance, 215
FTAO. *See* Fair Trade Advocacy Organization
Fuentes, Agustin, 196
'full-cost recovery' policies, with privatisation of water, 312

Galpin, Charles, 63
gender, 23, 89, 93, 403, 491. *See also* masculine identity
  privatisation of water and, 308
  relational agriculture, 96–7
  sustainable consumption by, 227
gender theory, masculine identity in, 104
  anti-green stereotyping in, feminine-coding in, 108
'Gene Revolution', 393
Generation Y, 242–3
genes
  centrism and, 345–6
  definition of, 345
  primacy of, 345–6
genetic variations, in genomic science, 343
genomic science
  biodiversity and, 343
  Crick and, 345
  critical developments in, 345–8
    CRISPR technology, 347–8
    epigenetics, 346–7
    'network thinking', 345–6
  data ownership and, 344
  ecological degradation and, 350–2
    of aquaculture, 351
    metabolism theories, 351
    'New Ecological Paradigm' and, 350
    of oceans, 350–2
  Encyclopedia of DNA Elements, 346
  environmental justice and, 352–4
  in environmental sociology, 344–5
  ethics and, 352–4
  genetic variation and, 343
  human exemptionalism and, 348
  Human Genome Project and, 342–3, 345–6
  in New Zealand, 353
    reconstruction of nature in, 350
  power of, 352–4
  reconstruction of nature through
    genomic editing, 349–50
    in New Zealand, 350
    'realist/materialist' approach, 348
    social constructivist approach, 348
    trait-based breeding, of animals, 349–50
  sociological approach to, 343
  theoretical approach to, 342–5
  Watson and, 345
geographic data. *See also* spatial analyses; spatial data
  isotropic map, 63
  for rural areas, 63
geographic information systems (GIS), 70
Georgescu-Roegan, N., 206, 208–9, 478
GHG emissions. *See* greenhouse gas emissions
Gilley, Brian, 88
GIS. *See* geographic information systems
global commons movement, 487–8
global fair trade movement
  alternative trade organisations, 460
  capitalism and, 469–70
  certification in, 467–9
  contemporary development of, 460–1
  environmental impacts from, 463–4
  ethical consumerism and, 461–2

excluded populations in, 463
future of, 471–2
in Global North, 462–3
in Global South, 462
goals of, 459
history of, 460–1
labeling in, 467–9
   competition in, 468–9
mainstreaming in, 460–1
non-governmental organisations in, 459
participants in, 464–7
   authorities, 465
   companies and brands, 465–6
   empowerment models for, 467
   multi-stakeholder initiatives for, 466
reforms in, 469–70
regulation mechanisms in
   private, 470
   state, 470
targets of, 461–2
in U.S., 463
Global North
   degrowth movement in, 207, 484
   environmental justice movements in, 508
   environmental sociology as discipline in, 14–15
   farming in, neoliberal globalisation and, 416
   food systems in, 396–401
      environmental consequences of, 393
      fair trade markets, 401–2
   global fair trade movement in, 462–3
   privatisation of water in, 309–12
   sociology in, 17
global precision farming, 377–8
Global Rural-Urban Mapping Project (GRUMP), 76–7
Global South
   CBR in, 47
   environmental justice in. *See* environmental justice
   farming in, neoliberal globalisation and, 416
   food systems in, 401–2
   global fair trade movement in, 462
   privatisation of water in, 306, 307
   sociology's capacity for self-transformation in, 16
global water crisis, 303
globalisation. *See also* alter-globalisation; neoliberal globalisation
   definition of, 412–13
Goffman, Erving, 430
The Good Society (Bellah), 29
Gorsuch, Anne, 104–5
governance. *See also* adaptive co-management; green governance models
   adaptive co-management and, 362–3
   climate, multi-level and, 285–7
   forestry and, 249–50
   global, carbon markets and, 269–73
   innovation in, 255–61
   sustainability cultures research as influence on policy, 244

Gray, Mary, 88
green governance models, sustainable consumption in, 224–6
   ABC models of behavior and, 225
   freeganism, 226
   role of individual in, 224–6
   social practice theories, 225–6
   voluntary simplifiers in, 226
green neoliberalism, 307
Green Revolution
   for food consumption and cultures, 393
   productivity gains of, 394–5
greenhouse gas (GHG) emissions
   in Brazil, 286–7
      by activity sector, 288–9
      increase of, 287–9
   carbon markets and, 273–4
growth-societies, 479–81
   Social Imaginary in, 479
   welfare states and, 480
GRUMP. *See* Global Rural-Urban Mapping Project
Guattari, Felix, 167

habitus concept, 238
Haining, Robert, 67
Hall, Matthew, 177
Hamer, Fannie Lou, 98
The Hansards, 325
Haraway, Donna J., 11, 169, 170
HAS. *See* human–animal studies
'having enough' concept, 215
Hawkey, Amos, 63
hegemony of male power, 103–6
   in hierarchies of categories of men, 105
Heidegger, Martin, 448
heterosexism, in sustainable agriculture movement, 94–5
The Hidden Life of Trees (Wohlleben), 179–80
hierarchism, 'white male' effect and, 108–9
Highly Indebted Poor Countries (HIPC), 312
Hochschild, Arlie, 93–4
homosexuality. *See also* LGBTQ populations
   masculine identity and, 107
household energy efficiency, in sustainability cultures, 237–8, 241–2, 243
Household Exposure Study, 129
How Forests Think (Kohn), 179–80
Howard, John, 90–1
Human Ecology (Hawkey), 63
Human Exceptionalism Paradigm, 161, 516
   genomic science and, 348
Human Exemptionalism Paradigm, 161
Human Genome Project, 342–3, 345–6
human health, ecological health and, 137
human nature, humanity influenced by, 19–20
human rights, water as, 304
human–animal relations, 196–9
   domesticatory practices, 196
   future of, 199–200

human–animal relations (cont.)
  human connections with animals, 188–9
human–animal studies (HAS), 188, 189
  conflict theory and, 189
  methodology of, 190
  multi-species ethnography and, 191, 199–200
  sociology in, 189–92, 199–200
  symbolic interactionism approach in, 189–90
human–environment interactions
  Chicago School on, 63
  extinction risks as result of, 521
  in spatial analyses
    interconnectedness of patterns and processes, 71–3
    social inequality elements in, 73–5
  in spatial data
    interconnectedness of patterns and processes, 71–3
    social inequality elements in, 73–5
humanity
  as causal agency, 21
  as concept, 19
  human nature and, 19–20
hydraulic fracturing. *See* fracking
hyper-masculine performances, male identity through, 103–6
  through masculine-coded performances, 104–5
  in patriarchal structures, 104–5
  of rationality, 109–11
  types of, 105

IARC. *See* International Agency for Research on Cancer
ICT. *See* information and communication technologies
identity-protective cognition, 'white male' effect and, 108–9
IMF. *See* International Monetary Fund
imperialism, capitalism and, 16
  in Western metaphysical tradition, 176–9
individualism, 'white male' effect and, 108–9
information and communication technologies (ICT), in sustainable consumption, 230, 231
International Agency for Research on Cancer (IARC), 140, 141–3
International Monetary Fund (IMF), 412
International Sociological Association, xix
'inverted quarantine', 400
ISEAL Alliance, 469
isotropic maps, 63

Johnson, Colin, 88
Johnson, E. Patrick, 90–1
justice. *See* environmental justice; social justice

Keller, Evelyn Fox, 345–6
Key, John, 349
Keywords (Williams), 381
Kim, Annette, 62
King, Leslie, 98
Klein, Melanie, 184
knowledge
  in adaptive co-management, gulf between types of, 364–5
  in environmental management policy, uncertainty of, 33
  as political, 15–16
  reverse-colonisation as result of, 15–16
knowledge economies, 178, 181–2, 183
Kohn, Eduardo, 179–80

labeling, in global fair trade movement, 467–9
land grabs
  alter-globalisation and, 420
  neoliberal globalisation, 417–18
landdyke movement. *See* lesbian land movement
Landless Rural Workers Movement, 88–9
learning. *See specific topics*
lesbian land movement (landdyke movement), 90
lesbian utopias, 91–2
Leslie, Isaac, 94–5
  relational agriculture, 96–7
leukemia. *See also* WebMD
  acute myeloid leukemia, 144, 149
  childhood, 120
  environmental causation perspective for
    contextual factors for, 147–8
    obscuring of causation, 143–4
  environmental health scholarship on, 140–3
  toxicant causes, 134–35
  International Agency for Research on Cancer and, 140
Levine, Adeline, 124
Lewin, Kurt, 47
LGBTQ populations
  in eco-queer movements, 94–7
    sexuality as factor in, 95
  environmental justice and, 97–9
    Critical Environmental Justice framework for, 98
    Food Justice Studies and, 99
    in queer intentional communities, 97–8
  National Resources Conservation Service campaigns, 92–3
  queer intentional communities
    environmental justice in, 97–8
    for farmers, in sustainable agriculture movement, 96
  in rural areas
    in conservative political contexts, 88–9
    lesbian land movement in, 90
    lesbian utopias, 91–2
    migration patterns to, 91
    oral histories of, 90–1
    as perceived threat, 89–92
    plural rural sexual approach, 91
    queer rural idylls and, as vacation destination, 91–2
    Rural Pride campaign, 87, 92–3
    same-sex marriage rights in, 89
  in sustainable agriculture movement, 94–7
    feminist agrifood systems theory, 94
    heterosexism in, 94–5

in queer-centered farming communities, 96
   relational agriculture, 96–7
   during Trump administration, 93
Li, Tania Murray, 168, 173
Limbaugh, Rush, 87, 92–4. *See also* Rural Pride campaign
Liu, John, 170–1
livestock. *See* antibiotic use
Llais Y Goedwig community, 256–7
Loch Arkaig Community Forest, 258
Loftus, Alex, 305
logos (articulation), 184
Love Canal, 124, 125
Luhmann, Niklas
   social systems and, 38
   social theory and, 37–9
Lukács, György, 186

MacGregor, Sherilyn, 103–4
male power. *See* hegemony of male power
manufacturing of consent, on fracking, 324–5
Māori, genomic science and, response to, 353
Marder, Michael, 171
Marx, Karl
   conflict theory and, 189
   on factory system, 163
   metabolism theories, 351
   on technological determinism, 447–8
masculine identity
   among African Americans, 109
   anti-green policing of, 105, 106–9
      feminine stereotyping in, 108
      in gender theory, 108
      as masculine overcompensation, 106–7
      'white male' effect in, 108–9
   in Cultural Cognition Project (CCP), 108–9
   environmental sustainability influenced by, 103
   gender binary incentivized by, 103–6
   in gender theory, 104
      anti-green stereotyping in, feminine-coding in, 108
   hegemony of male power, 103–6
      in hierarchies of categories of men, 105
   hyper-masculine performances incentivized by, 103–6
   through masculine-coded performance, 104–5
   in patriarchal structures, 104–5
   of rationality, 109–11
   types of, 105
   negative views on LGBTQs, 107
   overview of, 111–12
   rationality and
      ecomodern masculinity, 111
      as performative, 109–11
      rational actor model, 110
   'white male' effect and, 108–9
      hierarchism, 108–9
      identity-protective cognition, 108–9
      individualism, 108–9
Massey, Doreen, 62

material production of subjectivity, 483
material semiotics, 166
materiality, 163
MAUP. *See* modifiable area unit problem
Mauss, Marcell, 448, 487
McCarthy, Melissa, 221, 230
Mead, George Herbert, 430
medical information, environment in. *See also* human health; WebMD
   competence and confidence in, among physicians, 153
   disease framing from, environmental pollution in, 138–40
      ideological opposition to preventive approaches, 152–3
      in online medical publishing sites, 140
      reductionist approaches to, 139–40
      socio-economic implications of, 139
   environmental health scholarship and, on leukemia, 140–3
      toxicant causes, 134–35
   in mass media, 137–8
      on websites, 138
   in medical training, 152–3
   overview of, 153–4
   public ignorance as result of, 154
medical training, environmental causes of disease as discipline in, 152–3
Men Like That (Howard), 90–1
Mendes, Chico, 498–9, 501, 504
mental health effects, from environmental toxicity, 122–3
metabolism theories, genomic science and, 351
metamorphosing concept, for climate change, 15
The Metamorphosis of the World (Beck), 15
Mexico Migration Project, 72
migration patterns
   of LGBTQ populations, to rural areas, 91
   in spatial analyses, of geographic data, 71–3
Miller, Elaine, 176
Mills, C. Wright, 191
Mintz, Sidney, 164
mobility cultures, Generation Y and, 242–3
modes of production, 483
Modi, Narendra, 502
modifiable area unit problem (MAUP), 68–9
Mol, Annmarie, 170
monotechnics, 448–9
Monsanto, precision agriculture technologies and, 378–9
Moore, Jason, 399
moral economy, 430
moral orders, 432–5
   climate change and, 436
   with disasters, 436
   in Old West/New West, 433
   on population, 436
morality, sociology of, 433–7
   in agriculture, 438–40
      animal agriculture, 439–40

morality, sociology of (cont.)
    for farming, 438–9
    approach to, 429
    of carbon markets, meanings of, 273
    critique of, 429–30
    development of, 435–7
    environment and, 430–9
    environmental ethics and, 429–30
    environmental justice and, 435–6
    moral economy and, 430
    moral orders, 432–5
        climate change and, 436
        with disasters, 436
        in Old West/New West, 433
        on population, 436
    multiple natures and, 429
    natural conscience, 429, 430–2
        bourgeois and pagan, tensions between, 431
    technological, 437–8
'more-than-human' paradigm, in sociology, 11–12
Morton, Ted, 333–4
Morton, Timothy, 184
Moses, Robert, 386
Movement for the Survival of the Ogoni People (MOSOP), 500, 503–4
MST. *See* Brazilian Landless Movement
multi-species ethnography, 191, 199–200
Mumford, Lewis, 448, 455

Narmada Bacho Andolan (NBA), 499, 503, 506
National Comprehensive Cancer Network (NCCN), 151
National Environmental Protection Act, U.S., 118–19
National Policy on Climate Change, in Brazil, 291–2
National Resources Conservation Service (NRCS), 92–3
natural and organic foods movement, 397–9
    environmental justice and, 402
    fair trade markets for, 401–2
natural gas production, in Alberta, Canada, 330–2
    public need for, 332
    reserves for, 330–2
natural resource management. *See also* adaptive co-management
    creative experimentation in, 371
    DI approaches to, 31–3
        bureaucratic features of, 35–6
        complexity of, 32
        controversy over, 32–3
        'Solomon Trap' in, 33
        uncertainty of knowledge as element of, 33
        Weber on, 35–6
Natural Resource Wales Action Plan, 257
natural resources. *See also* natural resource management
    global economic impact of, 360
nature. *See also* human nature
    ecological feminism and, ontological politics of, 25
    as natural science, 22–3
    politics of, 22–4
        Anthropocene imaginary as influence on, 23

    reconstruction of, through genomics
        genomic editing, 349–50
        in New Zealand, 350
        'realist/materialist' approach, 348
        social constructivist approach, 348
        trait-based breeding, of animals, 349–50
NBA. *See* Narmada Bacho Andolan
NCCN. *See* National Comprehensive Cancer Network
NDVI. *See* Normal Difference Vegetation Index
neoliberal globalisation, 411–23. *See also* alter-globalisation
    contestation of, 418–22. *See also* alter-globalisation
    environment and, 412–13
    farming and, 416–18
        Farm Investment Management Organizations and, 417
        in Global North, 416
        in Global South, 416
        resource grabs, 417–18
        sovereign wealth funds and, 417
    food systems and, 414–16
        agri-food labour, 415
        alter-globalisation and, 419–20
        commodity index funds, 415–16
        corporate market concentration of, 414–15
        finance sector in, 415–16
        food waste and, 414
    International Monetary Fund and, 412
    overconsumption of commodities and, 413
    overview of, 422–3
    World Trade Organization, 412
neoliberalised agri-food systems, theoretical approach to, 411
neoliberalism
    definition of, 412
    degrowth movement and, 481–6
        environmental types of intervention, 482
        material production of subjectivity, 483
        modes of production and, 483
    Foucault on, 489
    green, 307
    mental infrastructure of, 486
    resilience and, 485
    self-valorisation and, 485
'network thinking', 345–6
New Dark Age (Bridle), 163–4
'New Ecological Paradigm', 350
'new materialism', 162–5
    actor-network theory and, 169–70
    assemblage theory and, 169–70
New Zealand
    genomic science in, 353
    reconstruction of nature in, 350
    Māori in, 353
non-governmental organisations (NGOs)
    carbon markets created though, 271
    environmental governance with, 251–2
    in global fair trade movement, 459

nonscalability, 217–18
Norgaard, Kari, 436
Normal Difference Vegetation Index (NDVI), 65–6, 72–3
NRCS. *See* National Resources Conservation Service

oceans, ecological degradation of, 350–2
Office of Environmental Health Hazard Assessment (OEHHA), 141
offset programs, for carbon emissions, criticisms of, 275, 276
Ogoni Uprising, 500, 507
oil production, in Alberta, Canada, 330
Okunoshima Island ('Rabbit Island'), 192–6
organic foods movement. *See* natural and organic foods movement
organisational logics, in sustainable consumption, 229–30
origin stories, in sociology, 14–15, 18–19

Paigen, Beverly, 130
Palin, Sarah, 104–5
Paris Agreement, 230–1
participation paradigms, 362
Patkar, Medha, 499, 501
PCBs. *See* polychlorinated biphenyls
Pellow, David N., 517
Pence, Mike, 89
Penniman, Leah, 96, 98, 518
Perdue, Sonny, 93
performativity approach, to carbon markets, 273
Perry, Rick, 230–1
philosophy, plants and
　collective constitution of plants, 179–80
　logos and, 184
　object-oriented ontology, 183
　'plant intelligence' and, 177–8
　　in critical plant studies, 178
　　as decentralized, 180
　　knowledge economies and, 178, 181–2, 183
　　plant thinking as distinct from, 182–3
　relationality theory, 184
　subject–object split in, 183–6
　task of, 183–7
　utopia model, 181
physical health effects, of environmental toxicity, 121–2
Pimentel, David, 390–1
plant thinking, 'plant intelligence' as distinct from, 182–3
plants
　collective constitution of, 179–80
　'intelligence' of, 177–8
　　in critical plant studies, 178
　　as decentralized, 180
　　knowledge economies and, 178, 181–2, 183
　　plant thinking as distinct from, 182–3
　philosophy and. *See* philosophy
　post-humanism and, 170–3
　in political economy systems, 171–2
Plants as Persons (Hall), 177
plural rural sexual approach, to LGBTQ populations, 91

Polanyian economic sociology, 277
policy coalitions, for carbon markets, 270–1
polychlorinated biphenyls (PCBs), 121–2
polytechnics, 449–50
popular education, 47
'popular' epidemiology, 118
post-humanism, 11–12
　actor-network theory, 165–7, 173
　　material semiotics and, 166
　　'new materialisms' and, 169–70
　　political economy systems and, 171–2
　　power and, 166–7
　assemblage theory, 167–9, 173
　　agencement, 167
　　governance and, 168
　　'new materialisms' and, 169–70
　　political economy systems and, 171–2
　　plants and, 170–3
　　in political economy systems, 171–2
　theoretical approach to, 161–2
post-traumatic stress disorder (PTSD), 123
potable water, access to, 303
poverty, CBR in communities in, 52–3
practice theory
　alternative technologies, 451–2
　social practice theories, in green governance models, 225–6
　sustainability cultures and, 238–9
pragmatism. *See* environmental pragmatism
precautionary principle, 127
precision agriculture technologies, 377–8
　active citizenship concept and, 381–6
　　word clouds and, 382, 383
　adoption rates, 379
　under Digital Millennium Copyright Act, 379, 384–5
　empirical approach to
　　big data industry, 380, 385, 386
　　Farm Hack, 380–1
　　by Monsanto, 378–9
　variable-rate input application technology, 379
prefigurative activism, 492
privatisation of resources, water, 305, 306–9, 312–13
　affordability crisis and, 311
　in Brazil, 309
　Chicago School approach, 306
　contestation of, 308–9
　environmental justice issues and, 311
　in EU, 311–12
　expansion of, 312
　financial efficiencies from, 308
　'full-cost recovery' policies, 312
　gendered effects of, 308
　global consensus on, 307
　in Global North, 309–12
　in Global South, 306, 307
　in Highly Indebted Poor Countries, 312
　market involvement in, 306–7
　public opposition to, 309

privatisation of resources (cont.)
    in public-private partnerships, 309, 310
    reversal of, 308–9
    in South Africa, 309, 315
    in U.K., 306, 312
    in U.S., 310–11
    through water concession contracts, 308–9
    through water justice activism, 308
'prosumers', 230
Pruitt, Lisa R., 89
Pruitt, Scott, 125
PTSD. *See* post-traumatic stress disorder
public health
    CBR and, 53
    environmental justice and, 53
public ownership, environmental governance through, 260–1
public sociology, xxii
public-private partnerships, water privatisation and, 309, 310

qualitative research approach, to CBR, 49
queer intentional communities
    environmental justice in, 97–8
    for farmers, in sustainable agriculture movement, 96
queer representation. *See* LGBTQ populations
queer rural idylls, 91–2
Queering the Countryside (Gilley, Gray, and Johnson, C.), 88

'Rabbit Island.' *See* Okunoshima Island
rabbits, 196–9. *See also* Okunoshima Island
race, sustainable consumption and, 227–8
racism, in rural areas, political identification as influence on, 89
radical ecological democracy, in India, 487
rational actor model, 110
rationality, masculine identity and
    ecomodern masculinity, 111
    as performative, 109–11
    rational actor model, 110
Reagan, Ronald, xxii
'realist/materialist' approach, to genomic science, 348
relational agriculture, 96–7
representative democracies, adaptive co-management in, 368–9
reproductive justice. *See* environmental reproductive justice
(re)productivity concept, 488
resilience, 485
resources. *See also* natural resource management; nature; water
    commodification of, 303
reverse-colonisation, 15–16
Rio Conference. *See* United Nations
Robbins, Paul, 170–1, 173
Rubber Tappers' Movement, 498–9, 502–3, 506
ruin economy, 170–1

rural areas
    'family values' myth in, 89–90
    geographic data for, 63
    in GRUMP, 76–7
    LGBTQ populations in. *See* LGBTQ populations
    political identification in, 88–9
        racist undertones as result of, 89
    social movements in
        Confederation Paysanne, in France, 88–9
        Landless Rural Workers Movement, in Brazil, 88–9
        lesbian land movement, 90
    Trump and, political support for, 89
Rural Pride campaign, 87, 92–3

Salatin, Joel, 440
same-sex marriage rights, for LGBTQ populations, in rural areas, 89
Sandel, Michael, 273
Sandilands, Catriona, 95
Saro-Wiwa, Ken, 500
Sbicca, Joshua, 95
Schneider, François, 481
Scotland, forestry management in, 258, 259–60
security mechanisms, 24–5
    État de siège, 24
self-valorisation, 485
semiotics. *See* material semiotics
Sherif, Muzafer, 39–40
Silent Spring (Carson), 108, 118–19
slavery, capitalism and, 16
Snow, John, 62
    epidemiology and, 117–18
social construction, of nature, 161–2, 197–8, 344
    as approach to genomics, 348
    realism and, 24–5, 348
    of scientific knowledge, 161–2
social Darwinism, 342
social exclusion, sustainable consumption as factor in, 229
Social Imaginary, 479
social justice
    in degrowth movement practices, 212–13
    environmental sociology and, 26–7
social movements. *See also* agricultural movements; anti-toxics movement; degrowth movement; sustainable agriculture movement
    emancipatory, 454–6
    in rural areas
        Confederation Paysanne, in France, 88–9
        Landless Rural Workers Movement, in Brazil, 88–9
        lesbian land movement, 90
    transnational, environmental governance and, 251
social oppression, ecological degradation and, 103–4
social practice theories, in green governance models, 225–6
social status, environmentalism and, 227
    in advertising, 221, 230

through sustainable consumption. *See* sustainable consumption
social systems, Luhmann and, 38
social theory, Luhmann and, 37–9
socialv construction1, of nature, 161–2, 197–8, 344
sociobiology, 342
sociology. *See also specific topics*
   carbon markets in, critical examination of, 279
   CBR in, 50
   climate change and. *See* climate change
   cosmopolitics of, 26–7
   critical construction tools in, 13
   definitions of, 16
   of environmental issues, xix–xx. *See also* environmental sociology
   epistemic hegemony of, 16–18
   Eurocentrism and, 16
   expansion of disciplines in, 14–15
   in Global North, relevancy in, 17
   in human–animal studies, 189–92, 199–200
   'modern' paradigms in, 16
   'more-than-human' paradigm in, 11–12
   origin stories in, 14–15, 18–19
   post-colonial critiques of, 18–19
   public, xxii
   theoretical approach to, 11–14
   'Western' paradigms in, 16
sociotechnical imaginaries, fracking and, in Canada, 325, 338
   in Alberta province, 329–34
      public concern as pillar of, 335
      safety concerns as pillar of, 333–4
      silencing of public voices as pillar of, 335–8
   conceptual background of, 325–7
   discourse as element of, 326
      silencing of, 335–8
   diversionary reframing of, 326–7
socio-technical systems theory, 238
'Solomon Trap', 33
South Africa, privatisation of water in, 309, 315
spatial analyses, of geographic data, 61, 66–70
   autocorrelation in, 67
   challenges with, 68–70
      with alternative spatial scales, 69
      with confidentiality, 69–70
   data quality as factor in, 68
   data structure as factor in, 68
   definition of, 62, 64
   environmental justice and, 76
   in environmental sociology, 70–5
      interconnectedness of human–environment patterns and processes, 71–3
      migration patterns and, 71–3
      NDVI and, 65–6, 72–3
      social inequality in molding human-environment interactions and, 73–5
      in Sub-Saharan Africa, 71, 74, 77
   exploratory spatial data analysis, 68
   future directions of, 75–8
   with GIS, 70
   historical development of, 67
   limitations of, 68–70
   modifiable area unit problem, 68–9
   pattern identification in, 66–7
   as quantitative, 62
   regional variation in, 66–7
   Tobler's First Law and, 67
   with VGI, 70
spatial data, in geographic data, 61, 62–6
   aggregation mechanisms for, 76–7
   availability of, 65–6
   continuous, 65
   data collection, 76
   data tools, 76
   definition of, 62, 65
   discrete, 65
   environmental justice and, 76
   in environmental sociology, 70–5
      interconnectedness of human–environment patterns and processes, 71–3
      migration patterns and, 71–3
      NDVI and, 65–6, 72–3
      social inequality in molding human–environment interactions and, 73–5
      in Sub-Saharan Africa, 71, 74, 77
   future directions of, 75–8
   historical view of, 63
   origins of, 65
   quality of, 65–6
   relational quality of, 62
   types of, 64–6
Spatial Data Analysis in the Social and Environmental Sciences (Haining), 67
Starr, Paul, 150
state hydraulic paradigm, for water paradigm, 303
Staying with the Trouble (Haraway), 11
stealth revolution, 479
Stein, Arlene, 93
Stivers, Richard, 437–8
The Stranger Next Door (Stein), 93
structuration theory, 238
subject–object split, in philosophy, 183–6
Sub-Saharan Africa, environmental sociology, 71, 74, 77
sustainability cultures
   components of, 237
      in 'cultures' framework, 240
   'cultures' framework, 239–41
      actor-centered methodology for, 240–1
      applications of, 241
      collective components of, 240
   origins of, 237–9
      for household energy efficiency, 237–8
   research on, 241–5
      on external influences, impact of, 242, 243
      governance policy influenced by, 244
      for household energy efficiency, 237–8, 241–2, 243

sustainability cultures (cont.)
  on mobility cultures, Generation Y and, 242–3
  on transformative change, 243–4
  sustainable outcomes as element of, 237
  theoretical approach to, 236–9
    application of, 245–6
    habitus concept in, 238
    practice theory in, 238–9
    socio-technical systems theory in, 238
    structuration theory in, 238
    systems approaches in, 238
sustainable agriculture movement, LGBTQ populations in, 94–7
  feminist agrifood systems theory, 94
  heterosexism in, 94–5
  in queer-centered farming communities, 96
  relational agriculture, 96–7
sustainable consumption, 221–30
  environmental issues and, 223–4
    causes, 223–4
    impacts of, 223–4
  future research on, 228–30
    information and communication technologies in, influence of, 230, 231
    organisational logics in, 229–30
    for pro-environmental behaviors, 228–9
    social exclusion factors in, 229
  gendered engagement in, 227
  in green governance models, 224–6
    ABC models of behavior and, 225
    freeganism, 226
    role of individual in, 224–6
    social practice theories, 225–6
    voluntary simplifiers in, 226
  problems with, 226–8
    material problem, 226–7
    symbolic problem, 226–7
  'prosumers', 230
  race and, 227–8
Sweet Tea (Johnson, E. P.), 90–1
symbolic interactionism approach, in human–animal studies, 189–90
systems
  in collaborative learning, 33
  communicative, 37–9
  in sustainability cultures, 238
Szasz, Andrew, 400

Tandon, Rajesh, 47
Tasting Food, Tasting Freedom (Mintz), 164
technological morality, 437–8
technology
  agricultural. *See also* precision agriculture technologies
    in 'Gene Revolution', 394
  alternative, 449–56
    adoption of, 454
    as emancipatory practice, 450–4
    emancipatory social movements and, 454–6
    as polytechnics, 449–50
    practices theory perspective, 451–2
  CRISPR, 347–8
  information and communication technologies, in sustainable consumption, 230, 231
  Marx on, 447–8
  modernism and, 163–4
  monotechnics, 448–9
  sociology and, historical relationship with, 447–8
  variable-rate input application, 379
Tester, Keith, 191
Thatcher, Margaret, 306, 490
Theatre of the Oppressed, 47
theoretical approach to, 176–9
theory of action, 451
A Thousand Plateaus, Capitalism and Schizophrenia (Deleuze and Guattari), 167
Till, John, 130–1
Tobler's First Law, 67
toxicity, environmental sources of. *See also* anti-toxics movement
  childhood leukemia, 120
  collaborations between academics and activist-researchers, 128–31
    advocacy biomonitoring, 129–30
    civic science mechanisms, 130
    in community-based participatory research, 128–9
    critical epidemiology approach, 130–1
    Household Exposure Study, 129
  contaminated communities
    Flint Water Crisis, 124
    Love Canal, 124, 125
  depression as result of, 122–3
  endocrine disrupter hypothesis, 121–2
  environmental health effects, 121–4
    biomarkers, 122
    community-wide effects, 123–4
    mental health effects, 122–3
    physical health effects, 121–2
  environmental health knowledge for, barriers to, 126–8
    data availability, 127–8
    among medical professionals, 128
    precautionary principle in, 127
    standards of proof, 126–7
  future directions for, in social science environmental health scholarship, 131–2
  'hormone havoc', 121–2
  hypervigilance and, 123
  polychlorinated biphenyls, 121–2
  post-traumatic stress disorder, 123
  public health protections, as government social responsibility, 124–6
  on WebMD, leukemia coverage in, 143–52
    articles on, 144–6
    disease framing in, 149
    environmental causation perspective, 143–4, 147–8

reductionist framework for, 147
social consequences of, 148–9
as tokenistic, on key webpages, 144
trait-based breeding, through genomic editing, 349–50
transgender populations. *See* LGBTQ populations
transnational food supply networks, 420–2
transnational social movements, environmental governance and, 251
treadmill of production theory, 22
Trump, Donald, 230–1
LGBTQ populations and, 93
rural support for, 89
Tsing, Anna, 26, 170–1
Tutuola, Amos, 164

U.K. *See* United Kingdom
UN. *See* United Nations
'uncooperative commodity,' water as, 314–15
United Kingdom (U.K.). *See also* Scotland; Wales
environmental governance in, for forestry management, with citizen and community forestry groups, 252, 253–4. *See also specific community groups*
privatisation of water in, 306, 312
United Nations (UN)
Conference on the Environment and Development, 249
on water as human right, 304
United States (U.S.)
carbon markets in, 279
Collaborative on Health and the Environment, 141
Digital Millennium Copyright Act, 379, 384–5
Emergency Planning and Community Right-to-Know Act, 125
Environmental Protection Agency, 75, 118–19
global fair trade movement in, 463
National Environmental Protection Act, 118–19
National Resources Conservation Service in, 92–3
Office of Environmental Health Hazard Assessment, 141
privatisation of water in, 310–11
water management in, 310–11
through emergency fiscal managers, 311
in Flint, Michigan, 124, 311
unsafe drinking water issues, 311
unsafe drinking water, 311
urban areas
Chicago School approach in, for urban and community development, 63
environmental justice in, through CBR, 54
in GRUMP, 76–7
U.S. *See* United States
utopia model, plants in, 181

variable-rate input application technology (VRT), 379
The Vegetative Soul (Miller), 176
VGI. *See* volunteered geographic information
La Via Campesina, 421–2

voluntary simplifiers, in green governance models, 226
volunteered geographic information (VGI), 70
volunteering, environmental governance and, 259–61
Voss, Paul, 71–2
VRT. *See* variable-rate input application technology

Wachter, Kenneth, 71–2
Wales, environmental governance in, 256–7, 259–60
water, as resource
accumulation of, 305
commodification of, 303, 304–5
'fictitious commodities', 304–5
as 'uncooperative commodity', 314–15
global water crisis, 303
as human right, 304
potable, access to, 303
privatisation of. *See* privatisation of resources
water concession contracts, 308–9
water justice activism, 308
water management
commodification of, 304
bottled water, 313–15
under Dublin Principles, 303
global approach to, 303–4
state hydraulic paradigm, 303
in U.S., 310–11
through emergency fiscal managers, 311
in Flint, Michigan, 124, 311
unsafe drinking water issues, 311
Watson, James, 345
WDSR Network. *See* Worker-Driven Social Responsibility Network
Weber, Max, on DI approaches, 33–34
to natural resource management, 35–6
nature of bureaucracies in, 35–6
WebMD, 138, 140
coverage of toxicants, as cause of leukemia, 143–52
articles on, 144–6
disease framing in, 149
environmental causation perspective, 143–4, 147–8
reductionist framework for, 147
social consequences of, 148–9
tokenistic, on key webpages, 144
disease framing, 149–53
of coverage of toxicants, as cause of leukemia, 149
financial interests as factor in, 150–2
political economy approach, 149–50
public health interests in, prioritisation over, 150–2
environmental causation perspective, for leukemia
contextual factors for, 147–8
obscuring of, 143–4
'Western' paradigms, in sociology, 16
Weston, Kath, 90
WFTO. *See* World Fair Trade Organization
Wheeler, Andrew, 230–1
When Species Meet (Haraway), 170
White, Monica, 98
White, Richard, 433

'white male' effect, 108–9
  hierarchism, 108–9
  identity-protective cognition, 108–9
  individualism, 108–9
Williams, Raymond, 381
Winner, Landon, 449–50
Wohlleben, Peter, 179–80
women. *See also* feminine identity; feminism; feminist sociology; gender theory
  in anti-toxics movement, 120–1
  sustainable consumption by, 227
Woodlland Trust, 258
Woodmatters, 258–9

Worker-Driven Social Responsibility (WDSR) Network, 463, 467
World Fair Trade Organization (WFTO), 462–3, 465
World Trade Organization (WTO), 412
Worldwide Opportunities on Organic Farms (WWOOF) movement, 216–17
WTO. *See* World Trade Organization
WWOOF. *See* Worldwide Opportunities on Organic Farms movement

Zapatista Movement, 499, 503
Zelizer, Viviana, 273

CPSIA information can be obtained
at www.ICGtesting.com
Printed in the USA
LVHW061113280321
PP16692800003B/1